Lecture Notes in Computer Science   12126

Advanced Research in Computing and Software Science

Subline of Lecture Notes in Computer Science

More information about this series at http://www.springer.com/series/7407

Leszek Gąsieniec · Ralf Klasing ·
Tomasz Radzik (Eds.)

Combinatorial Algorithms

31st International Workshop, IWOCA 2020
Bordeaux, France, June 8–10, 2020
Proceedings

 Springer

Editors
Leszek Gąsieniec
University of Liverpool
Liverpool, UK

Augusta University
Augusta, GA, USA

Tomasz Radzik
King's College London
London, UK

Ralf Klasing
CNRS and University of Bordeaux
Talence, France

ISSN 0302-9743   ISSN 1611-3349   (electronic)
Lecture Notes in Computer Science
ISBN 978-3-030-48965-6   ISBN 978-3-030-48966-3   (eBook)
https://doi.org/10.1007/978-3-030-48966-3

LNCS Sublibrary: SL1 – Theoretical Computer Science and General Issues

This Springer imprint is published by the registered company Springer Nature Switzerland AG
The registered company address is: Gewerbestrasse 11, 6330 Cham, Switzerland

Preface

The 31st International Workshop on Combinatorial Algorithms (IWOCA 2020) was originally scheduled to take place during June 8–10, 2020, in Bordeaux, France. Due to the COVID-19 pandemic, those original arrangements had to be changed. The symposium was run online on the originally set dates of 8–10 June 2020, organized and coordinated from Bordeaux by the IWOCA 2020 Organizing Committee.

Since its inception in 1989 as AWOCA (Australasian Workshop on Combinatorial Algorithms), IWOCA has provided an annual forum for researchers who design algorithms for the myriad combinatorial problems that underlie computer applications in science, engineering, and business. Previous IWOCA and AWOCA meetings have been held in Australia, Canada, Czech Republic, Finland, France, Indonesia, India, Italy, Japan, Singapore, South Korea, the UK, and the USA.

The Program Committee (PC) of IWOCA 2020 received 62 submissions. Each submission was reviewed by at least three PC members and some trusted external referees, and evaluated on its quality, originality, and relevance to the symposium. The PC selected 30 papers for presentation at the symposium and inclusion in the proceedings.

Three invited talks were given at IWOCA 2020, by *Dan Alistarh* (Institute of Science and Technology, Austria), *Sándor Fekete* (Technische Universität Braunschweig, Germany), and *Tatiana Starikovskaya* (École normale supérieure, Paris, France). This volume contains the abstracts or the extended versions of the three invited talks.

We thank the Steering Committee for giving us the opportunity to serve as program chairs of IWOCA 2020, and for the responsibilities of selecting the PC, the conference program, and publications.

The PC selected three contributions for the best paper and the best student paper awards, sponsored by Springer.

The Best Paper Award was given to:

- *Peter Damaschke* for his paper "Two Robots Patrolling on a Line: Integer Version and Approximability"

The Best Student Paper Award was shared between two papers:

- *Pratibha Choudhary* for her paper "Polynomial Time Algorithms for Tracking Path Problems"
- *Florent Foucaud, Benjamin Gras, Anthony Perez, and Florian Sikora* for their paper "On the complexity of Broadcast Domination and Multipacking in digraphs"

We gratefully acknowledge additional financial support from the following institutions: the University of Bordeaux, LaBRI, CNRS, Bordeaux INP, IDEX Bordeaux, Cluster SysNum, and the French National Research Agency (ANR).

We would like to thank all the authors who responded to the call for papers, the invited speakers, the members of the PC, the external referees, and – last but not least – the members of the Organizing Committee.

We would like to thank Springer for publishing the proceedings of IWOCA 2020 in their ARCoSS/LNCS series and for their support.

Finally, we acknowledge the use of the EasyChair system for handling the submission of papers, managing the review process, and generating these proceedings.

June 2020

Leszek Gąsieniec
Ralf Klasing
Tomasz Radzik

Organization

Steering Committee

Maria Chudnovsky	Princeton University, USA
Charles Colbourn	Arizona State University, USA
Costas Iliopoulos	King's College London, UK
Bill Smyth	McMaster University, Canada; Murdoch University, Australia; and King's College London, UK

Program Committee

Amihood Amir	Bar Ilan University, Israel
Petra Berenbrink	University of Hamburg, Germany
Hans L. Bodlaender	Utrecht University, The Netherlands
Hans-Joachim Böckenhauer	ETH Zürich, Switzerland
Marthe Bonamy	CNRS and University of Bordeaux, France
Arnaud Casteigts	University of Bordeaux, France
Marek Chrobak	University of California, Riverside, USA
Charles Colbourn	Arizona State University, USA
Anne Driemel	University of Bonn, Germany
Leah Epstein	University of Haifa, Israel
Thomas Erlebach	University of Leicester, UK
Paola Flocchini	University of Ottawa, Canada
Florent Foucaud	University of Bordeaux, France
Pierre Fraigniaud	CNRS and Paris Diderot University, France
Luisa Gargano	University of Salerno, Italy
Leszek Gąsieniec (Co-chair)	The University of Liverpool, UK, and Augusta University, USA
Juraj Hromkovič	ETH Zürich, Switzerland
Sun-Yuan Hsieh	National Cheng Kung University, Taiwan
Ling-Ju Hung	National Taipei University of Business, Taiwan
Ralf Klasing (Co-chair)	CNRS and University of Bordeaux, France
Tomasz Kociumaka	Bar Ilan University, Israel
Dennis Komm	ETH Zürich, Switzerland
Rastislav Královič	Comenius University, Slovakia
Thierry Lecroq	University of Rouen, France
Dimitrios Letsios	King's College London, UK
Andrea Marino	University of Florence, Italy
Tobias Mömke	Saarland University, Germany
Xavier Muñoz	Universitat Politècnica de Catalunya, Spain
Ian Munro	University of Waterloo, Canada
Rolf Niedermeier	Technical University of Berlin, Germany

Aris Pagourtzis	National Technical University of Athens, Greece
Marina Papatriantafilou	Chalmers University of Technology, Sweden
Tomasz Radzik (Co-chair)	King's College London, UK
Sohel Rahman	Bangladesh University of Engineering and Technology, Bangladesh
Peter Rossmanith	RWTH Aachen University, Germany
Joe Ryan	The University of Newcastle, Australia
Alex Schwarzmann	Augusta University, USA
Bill Smyth	McMaster University, Canada; Murdoch University, Australia; and King's College London, UK
Paul Spirakis	The University of Liverpool, UK, and University of Patras, Greece
Walter Unger	RWTH Aachen University, Germany
Przemysław Uznański	University of Wrocław, Poland
Yukiko Yamauchi	Kyushu University, Japan

Organizing Committee

Marthe Bonamy	CNRS and University of Bordeaux, France
Arnaud Casteigts	University of Bordeaux, France
Auriane Dantes	University of Bordeaux, France
Florent Foucaud (Co-chair)	University of Bordeaux, France
Isabelle Garcia	University of Bordeaux, France
Shih-Shun Kao	University of Bordeaux, France, and National Cheng Kung University, Taiwan
Ralf Klasing (Co-chair)	CNRS and University of Bordeaux, France
Dimitri Lajou	University of Bordeaux, France
Alessia Milani	Bordeaux INP, France
Jonathan Narboni	University of Bordeaux, France
Jason Schoeters	University of Bordeaux, France

External Reviewers

Abels, Andreas	Chang, Shun-Chieh
Afshani, Peyman	Colini Baldeschi, Riccardo
Akrida, Eleni C.	Cordasco, Gennaro
Abels, Andreas	Coudert, David
Afshani, Peyman	Dailly, Antoine
Akrida, Eleni C.	Damaschke, Peter
Benter, Matthias	de Castro Mendes Gomes, Guilherme
Bergognoux, Benjamin	Deligkas, Argyrios
Brejová, Broňa	Dereniowski, Dariusz
Burjons, Elisabet	Dobrev, Stefan
Cames van Batenburg, Wouter	Dreier, Jan
Chakraborty, Dibyayan	Duvignau, Romaric
Chang, Jou-Ming	Faliszewski, Piotr

Fischer, Dennis
Frascaria, Dario
Frei, Fabian
Froese, Vincent
Fuchs, Janosch
Fukunaga, Takuro
Fürst, Maximilian
Georgiou, Konstantinos
Gkikas, Angelos
Habib, Mursalin
Hanaka, Tesshu
Hartmann, Tim A.
Hassan, Shahriar
Heeger, Klaus
Himmel, Anne-Sophie
Hoefer, Martin
Hsu, Daniel
Kao, Mong-Jen
Kauers, Manuel
Kellerhals, Leon
Kijima, Shuji
Kim, Eun Jung
Knop, Dušan
Kobayashi, Yasuaki
Koumoutsos, Grigorios
Kranakis, Evangelos
La, Hoang
Lajou, Dimitri
Lee, Chuan-Min
Levin, Asaf
Lin, Chuang-Chieh
Lopes, Raul
Lotze, Henri

Lucarelli, Giorgio
Łukasiewicz, Aleksander
Lukoťka, Robert
Mautner, Stefan
Mc Inerney, Fionn
Mertzios, George
Michail, Othon
Mnich, Matthias
Mobin, Jaiaid
Nakos, Vasileios
Narboni, Jonathan
Novotna, Jana
Okrasa, Karolina
Ordyniak, Sebastian
Pandit, Supantha
Prezza, Nicola
Psarros, Ioannis
Punzi, Giulia
Rashid, Syed Md. Mukit
Ravindran Vijayalakshmi, Vipin
Rescigno, Adele
Ruangwises, Suthee
Sadakane, Kunihiko
Sajenko, Andrej
Santoro, Nicola
Serna, Maria
Shamil, Md. Salman
Silva, Ana
Suzuki, Akira
Tauer, Björn
Vaccaro, Ugo
Zamaraev, Viktor
Zhang, Jingru

Sponsors

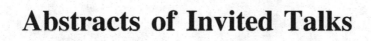

Abstracts of Invited Talks

Optimization by Population: Large-Scale Distributed Optimization Via Population Protocols

Dan Alistarh

Institute of Science and Technology, Austria
dan.alistarh@ist.ac.at

The population model is a standard way to represent large-scale decentralized distributed systems, in which agents with limited computational power interact in randomly chosen pairs, in order to collectively solve global computational tasks. In contrast with synchronous gossip models, nodes are anonymous, lack a common notion of time, and have no control over their scheduling. In this talk, I will describe recent work examining whether large-scale distributed optimization can be performed in this extremely restrictive setting.

I will introduce and analyze natural decentralized variants of the classical stochastic gradient descent (SGD) procedure, in which every node maintains a local estimate of the optimal set of parameters, and is able to compute stochastic gradients with respect to this parameter. Every pair-wise node interaction performs a stochastic gradient step at each agent, followed by averaging of the two models. I will show that, under standard assumptions, SGD can converge even in this extremely loose, decentralized setting. Moreover, surprisingly, the algorithm can achieve linear speedup in the number of nodes n. In addition, I will show experimental results showing that this algorithm can achieve convergence and speedup for large-scale distributed learning tasks.

Coordinating Swarms of Objects at Extreme Dimensions

Sándor P. Fekete

Department of Computer Science, TU Braunschweig, Germany
s.fekete@tu-bs.de

We describe a variety of algorithmic challenges arising from coordination and reconfiguration of swarms of potentially many objects, ranging in size from minuscule particles all the way to far-away satellite swarms. Particular results include methods for coordinating the motion of vehicles in traffic in order to avoid inefficient stop-and-go congestions; using uniform global forces for controlling particle swarms; online triangulation and structured exploration; cohesive control for swarms of robots with only local communication; coordinated motion planning for efficiently reconfiguring an arrangement of robots; and constructing and reconfiguring large-scale structures by finite automata. All presented work is based on collaborations with a variety of authors.

Algorithms for String Processing
in Restricted-Access Models of Computation

Tatiana Starikovskaya

DIENS, École normale supérieure, PSL Research University, Paris, France
tat.starikovskaya@gmail.com

Many classical algorithms for string processing assume that the input can be accessed in full via constant-time random access, which poses a serious limitation in the modern era of data deluge. In this talk, we will consider two novel models of computation that avoid this assumption: streaming and property testing. In this talk, we will discuss recent developments related to these two models and will show that they allow developing ultra-efficient approaches to string processing.

In the streaming model of computation, we assume that the input arrives as a stream, one character at a time, which captures a situation when the data are sequential measurements or an output of an algorithm. In this model, the space complexity is defined as all the space used, including the space used to store any information about the input. The goal is to develop an algorithm which uses sublinear (preferably, polylogarithmic) space. The first streaming algorithm for pattern matching was presented in the seminal paper of Porat and Porat in FOCS 2009. For a pattern of length m, the algorithm uses only $\mathcal{O}(\log m)$ space, while any classical algorithm requires $\Omega(m)$ space. This result served as a foundation of the area of streaming algorithms for pattern matching, which we will survey in the first part of the talk.

In the property testing model, our task is to decide whether the input string has a particular property $\mathcal{P}$. As opposed to streaming algorithms, property testers have random access to their input. However, we pose a restriction on property testers and require them to access as few characters of the input as possible, which allows developing very fast algorithms. In order to make the decision task possible, often two relaxations are allowed. First, we allow probabilistic algorithms. Second, we only ask the tester to distinguish between two cases: the case where the input satisfies $\mathcal{P}$; and the case where we must change at least an ε-fraction of the characters so that the input satisfies $\mathcal{P}$. In FOCS 1999, Alon et al. showed that any regular language can be tested by accessing only $\mathcal{O}(\log^3(1/\varepsilon)/\varepsilon)$ characters of the input. Since this fundamental result, the literature has seen a series of property testers for formal languages, and we will discuss some of them in the second part of the talk.

Algorithms for String Processing
in Restricted-Access Models of Computation

Tatiana Starikovskaya

DIENS, École normale supérieure, PSL Research University, Paris, France

Contents

Invited Paper

Coordinating Swarms of Objects
at Extreme Dimensions

Sándor P. Fekete(✉)

Department of Computer Science, TU Braunschweig, Braunschweig, Germany
s.fekete@tu-bs.de

Abstract. We describe a variety of algorithmic challenges arising from coordination and reconfiguration of swarms of potentially many objects, ranging in size from minuscule particles all the way to far-away satellite swarms. Particular results include methods for coordinating the motion of vehicles in traffic in order to avoid inefficient stop-and-go congestions; using uniform global forces for controlling particle swarms; online triangulation and structured exploration; cohesive control for swarms of robots with only local communication; coordinated motion planning for efficiently reconfiguring an arrangement of robots; and constructing and reconfiguring large-scale structures by finite automata. All presented work is based on collaborations with a variety of authors, who are named in the respective sections of this overview.

1 Introduction

Problems of coordinating and reconfiguring arrangements of objects pose important questions at various dimensions, ranging from tiny particles all the way to far-away satellite swarms. Ironically, systems at these very small and very large distances share a fundamental property: It becomes difficult to use "external" computation, in which a powerful central computing device is provided with input about the system, and output is fed back into the system. Instead, it becomes important to consider "internal" computation, in which algorithms and execution remain within the system itself, even if that comes at the expense of processing power.

In this overview we provide a number of different contexts that require dealing with computation, communication and coordination of swarms of robots at extreme dimensions. These dimensions may correspond to extremely small or large sizes or extremely large numbers; both give rise to a wide range of algorithmic challenges. This overview provides pointers to more detailed contexts and references, and gives credit to the numerous collaborators and their contributions.

2 Traffic

How can we coordinate the motion of many autonomous vehicles, such that the overall traffic flow is smooth and efficient? (Fig. 1).

© Springer Nature Switzerland AG 2020
L. Gąsieniec et al. (Eds.): IWOCA 2020, LNCS 12126, pp. 3–13, 2020.
https://doi.org/10.1007/978-3-030-48966-3_1

Fig. 1. Typical stop-and-go traffic.

We describe a distributed and self-regulated approach for the self-organization of a large system of many self-driven, mobile objects, i.e., cars in traffic. Based on methods for mobile ad-hoc networks using short-distance communication between vehicles, and ideas from distributed algorithms, we consider reactions to specific traffic structures (e.g., traffic jams). Building on models from traffic physics, we are able to develop strategies that significantly improve the flow of congested traffic. Results include fuel savings up to 40% for cars in stop-and-go traffic; we present a number of simulation results illustrating the efficacy of the underlying mechanisms (Fig. 2).

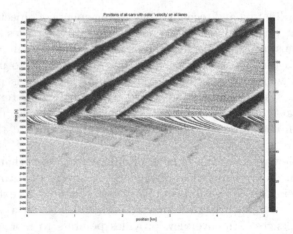

Fig. 2. A simulation of dense traffic. Time proceeds from top to bottom, vehicles move from left to right, so an individual vehicle follows a trajectory from top left to bottom right, with color indicating momentary vehicle speeds. Observe the stop-and-go density fluctuations in the upper part of the diagram, which are dissolved in the bottom part by switching on our developed mechanism, resulting in a better flow of traffic. (Color figure online)

The results of this section summarize joint work with Sebastian Ebers, Stefan Fischer, Horst Hellbrück, Björn Hendriks, Christopher Tessars, and Axel Wegener. See [13,16,19] for further details.

3   Uniform Global Control for Particle Swarms

How can we rearrange a potentially large swarm of particles that do not have their own energy supply? (Fig. 3).

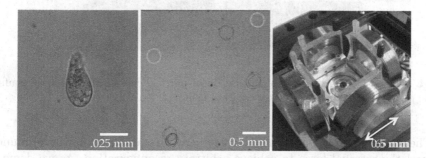

Fig. 3. After feeding iron particles to ciliate eukaryons (*Tetrahymena pyriformis*) and magnetizing the particles with a permanent magnet, the cells can be turned by changing the orientation of an external magnetic field (see colored paths in the center image). (Right) Using two orthogonal Helmholz electromagnets, Becker et al. [11] demonstrated steering many living magnetized *T. pyriformis* cells. All cells are steered by the same global field. (Color figure online)

We consider algorithmic control of a large swarm of mobile particles, such as robots, sensors, or building material. The objective is to achieve arbitrary reconfiguration, even if the particles are too small to carry their own energy supply. Instead, they are moved around with the help of external forces, such as a magnetic field or gravity. Upon actuation, each object is pushed in the same direction until it collides with an obstruction. This concept can be used for a wide range of applications in which particles follow a uniform global signal. In an open workspace, this system model is of limited use, because all particles receive the same inputs and move uniformly. Thus, a crucial challenge for achieving any desired target configuration is breaking global symmetry in a controlled fashion.

We provide two different methods for this objective. The first is to add a maze of obstacles to the environment, which can make the system drastically more complex but also more useful. We provide a variety of results for a wide range of questions. These can be subdivided into *external* algorithmic problems, in which particle configurations serve as input for computations that are performed elsewhere, and *internal* logic problems, in which the particle configurations themselves are used for carrying out computations (Fig. 4).

Fig. 4. Gravity-fed hardware implementation of particle computation. This reconfigurable prototype is set up as a FAN-OUT gate using a 2×1 robot (white)

The second approach uses the interplay between static friction with a boundary and the external force to achieve arbitrary reconfiguration. As we demonstrate, it is possible to determine a precise theoretical characterization of the critical coefficient of friction that is sufficient for rearranging two particles in triangles, convex polygons, and regular polygons. We also illustrate a method for reconfiguring multiple particles in rectangular workspaces, and deriving practical algorithms for these rearrangements.

The results of this section highlight a broad spectrum of joint work with Victor Baez, Aaron Becker, Erik Demaine, Golnaz Habibi, Li Huang, Phillip Keldenich, Linda Kleist, Dominik Krupke, Jarrett Lonsford, Arun Mahadev, Sheryl Manzoor, James McLurkin, Rose Morris-Wright, Hamed Mohtasham Shad, Christian Rieck, Christian Scheffer, and Arne Schmidt; see [3, 4, 6–8, 10, 20, 23, 24, 26, 27] for further details. The shown video [2] is available at https://www.ibr.cs.tu-bs.de/users/fekete/Videos/SoCG/2020/Friction_SoCG.mp4; an earlier video with further theoretical results [5] can be found at https://www.ibr.cs.tu-bs.de/users/fekete/Videos/SoCG/2015/TiltforCompGeom100mb.mp4.

4 Online Triangulation and Structured Exploration

How can we allow a swarm of relatively simple robots to cooperate in exploring an unknown environment? (Figs. 5 and 6).

We consider a fundamental framework for organizing exploration, coverage, and surveillance by a swarm of robots with limited individual capabilities, based on triangulating an unknown environment with a multi-robot system. Locally, an individual triangle is easy for a single robot to manage and covers a small area; globally, the topology of the triangulation approximately captures the geometry of the entire environment. Combined, a multi-robot system can explore,

Fig. 5. A swarm of small robots building an expanding triangulation for carrying out a collective online exploration algorithm.

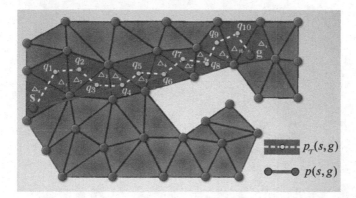

Fig. 6. Using a dual path for routing in a triangulated environment: a shortest path (shown in red) is approximated by a minimum-hop path (shown in yellow), achieving constant stretch. (Color figure online)

map, navigate, and patrol. Algorithms can store information in triangles that the robots can read and write as they run their algorithms. This creates a physical data structure (PDS) that is both robust and versatile. We study distributed approaches to triangulating an unknown, two-dimensional Euclidean space using a multi-robot network. The resulting PDS is a compact representation of the workspace, contains distributed knowledge of each triangle, encodes the dual graph of the triangulation, and supports reads and writes of auxiliary data. The ability to store and process this auxiliary information enables the simple robots to solve complex problems. This leads to distributed algorithms for dual-graph navigation, patrolling, construction of a topological Voronoi tessellation, and location of the geodesic centers in non-convex regions, making it possible to provide theoretical performance guarantees for the quality of constructed triangulation and the connectivity of a dual graph in the triangulation. In addition, the path lengths of the physical navigation are within a constant factor of the shortest-path Euclidean distance. These theoretical results were also practically validated with simulations and experiments with dozens of robots.

The results of this section summarize joint work with Aaron Becker, Tom Kamphans, Alexander Kröller, Seoung Kyou Lee, James McLurkin, Joe Mitchell, and Christiane Schmidt. See our papers [17,22] for further details; the shown video [18] is available at https://www.ibr.cs.tu-bs.de/users/fekete/Videos/SoCG/2013/MATP-Video.mov.

5   Cohesive Control

How can we enable a swarm of simple robots to maintain connectivity, even if it is being pulled apart by external forces? (Fig. 7).

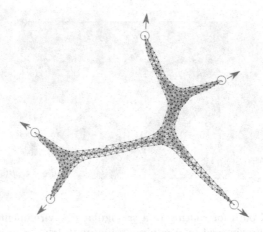

Fig. 7. A robust robot swarm emulating a Steiner tree between diverging leader robots.

Consider a swarm of robots that needs to remain connected. There is no central control and no knowledge of the overall environment. This environment is hostile: The swarm is being pulled apart by external forces, stretching it into a number of different directions, so it is in danger of breaking up. Individual robots are weak, with limited sensing, limited communication, and limited connectivity; even worse, each robot's expected lifetime is limited by random, permanent failures, which may destroy connectedness and functioning of the swarm as a whole. The objective is to achieve coordinated dynamic swarm behavior without centralized coordination, employing each robot as much as possible, without depending on it if it fails, and balancing overall flexibility and robustness to deal with the hostile environment.

We propose a set of local continuous algorithms that together produce a generalization of a Euclidean Steiner tree. At any stage, the resulting overall shape achieves a good compromise between local thickness, global connectivity, and flexibility to further continuous motion of the terminals. The resulting swarm behavior scales well, is robust against node failures, and performs close to the best known approximation bound for a corresponding centralized static optimization problem.

The results of this section summarize joint work with Dominik Krupke, Maximilian Ernestus, and Michael Hemmer. See our paper [21] for further details.

6 Coordinated Motion Planning

How can we coordinate the collision-free motion of many robots, vehicles, aircraft, or people, such that each one reaches its destination as quickly as possible? (Fig. 8).

Fig. 8. Coordinated motion planning: (Left) A start configuration of labeled robots. (Center) A feasible move, coordinating parallel relocation of many robots. (Right) The desired target configuration.

We develop constant-factor approximation algorithms for minimizing the execution time of a coordinated parallel motion plan for a relatively dense swarm of homogeneous robots in the absence of obstacles. In our first model, each robot has a specified start and destination on the square grid, and in each round of coordinated parallel motion, every robot can move to any adjacent position that is either empty or simultaneously being vacated by another robot. In this model, our algorithm achieves a *constant stretch factor*: If every robot starts at a distance of at most d from its destination, then the total duration of the overall schedule is $O(d)$, which is optimal up to constant factors. Our result holds for distinguished robots (each robot has a specific destination), identical (unlabeled) robots, and most generally, classes of different robot types (where each destination specifies a required type of robot). We also show that finding the optimal coordinated parallel motion plan is NP-hard, justifying approximation algorithms.

In our second model, each robot is a unit-radius disk in the plane, and robots can translate continuously in parallel subject to not intersecting, i.e., having disk centers at L_2-distance at least 2. We prove the same result—constant-factor approximation algorithm to minimizing execution time via constant stretch factor—when the pairwise L_∞-distance between disk centers is at least $2\sqrt{2} = 2.8284\ldots$. On the other hand, for N densely packed disks at distance at most $2 + \delta$ for a sufficiently small $\delta > 0$, we prove that a stretch factor of $\Omega(N^{1/4})$ is sometimes necessary (when densely packed), while a stretch factor of $\mathcal{O}(N^{1/2})$ is always possible.

The results of this section summarize joint work with Aaron Becker, Erik Demaine, Phillip Keldenich, Lilian Li, and Henk Meijer. See our paper [12] for further details; the shown video [9] is available at https://www.ibr.cs.tu-bs.de/users/fekete/Videos/SoCG/2018/CoordinatedMotionPlanning.mp4.

7   Constructing and Reconfiguring Large-Scale Structures

How can we use simple robots to construct large-scale structures, such as space stations? (Fig. 9).

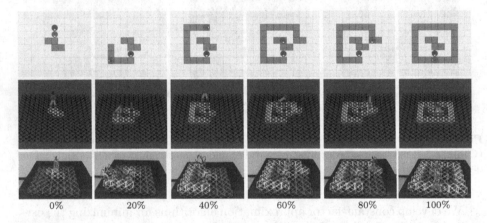

Fig. 9. Snapshots from building a bounding box for a Z-shaped polyomino using 2D simulator, 3D simulator, and staged hardware robots, synchronized so all are shown at steps $\{0, 24, 48, 72, 96, 120\}$.

We consider recognition and reconfiguration of lattice-based cellular structures by very simple robots with only basic functionality. The underlying motivation is the construction and modification of space facilities of enormous dimensions, where the combination of new materials with extremely simple robots promises structures of previously unthinkable size and flexibility. We present algorithmic methods that are able to detect and reconfigure arbitrary polyominoes, based on finite-state robots, while also preserving connectivity of a structure during reconfiguration. Specific results include methods for determining a bounding box, scaling a given arrangement, and adapting more general algorithms for transforming polyominoes.

The results of this section summarize joint work with Amira Abdel-Rahman, Aaron Becker, Daniel Biediger, Kenny Cheung, Neil Gershenfeld, Sabrina Hugo, Ben Jenett, Phillip Keldenich, Eike Niehs, Christian Rieck, Arne Schmidt, Christian Scheffer, and Michael Yannuzzi. See our papers [15,25] for further details; the shown video [1] is available at https://www.ibr.cs.tu-bs.de/users/fekete/ Videos/SoCG/2020/SpaceAnts_SoCG.mp4.

8 Conclusion

Many of the presented topics are still subject to ongoing work; see the recent survey article [14] for more details on context, content and technical details, as well as additional references for some of the described topics. We are confident that more progress on a wide range of related problems is imminent.

References

1. Abdel-Rahman, A., et al.: Space ants: constructing and reconfiguring large-scale structures with finite automata. In: Symposium on Computational Geometry (SoCG), pp. 73:1–73:7 (2020). https://www.ibr.cs.tu-bs.de/users/fekete/Videos/ SoCG/2020/SpaceAnts_SoCG.mp4
2. Baez, V.M., Becker, A.T., Fekete, S.P., Schmidt, A.: Coordinated particle relocation with global signals and local friction. In: Symposium on Computational Geometry (SoCG), pp. 72:1–72:8 (2020). https://www.ibr.cs.tu-bs.de/users/ fekete/Videos/SoCG/2020/Friction_SoCG.mp4
3. Becker, A., Demaine, E.D., Fekete, S.P., Habibi, G., McLurkin, J.: Reconfiguring massive particle swarms with limited, global control. In: Flocchini, P., Gao, J., Kranakis, E., Meyer auf der Heide, F. (eds.) ALGOSENSORS 2013. LNCS, vol. 8243, pp. 51–66. Springer, Heidelberg (2014). https://doi.org/10.1007/978-3-642-45346-5_5
4. Becker, A., Demaine, E.D., Fekete, S.P., McLurkin, J.: Particle computation: designing worlds to control robot swarms with only global signals. In: IEEE International Conference on Robotics and Automation (ICRA), pp. 6751–6756 (2014)
5. Becker, A., Demaine, E.D., Fekete, S.P., Shad, S.H.M., Morris-Wright, R.: Tilt: the video. Designing worlds to control robot swarms with only global signals. In: Symposium on Computational Geometry (SoCG), pp. 16–18 (2015). https://www. ibr.cs.tu-bs.de/users/fekete/Videos/SoCG/2015/TiltforCompGeom100mb.mp4
6. Becker, A., Morris-Wright, R., Demaine, E.D., Fekete, S.P.: Particle computation: device fan-out and binary memory. In: IEEE International Conference on Robotics and Automation (ICRA), pp. 5384–5389 (2015)
7. Becker, A.T., Demaine, E.D., Fekete, S.P., Lonsford, J., Morris-Wright, R.: Particle computation: complexity, algorithms, and logic. Nat. Comput. 18(1), 181–201 (2017). https://doi.org/10.1007/s11047-017-9666-6
8. Becker, A.T., et al.: Targeted drug delivery: algorithmic methods for collecting a swarm of particles with uniform, external forces. In: IEEE International Conference on Robotics and Automation (ICRA), (2020). https://www.ibr.cs.tu-bs.de/users/ fekete/hp/publications/PDF/2020-Gathering_ICRA.pdf

9. Becker, A.T., Fekete, S.P., Keldenich, P., Konitzny, M., Lin, L., Scheffer, C.: Coordinated motion planning: the video. In: Symposium on Computational Geometry (SoCG), vol. 99, pp. 74:1–74:6 (2018). https://www.ibr.cs.tu-bs.de/users/fekete/Videos/SoCG/2018/CoordinatedMotionPlanningDefinitiveVersion.mp4
10. Becker, A.T., et al.: Tilt assembly: algorithms for micro-factories that build objects with uniform external forces. Algorithmica **82**(2), 165–187 (2020)
11. Becker, A.T., Ou, Y., Kim, P., Kim, M.J., Julius, A.: Feedback control of many magnetized *Tetrahymena pyriformis* cells by exploiting phase inhomogeneity. In: IEEE/RSJ International Conference on Intelligent Robots and Systems (IROS), pp. 3317–3323 (2013)
12. Demaine, E.D., Fekete, S.P., Keldenich, P., Scheffer, C., Meijer, H.: Coordinated motion planning: reconfiguring a swarm of labeled robots with bounded stretch. SIAM J. Comput. **48**, 1727–1762 (2019)
13. Fekete, S., Tessars, C., Schmidt, C., Wegener, A., Fischer, S., Hellbrück, H.: Verfahren und Vorrichtung zur Ermittlung einer Fahrstrategie. Patentnummer DE 10, 047 (2008)
14. Fekete, S.P.: Geometric aspects of robot navigation: from individual robots to massive particle swarms. In: Flocchini, P., Prencipe, G., Santoro, N. (eds) Distributed Computing by Mobile Entities. LNCS, vol. 11340. Springer, Cham (2019). https://doi.org/10.1007/978-3-030-11072-7_21
15. Fekete, S.P., Gmyr, R., Hugo, S., Keldenich, P., Scheffer, C., Schmidt, A.: Cadbots: algorithmic aspects of manipulating programmable matter with finite automata. In: Algorithmic Foundations of Robotics (WAFR 2018), Advanced Tracts in Robotics. Springer, (2020). https://www.ibr.cs.tu-bs.de/users/fekete/hp/publications/PDF/2020-CADBots_WAFR.pdf
16. Fekete, S.P., et al.: Methods for improving the flow of traffic. In: Müller-Schloer, C., Schmeck, H., Ungerer, T. (eds) Organic Computing – A Paradigm Shift for Complex Systems. Autonomic Systems, vol. 1, pp. 447–460. Springer, Basel (2011). https://doi.org/10.1007/978-3-0348-0130-0_29
17. Fekete, S.P., Kamphans, T., Kröller, A., Mitchell, J.S.B., Schmidt, C.: Exploring and triangulating a region by a swarm of robots. In: Goldberg, L.A., Jansen, K., Ravi, R., Rolim, J.D.P. (eds.) APPROX/RANDOM -2011. LNCS, vol. 6845, pp. 206–217. Springer, Heidelberg (2011). https://doi.org/10.1007/978-3-642-22935-0_18
18. Fekete, S.P., Kröller, A., Kyou, L., Schmidt, J.M.C.: Triangulating unknown environments using robot swarms. In: Symposium on Computational Geometry (SoCG), pp. 345–346 (2013). https://www.ibr.cs.tu-bs.de/users/fekete/Videos/SoCG/2013/MATP-Video.mov
19. Fekete, S.P., Schmidt, C., Wegener, A., Hellbrück, H., Fischer, S.: Empowered by wireless communication: distributed methods for self-organizing traffic collectives. ACM Trans. Auton. Adapt. Syst. **5**(3), 1–30 (2010)
20. Keldenich, P., et al.: On designing 2D discrete workspaces to sort or classify polynminoes. In: IEEE/RSJ International Conference on Intelligent Robots and Systems (IROS), pp. 1–9 (2018)
21. Krupke, D.M., Ernestus, M., Hemmer, M., Fekete, S.: Distributed cohesive control for robot swarms: maintaining good connectivity in the presence of exterior forces. In: IEEE/RSJ International Conference on Intelligent Robots and Systems (IROS), pp. 413–420 (2015)
22. Lee, S.K., Fekete, S.P., McLurkin, J.: Structured triangulation in multi-robot systems: coverage, patrolling, voronoi partitions, and geodesic centers. Int. J. Robot. Res. **9**(35), 1234–1260 (2016)

23. Mahadev, A.V., Krupke, D., Fekete, S.P., Becker, A.: Mapping, foraging, and coverage with a particle swarm controlled by uniform inputs. In: IEEE/RSJ International Conference on Intelligent Robots and Systems (IROS), pp. 1097–1104 (2017)
24. Mahadev, A.V., Krupke, D., Reinhardt, J.-M., Fekete, S.P., Becker, A.: Collecting a swarm in a 2D environment using shared, global inputs. In: Conference on Automation Science and Engineering (CASE), pp. 1231–1236 (2016)
25. Niehs, E., et al.: Recognition and reconfiguration of lattice-based cellular structures by simple robots. In: IEEE International Conference on Robotics and Automation (ICRA) (2020). https://www.ibr.cs.tu-bs.de/users/fekete/hp/publications/PDF/2020-Automata_ICRA.pdf
26. Schmidt, A., Manzoor, S., Huang, L., Becker, A., Fekete, S.P.: Efficient parallel self-assembly under uniform control inputs. Robot. Autom. Lett. **3**, 3521–3528 (2018)
27. Schmidt, A., Montano, V., Becker, A., Fekete, S.P.: Coordinated particle relocation using finite static friction with boundary walls. Robot. Autom. Lett. **2**, 985–992 (2020)

Contributed Papers

A Family of Tree-Based Generators for Bubbles in Directed Graphs

Vicente Acuña[1], Leandro Lima[2], Giuseppe F. Italiano[3,5], Luca Pepè Sciarria[4],

Marie-France Sagot[5,6], and Blerina Sinaimeri[5,6(✉)]

[1] Center for Mathematical Modeling (UMI 2807 CNRS), University of Chile, Santiago, Chile
[2] European Bioinformatics Institute, Cambridge, UK
[3] LUISS University, Roma, Italy
[4] University of Rome Tor Vergata, Rome, Italy
[5] Erable, INRIA Grenoble Rhône-Alpes, Montbonnot-Saint-Martin, France
blerina.sinaimeri@inria.fr
[6] Université de Lyon, Université Lyon 1, Laboratoire de Biométrie et Biologie Evolutive,
UMR 5558, Villeurbanne, France

Abstract. Bubbles are pairs of internally vertex-disjoint (s, t)-paths in a directed graph. In de Bruijn graphs built from reads of RNA and DNA data, bubbles represent interesting biological events, such as alternative splicing (AS) and allelic differences (SNPs and indels). However, the set of all bubbles in a de Bruijn graph built from real data is usually too large to be efficiently enumerated and analysed in practice. In particular, despite significant research done in this area, listing bubbles still remains the main bottleneck for tools that detect AS events in a reference-free context. Recently, in [1] the concept of a bubble generator was introduced as a way for obtaining a compact representation of the bubble space of a graph. Although this generator was quite effective in finding AS events, preliminary experiments showed that it is about 5 times slower than state-of-art methods. In this paper we propose a new family of bubble generators which improve substantially on the previous generator: generators in this new family are about two orders of magnitude faster and are still able to achieve similar precision in identifying AS events. To highlight the practical value of our new generators, we also report some experimental results on a real dataset.

Keywords: Bubble generator · Directed graphs · Alternative splicing

1 Introduction

The advent of sequencing technologies has revolutionised the study of DNA and RNA data. The information contained in the reads coming from genome or transcriptome

V. Acuña is supported by Fondecyt 1140631, PIA Fellowship AFB170001 and Center for Genome Regulation FONDAP 15090007. G. F. Italiano is partially supported by MIUR, the Italian Ministry for Education, University and Research, under PRIN Project AHeAD (Efficient Algorithms for HArnessing Networked Data). B. Sinaimeri and M.-F. Sagot are partially funded by the French ANR project Aster (2016–2020). Part of this work was done while G. F. Italiano was visiting Université de Lyon and B. Sinaimeri and M.-F. Sagot were visiting LUISS University in Rome.

© Springer Nature Switzerland AG 2020
L. Gąsieniec et al. (Eds.): IWOCA 2020, LNCS 12126, pp. 17–29, 2020.
https://doi.org/10.1007/978-3-030-48966-3_2

sequencing is usually represented by a de Bruijn graph (see *e.g.,* [18,20]). In this graph *bubbles, i.e.,* pairs of internally vertex-disjoint (s, t)-paths, play an important role in the study of genetic variations, which include Alternative Splicing (AS) in RNA-data [16,20–22] and SNPs (Single Nucleotide Polymorphism), and indels in DNA-data [10,24,25]. Since bubbles can be associated to such biologically relevant events, in recent years there have been several theoretical studies on bubbles (see *e.g.,* [3,4,19,21,23]), and in particular there has been a growing interest in algorithms for listing all bubbles in a directed graph. However, in real data graphs the number of bubbles can be exponential in the size of the graph. As a consequence, in practice current algorithms are able to list only a subset of the bubble space, thus losing the information related to the bubbles that are left unexplored. Furthermore, not every bubble corresponds to a biological event. Indeed, a significant number of these bubbles can be false positives (*i.e.,* they are not biologically relevant events), and are produced as artifacts of the underlying construction of the de Bruijn graph. In this framework, the main question is how to find a subset of bubbles that can be efficiently computed in practice and that correspond to relevant biological events.

To tackle this question, the notion of bubble generator was first introduced in [1]. Intuitively, a bubble generator is a subset of bubbles of polynomial size, from which all the other bubbles in the graph can be obtained through a suitable application of a specific symmetric difference operator. In particular, the generator proposed in [1] contains at most $m \cdot n$ bubbles, where m and n denote respectively the number of edges and vertices in the input graph. Furthermore, the authors of [1] provided an algorithm that, given any bubble B in the graph, is able to find in $O(n^3)$ time the bubbles of the generator that can be combined to produce B through a symmetric difference operator. To test its practical value, the generator was used to find AS events in a real dataset. As reported in [1], this generator was able to achieve about the same precision in identifying AS events as the state-of-art-algorithm KISSPLICE [16,20], but unfortunately building the generator was about 5 times slower than finding AS events with KISSPLICE. Despite its great theoretical value, this poses a serious limitation on the practical application of this generator to large-scale datasets, which are typical of biological applications.

To address this issue, in this paper we present a new family of bubble generators which improves substantially on the generator of [1]. In particular, in the same RNA dataset used in [1], generators in our family are about two orders of magnitude faster in practice than the generator in [1], and improve the precision in identifying AS events from 77.3% to 90%. When compared to the state-of-the-art algorithm for identifying AS events, our generators are also much faster than KISSPLICE [16,20], have similar precision, and find AS events that KISSPLICE cannot find. In the experiments, we observed that our new generators also contain many bubbles that correspond to a particular type of AS event, namely *intron retention* (IR), which is usually considered a hard-to-find event. We believe that our experimental findings make the new generators the method of choice for finding AS events in a reference-free context, especially in large-scale data sets.

From the theoretical viewpoint, our new generators are of minimum size (*i.e.* size $m - n + 1$) for flow graphs, *i.e.,* graphs in which there exists a vertex that can reach all other vertices. In case of general graphs, their size is bounded by $|S|(m - n + 1)$, where S is the source set, *i.e.,* a minimum set of vertices that can reach every other vertex in the graph. Although in the worst case this is asymptotically equivalent to the size of the generator in [1], in our experiments the new generators had a much smaller size in practice. Furthermore, the new generators have a much faster decomposition algorithm: given a bubble B it is possible to compute in $O(n)$ time the set of bubbles in the new generators from which B can be composed, while the bubble decomposition algorithm of [1] required as much as $O(n^3)$ time for this task.

To design our new family of generators, we find a way to exploit some connections with cycle bases. We observe that the techniques developed for cycle bases (both in undirected and in directed graphs) cannot be applied directly to bubble generators. Indeed, as reported in [1], the main difference with cycle bases is that in our problem, in order to have biological relevance the following two properties are needed:

($\mathcal{P}_1$) A bubble generator for a directed graph G *must contain only bubbles*;
($\mathcal{P}_2$) Each bubble of G should be decomposed into bubbles of the generator, so that *only bubbles are generated at each step of this decomposition*.

We remark that ensuring properties ($\mathcal{P}_1$) and ($\mathcal{P}_2$) for cycles (in place of bubbles) is already non-trivial. Indeed, Gleiss *et al.* [8] have shown that it is possible to find a basis composed of directed cycles if the graph is strongly connected. However, this is not known in the case of general directed graphs. On the other side, Property ($\mathcal{P}_2$) is somewhat reminiscent of the notion of *cyclically robust cycle bases* which allows one to generate all cycles of a given graph by iteratively adding cycles of the basis [11, 15]. Unfortunately, not all graphs have a cyclically robust cycle basis [9] and understanding for which graph classes such a basis can be found is still an important open problem (see *e.g.,* [15]). Despite all these difficulties, we prove that a bubble generator based on spanning trees of the input graph satisfies properties ($\mathcal{P}_1$) and ($\mathcal{P}_2$). Since our bubble generators are identified from a chosen spanning tree, we also investigate the influence of the choice of spanning tree on the resulting generator.

The remainder of this paper is organised as follows. Section 2 presents some definitions that will be used throughout the paper. Section 3 introduces our family of bubble generators for flow graphs and for arbitrary graphs and we prove that it satisfies properties ($\mathcal{P}_1$) and ($\mathcal{P}_2$). Section 4 presents our experimental results: we first provide an empirical analysis of the characteristics of our new bubble generators based on the choice of the spanning tree (Subsect. 4.1) and then we show an application of our new bubble generators in processing and analysing RNA data (Subsect. 4.2). Finally, we conclude with some open problems in Sect. 5.

2 Preliminaries

Throughout the paper, we assume that the reader is familiar with the standard graph terminology, as contained for instance in [6]. A graph is a pair $G = (V, E)$, where V is the set of vertices, and $E \subseteq V \times V$ is the set of edges. For convenience, we may also denote the set of vertices V of G by $V(G)$ and its set of edges E by $E(G)$. We further set $n = |V(G)|$ and $m = |E(G)|$. A graph may be *directed* or *undirected*, depending on whether its edges are directed or undirected. In this paper, we deal with graphs that are directed, unweighted, finite and without parallel edges. An edge $e = (u, v)$ is said to be *incident* to the vertices u and v, and u and v are said to be the endpoints of $e = (u, v)$. For a directed graph, edge $e = (u, v)$ is said to be leaving vertex u and entering vertex v. Alternatively, $e = (u, v)$ is an outgoing edge for u and an incoming edge for v. The *in-degree* of a vertex v is given by the number of edges entering v, while the *out-degree* of v is the number of edges leaving v. The *degree* of v is the sum of its in-degree and out-degree.

We say that a graph $G' = (V', E')$ is a *subgraph* of a graph $G = (V, E)$ if $V' \subseteq V$ and $E' \subseteq E$. Given a subset of vertices $V' \subseteq V$, the subgraph of G *induced* by V', denoted by $G_{V'}$, has V' as vertex set and contains all edges of G that have both endpoints in V'. Given a subset of edges $E' \subseteq E$, the subgraph of G *induced* by E', denoted by $G_{E'}$, has E' as edge set and contains all vertices of G that are endpoints of edges in E'. Given two subgraphs G and H, their union $G \cup H$ is the graph F for which $V(F) = V(G) \cup V(H)$ and $E(F) = E(G) \cup E(H)$. Their intersection $G \cap H$ is the graph F for which $V(F) = V(G) \cap V(H)$ and $E(F) = E(G) \cap E(H)$.

Let s, t be any two vertices in G. A (*directed*) *path* from s to t in G, denoted as $s \rightsquigarrow t$, is a sequence of vertices and edges $s = v_1, e_1, v_2, e_2, \ldots, v_{k-1}, e_{k-1}, v_k = t$, such that $e_i = (v_i, v_{i+1})$ for $i = 1, 2, \ldots, k - 1$. Since there is no danger of ambiguity, in the remainder of the paper we will also denote a path simply as $s = v_1, v_2, \ldots, v_{k-1}, v_k = t$ (*i.e.*, as a sequence of vertices). A path is *simple* if it does not contain repeated vertices, except possibly for the first and the last vertex. Throughout this paper, all the paths considered will be simple and referred to as paths. A path from s to t is also referred to as an (s, t)-path.

A directed graph G is a *flow graph* if there is one vertex s (referred to as the *start vertex*) which can reach all other vertices. Given a graph G, a rooted spanning tree T of G is a tree where each leaf is reachable from the root by a directed path. Notice that any flow graph has a spanning tree rooted at the start vertex through a graph visit.

Definition 1. *Given a directed graph G and two (not necessarily distinct) vertices $s, t \in V(G)$, an (s, t)-bubble consists of two directed (s, t)-paths that are internally vertex disjoint. Vertex s is the source and t is the target of the bubble. If $s = t$ then exactly one of the paths of the bubble has length 0, and therefore B corresponds to a directed cycle. In this case, we say that B is a degenerate bubble.*

Let G be an undirected graph. Two subgraphs G_1, G_2 of G can be combined by the operator Δ that simply consists in the symmetric difference of the set of edges. More formally, $G_1 \Delta G_2 = (G_1 \cup G_2) \setminus (E(G_1) \cap E(G_2))$ where $E(G_i)$ is the set of edges of G_i. If $G_3 = G_1 \Delta G_2$ we say that G_3 is *generated* by G_1 and G_2. With this operation, it can be shown that the space of all Eulerian subgraphs of G (called the *cycle space* of G) is a vector space [8, 12, 13, 17].

It is known that a cycle basis for a connected undirected graph G, denoted by $C(G)$, has dimension $m - n + 1$. If the graph G is not connected this is generalised to $m - n + c$, where c is the number of connected components (see, *e.g.*, [8, 12, 13, 17]). For a given graph G and a spanning tree T on it, the insertion of one further edge e of the graph to this tree produces a unique cycle $C(T, e)$. Given a spanning tree T of G, the set $C(G) = \{C(T, e) | e \in E(G) \setminus E(T)\}$ is called *Kirchhoff* cycle basis [14].

Let $\mathcal{B}$ be a set of bubbles in G. $\mathcal{B}$ is a *bubble generator* if each bubble in G can be generated by a subset of bubbles in $\mathcal{B}$. A generator is *minimal* if it does not contain a proper subset that is also a generator; and a generator is *minimum* if it has the minimum cardinality. We say that B has *a tree decomposition* in $\mathcal{B}$, if B can be decomposed in a binary-tree-like-fashion where the leaves correspond to bubbles in $\mathcal{B}$ and the internal nodes are bubbles. Notice that a bubble generator satisfies Property $\mathcal{P}_2$ if every bubble of the graph has a tree-decomposition in $\mathcal{B}$.

3   Defining a Bubble Generator from a Spanning Tree

In this section, we define a bubble generator that satisfies properties $(\mathcal{P}_1)$ and $(\mathcal{P}_2)$ starting from a spanning tree of the input graph. We consider first flow graphs and then we extend our results to general graphs. Given a flow graph G with start vertex s, we find a rooted spanning tree T of G, by performing any graph visit starting from s. In the experimental results in Sect. 4 we consider different types of visits, such as Depth-First Search, Breadth-First Search and Scan-First Search [5].

Every non-tree edge $e = (u, v)$ encountered during this visit defines a bubble. The source of this bubble is the least common ancestor w of u and v, and its target is v. The two paths of this bubble are the tree path from w to v and the tree path from w to u followed by the edge (u, v). We denote by $B_T(G)$ the set of bubbles obtained in this way for the flow graph G.

Theorem 1. *Let G be a flow graph with start vertex s, and let $B_T(G)$ be the set of bubbles identified by a tree T obtained through a visit starting from s. Then each bubble in G can be generated starting from the bubbles in $B_T(G)$ (with a symmetric difference operator), and $|B_T(G)| = m - n + 1$.*

Proof. Let T be a rooted spanning tree of G obtained by a visit starting from s and let $B_T(G)$ be the set of bubbles identified by the non-tree edges of T. Consider the undirected graph G' obtained by ignoring the direction of edges in G. We now consider two cases, depending on whether there are parallel edges in G' or not.

Assume first that there are no parallel edges in G'. Note that there is a one-to-one mapping between (undirected) cycles in G' and bubbles in G, and that the spanning tree T found in G is trivially a spanning tree for G'. It is well-known (see for example [13]) that, given an undirected graph G' without parallel edges, taking the cycles formed by the combination of a path in the spanning tree and a single edge outside the tree yields a cycle basis in G' (with a symmetric difference operator). Consider any bubble B in G and let $B_1, \ldots, B_k$ be the bubbles in $B_T(G)$ identified by the non-tree edges of B. If we ignore the directions of the edges, the above property implies that $B \Delta B_1 \Delta \ldots \Delta B_k$ is empty. Consider now the directed graph G notice that $B \Delta B_1 \Delta \ldots \Delta B_k$ is again empty as each edge in G appears in exactly one direction. Hence, each bubble in G can be generated starting from the bubbles in $B_T(G)$. Since there are $m - (n - 1)$ non-tree edges, $|B_T(G)| = m - n + 1$.

If G' has parallel edges, the previous argument cannot be applied directly. However, in this case a simple reduction will work. Note that in G' there can be at most two parallel edges between any two vertices u and v, corresponding to the two edges (u, v) and

(v, u) in the original directed graph G. To deal with this, we transform G into another directed graph G_o as follows: if there are two edges (u, v) and (v, u) in G, we subdivide one of them, say (u, v), by adding a new vertex x_{uv}, by removing the edge (u, v) and by adding two new edges $(u, x_{uv}), (x_{uv}, v)$. Note that there is a one-to-one mapping between bubbles in G and bubbles in G_o: for any vertex x_{uv} in G_o, $(u, x_{uv}), (x_{uv}, v)$ belong to a bubble B_o in G_o if and only if (u, v) belongs to a corresponding bubble B in G. Furthermore, let G'_o be the undirected graph obtained by ignoring the direction of edges in G_o. Since G'_o has no parallel edges, each bubble of G_o can be generated starting from the bubbles in $B_T(G_o)$. Due to the one-to-one mapping between bubbles of G and bubbles of G_o, this implies that each bubble of G can be generated starting from the bubbles in $B_T(G_o)$. Let k be the number of new vertices x_{uv} added to G_o: note that for each new vertex added to G_o, the number of edges of G_o increases by one. This implies that $B_T(G) = B_T(G_o) = (m + k) - (n + k) + 1 = m - n + 1$ and yields the theorem. ∎

Let G be a flow graph with start vertex s and let T be a spanning tree from s. Since each non-tree edge (u, v) is contained exactly in one bubble of $B_T(G)$, Theorem 1 implies that, in order to decompose a generic bubble B into the bubbles of $B_T(G)$, one needs to consider all and only the bubbles of $B_T(G)$ identified by the non-tree edges of B (with respect to T). Moreover, the set $B_T(G)$ can be found efficiently by simply performing a visit from the start vertex s and by returning the non-tree edges.

It is worth mentioning that Theorem 1 can be extended to general graphs as follows. Let G be an arbitrary directed graph G. Let S be a minimum set of vertices from which every vertex of G can be reached. We denoted by S a *source set* of G. Note that in the worst case, $|S| = O(n)$. For each $s \in S$, let $B_T(G, s)$ be the set of bubbles identified by a visit starting from the vertex s of G. Consider the set $B(G, S) = \cup_{s \in S} B_T(G, s)$. Observe that the source of any bubble B in G can be reached by at least one vertex s in S. Thus B belongs to a subgraph of G, which is a flow graph rooted in s, and hence can be expressed as a composition of bubbles in $B_T(G, s)$. This can be summarised by the following theorem.

Theorem 2. *Let G be a directed graph and let S be its source set. Then there is a set of bubbles $\mathcal{B}$, such that each bubble in G can be generated starting from the bubbles in $\mathcal{B}$ (with a symmetric difference operator), and $|\mathcal{B}| \leq |S|(m - n + 1)$.*

Notice that for general graphs, our generator can reach the size of the generator proposed in [1]. However, it will be shown in Sect. 4 that in practice the size of our generator is much smaller. Finally, we show that our generator ensures a tree-like decomposition and thus satisfies Property $\mathcal{P}_2$. In other words, we show that each bubble B in G has a tree decomposition using a subset of bubbles in $\mathcal{B}_T$ and such that in each step we combine only bubbles. To prove this we need first two propositions.

Given a bubble B and two *distinct* vertices u, v in B (not necessarily distinct from s, t), an (u, v)-*chord* of B is a directed path from u to v that is internally vertex disjoint with B (i.e. except for u and v, the path $u \rightsquigarrow v$ has no other vertex in common with B).

Proposition 1. *Given a non-degenerate (s, t)-bubble B and an (u, v)-chord of B such that either there is no directed path $v \rightsquigarrow u$ in B or $\{u, v\} \cap \{s, t\} \neq \emptyset$, then the chord defines two bubbles B_1 and B_2 such that $B = B_1 \triangle B_2$.*

Proof. If u and v are on different legs of B, then we define B_1 to be the bubble with source u and target t and B_2 to be the bubble with source s and target v. Notice that if at least one of u and v coincides with s or t, they can be considered to be in different legs as s and t belong to both legs of B. It is easy to see that $B = B_1 \vartriangle B_2$. These cases are depicted in Fig. 1(a)–(d). If u and v are on the same leg of B then we define B_1 to be the bubble with source u and target v and B_2 to be the bubble with source s and target t. However, if there exists a path from $v \rightsquigarrow u$ in B (see Fig. 1(e_2)) then it is not possible to define the two bubbles B_1 and B_2. Notice that this is the only case where the (u, v)-chord does not allow to define the two bubbles for which $B = B_1 \vartriangle B_2$.   ∎

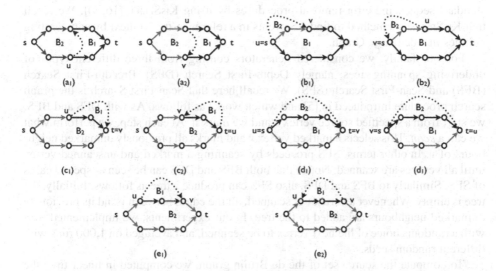

Fig. 1. All the possible cases considered in Proposition 1. In dotted line we have the edges of the (u, v)-chord, the bubble B is composed by the black and grey edges, the bubble B_1 is composed by the black and the dotted line edges and the bubble B_2 by the grey and the dotted line edges.

Proposition 2. *Given a degenerate bubble B then any (u, v)-chord of B defines two bubbles B_1 and B_2 such that $B = B_1 \vartriangle B_2$.*

Proof. The proof follows straightforwardly by observing that every vertex in a directed cycle C has in-degree and out-degree equal to one. After adding the edges of the (u, v)-chord, u has out-degree equal to 2 and v has in-degree 2. Thus the directed cycle C can be written as the sum of B_1 that is the non-degenerate bubble with source u and target v and B_2 that is the degenerate bubble with source and target u (or v).   ∎

Propositions 1 and 2 can be used to prove the following theorem. For lack of space, its proof is deferred to a full version of this paper. Moreover, using the same arguments as for Theorem 2, we can extend it to general graphs.

Theorem 3. *Let G be a flow graph with start vertex r, and let $B_T(G)$ be the set of bubbles identified by a spanning tree T rooted in r. Then any bubble B in G can be decomposed in $O(n)$ time in bubbles in $B_T(G)$ in a tree-like fashion.*

4   Experimental Results

To test the usefulness of our family of generators in practice, we applied it to the identification of AS events in RNA data in a reference-free context. In order to compare our generators to both the state-of-art algorithm KISSPLICE [16, 20] and to the generator defined in [1], we used in our experiments exactly the same dataset as in [1]. This dataset is constructed by selecting the reads corresponding to chromosome 10 from the set of 58 million RNA-seq Illumina paired-end reads extracted from the mouse brain tissue (available in the ENA repository under the following study: PRJEB25574). This leads to a set of 4,932,572 reads. We built the de Bruijn graph from these reads and applied standard sequencing-error-removal procedures by using KISSPLICE [16, 20]. We recall that KISSPLICE is a method to find AS events in a reference-free context by enumerating bubbles in a de Bruijn Graph.

For our family, we considered generators coming from three different types of underlying spanning trees, namely Depth-First Search (DFS), Breadth-First Search (BFS) and Scan-First Search (SFS). We recall here that Scan-First Search is the graph search procedure introduced in [5] and which works as follows. As with DFS and BFS, we start from a specified source vertex s and we mark it. At each step, we perform what we call a *scan*. This selects a marked vertex v and marks all previously unmarked neighbours of v. In other terms, SFS proceeds by scanning a marked and unscanned vertex until all vertices are scanned. Notice that both BFS and DFS can be seen as special cases of SFS. Similarly to BFS and DFS, also SFS can produce a tree as follows. Initially, the tree is empty. Whenever a vertex v is scanned, all the edges between v and its previously unmarked neighbours are added to the tree. In our experiments, we implemented SFS with a random choice of the next vertex to be scanned, and averaged on 1,000 runs with different random seeds.

To compute the source set of the de Bruijn graph, we computed in linear time the DAG of its strongly connected components and chose a vertex from each source. The de Bruijn graph corresponding to our dataset had a total of 83,400 vertices, 99,038 edges and 18,385 source vertices.

Finally, we recall that for general graphs, our new generators are not necessarily minimal. In order to avoid producing duplicates of the same bubble, we discarded a bubble whenever its source was already contained in a tree previously computed from another start vertex. Notice that this does not guarantee the minimality of the generator as there can still be bubbles that can be composed from bubbles that were already present in the generator. For this reason, in general graphs we expect that the size of the generator may vary substantially, depending on the underlying tree chosen.

All our experiments were carried out on a 64-bit machine running Ubuntu 16.04 LTS, equipped with a 2.30 GHz processor Intel(R) Xeon(R) Gold 511, 192 GB of RAM, 16 MB of L3 cache and 1 MB of L2 cache.

4.1   An Empirical Analysis of the Characteristics of the Bubble Generator Based on the Choice of the Spanning Tree

We first explore experimentally some characteristics of bubble generators in our family, depending on the choice of the underlying spanning tree. The parameters we consider are: (i) the size of the generator, (ii) the number of degenerate bubbles (cycles), (iii) the average length of the longest leg, (iv) the average length of the shortest leg, (v) the number of branching bubbles (a branching bubble is a bubble containing more than 5 vertices of in-degree or out-degree greater than 1 [16,20]).

Table 1 shows the main characteristics of generators in our family. We also include the time required to compute each generator. We do not include in this running time the pre-processing time spent in creating the de Bruijn graph, which is exactly the same for all generators. We refer to a generator in our family simply by the graph search used to generate it and we denote by SP-Gen the generator defined in [1].

Table 1. Characteristics of the generators in our family. The columns represent: the size of the generator, $\#ND_{Bubbles}$ the number of non degenerate bubbles found, $\#D_{Bubbles}$ the number of degenerate bubbles (*i.e.* cycles), AvgLong and AvgShort the average length of the longest and shortest leg, respectively, and the time the algorithm spent in seconds. Notice that for Scan-First search trees (SFS) we report the mean and the standard deviation of 1000 different runs.

Generator		Size	$\#ND_{Bubbles}$	$\#D_{Bubbles}$	AvgLong	AvgShort	Time (s)
DFS		12175	11792	383	90.53	40.5	3
BFS		42324	41959	365	33.57	21.23	3
SFS	Mean	41388	41187	201	56.58	41.47	3
	STD	1102.8	1096	6.8	0.3	0.32	0.09
SP-Gen [1]		91486	80108	11378	70.12	31.31	380

As illustrated in Table 1, the size of all our new generators, independently of the underlying spanning tree, is much smaller than the size of SP-Gen [1]. Furthermore, all our new generators can be computed two orders of magnitude faster than SP-Gen. Furthermore, compared to BFS and SFS, the DFS generator usually has smaller size and its bubbles have longer legs. We also observe that, compared to SP-Gen, the percentage of cycles significantly drops in our new generators: from 12.4% for SP-Gen to 3.1% for DFS, 0.8% for BFS and 0.5% for SFS. This is desirable as cycles are degenerate bubbles that do not correspond to AS events, and thus generators that avoid cycles are preferable.

4.2 Application of the Bubble Generator to the Identification of AS Events in RNA-seq Data

As already mentioned in the introduction, identifying AS events in the absence of a reference genome remains a challenging problem. Local assemblers such as KISSPLICE [16] are faced with a dramatically large (and often practically unfeasible) running time due to the exponentially large number of bubbles present, most of which are false positives, *i.e.* they are artificial bubbles not associated with biological events. Indeed, a significantly large number of such artificial bubbles comes from complex subgraphs created by the presence of approximate repeats in the transcriptomic sequence. Thus, tools such as KISSPLICE use heuristics in order to avoid dealing with large portions of a de Bruijn graph containing such complex subgraphs. Here we show how the set of bubbles belonging to generators in our family can be used to predict AS events. Notice that our method is reference-free; however, in order to evaluate it, we make use of annotated reference genomes to assess if our predictions are correct.

To estimate the precision of our new generators in predicting AS events we proceed as follows. We consider the whole set of bubbles belonging to the generator. We then apply the same filter (based on the length of the legs) as in KISSPLICE to extract the bubbles that can be considered as putative AS events. To determine the true AS events, we map the putative bubbles to the *Mus musculus* reference genome and annotations (Ensemble release 94) using STAR [7], which are then analysed by KISSPLICE2REFGENOME [2]. Following [16], a bubble corresponds to a true AS event (or a true positive (TP)) if one leg matches the inclusion isoform and the other the exclusion isoform. Otherwise, the bubble is classified as a false positive. The precision of the method is defined as $TP/(TP + FP)$.

The results for DFS/BFS/SFS and SP-Gen are reported in Table 2. The results show that the number of true AS events found by our generators is comparable to the number of true AS events found by SP-Gen whereas the number of false positives is significantly smaller. Indeed, our generators have a precision between 87.7% and 91.6%, compared to 77.3% for the SP-Gen. An interesting aspect of SP-Gen was that it contained many bubbles that were classified as Intron Retention (IR), which is a type of AS event that is generally particularly hard to identify. As shown in Table 2, the number of IR for our generators remains similar to the one found by SP-Gen.

Table 2. Precision of the generators in our family. The columns represent: number of putative AS events, number of true AS events, precision and number of intron retention events.

Algorithm	#putatitve AS events	#true AS events	Precision	#IR
BFS	1046	959	(91.6%)	319
DFS	1178	1034	(87.7%)	392
SFS	1163	1053	(90.5%)	391
SP-Gen [1]	1403	1085	(77.3%)	377

Since the computation of generators in our family is truly fast in practice, we combined them by taking the union of bubbles coming from different generators and tested whether this would increase the number of AS events found. Notice that the same bubble could be found in two different generators in our family, and thus we eliminated duplicate bubbles in this process. In Table 3 we report the results of different unions of generators in our family (DFS, BFS and 10 randomly chosen runs of SFS), together with the results of SP-Gen and KISSPLICE. As can be seen, the union of different generators in our family allows us to find more true AS events than both SP-Gen and KISSPLICE.

Finally, in [1] it was shown that SP-Gen was able to identify some AS events that will certainly be lost by KISSPLICE. Indeed, the heuristic used by KISSPLICE does not generate bubbles containing a number of branching vertices (*i.e.,* vertices with in-degree or

Table 3. Combining different generators in our family. The columns represent: number of putative AS events, number of true AS events and precision.

Algorithm	#putatitve AS events	#true AS events	Precision
BFS + DFS	1245	1099	88.3%
10-SFS	1622	1179	72.7%
BFS + DFS + 10-SFS	1677	1196	71%
SP-Gen [1]	1403	1085	77.3%
KISSPLICE	1293	1159	89.63%

out-degree at least 2) higher than some threshold. In KISSPLICE, the default value for this branching threshold is 5. Increasing the value of this threshold will increase exponentially the running time of the algorithm and thus a large branching threshold is unfeasible in practice. As reported in [1], around 27 true AS events in SP-Gen have a branching number higher than 5, and are lost by KISSPLICE. For the family of our generators, we have that the number of true AS events that are certainly lost by KISSPLICE is: (a) 16 for the BFS, (b) 77 for the DFS, and (c) an average of 80 for SFS (averaged over different choices of the random seed).

5   Conclusions and Open Problems

In this paper, we have proposed a new family of bubble generators which improves substantially on the previous generator (SP-Gen [1]): generators in the new family are much faster, *i.e.,* about two orders of magnitude faster than SP-Gen, and they are still able to achieve similar (and sometimes higher) precision in identifying AS events.

Our work raises several new and perhaps intriguing questions. First, we notice that while for flow graphs our family produces minimum generators, for general graphs it is still open to find a minimum bubble generator. Second, the fast computation of our new generators opens the way to the design of algorithms that efficiently combine the bubbles of a generator in order to find more AS events. Third, we believe that the number of false positives could be reduced by adding more biologically motivated constraints. An example of constraint that can be introduced toward this aim is to give a weight to each edge of the de Bruijn graph based on the reads coverage. A true AS event would then correspond to bubbles in which the edges inside a leg must have similar weights (but different legs may have different coverage). Fourth, when constructing a de Bruijn graph from RNA-seq reads, some filters are applied that are meant to eliminate sequencing errors. These filters remove vertices and edges whose coverage by the set of reads is below some given thresholds. Changing those thresholds has a significant impact on the resulting de Bruijn graph, and hence on the set of solutions. Is it possible to compute in a dynamic fashion a bubble generator when this coverage threshold is changing, without having to recompute everything from scratch?

References

1. Acuña, V., et al.: On bubble generators in directed graphs. In: Bodlaender, H.L., Woeginger, G.J. (eds.) WG 2017. LNCS, vol. 10520, pp. 18–31. Springer, Cham (2017). https://doi.org/10.1007/978-3-319-68705-6_2. Announced at WG 2017
2. Benoit-Pilven, C., et al.: Complementarity of assembly-first and mapping-first approaches for alternative splicing annotation and differential analysis from RNAseq data. Sci. Rep. **8**(1), 1–13 (2018)
3. Birmelé, E., et al.: Efficient bubble enumeration in directed graphs. In: Calderón-Benavides, L., González-Caro, C., Chávez, E., Ziviani, N. (eds.) SPIRE 2012. LNCS, vol. 7608, pp. 118–129. Springer, Heidelberg (2012). https://doi.org/10.1007/978-3-642-34109-0_13
4. Brankovic, L., Iliopoulos, C.S., Kundu, R., Mohamed, M., Pissis, S.P., Vayani, F.: Linear-time superbubble identification algorithm for genome assembly. Theoret. Comput. Sci. **609**, 374–383 (2016)
5. Cheriyan, J., Kao, M.-Y., Thurimella, R.: Scan-first search and sparse certificates: an improved parallel algorithm for k-vertex connectivity. SIAM J. Comput. **22**(1), 157–174 (1993)
6. Cormen, T.H., Leiserson, C.E., Rivest, R.L., Stein, C.: Introduction to Algorithms, 3rd edn. The MIT Press, Cambridge (2009)
7. Dobin, A., et al.: STAR: ultrafast universal RNA-seq aligner. Bioinformatics **29**(1), 15–21 (2013)
8. Gleiss, P.M., Leydold, J., Stadler, P.F.: Circuit bases of strongly connected digraphs. Discuss. Math. Graph Theory **23**(2), 241–260 (2003)
9. Hammack, R.H., Kainen, P.C.: Robust cycle bases do not exist for $K_{n,n}$ if $n \geq 8$. Discret. Appl. Math. **235**, 206–211 (2018)
10. Iqbal, Z., Caccamo, M., Turner, I., Flicek, P., McVean, G.: *De novo* assembly and genotyping of variants using colored de Bruijn graphs. Nat. Genet. **44**(2), 226–232 (2012)
11. Kainen, P.C.: On robust cycle bases. Electron. Notes Discret. Math. **11**, 430–437 (2002). The Ninth Quadrennial International Conference on Graph Theory. Combinatorics, Algorithms and Applications
12. Kavitha, T., et al.: Cycle bases in graphs characterization, algorithms, complexity, and applications. Comput. Sci. Rev. **3**(4), 199–243 (2009)
13. Kavitha, T., Mehlhorn, K.: Algorithms to compute minimum cycle bases in directed graphs. Theory Comput. Syst. **40**(4), 485–505 (2007)
14. Kirchhoff, G.: Ueber die auflösung der gleichungen, auf welche man bei der untersuchung der linearen vertheilung galvanischer ströme geführt wird. Ann. Phys. **148**(12), 497–508 (1847)
15. Klemm, K., Stadler, P.F.: A note on fundamental, non-fundamental, and robust cycle bases. Discret. Appl. Math. **157**(10), 2432–2438 (2009). Networks in Computational Biology
16. Lima, L., et al.: Playing hide and seek with repeats in local and global de novo transcriptome assembly of short RNA-seq reads. Algorithms Mol. Biol. **12**, 2 (2017)
17. MacLane, S.: A combinatorial condition for planar graphs. Fundamenta Mathematicae **28**, 22–32 (1937)
18. Miller, J.R., Koren, S., Sutton, G.: Assembly algorithms for next-generation sequencing data. Genomics **95**(6), 315–327 (2010)
19. Onodera, T., Sadakane, K., Shibuya, T.: Detecting superbubbles in assembly graphs. In: Darling, A., Stoye, J. (eds.) WABI 2013. LNCS, vol. 8126, pp. 338–348. Springer, Heidelberg (2013). https://doi.org/10.1007/978-3-642-40453-5_26
20. Sacomoto, G., et al.: KISSPLICE: de-novo calling alternative splicing events from RNA-seq data. BMC Bioinform. **13**(S–6), S5 (2012). https://doi.org/10.1186/1471-2105-13-S6-S5

21. Sacomoto, G., Lacroix, V., Sagot, M.-F.: A polynomial delay algorithm for the enumeration of bubbles with length constraints in directed graphs and its application to the detection of alternative splicing in RNA-seq data. In: Darling, A., Stoye, J. (eds.) WABI 2013. LNCS, vol. 8126, pp. 99–111. Springer, Heidelberg (2013). https://doi.org/10.1007/978-3-642-40453-5_9

22. Sammeth, M.: Complete alternative splicing events are bubbles in splicing graphs. J. Comput. Biol. **16**(8), 1117–1140 (2009)

23. Sung, W.-K., Sadakane, K., Shibuya, T., Belorkar, A., Pyrogova, I.: An $O(m \log m)$-time algorithm for detecting superbubbles. IEEE/ACM Trans. Comput. Biol. Bioinform. **12**(4), 770–777 (2015)

24. Uricaru, R., et al.: Reference-free detection of isolated SNPs. Nucleic Acids Res. **43**(2), e11 (2015)

25. Younsi, R., MacLean, D.: Using $2k + 2$ bubble searches to find single nucleotide polymorphisms in k-mer graphs. Bioinformatics **31**(5), 642–646 (2015)

The Micro-world of Cographs

Bogdan Alecu[1]([✉]), Vadim Lozin[1], and Dominique de Werra[2]

[1] Mathematics Institute, University of Warwick, Coventry CV4 7AL, UK
{B.Alecu,V.Lozin}@warwick.ac.uk
[2] Institute of Mathematics, EPFL, 1015 Lausanne, Switzerland
dominique.dewerra@epfl.ch

Abstract. Cographs constitute a small point in the atlas of graph classes. However, by zooming in on this point, we discover a complex world, where many parameters jump from finiteness to infinity. In the present paper, we identify several milestones in the world of cographs and create a hierarchy of graph parameters grounded on these milestones.

1 Introduction

Large things are seen from a distance, but to examine small things, one needs to look up-close. Cographs constitute a small class and in this paper we analyse it with a "magnifying glass", trying to spot the details. With a closer look at this class we discover a complex world and observe that many important parameters can be arbitrarily large within cographs. This is the case, for instance, for chromatic number, co-chromatic number, matching number, tree-width, linear clique-width and many others. Moreover, such parameters jump to infinity on specific subclasses of cographs. This is due to the fact that the class of cographs is well-quasi-ordered under the induced subgraph relation [8], and therefore, for every parameter p which is unbounded in the class of cographs, there exists a finite collection $M(p)$ of inclusion-wise minimal hereditary subclasses of cographs, where p can be arbitrarily large. This observation suggests a simple way of comparing two parameters: a parameter p_1 is stronger than a parameter p_2 if for every class $X \in M(p_1)$ there exists a class $Y \in M(p_2)$ such that $Y \subseteq X$. In other words, p_1 is stronger than p_2 if the family of cograph subclasses where p_1 is bounded contains the family of cograph subclasses where p_2 is bounded.

For some parameters, identifying minimal classes is an easy task. For instance, since cographs are perfect, the chromatic number is bounded if and only if the clique number is bounded and hence the class of complete graphs is the only minimal hereditary subclass of cographs where the chromatic number is unbounded. However, in general, identifying minimal classes is far from being trivial, as the example of linear clique-width shows. The authors of [5] develop a sophisticated approach to show that there exist precisely two minimal hereditary subclasses of cographs where linear clique-width is unbounded: the class of (P_4, C_4)-free graphs, also known as the quasi-threshold [21] or trivially perfect [15] graphs, and the class of their complements.

L. Gąsieniec et al. (Eds.): IWOCA 2020, LNCS 12126, pp. 30–42, 2020.
https://doi.org/10.1007/978-3-030-48966-3_3

In the present paper, we characterise a variety of other graphs parameters in terms of minimal hereditary subclasses of cographs where these parameters are unbounded, which is the content of Sects. 3 and 4. In Sect. 2, we introduce basic terminology and notation used throughout the paper.

2 Preliminaries

All graphs in this paper are simple, i.e., finite, undirected, without loops and without multiple edges. The vertex set and the edge set of a graph G are denoted by $V(G)$ and $E(G)$, respectively. As usual, P_n, C_n, K_n denote a chordless path, a chordless cycle and a complete graph with n vertices, respectively. Also, $K_{n,m}$ is a complete bipartite graph with parts of size n and m.

The complement of a graph G is denoted by $\overline{G}$. Given two graphs G and H, we denote by $G \cup H$ the disjoint union of G and H and by $G \times H$ the join of G and H, i.e., the graph obtained from $G \cup H$ by adding all possible edges between G and H. Two sets $A, B \subseteq V(G)$ are said to be complete to each other if every possible edge between them appears in G, and anticomplete to each other if they are complete to each other in $\overline{G}$. The disjoint union of p copies of G will be denoted by pG.

A *clique* in a graph is a subset of pairwise adjacent vertices and an *independent set* is a subset of pairwise non-adjacent vertices. We say that a graph G is H-free if G does not contain a copy of H as an induced subgraph.

A class of graphs is *hereditary* if it is closed under taking induced subgraphs. It is well-known (and not difficult to see) that a class is hereditary if and only if it can be characterised in terms of minimal forbidden induced subgraphs.

The class of cographs is the class of graphs that can be obtained from K_1 by taking complements and disjoint unions. In particular, every cograph with at least two vertices can be represented either as $G \cup H$ or as $G \times H$ for two non-empty graphs G and H. It is well known that the class of cographs is precisely the class of P_4-free graphs.

Since the complement of a cograph is again a cograph, with every subclass $\mathcal{X}$ of cographs we associate the subclass $\overline{\mathcal{X}}$ of complements of graphs in $\mathcal{X}$. The following subclasses of cographs will play a critical role in our study:

$\mathcal{Q}$ the class of *quasi-threshold graphs*, i.e., (P_4, C_4)-free graphs,

$\mathcal{T}$ the class of *threshold graphs*. This is the class of $(P_4, C_4, 2K_2)$-free graphs, i.e., the intersection of $\mathcal{Q}$ and $\overline{\mathcal{Q}}$.

$\mathcal{U}$ the class of P_3-free graphs, i.e., graphs every connected component of which is a clique.

$\mathcal{K}$ the class of complete graphs.

$\mathcal{F}$ the class of star forests, i.e., graphs every connected component of which is a star. This is the class of (P_4, C_4, K_3)-free graphs, i.e., the class of bipartite graphs in $\mathcal{Q}$.

$\mathcal{M}$ the class of graphs of vertex degree at most 1. This is the class of (P_3, K_3)-free graphs, i.e., the class of bipartite graphs in $\mathcal{U}$.

$\mathcal{B}$ the class of complete bipartite graphs (an edgeless graph is counted as complete bipartite with one part being empty). This is the class of $(\overline{P}_3, K_3)$-free graphs, i.e., the class of bipartite graphs in $\overline{\mathcal{U}}$.

$\mathcal{S}$ the class of *stars*, i.e., graphs of the form $K_{1,n}$ and their induced subgraphs.

The Ramsey number $R(a, b)$ is the smallest natural number such that any graph with $R(a, b)$ vertices contains a clique of size a or an independent set of size b.

3 Graph Parameters

We start by reporting some known results or results that readily follows from known results. In particular, directly from Ramsey's Theorem we derive the following conclusion:

Proposition 1. *The class $\mathcal{K}$ of complete graphs and the class of $\mathcal{S}$ of stars are the only two minimal hereditary classes of graphs of unbounded maximum vertex degree.*

To report more results, we denote by

$\alpha(G)$ the *independence number* of G, i.e., the size of a maximum independent set in G,

$\omega(G)$ the *clique number* of G, i.e., the size of a maximum clique in G,

$\chi(G)$ the *chromatic number* of G, i.e., the minimum number of subsets in a partition of $V(G)$ such that each subset is an independent set,

$y(G)$ the *clique partition* (also known as *clique cover*) *number*, i.e., the minimum number of subsets in a partition of $V(G)$ such that each subset is a clique.

Clearly, the class $\mathcal{K}$ of complete graphs is the only minimal hereditary class of unbounded clique number, i.e., by forbidding a complete graph we obtain a class of bounded clique number. Also, it is not difficult to see that $\mathcal{K}$ is a minimal hereditary class of unbounded chromatic number. However, it is not the only minimal hereditary class of unbounded chromatic number, i.e., forbidding a complete graph does not guarantee a bound on the chromatic number. Moreover, as shown by Erdős [10] chromatic number is unbounded even in the class of $(C_3, C_4, \ldots, C_k)$-free graphs for any value of k, which means that in the universe of hereditary classes chromatic number cannot be characterised by means of minimal classes where this parameter is unbounded. On the other hand, when we restrict ourselves to cographs such a characterization is possible, which is due to the fact that cographs are perfect, and hence $\omega(G) = \chi(G)$ for any cograph G. As a result, we obtain the following conclusion.

Proposition 2. *The class $\mathcal{K}$ of complete graphs is the only minimal hereditary subclass of cographs of unbounded clique number and chromatic number.*

The *degeneracy* of a graph G is the smallest value of k such that every induced subgraph of G has a vertex of degree at most k. It is not difficult to see that the class $\mathcal{K}$ of complete graphs and the class of $\mathcal{B}$ of complete bipartite graphs are minimal hereditary classes of unbounded degeneracy. However, these are not the only minimal classes, because forbidding a complete graph and a complete bipartite graph does not guarantee a bound on the degeneracy. To explain this, we observe that the degeneracy of G is bounded from below by $\chi(G) - 1$ and from above by the tree-width of G. Therefore, degeneracy and tree-width are unbounded in the class of $(C_3, C_4, \ldots, C_k)$-free graphs for any value of k, and for $k \geq 4$ the set of forbidden induced subgraphs include both a complete graph C_3 and a complete bipartite graph C_4. This discussion shows that, similarly to chromatic number, in the universe of all hereditary classes neither degeneracy nor tree-width admit a characterization in terms of minimal classes where these parameters are unbounded. On the other hand, again similarly to chromatic number, such a characterization is possible when restricting to cographs, and it is presented in the next claim.

Proposition 3. *The class $\mathcal{K}$ of complete graphs and the class of $\mathcal{B}$ of complete bipartite graphs are the only two minimal hereditary subclasses of cographs of unbounded degeneracy and tree-width.*

Proof. To prove the claim, it suffices to show that for any s and p, the tree-width of $(P_4, K_s, K_{p,p})$-free graphs is bounded by a constant. For this, we refer the reader to the following result from [1]: for every t, p, s, there exists a $z = z(t, p, s)$ such that every graph with a (not necessarily induced) path of length at least z contains either an induced P_t or an induced $K_{p,p}$ or a clique of size s. From this result it follows that $(P_4, K_s, K_{p,p})$-free graphs do not contain (not necessarily induced) paths of length $z(4, p, s)$. It is well known (see, e.g., [12]) that graphs of bounded path number (the length of a longest path) have bounded tree-width.
□

The *matching number* of a graph G is the size of a maximum matching in G. The following result was proved in [7].

Lemma 1. *For any natural numbers s, t and p, there is a number $N(s, t, p)$ such that every graph with a matching of size at least $N(s, t, p)$ contains either a clique K_s or an induced bi-clique $K_{t,t}$ or an induced matching pK_2.*

A natural corollary from this result is the following characterization of the matching number in terms of minimal hereditary classes where this parameter is unbounded.

Theorem 1. *$\mathcal{M}$, $\mathcal{B}$ and $\mathcal{K}$ are the only three minimal hereditary classes of graphs of unbounded matching number.*

The *vertex cover number* of a graph G is the size of a minimum vertex cover in G. It is well known that the vertex cover number is never smaller than the matching number and never larger than twice the matching number. Therefore, the characterization of matching number given in Theorem 1 applies to the vertex cover number as well.

Theorem 2. $\mathcal{M}$, $\mathcal{B}$ and $\mathcal{K}$ *are the only three minimal hereditary classes of graphs of unbounded vertex cover number.*

The *neighbourhood diversity* of a graph was introduced in [16] and can be defined as follows.

Definition 1. *Let us say that two vertices x and y are similar if there is no vertex z distinguishing them (i.e., if there is no vertex z adjacent to exactly one of x and y). Vertex similarity is an equivalence relation. We denote by $nd(G)$ the number of similarity classes in G and call it the neighbourhood diversity of G.*

Neighbourhood diversity was characterised in [17] by means of nine minimal hereditary classes of graphs where this parameter is unbounded. Six of these minimal classes contain a P_4. Therefore, when restricted to cographs, neighbourhood diversity can be characterised by three minimal classes as follows.

Theorem 3. $\mathcal{M}$, $\overline{\mathcal{M}}$, *and* $\mathcal{T}$ *are the only three minimal hereditary subclasses of cographs of unbounded neighbourhood diversity.*

3.1 Co-chromatic Number

The *co-chromatic number* of G, denoted $z(G)$, is the minimum number of subsets in a partition of $V(G)$ such that each subset is either a clique or an independent set [11]. It is not difficult to see that the co-chromatic number can be arbitrarily large in the class of P_3-free graphs, where each graph is a disjoint union of cliques. Therefore, it is also unbounded in the complements of P_3-free graphs, also known as complete multipartite graphs. In what follows, we show that these are the only two minimal subclasses of cographs of unbounded co-chromatic number.

Lemma 2. *Let n, m, t be positive integers with $t \geq 2$. If G is a $(nK_t, \overline{mK_t})$-free cograph, then $z(G) \leq 2^{m+n-1}(t-1)$.*

Proof. Call a partition of $V(G)$ *good* if it contains at least $t-1$ cliques and $t-1$ independent sets (empty sets in the partition may count as either). We prove by induction on $m+n$ that G admits a good partition into $2^{m+n-1}(t-1)$ sets, each of which is a clique or an independent set.

If $m+n = 2$ ($n = m = 1$), then G is K_t-free. Hence $\chi(G) = \omega(G) \leq t-1$; we add empty sets to the partition until we reach $2(t-1)$ sets in total. This makes the partition good, and we have proved the basis for the induction. In general, put $G' := G$. We are in one of the following three cases:

(a) $G' = G_1 \cup G_2$, and both G_1 and G_2 are K_t-free, OR $G' = G_1 \times G_2$, and both G_1 and G_2 are $\overline{K_t}$-free.

(b) $G' = G_1 \cup G_2$, and both G_1 and G_2 contain a K_t, OR $G' = G_1 \times G_2$, and both G_1 and G_2 contain a $\overline{K_t}$.

(c) $G' = G_1 \cup G_2$, G_1 contains a K_t and G_2 is K_t-free, OR $G' = G_1 \times G_2$, G_1 contains a $\overline{K_t}$ and G_2 is $\overline{K_t}$-free.

As long as we are in case (c), iteratively put $G' := G_1$. We end up with a graph G' in either case (a) or (b). Note first that any good partition of G' extends to a good partition of G without increasing the number of sets. Indeed, at each step, G_2 was either K_t-free and anticomplete to the rest of the graph or $\overline{K_t}$-free and complete to the rest of the graph. The disjoint union of all K_t-free G_2s is again K_t-free and hence can be partitioned into at most $t-1$ independent sets, and we take the union of each of these sets with one of the independent sets in the good partition of G' injectively. Similarly, the join of the $\overline{K_t}$-free G_2s can be partitioned into at most $t-1$ cliques, each of which we join to one of the cliques in the good partition of G' injectively.

Now, if G' is in case (a), then G' is K_t-free or $\overline{K_t}$-free and we act like in the base case to obtain a good partition of G' (and therefore of G) in $\overline{2(t-1)}$ sets. If G' is in case (c), then G_1 and G_2 are both either $(n-1)K_t$-free or $\overline{(m-1)K_t}$-free. In either case, the inductive hypothesis applies, and we have a good partition of G' of size at most

$$2^{m+n-2}(t-1) + 2^{m+n-2}(t-1) = 2^{m+n-1}(t-1).$$

Like before, this extends to a partition of G, concluding the proof. □

Lemma 2 naturally leads to the following conclusion.

Theorem 4. *The class $\mathcal{U}$ of P_3-free graphs and the class $\overline{\mathcal{U}}$ of $\overline{P}_3$-free graphs are the only two minimal hereditary subclasses of cographs of unbounded co-chromatic number.*

3.2 Lettericity

The notion of letter graphs was introduced in [19] and can be defined as follows.

Let A be a finite alphabet, $D \subseteq A^2$ and $w = w_1 w_2 \ldots w_n$ a word over A (repetitions allowed). The letter graph $G(D, w)$ associated to w has $\{1, 2, \ldots, n\}$ as its vertex set, and two vertices $i < j$ are adjacent if and only if the ordered pair (w_i, w_j) belongs to D. A graph G is said to be a letter graph if there exist an alphabet A, a subset $D \subseteq A^2$ and a word $w = w_1 w_2 \ldots w_n$ over A such that G is isomorphic to $G(D, w)$.

The role of D is to decode (transform) a word into a graph and therefore we refer to D as a decoder. Every graph G is trivially a letter graph over the alphabet $A = V(G)$ with the decoder $D = \{(v, w), (w, v) : \{v, w\} \in E(G)\}$. The lettericity of G, denoted $\ell(G)$, is the minimum k such that G is representable as a letter graph over an alphabet of k letters.

To give a less trivial example, consider the alphabet $A = \{a, b\}$ and the decoder $D = \{(a, a), (a, b)\}$. Then the word $abababab$ describes the graph represented in Fig. 1. This graph can be constructed from a single vertex by means of two operations: adding a dominating vertex (corresponds to adding letter a as a prefix) or adding an isolated vertex (corresponds to adding letter b as a prefix). The class of all graphs that can be constructed by means of these two operations coincides with the class of threshold graphs defined in Sect. 2

as $(2K_2, C_4, P_4)$-free graphs [18]. The above discussion shows that a graph is threshold if and only if it is a letter graph over the alphabet $A = \{a, b\}$ with the decoder $D = \{(a, a), (a, b)\}$.

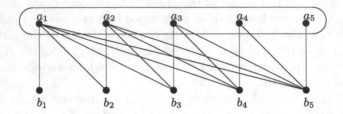

Fig. 1. The letter graph of the word *abababab* (the oval represents a clique). We use indices to indicate in which order the a-letters and the b-letters appear in the word.

Lemma 3. $\ell(nK_2) = n$.

Proof. First, it is not difficult to see that $\ell(nK_2) \leq n$, since n letters suffice (one letter per edge). Assume $\ell(nK_2) < n$, then there must exist a letter a representing at least 3 vertices of the graph. Clearly, $(a, a) \notin D$, since otherwise a triangle arises. Then the neighbour of the middle a is different from a, say b. If this neighbour appears before the middle a, it must also be adjacent to the last a. If it appears after the middle a, it must also be adjacent to the first a. In both case, b has at least two neighbours. Therefore, $\ell(nK_2) \geq n$. $\square$

The above theorem shows that the lettericity is unbounded in the class $\mathcal{M}$ of graphs of vertex degree at most 1. Therefore, it is also unbounded in the class $\overline{\mathcal{M}}$, since $\ell(G) = \ell(\overline{G})$.

Theorem 5. $\mathcal{M}$ and $\overline{\mathcal{M}}$ *are the only two minimal hereditary subclasses of cographs of unbounded lettericity.*

Proof. To prove the theorem, we will show that for any natural numbers $p, t \geq 2$, the lettericity of a $(P_4, pK_2, \overline{tK}_2)$-free graph G is at most 2^{p+t-3}. This will be shown by induction on $p + t$. Moreover, we will show that G can be represented with a decoder D containing a source letter, i.e., a letter a such that $(a, b) \in D$ for any letter b, and a sink letter, i.e., a letter b such that $(b, a) \notin D$ for any letter a.

If $p = t = 2$, then G is a threshold graph and its lettericity is at most 2, because any threshold graph can be represented over the decoder $D = \{(a, a), (a, b)\}$. In this decoder, a is a source letter and b is a sink letter.

Assume that every $(P_4, pK_2, \overline{tK}_2)$-free graph with $p + t \leq k$ can be represented as a letter graph over an alphabet of at most 2^{p+t-3} letters with a decoder containing a source vertex a and a sink vertex b. Consider now a $(P_4, pK_2, \overline{tK}_2)$-free graph G with $p + t = k + 1$.

The presence of source and sink letters in the decoder allows us to assume that G has neither dominating nor isolated vertices. Indeed, if v is dominating, then a word for G can be constructed from a word for $G - v$ by adding a source letter as a prefix, and if v is isolated, then a word for G can be constructed from a word for $G - v$ by adding a sink letter as a prefix. Therefore, in the rest of the proof we assume that G has neither isolated nor dominating vertices.

Case 1: G is disconnected. Denote by G_1 a connected component of G and by G_2 the rest of the graph. Observe that each of G_1 and G_2 contains a K_2, since otherwise G has an isolated vertex. Therefore, each of G_1 and G_2 is $(p-1)K_2$-free and hence we can apply induction to each of G_1 and G_2. In other words, G_1 can be represented by a word ω_1 over an alphabet A_1 of size at most 2^{p+t-4} with a decoder containing a source vertex a_1 and a sink vertex b_1, and G_1 can be represented by a word ω_2 over an alphabet A_2 of size at most 2^{p+t-4} with a decoder containing a source vertex a_2 and a sink vertex b_2 (we assume that A_1 and A_2 are disjoint). Then the word $\omega = \omega_1 \omega_2$ represents G over the alphabet $A_1 \cup A_2$ of size at most 2^{p+t-3} with the decoder $D = D_1 \cup D_2$. In this decoder, vertex b_2 is a sink vertex. To guarantee the presence of a source vertex, we add to D the pair (a_2, c) for every vertex $c \in A_1$. This extension transforms a_2 into a source vertex and does not change the graph represented by the word ω, since every letter from A_1 appears in ω before any appearance of a_2.

Case 2: G is connected. In this case, $\overline{G}$ is disconnected and $(P_4, tK_2, \overline{pK_2})$-free. A similar argument as above gives a representation for $\overline{G}$ with at most 2^{p+t-3} letters, and complementing the corresponding decoder produces one for G (note that when doing that, sink letters become source letters and vice-versa).   □

3.3 Boxicity

The *boxicity* box(G) of a graph G is the minimum dimension in which G can be represented as an intersection graph of hyper-rectangles. Equivalently, it is the smallest number of interval graphs on the same set of vertices whose intersection is G. The next lemma was shown in [20]; we give here a proof for the sake of completeness.

Lemma 4. box$(\overline{nK_2}) = n$.

Proof. To see that box$(\overline{nK_2}) \leq n$, note that K_{2n} without an edge is an interval graph, and $\overline{nK_2}$ is the intersection of n such graphs. Conversely, note that two different matched non-edges in $\overline{nK_2}$ cannot belong to the same interval graph (since the corresponding four vertices would induce a C_4, which is not an interval graph). Hence we need at least n interval graphs to obtain $\overline{nK_2}$ as an intersection.   □

Lemma 5. *Let G_1 and G_2 be two graphs. Then*

$$\text{box}(G_1 \cup G_2) \leq \max(\text{box}(G_1), \text{box}(G_2)) \text{ and } \text{box}(G_1 \times G_2) \leq \text{box}(G_1) + \text{box}(G_2).$$

Moreover, if G_2 is a clique, then box$(G_1 \times G_2) = $ box(G_1).

Proof. Suppose $G_1 = \bigcap_{i=1}^{s} A_i$ and $G_2 = \bigcap_{i=1}^{t} B_i$ where the A_i and B_i are interval graphs, and assume without loss of generality that $s \geq t$. Put $C_i = A_i \cup B_i$ for $1 \leq i \leq t$ and $C_i = A_i \cup K_{|V(G_2)|}$ for $t < i \leq s$. Put $D_i = A_i \times K_{|V(G_2)|}$ for $1 \leq i \leq s$ and $D_i = K_{|V(G_1)|} \times B_{i-s}$ for $s < i \leq s+t$.

The C_i and D_i are interval graphs, and with the obvious labellings of C_i and D_i, we have $G_1 \cup G_2 = \bigcap_{i=1}^{s} C_i$ and $G_1 \times G_2 = \bigcap_{i=1}^{s+t} D_i$.

For the final claim, if $G_2 = K_{|V(G_2)|}$ is a clique, then $G_1 \times G_2 = \bigcap_{i=1}^{s} (A_i \times K_{|V(G_2)|})$, and each of those is an interval graph. $\square$

Theorem 6. $\overline{\mathcal{M}}$ *is the only minimal hereditary subclass of cographs of unbounded boxicity.*

Proof. Let $n \geq 2$. We prove by induction on n that $(P_4, \overline{nK_2})$-free graphs have boxicity at most 2^{n-2}. The result is true for $n = 2$, since (P_4, C_4)-free graphs are known to be interval graphs (see, e.g., [4]).

For the induction step, suppose the result is true for some $n \geq 2$, and let G be a cograph that is $\overline{(n+1)K_2}$-free. By Lemma 5, we may assume that G is connected, and in particular that $G = G_1 \times G_2$ where neither of the cographs G_1 or G_2 is a clique. But then G_1 and G_2 each have a $\overline{K_2}$, and so they are both $\overline{nK_2}$-free. The induction hypothesis applies, and another application of Lemma 5 gives us that $\operatorname{box}(G) \leq \operatorname{box}(G_1) + \operatorname{box}(G_2) \leq 2^{n-2} + 2^{n-2} = 2^{n-1}$ as required. $\square$

3.4   H-Index

The H-index $h(G)$ of a graph G is the largest $k \geq 0$ such that G has k vertices of degree at least k. This parameter is important in the study of dynamic algorithms [9]. Clearly, H-index is unbounded for cographs, since it is unbounded for complete graphs. To characterise this parameter in terms of minimal subclasses of cographs with unbounded H-index, we start with a helpful lemma.

Lemma 6. *Let $G_1, \dots, G_t$ be graphs. Then*

$$h\left(\bigcup_{i=1}^{t} G_i\right) \leq \sum_{i=1}^{t} h(G_i), \quad \text{and} \quad h(G_1 \times G_2) \leq \min(h(G_1) + |V(G_2)|, h(G_2) + |V(G_1)|).$$

Proof. For the first bound, note that for any j, $1 + \sum_i h(G_i) > h(G_j)$. In particular, by definition of the H-index, each G_j has at most $h(G_j)$ vertices of degree $1 + \sum_i h(G_i)$ or more, and so $\bigcup_j G_j$ has at most $\sum_j h(G_j)$ vertices of degree at least $1 + \sum_i h(G_i)$, from which the claim follows.

For the other bound, note that $G_1 \times G_2$ has at most $|V(G_2)|$ vertices of degree at least $h(G_1) + |V(G_2)| + 1$ coming from G_2, and at most $h(G_1)$ coming from G_1, since[1] $\deg_{G_1 \times G_2}(v) = \deg_{G_1}(v) + |V(G_2)|$ for any $v \in G_1$, and G_1 does not have

---

[1] When a vertex v appears in more than one graph, we write $\deg_G(v)$ for the degree of v in graph G.

more than $h(G_1)$ vertices of degree $h(G_1) + 1$. By definition of the H-index, we obtain that $h(G_1 \times G_2) \leq h(G_1) + |V(G_2)|$, and the claim follows by symmetry.

□

Theorem 7. $\mathcal{K}$, $\mathcal{B}$ and the class $\mathcal{F}$ of star forests are the only minimal hereditary subclasses of cographs of unbounded H-index.

Proof. One can check that those are, indeed, minimal hereditary classes of unbounded H-index. To see they are the only ones, let $p, q, r, s \geq 1$. We will show by induction on $p + r$ that if G avoids K_p, $K_{q,q}$ and $rK_{1,s}$, then the H-index of G is bounded by a constant $H(p, q, r, s)$. For the base case, note that if $p = 1$, this is trivial, and if $r = 1$, then G is $(K_p, K_{1,s})$-free and therefore the maximum vertex degree in G is bounded by $R(p, s)$. This in turn implies that $h(G) \leq R(p, s)$. We may thus assume $p, r \geq 2$.

If $G = G_1 \times G_2$ is a join of non-empty graphs, then not both G_1 and G_2 have more than $R(p, q)$ vertices. Indeed, if both do, then either one of them contains a clique of size p, which is forbidden, or they both have independent sets of size q, which again cannot happen since $K_{q,q}$ is forbidden. Without loss of generality, we may assume that $|V(G_2)| \leq R(p, q)$. In this case, by Lemma 6, $h(G) \leq h(G_1) + R(p, q)$. Since $|V(G_2)| \geq 1$, G_1 is K_{p-1}-free, so by the induction hypothesis, $h(G_1)$ is bounded by $H(p - 1, q, r, s)$.

If $G = \bigcup_{i=1}^{t} G_i$ is a union of connected graphs, we may write $G = G_1 \cup \ldots G_l \cup G'$, where $G_1, \ldots, G_l$ each have a $K_{1,s}$, and G' is $K_{1,s}$-free (we may have $l = 0$). Since K_p and $K_{1,s}$ are forbidden for G', the maximum vertex degree, and hence the H-index of G', is bounded by $R(p, s)$. Moreover, if $l \geq 2$ and so two of the components of G do have a $K_{1,s}$, then we may write G as the union of two graphs that are $(r - 1)K_{1,s}$-free, and by Lemma 6, $h(G) \leq 2H(p, q, r - 1, s)$. Finally, if only one component has a $K_{1,s}$, then that component is a join of non-empty graphs and we obtain, again by Lemma 6 and from the previous paragraph, $h(G) \leq H(p - 1, q, r, s) + R(p, q) + R(p, s)$.

Combining the above, we obtain

$$H(p, q, r, s) \leq \max(H(p - 1, q, r, s) + R(p, q) + R(p, s), 2H(p, q, r - 1, s)).$$

□

3.5   Achromatic Number

A *complete k-colouring* is a partition of G into k independent sets (the "colour classes") such that any two independent sets in the partition have at least one edge between them. The *achromatic number* $\psi(G)$ of a graph G is the maximum number k such that G admits a complete k-colouring. Computing this parameter is a difficult task even for cographs and interval graphs [3].

Note that the class $\mathcal{K}$ of complete graphs and the class $\mathcal{M}$ of matchings have unbounded achromatic number. Indeed, this is clear for complete graphs, and we note that $\binom{n}{2}K_2$ admits a complete n-colouring where each edge of the matching joins two of the colour classes. We claim that among cographs, those are the only minimal classes of unbounded achromatic number. To show this, we start with a short lemma.

Lemma 7. *Let r, $s \in \mathbb{N}$. The class of (K_r, sK_2, P_4)-free graphs has bounded neighbourhood diversity.*

Proof. From Theorem 3, the only minimal subclasses of cographs where neighbourhood diversity is unbounded are $\mathcal{M}$, $\overline{\mathcal{M}}$ and $\mathcal{T}$. K_r belongs to both $\overline{\mathcal{M}}$ and $\mathcal{T}$, while sK_2 belongs to $\mathcal{M}$. □

We are now ready to prove the main result of this section.

Theorem 8. *$\mathcal{K}$ and $\mathcal{M}$ are the only minimal hereditary subclasses of cographs of unbounded achromatic number.*

Proof. It suffices to show that for any r, $s \in \mathbb{N}$, the class of (K_r, sK_2, P_4)-free graphs has bounded achromatic number. Let G be a graph in this class. By Lemma 7, the class has bounded neighbourhood diversity. In other words, there is a constant k (independent of G) such that the vertex set of G can be partitioned into k similarity classes, each similarity class being a clique or an independent set. Moreover, since the size of cliques is bounded by r, we may further assume that each of these similarity classes is an independent set. Let G' be the quotient of G by this partition, i.e., the graph whose vertices are the independents sets, with two vertices being adjacent if and only if the corresponding sets are complete to each other.

Now consider a t-colouring of G, and interpret the colours as vertices of the complete graph K_t. From each edge e of G', we obtain a complete bipartite subgraph of K_t as follows: if the edge e in G' joins independent sets A_1 and A_2, then the two sets are complete to each other, so the sets of colours $I_1, I_2 \subseteq V(K_t)$ appearing in A_1 and A_2 respectively are disjoint. The complete bipartite graph B^e corresponding to e has I_1 and I_2 as its parts. With this set-up, the t-colouring is complete if any only if the edges of the graphs $B^e{}_{e \in E(G')}$ cover the edges of K_t. From [13], we need at least $\lceil \log_2(t) \rceil$ complete bipartite graphs to cover K_t. It follows that $t \leq 2^{|E(G')|} \leq 2^{\binom{k}{2}}$, as required. □

4   The Hierarchy

In this section, we bring together the different pieces of our analysis and draw a hierarchy of the parameters studied in this paper. Each parameter p is presented in Fig. 2 together with a collection $M(p)$ of minimal hereditary subclasses of cographs where p is unbounded. We say that a parameter p_1 is stronger than a parameter p_2 if the family of classes where p_1 is bounded contains the family of classes where p_2 is bounded. It is not difficult to see that p_1 is stronger than p_2 if for every class $X \in M(p_1)$ there exists a class $Y \in M(p_2)$ such that $Y \subseteq X$.

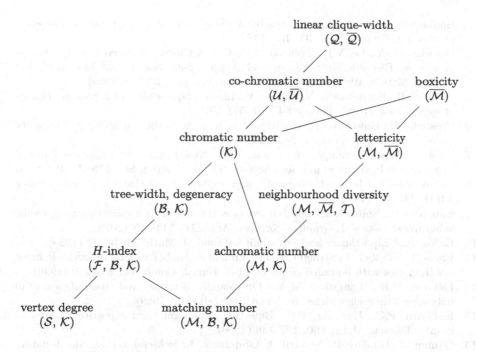

Fig. 2. A Hasse diagram of graph parameters within the universe of cographs

5   Conclusion and Open Problems

There are many other interesting parameters that are unbounded in the class of cographs, such as linearity [6], shrub-depth [14] or distinguishing number [2]. However, surprisingly, there are not so many "interesting" subclasses of cographs that appear in the characterization of those parameters. For instance, shrub-depth and distinguishing number can be characterised without extending the set of classes studied in this paper. Understanding this phenomenon is a challenging research problem.

As we observed earlier, computing the achromatic number is an NP-complete problem for cographs, and again due to well-quasi-orderability of cographs there must exist a finite collection of minimal hereditary subclasses of cographs, where the problem is NP-complete. Identifying this collection is one more open problem.

References

1. Atminas, A., Lozin, V.V., Razgon, I.: Linear time algorithm for computing a small biclique in graphs without long induced paths. In: Fomin, F.V., Kaski, P. (eds.) SWAT 2012. LNCS, vol. 7357, pp. 142–152. Springer, Heidelberg (2012). https://doi.org/10.1007/978-3-642-31155-0_13
2. Atminas, A., Brignall, R.: Well-quasi-ordering and finite distinguishing number. J. Graph Theory. https://doi.org/10.1002/jgt.22523

3. Bodlaender, H.L.: Achromatic number is NP-complete for cographs and interval graphs. Inf. Process. Lett. **31**, 135–138 (1989)
4. Brandstädt, A., Le, V.B., Spinrad, J.P.: Graph Classes: A Survey. SIAM Monographs on Discrete Mathematics and Applications, Society for Industrial and Applied Mathematics (SIAM), Philadelphia, PA, pp. xii+304 (1999)
5. Brignall, R., Korpelainen, N., Vatter, V.: Linear clique-width for hereditary classes of cographs. J. Graph Theory **84**, 501–511 (2017)
6. Crespelle, C., Gambette, P.: (Nearly-)tight bounds on the contiguity and linearity of cographs. Theor. Comput. Sci. **522**, 1–12 (2014)
7. Dabrowski, K., Demange, M., Lozin, V.V.: New results on maximum induced matchings in bipartite graphs and beyond. Theor. Comput. Sci. **478**, 33–40 (2013)
8. Damaschke, P.: Induced subgraphs and well-quasi-ordering. J. Graph Theory **14**(4), 427–435 (1990)
9. Eppstein, D., Spiro, E.S.: The h-index of a graph and its application to dynamic subgraph statistics. J. Graph Algorithms Appl. **16**, 543–567 (2012)
10. Erdős, P.: Graph theory and probability. Canad. J. Math. **11**, 34–38 (1959)
11. Erdős, P., Gimbel, J., Straight, H.J.: Chromatic number versus cochromatic number in graphs with bounded clique number. Eur. J. Comb. **11**, 235–240 (1990)
12. Fellows, M.R., Langston, M.A.: On search, decision and the efficiency of polynomial-time algorithms. In: STOC, pp. 501–512 (1989)
13. Fishburn, P.C., Hammer, P.L.: Bipartite dimensions and bipartite degrees of graphs. Discrete Math. **160**, 127–148 (1996)
14. Ganian, R., Hliněný, P., Nešetřil, J., Obdržálek, J., de Mendez, P.O.: Shrub-depth: capturing height of dense graphs. In: Logical Methods in Computer Science, vol. 15, pp. 7:1–7:25 (2019)
15. Golumbic, M.C.: Trivially perfect graphs. Discrete Math. **24**, 105–107 (1978)
16. Lampis, M.: Algorithmic meta-theorems for restrictions of treewidth. Algorithmica **64**, 19–37 (2012). https://doi.org/10.1007/s00453-011-9554-x
17. Lozin, V.: Graph parameters and ramsey theory. In: Brankovic, L., Ryan, J., Smyth, W.F. (eds.) IWOCA 2017. LNCS, vol. 10765, pp. 185–194. Springer, Cham (2018). https://doi.org/10.1007/978-3-319-78825-8_15
18. Mahadev, N.V.R., Peled, U.N.: Threshold graphs and related topics. In: Annals of Discrete Mathematics, pp. xiv+543. 56. North-Holland Publishing Co., Amsterdam (1995)
19. Petkovšek, M.: Letter graphs and well-quasi-order by induced subgraphs. Discrete Math. **244**, 375–388 (2002)
20. Roberts, F.S.: On the boxicity and cubicity of a graph. In: Recent Progress in Combinatorics, pp. 301–310. Academic Press, Cambridge (1969)
21. Yan, J.-H., Chen, J.-J., Chang, G.J.: Quasi-threshold graphs. Discrete Appl. Math. **69**, 247–255 (1996)

Parameterized Complexity of (A, ℓ)-Path Packing

Rémy Belmonte[1], Tesshu Hanaka[2], Masaaki Kanzaki[3], Masashi Kiyomi[4],
Yasuaki Kobayashi[5], Yusuke Kobayashi[5], Michael Lampis[6],
Hirotaka Ono[7], and Yota Otachi[7(✉)]

[1] The University of Electro-Communications, Chofu, Tokyo, Japan
remybelmonte@gmail.com
[2] Chuo University, Bunkyo-ku, Tokyo, Japan
hanaka.91t@g.chuo-u.ac.jp
[3] Kumamoto University, Kumamoto 860-8555, Japan
c5744@st.cs.kumamoto-u.ac.jp
[4] Yokohama City University, Yokohama, Japan
masashi@yokohama-cu.ac.jp
[5] Kyoto University, Kyoto, Japan
kobayashi@iip.ist.i.kyoto-u.ac.jp, yusuke@kurims.kyoto-u.ac.jp
[6] Université Paris-Dauphine, PSL University, CNRS, LAMSADE,
75016 Paris, France
michail.lampis@lamsade.dauphine.fr
[7] Nagoya University, Nagoya 464-8601, Japan
{ono,otachi}@nagoya-u.jp

Abstract. Given a graph $G = (V, E)$, $A \subseteq V$, and integers k and ℓ, the (A, ℓ)-PATH PACKING problem asks to find k vertex-disjoint paths of length ℓ that have endpoints in A and internal points in $V \setminus A$. We study the parameterized complexity of this problem with parameters $|A|$, ℓ, k, treewidth, pathwidth, and their combinations. We present sharp complexity contrasts with respect to these parameters. Among other results, we show that the problem is polynomial-time solvable when $\ell \leq 3$, while it is NP-complete for constant $\ell \geq 4$. We also show that the problem is W[1]-hard parameterized by pathwidth $+ |A|$, while it is fixed-parameter tractable parameterized by treewidth $+ \ell$.

Keywords: A-path packing · Fixed-parameter tractability · Treewidth

1 Introduction

Let $G = (V, E)$ be a graph and $A \subseteq V$. A path P in G is an A path if the first and the last vertices of P belong to A and all other vertices of P belong to $V \setminus A$. Given

Partially supported by PRC CNRS JSPS project PARAGA, by JSPS KAKENHI Grant Numbers JP16K16010, JP17H01698, JP17H01788, JP18H05291, JP18K11157, JP18K11168, JP18K11169, JP18H04091, JP19K21537, and by JST CREST JPMJCR 1401. The authors thank Tatsuya Gima for helpful discussions.

L. Gąsieniec et al. (Eds.): IWOCA 2020, LNCS 12126, pp. 43–55, 2020.
https://doi.org/10.1007/978-3-030-48966-3_4

G and A, A-PATH PACKING is the problem of finding the maximum number of vertex-disjoint A-paths in G. The A-PATH PACKING problem is well studied and even some generalized versions are known to be polynomial-time solvable (see e.g., [5,6,11,15,18,19]). Note that A-PATH PACKING is a generalization of MAXIMUM MATCHING since they are equivalent when $A = V$.

In this paper, we study a variant of A-PATH PACKING that also generalizes MAXIMUM MATCHING. An A-path of length ℓ is an (A, ℓ)-*path*, where the length of a path is the number of edges in the path. Now our problem is defined as follows:

(A, ℓ)-PATH PACKING (ALPP)
Input: A tuple (G, A, k, ℓ), where $G = (V, E)$ is a graph, $A \subseteq V$, and k and ℓ are positive integers.
Question: Does G contain k vertex-disjoint (A, ℓ)-paths?

To the best of our knowledge, this natural variant of A-PATH PACKING was not studied in the literature. Our main motivation of studying ALPP is to see theoretical differences from the original A-PATH PACKING, but practical motivations of having the length constraint may come from some physical restrictions or some fairness requirements. Note that if $\ell = 1$, then ALPP is equivalent to MAXIMUM MATCHING. Another related problem is ℓ-PATH PARTITION [16,20,21], which asks for vertex-disjoint paths of length ℓ (without specific endpoints).

In the rest of paper, we assume that $k \leq |A|/2$ in every instance as otherwise the instance is a trivial no-instance. The restricted version of the problem where the equality $k = |A|/2$ is forced is also of our interest as that version corresponds to a "full" packing of A-paths. We call this version FULL (A, ℓ)-PATH PACKING (Full-ALPP, for short). In this paper, all our positive results showing tractability of some cases will be on the general ALPP, while all our negative (or hardness) results will be on the possibly easier Full-ALPP.

We assume that the reader is familiar with terminologies in the parameterized complexity theory. See the textbook by Cygan et al. [8] for standard definitions.

Our Results

In summary, we show that ALPP is intractable even on very restricted inputs, while it has some nontrivial cases that admit efficient algorithms. (See Fig. 1.)

We call $|A|$, k, and ℓ the *standard parameters* of ALPP as they naturally arise from the definition of the problem. We determine the complexity of ALPP with respect to all standard parameters and their combinations. We first observe that Full-ALPP is NP-complete for any constant $|A| \geq 2$ (Observation 3.1) and for any constant $\ell \geq 4$ (Observation 3.2), while it is polynomial-time solvable when $\ell \leq 3$ (Theorem 3.3). On the other hand, ALPP is fixed-parameter tractable when parameterized by $k + \ell$ and thus by $|A| + \ell$ as well (Theorem 3.5). We later strengthen Observation 3.2 by showing that NP-complete for every fixed $\ell \geq 4$ even on grid graphs (Theorem 5.1).

We then study structural parameters such as treewidth and pathwidth in combination with the standard parameters. We first observe that ALPP can be solved in time $n^{O(\mathsf{tw})}$ (Theorem 4.1), where n and tw are the number of vertices and the treewidth of the input graph, respectively. Furthermore, we show that ALPP parameterized by $\mathsf{tw} + \ell$ is fixed-parameter tractable (Theorem 4.2). We finally show that Full-ALPP parameterized by $\mathsf{pw} + |A|$ is W[1]-hard (Theorem 4.5), where pw is the pathwidth of the input graph.

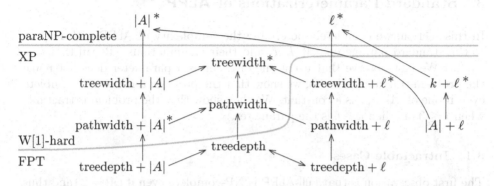

Fig. 1. Summary of the results. An arrow $\alpha \to \beta$ indicates that there is a function f such that $\alpha \geq f(\beta)$ for every instance of ALPP. Some possible arrows are omitted to keep the figure readable. The results on the parameters marked with $*$ are explicitly shown in this paper, and the other results follow by the hierarchy of the parameters. We have a bidirectional arrow treedepth $\leftrightarrow$ treedepth $+ \ell$ because the maximum length of a path in a graph is bounded by a function of treedepth [17, Section 6.2].

2 Preliminaries

A graph $G = (V, E)$ is a *grid graph* if V is a finite subset of $\mathbb{Z}^2$ and $E = \{\{(r, c), (r', c')\} \mid |r - r'| + |c - c'| = 1\}$. From the definition, all grid graphs are planar, bipartite, and of maximum degree at most 4. To understand the intractability of a graph problem, it is preferable to show hardness on a very restricted graph class. The class of grid graphs is one of such target classes.

A *tree decomposition* of a graph $G = (V, E)$ is a pair $(\{X_i \mid i \in I\}, T = (I, F))$, where $X_i \subseteq V$ for each i and T is a tree such that

- for each vertex $v \in V$, there is $i \in I$ with $v \in X_i$;
- for each edge $\{u, v\} \in E$, there is $i \in I$ with $u, v \in X_i$;
- for each vertex $v \in V$, the induced subgraph $T[\{i \mid v \in X_i\}]$ is connected.

The *width* of a tree decomposition $(\{X_i \mid i \in I\}, T)$ is $\max_{i \in I} |X_i| - 1$, and the *treewidth* of a graph G, denoted $\mathsf{tw}(G)$, is the minimum width over all tree decompositions of G.

The *pathwidth* of a graph G, denoted $\mathsf{pw}(G)$, is defined by restricting the trees T in tree decompositions to be paths. We call such decompositions *path decompositions*. It is easy to observe that pathwidth does not change significantly by subdividing some edges and attaching paths to some vertices.

Corollary 2.1 ($\bigstar^1$). *Let $G = (V, E)$ be a graph without isolated vertices. If G' is a graph obtained from G by subdividing a set of edges $F \subseteq E$ an arbitrary number of times, and attaching a path of arbitrary length to each vertex in a set $U \subseteq V$, then $\mathsf{pw}(G') \leq \mathsf{pw}(G) + 2$.*

3 Standard Parameterizations of ALPP

In this section, we completely determine the complexity of ALPP with respect to the standard parameters $|A|$, k, ℓ, and their combinations. (Recall that $k \leq |A|/2$.) We first observe that using one of them as a parameter does not make the problem tractable. That is, we show that the problem remains NP-complete even if one of $|A|$, k, ℓ is a constant. We then show that the problem is tractable when $\ell \leq 3$ or when $k + \ell$ is the parameter.

3.1 Intractable Cases

The first observation is that Full-ALPP is NP-complete even if $|A| = 2$ (and thus $k = 1$). This can be shown by an easy reduction from HAMILTONIAN CYCLE [12]. This observation is easily extended to every fixed even $|A|$.

Observation 3.1 ($\bigstar$). *For every fixed even number $\alpha \geq 2$, Full-ALPP on grid graphs is NP-complete even if $|A| = \alpha$.*

The NP-hardness of Full-ALPP for fixed ℓ can be shown also by an easy reduction from a known NP-hard problem, but in this case only for $\ell \geq 4$. This is actually tight as we see later that the problem is polynomial-time solvable when $\ell \leq 3$ (see Theorem 3.3).

Observation 3.2 ($\bigstar$). *For every fixed $\ell \geq 4$, Full-ALPP is NP-complete.*

We can strengthen Observation 3.2 to hold on grid graphs by constructing an involved reduction from scratch. As the proof is long and the theorem does not really fit the theme of this section, we postpone it to Sect. 5.

3.2 Tractable Cases

Theorem 3.3. *If $\ell \leq 3$, then ALPP can be solved in polynomial time.*

Proof. Let (G, A, k, ℓ) with $G = (V, E)$ be an instance of ALPP with $\ell \leq 3$.

If $\ell = 1$, then the problem can be solved by finding a maximum matching in $G[A]$. Since a maximum matching can be found in polynomial time [9], this case is polynomial-time solvable.

Consider the case where $\ell = 2$. We reduce this case to the case of $\ell = 3$. We can assume that $G[A]$ and $G[V \setminus A]$ do not contain any edges as such edges are not included in any $(A, 2)$-path. New instance $(G', A, k, 3)$ is constructed by

---

[1] A star $\bigstar$ means that the proof is omitted.

adding a true twin v' to each vertex $v \in V \setminus A$; i.e., $V(G') = V \cup \{v' \mid v \in V \setminus A\}$ and $E(G') = E \cup \{\{v, v'\} \mid v \in V \setminus A\} \cup \{\{u, v'\} \mid u \in A, v \in V \setminus A, \{u, v\} \in E\}$. Clearly, $(G, A, k, 2)$ is a yes-instance if and only if so is $(G', A, k, 3)$.

For the case of $\ell = 3$, we construct an auxiliary graph $G' = (A \cup V_1 \cup V_2, E_{A,1} \cup E_{1,2} \cup E_{2,2})$ as follows (see Fig. 2):

$$V_i = \{v_i \mid v \in V \setminus A\} \text{ for } i \in \{1, 2\},$$
$$E_{A,1} = \{\{u, v_1\} \mid u \in A, v \in V \setminus A, \{u, v\} \in E\},$$
$$E_{1,2} = \{\{v_1, v_2\} \mid v \in V\},$$
$$E_{2,2} = \{\{u_2, v_2\} \mid u, v \in V \setminus A, \{u, v\} \in E\}.$$

We show that $(G, A, k, 3)$ is a yes-instance if and only if G' has a matching of size $k + |V \setminus A|$, which implies that the problem can be solved in polynomial time.

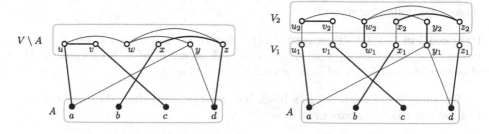

Fig. 2. The construction of G' (right) from G (left).

To prove the only-if direction, let $P_1, \ldots, P_k$ be k vertex-disjoint $(A, 3)$-path in G. We set $M = M_{A,1} \cup M_{1,2} \cup M_{2,2}$, where

$$M_{A,1} = \{\{u, v_1\} \in E_{A,1} \mid \text{edge } \{u, v\} \text{ appears in some } P_i\},$$
$$M_{1,2} = \{\{v_1, v_2\} \in E_{1,2} \mid \text{vertex } v \text{ does not appear in any } P_i\},$$
$$M_{2,2} = \{\{u_2, v_2\} \in E_{2,2} \mid \text{edge } \{u, v\} \text{ appears in some } P_i\}.$$

Since the $(A, 3)$-paths $P_1, \ldots, P_k$ are pairwise vertex-disjoint, M is a matching. We can see that $|M| = k + |V \setminus A|$ as $|M_{2,2}| = k$ and $|M_{A,1}| + |M_{1,2}| = |V_1| = |V \setminus A|$.

To prove the if direction, assume that G' has a matching of size $k + |V \setminus A|$. Let M be a maximum matching of G' that includes the maximum number of vertices in $V_1 \cup V_2$ among all maximum matchings of G'. We claim that M actually includes all vertices in $V_1 \cup V_2$. Suppose to the contrary that v_1 or v_2 is not included in M for some $v \in V \setminus A$. Now, since M is maximum, exactly one of v_1 and v_2 is included in M.

Case 1: $v_1 \in V(M)$ and $v_2 \notin V(M)$. There is a vertex $u \in A$ such that $\{u, v_1\} \in M$. The set $M - \{u, v_1\} + \{v_1, v_2\}$ is a maximum matching that uses more vertices in $V_1 \cup V_2$ than M. This contradicts how M was selected.

Case 2: $v_1 \notin V(M)$ and $v_2 \in V(M)$. There is a vertex $w_2 \in V_2$ such that $\{v_2, w_2\} \in M$. The edge set $M' := M - \{v_2, w_2\} + \{v_1, v_2\}$ is a maximum matching that uses the same number of vertices in $V_1 \cup V_2$ as M. Since M' is maximum and w_2 is not included in M', the vertex w_1 has to be included in M', but such a case leads to a contradiction as we saw in Case 1.

Now we construct k vertex-disjoint $(A, 3)$-paths from M as follows. Let $\{u_2, v_2\} \in M \cap E_{2,2}$. Since M includes all vertices in V_1, it includes edges $\{u_1, x\}$ and $\{v_1, y\}$ for some $x, y \in A$. This implies that G has an $(A, 3)$-path (x, u, v, y). Let (x', u', v', y') be the $(A, 3)$-path constructed in the same way from a different edge in $M \cap E_{2,2}$. Since M is a matching, these eight vertices are pairwise distinct, and thus (x, u, v, y) and (x', u', v', y') are vertex-disjoint $(A, 3)$-paths. Since $|M| \geq k + |V \setminus A|$ and each edge in $E_{A,1} \cup E_{1,2}$ uses one vertex of V_1, M includes at least k edges in $E_{2,2}$. By constructing an $(A, 3)$-path for each edge in $M \cap E_{2,2}$, we obtain a desired set of k vertex-disjoint $(A, 3)$-paths.   □

In their celebrated paper on *Color-Coding* [1], Alon, Yuster, and Zwick showed the following result.

Proposition 3.4 ([1, **Theorem 6.3**]).   *Let H be a graph on h vertices with treewidth t. Let G be a graph on n vertices. A subgraph of G isomorphic to H, if one exists, can be found in time $O(2^{O(h)} \cdot n^{t+1} \log n)$.*

By using Proposition 3.4 as a black box, we can show that ALPP parameterized by $k + \ell$ is fixed-parameter tractable.

Theorem 3.5. *ALPP on n-vertex graphs can be solved in $O(2^{O(k\ell)} n^6 \log n)$ time.*

Proof. Let (G, A, k, ℓ) be an instance of ALPP. Observe that the problem ALPP can be seen as a variant of the SUBGRAPH ISOMORPHISM problem as we search for $H = kP_{\ell+1}$ in G as a subgraph with the restriction that each endpoint of $P_{\ell+1}$ in H has to be mapped to a vertex in A, where $P_{\ell+1}$ denotes an $(\ell+1)$-vertex path (which has length ℓ) and $kP_{\ell+1}$ denotes the disjoint union of k copies of $P_{\ell+1}$. We reduce this problem to the standard SUBGRAPH ISOMORPHISM problem [12].

Let G' and H' be the graphs obtained from G and H, respectively, by subdividing each edge once. The graphs G' and $H' = kP_{2\ell+1}$ are bipartite. We then construct G'' from G' by attaching a triangle to each vertex in A; that is, for each vertex $u \in A$ we add two new vertices v, w and edges $\{u, v\}$, $\{v, w\}$, and $\{w, u\}$. Similarly, we construct H'' from H' by attaching a triangle to each endpoint of each $P_{2\ell+1}$. Note that $|V(G'')| \in O(n^2)$, $|V(H'')| = k(2\ell + 1)$, and $\mathsf{tw}(H'') = 2$. Thus, by Proposition 3.4, it suffices to show that (G, A, k, ℓ) is a yes-instance of ALPP if and only if G'' has a subgraph isomorphic to H''.

To show the only-if direction, assume that G has k vertex-disjoint (A, ℓ)-paths $P_1, \ldots, P_k$. In G'', for each P_i, there is a unique path Q_i of length 2ℓ plus triangles attached to the endpoints; that is, Q_i consists of the vertices of P_i, the new vertices and edges introduced by subdividing the edges in P_i, and the triangles attached to the endpoints of the subdivided path. Furthermore, since the paths P_i are pairwise vertex-disjoint, the subgraphs Q_i of G'' are pairwise vertex-disjoint. Thus, G'' has a subgraph isomorphic to $H'' = \bigcup_{1 \leq i \leq k} Q_i$.

To prove the if direction, assume that G has a subgraph H' isomorphic to H. Let $R_1, \ldots, R_k$ be the connected components of H'. Each R_i is isomorphic to a path of length 2ℓ with a triangle attached to each endpoint. Let $u, v \in V(R_i)$ be the degree-3 vertices of R_i. Since G'' is obtained from the triangle-free graph G' by attaching triangles at the vertices in A, we have $u, v \in A$. Since the u-v path of length 2ℓ in R_i is obtained from a u-v path of length ℓ in G by subdividing each edge once, the graph $G[V(R_i) \cap V(G)]$ contains an (A, ℓ)-path. Since $V(R_1), \ldots, V(R_k)$ are pairwise disjoint, G contains k vertex-disjoint (A, ℓ)-paths. □

4   Structural Parameterizations

In this section, we study structural parameterizations of ALPP. First we present XP and FPT algorithms parameterized by tw and tw $+ \ell$, respectively.

The XP-time algorithm parameterized by tw is based on an efficient algorithm for computing a tree decomposition [4] and a standard dynamic-programming over *nice tree decompositions* [14]. The FPT algorithm parameterized tw $+ \ell$ is achieved by expressing the problem in the *monadic second-order logic* (MSO₂) of graphs [2,3,7]. The proofs of them are omitted.

Theorem 4.1 (★). *ALPP can be solved in time* $n^{O(\text{tw})}$.

Theorem 4.2 (★). *ALPP parameterized by* tw $+ \ell$ *is fixed-parameter tractable.*

Now we show that Full-ALPP is W[1]-hard parameterized by pathwidth (and hence also by treewidth), even if we also consider $|A|$ as an additional parameter. We present a reduction from a W[1]-complete problem k-MULTI-COLORED CLIQUE (k-MCC) [10], which goes through an intermediate version of our problem. Specifically, we will consider a version of Full-ALPP with the following modifications: the graph has (positive integer) edge weights, and the length of a path is the sum of the weights of its edges; the set A is given to us partitioned into pairs indicating the endpoints of the sought A-paths; for each such pair the value of ℓ may be different.

More formally, Extended-ALPP is the following problem: we are given a graph $G = (V, E)$, a weight function $w : E \to \mathbb{Z}^+$, and a sequence of r triples $(s_1, t_1, \ell_1), \ldots, (s_r, t_r, \ell_r)$, where all the $s_i, t_i \in V$ are distinct vertices and $\ell_i \in \mathbb{Z}^+$ for all $i \in [r]^2$. We are asked if there exists a set of r vertex-disjoint paths in G such that for all $i \in [r]$ the i-th path in this set has endpoints s_i, t_i and the sum of the weights of its edges is ℓ_i. We first show that establishing that this variation of the problem is hard implies also the hardness of Full-ALPP.

Lemma 4.3. *There exists an algorithm which, given an instance of Extended-ALPP on an n-vertex graph G with r triples and maximum edge weight W,*

---

[2] For a positive integer r, we denote the set $\{1, 2, \ldots, r\}$ by $[r]$.

constructs in time polynomial in $n + W$ *an equivalent instance* $(G', A, |A|/2, \ell)$ *of Full-ALPP with the properties: (i)* $|A| = 2r$*, (ii)* $\mathsf{pw}(G') \leq \mathsf{pw}(G) + 2$.

Proof. First, we simplify the given instance of Extended-ALPP by removing edge weights: for every edge $e = \{u, v\} \in E(G)$ with $w(e) > 1$, we remove this edge and replace it with a path from u to v with length $w(e)$ going through new vertices (in other words we subdivide e $w(e) - 1$ times). It is not hard to see that we have an equivalent instance of Extended-ALPP on the new graph, which we call G_1, where the weight of all edges is 1 and $|V(G_1)| \leq n^2 W$. We now give a polynomial-time reduction from this new instance of Extended-ALPP to Full-ALPP.

Let $n_1 = |V(G_1)|$ and $\ell = n_1^3$. For each $i \in [r]$ we do the following: we construct a new vertex s_i' and connect it to s_i using a path of length $i \cdot n_1^2$ going through new vertices; we construct a new vertex t_i' and connect it to t_i using a path of length $(n_1 - i) \cdot n_1^2 - \ell_i$ through new vertices. We set A to contain all the vertices s_i', t_i' for $i \in [r]$. This completes the construction and it is clear that $|A| = 2r$ (because the s_i, t_i vertices are distinct), the new graph G' has order at most $n_1^5 \leq n^{10} \cdot W^5$ and can be constructed in time polynomial in $n + W$.

We claim that the new graph G' has $|A|/2$ vertex-disjoint (A, ℓ)-paths if and only if the Extended-ALPP instance of G_1 has a positive answer. Indeed, if there exists a collection of r vertex-disjoint paths in G_1 such that the i-th path has endpoints s_i, t_i and length ℓ_i, we add to this path the paths from s_i' to s_i and from t_i to t_i' and this gives a path of length $\ell = n_1^3$ with endpoints in A. Observe that all these paths are vertex-disjoint, so we obtain a yes-certificate of Full-ALPP. For the converse direction, suppose that G' has a set $\mathcal{A}$ of $|A|/2$ vertex-disjoint (A, ℓ)-paths. If $\mathcal{A}$ contains a path P with endpoints s_i' and s_j', then considering the length of P we get $(i + j) \cdot n_1^2 + 1 \leq n_1^3 \leq (i + j) \cdot n_1^2 + n_1 - 1$. The first inequality implies $i + j \leq n_1 - 1$, but then this implies $(i + j) \cdot n_1^2 + n_1 - 1 \leq n_1^3 - n_1^2 + n_1 - 1 < n_1^3$, a contradiction. Also, there cannot be a path in $\mathcal{A}$ with endpoints t_i' and t_j', since existence of such a path implies, by the pigeon hole principle, that there is a path in $\mathcal{A}$ with endpoints s_p' and s_q'. Assume that $\mathcal{A}$ contains a path with endpoints s_i' and t_j'. Then the length of this path is at least $i \cdot n_1^2 + (n_1 - j) \cdot n_1^2 - \ell_j + 1$ and at most $i \cdot n_1^2 + (n_1 - j) \cdot n_1^2 - \ell_j + n_1 - 1$. Therefore, if this path has length exactly $\ell = n_1^3$, it must be the case that $i = j$. Furthermore, if $i = j$ we infer that the length of the part of the path from s_i to t_i is exactly ℓ_i. We therefore obtain a solution to the Extended-ALPP instance.

Finally, observe that the only modifications we have done on G is to subdivide some edges and to attach paths to some vertices. By Corollary 2.1, the pathwidth is increased only by at most 2. $\qquad \square$

We can now reduce the k-MCC problem to Extended-ALPP.

Lemma 4.4. *There exists a polynomial-time algorithm which, given an instance of k-MCC on a graph G with n vertices, produces an equivalent instance of Extended-ALPP on a graph G', with $r \in O(k^2)$ triples, $\mathsf{pw}(G') \in O(k^2)$, and maximum edge weight $W \in n^{O(1)}$.*

Proof. We are given a graph $G = (V, E)$ with V partitioned into k sets $V_1, \ldots, V_k$, and are asked for a clique of size k that contains one vertex from each set. To ease notation, we will assume that n is even and $|V_i| = n$ for $i \in [k]$ (so the graph has kn vertices in total) and that the vertices of V_i are numbered $1, \ldots, n$. We define two lengths $L_1 = n^3 + (k+1)(2n+2)$ and $L_2 = n^6$.

For $i \in [k]$ we construct a vertex-selection gadget as follows (see Fig. 3): we make $2n + 3$ paths of length k, call them $P_{i,j}$, where $j \in [2n + 3]$. Let $a_{i,j}, b_{i,j}$ be the first and last vertex of path $P_{i,j}$ respectively. We label the remaining vertices of the path $P_{i,j}$ as $x_{i,j,i'}$ for $i' \in \{1, \ldots, k\} \setminus \{i\}$ in some arbitrary order. Then for each $j \in [2n + 2]$ we connect $a_{i,j}$ to $a_{i,j+1}$ and $b_{i,j}$ to $b_{i,j+1}$. All edges constructed so far have weight 1. We add two vertices s_i, t_i, connect s_i to $a_{i,1}$ with an edge of weight $n^3/2$ and t_i to $a_{i,2n+3}$ also with an edge of weight $n^3/2$. We add to the instance the triple (s_i, t_i, L_1).

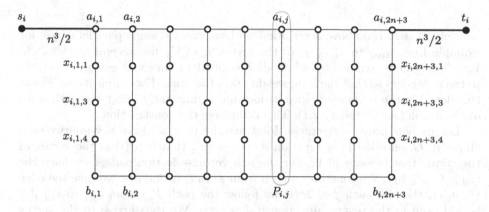

Fig. 3. An example of the vertex-selection gadget for $n = 3$, $k = 4$, and $i = 2$.

We now need to construct an edge-verification gadget as follows (see Fig. 4): for each $i_1, i_2 \in [k]$ with $i_1 < i_2$ we construct three vertices $s_{i_1,i_2}, t_{i_1,i_2}, p_{i_1,i_2}$. For each edge e of G between V_{i_1} and V_{i_2} we do the following: suppose e connects vertex j_1 of V_{i_1} to vertex j_2 of V_{i_2}. We add the following four edges:

1. An edge from s_{i_1,i_2} to $x_{i_1,2j_1,i_2}$. This edge has weight $L_2/4 + j_1 n^4 + j_2 n^2$.
2. An edge from $x_{i_1,2j_1,i_2}$ to p_{i_1,i_2}. This edge has weight $L_2/4$.
3. An edge from p_{i_1,i_2} to $x_{i_2,2j_2,i_1}$. This edge has weight $L_2/4$.
4. An edge from $x_{i_2,2j_2,i_1}$ to t_{i_1,i_2}. This edge has weight $L_2/4 - j_1 n^4 - j_2 n^2$.

We call the edges constructed in the above step heavy edges, since their weight is close to $L_2/4$. We add the $k(k-1)/2$ triples $(s_{i_1,i_2}, t_{i_1,i_2}, L_2)$ to the instance, for all $i_1, i_2 \in [k]$, with $i_1 < i_2$.

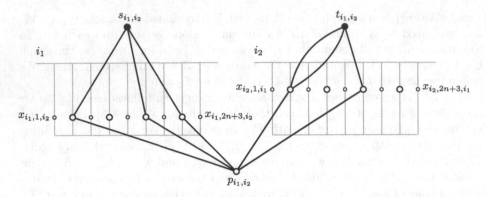

Fig. 4. An example of the edge-verification gadget for V_{i_1} and V_{i_2} ($i_1 < i_2$). In this example, there are exactly three edges between V_{i_1} and V_{i_2}.

Note that in the above description we have created some parallel edges, for example from s_{i_1,i_2} to $x_{i_1,2j_1,i_2}$ (if the vertex j_1 of V_{i_1} has several neighbors in V_{i_2}). This can be avoided by subdividing such edges once and assigning weights to the new edges so that the total weight stays the same. For simplicity we ignore this detail in the remainder since it does not significantly affect the pathwidth of the graph (see Corollary 2.1). This completes the construction.

Let us now prove correctness. First assume that we have a k-multicolored clique in G, encoded by a function $\sigma\colon [k] \to [n]$, that is, $\sigma(i)$ is the vertex of the clique that belongs in V_i. For the i-th vertex-selection gadget we have the triple (s_i, t_i, L_1). We construct a path from s_i to t_i as follows: we take the edge $(s_i, a_{i,1})$, then for each $j < 2\sigma(i)$ we follow the path $P_{i,j}$ from $a_{i,j}$ to $b_{i,j}$ if j is odd, and in the reverse direction if j is even. We thus arrive to the vertex $b_{i,2\sigma(i)-1}$. We then skip the path $P_{i,\sigma(i)}$, proceed through $b_{i,2\sigma(i)}$ to the vertex $b_{i,2\sigma(i)+1}$ and traverse the paths by reversing our parity rule: for $j > 2\sigma(i)$ we traverse $P_{i,j}$ from $b_{i,j}$ to $a_{i,j}$ if j is odd, and in the reverse direction otherwise. Hence, the last vertex of this traversal is $a_{i,2n+3}$, after which we reach t_i. The first and last edge of this path have total cost n^3; we have traversed $2n+2$ paths $P_{i,j}$, each of which has k edges; we have also traversed $2n+2$ edges connecting adjacent paths. The total length is therefore, $n^3 + (2n+2)k + 2n+2 = L_1$. In this way we have satisfied all the k triples (s_i, t_i, L_1) and have not used the vertices $x_{i,2\sigma(i),i'}$ for any $i' \neq i$.

Consider now a triple $(s_{i_1,i_2}, t_{i_1,i_2}, L_2)$, for $i_1 < i_2$. Because we have selected a clique, there exists an edge between vertex $\sigma(i_1)$ of V_{i_1} and $\sigma(i_2)$ of V_{i_2}. For this edge we have constructed four edges in our new instance, linking s_{i_1,i_2} to t_{i_1,i_2} with a total weight of L_2. We use these paths to satisfy the $\binom{k}{2}$ triples $(s_{i_1,i_2}, t_{i_1,i_2}, L_2)$. These paths are disjoint from each other: when $i_1 < i_2$, $x_{i_1,2\sigma(i_1),i_2}$ is only used in the path from s_{i_1,i_2} to t_{i_1,i_2} and when $i_1 > i_2$, $x_{i_1,2\sigma(i_1),i_2}$ is only used in the path from s_{i_2,i_1} to t_{i_2,i_1}. Furthermore, these paths are disjoint from the paths in the vertex-selection gadgets, as we observed that

$x_{i,2\sigma(i),i'}$ are not used by the path connecting s_i to t_i. We thus have a valid solution. See Fig. 5.

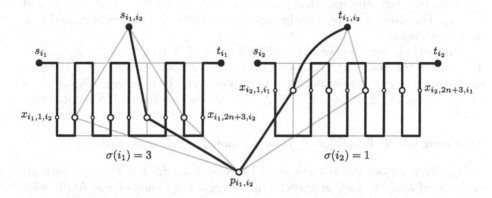

Fig. 5. Construction of paths from σ.

For the converse direction, suppose we have a valid solution for the Extended-ALPP instance. First, consider the path connecting s_i to t_i. This path has length L_1, therefore it cannot be using any heavy edges, since these edges have cost at least $L_2/4 - n^5 - n^3 > L_1$. Inside the vertex-selection gadget, the path may use either all of the edges of a path $P_{i,j}$ or none. Let us now see how many $P_{i,j}$ are unused. First, a simple parity argument shows that, because s_i, t_i are both connected to an $a_{i,j}$ vertex, the number of paths traversed in the $a_{i,j} \to b_{i,j}$ direction is equal to those traversed in the opposite direction, so the total number of used paths is even. Since we have an odd number of paths in total, at least one path is not used. We conclude that exactly one $P_{i,j}$ is not used, otherwise the path from s_i to t_i would be too short. Let $\sigma(i)$ be defined as the index j such that the internal vertices of $P_{i,j}$ are not used in the $s_i \to t_i$ path of the solution. We define a clique in G by selecting for each i the vertex $\lfloor \sigma(i)/2 \rfloor$.

Let us argue why this set induces a clique. Let j_1, j_2 be the vertices selected in V_{i_1}, V_{i_2} respectively, with $i_1 < i_2$, and consider the triple $(s_{i_1,i_2}, t_{i_1,i_2}, L_2)$. This triple must be satisfied by a path that uses exactly four heavy edges, since each heavy edge has weight strictly larger than $L_2/5$ and strictly smaller than $L_2/3$ and all other edges together are either incident on another terminal or have weight smaller than $L_2/5n^2$. Hence, every such path is using at least two internal vertices of some $P_{i,j}$ because every heavy edge is incident on such a vertex. But, by our previous reasoning, the paths that satisfy the (s_i, t_i, L_1) triples have used all such vertices except for one path $P_{i,j}$ for each i. There exist therefore exactly $k(k-1)$ such vertices available, so each of the $k(k-1)/2$ triples $(s_{i_1,i_2}, t_{i_1,i_2}, L_2)$ has a path using exactly two of these vertices. Hence, each such path consists of four heavy edges and no other edges.

Such a path must therefore be using one edge incident on s_{i_1,i_2}, one edge incident on t_{i_1,i_2} and two edges incident on p_{i_1,i_2}. The used edge incident on

s_{i_1,i_2} must have as other endpoint $x_{i_1,2j_1,i_2}$, which implies that its weight is $L_2/4 + j_1 n^4 + j_2' n^2$, for some j_2'. Similarly, the edge incident on t_{i_1,i_2} must have weight $L_2/4 - j_1' n^4 - j_2 n^2$, as its other endpoint is necessarily $x_{i_2,2j_2,i_1}$. We conclude that the only way that the length of this path is L_2 is if $j_1 = j_1'$ and $j_2 = j_2'$. Therefore, we have an edge between the two selected vertices, and as a result a k-clique.

To conclude we observe that deleting the $O(k^2)$ vertices $s_{i_1,i_2}, p_{i_1,i_2}, t_{i_1,i_2}$ disconnects the graph into components that correspond to the vertex gadgets. Each vertex gadget has pathwidth at most 4 as it can be seen as a subgraph of a subdivision of the $2 \times (2n + 4)$ grid. As a result the whole graph has pathwidth $O(k^2)$. □

Theorem 4.5. *Full-ALPP is W[1]-hard parameterized by* pw $+ |A|$.

Proof. We compose the reductions of Lemmas 4.3 and 4.4. Starting with an instance of k-MCC with n vertices this gives an instance of Full-ALPP with $n^{O(1)}$ vertices, $|A| = O(k^2)$, and pathwidth $O(k^2)$. □

5   Hardness on Grid Graphs

We first reduce PLANAR CIRCUIT SAT to Full-ALPP on planar bipartite graphs of maximum degree at most 4. We then modify the instance by subdividing edges and adding terminal vertices in a appropriate way, and have an equivalent instance on grid graphs. All proofs in this section are omitted.

Theorem 5.1 (★). *For every fixed $\ell \geq 4$, Full-ALPP is NP-complete on grid graphs.*

6   Concluding Remarks

In this paper, we have introduced a new problem (A, ℓ)-PATH PACKING and showed tight complexity results. One possible future direction would be the parameterization by clique-width cw, a generalization of treewidth (see [13]). In particular, we ask the following two questions.

– Does ALPP admit an algorithm of running time $O(n^{\mathsf{cw}})$?
– Is ALPP fixed-parameter tractable parameterized by cw $+ \ell$?

References

1. Alon, N., Yuster, R., Zwick, U.: Color-coding. J. ACM **42**(4), 844–856 (1995). https://doi.org/10.1145/210332.210337
2. Arnborg, S., Lagergren, J., Seese, D.: Easy problems for tree-decomposable graphs. J. Algorithms **12**(2), 308–340 (1991). https://doi.org/10.1016/0196-6774(91)90006-K

3. Bodlaender, H.L.: A linear-time algorithm for finding tree-decompositions of small treewidth. SIAM J. Comput. **25**(6), 1305–1317 (1996). https://doi.org/10.1137/S0097539793251219

4. Bodlaender, H.L., Drange, P.G., Dregi, M.S., Fomin, F.V., Lokshtanov, D., Pilipczuk, M.: A $c^k n$ 5-approximation algorithm for treewidth. SIAM J. Comput. **45**(2), 317–378 (2016). https://doi.org/10.1137/130947374

5. Chudnovsky, M., Cunningham, W.H., Geelen, J.: An algorithm for packing non-zero a-paths in group-labelled graphs. Combinatorica **28**(2), 145–161 (2008). https://doi.org/10.1007/s00493-008-2157-8

6. Chudnovsky, M., Geelen, J., Gerards, B., Goddyn, L.A., Lohman, M., Seymour, P.D.: Packing non-zero a-paths in group-labelled graphs. Combinatorica **26**(5), 521–532 (2006). https://doi.org/10.1007/s00493-006-0030-1

7. Courcelle, B.: The monadic second-order logic of graphs III: tree-decompositions, minor and complexity issues. Theor. Inf. Appl. **26**, 257–286 (1992). https://doi.org/10.1051/ita/1992260302571

8. Cygan, M., et al.: Parameterized Algorithms. Springer, Cham (2015). https://doi.org/10.1007/978-3-319-21275-3

9. Edmonds, J.: Paths, trees, and flowers. Can. J. Math. **17**, 449–467 (1965). https://doi.org/10.4153/CJM-1965-045-4

10. Fellows, M.R., Hermelin, D., Rosamond, F.A., Vialette, S.: On the parameterized complexity of multiple-interval graph problems. Theor. Comput. Sci. **410**(1), 53–61 (2009). https://doi.org/10.1016/j.tcs.2008.09.065

11. Gallai, T.: Maximum-Minimum Sätze und verallgemeinerte Faktoren von Graphen. Acta Mathematica Academiae Scientiarum Hungarica **12**, 131–173 (1961). https://doi.org/10.1007/BF02066678

12. Garey, M.R., Johnson, D.S.: Computers and Intractability: A Guide to the Theory of NP-Completeness. Freeman W. H., New York (1979)

13. Hliněný, P., Oum, S., Seese, D., Gottlob, G.: Width parameters beyond tree-width and their applications. Comput. J. **51**(3), 326–362 (2008). https://doi.org/10.1093/comjnl/bxm052

14. Kloks, T. (ed.): Treewidth, Computations and Approximations. LNCS, vol. 842. Springer, Heidelberg (1994). https://doi.org/10.1007/BFb0045375

15. Mader, W.: Über die maximalzahl kreuzungsfreier H-wege. Archiv der Mathematik (Basel) **31**, 387–402 (1978). https://doi.org/10.1007/BF01226465

16. Monnot, J., Toulouse, S.: The path partition problem and related problems in bipartite graphs. Oper. Res. Lett. **35**(5), 677–684 (2007). https://doi.org/10.1016/j.orl.2006.12.004

17. Nešetřil, J., Ossona de Mendez, P.: Sparsity - Graphs, Structures, and Algorithms. AC, vol. 28. Springer, Heidelberg (2012). https://doi.org/10.1007/978-3-642-27875-4

18. Pap, G.: Packing non-returning a-paths. Combinatorica **27**(2), 247–251 (2007). https://doi.org/10.1007/s00493-007-0056-z

19. Pap, G.: Packing non-returning a-paths algorithmically. Discrete Math. **308**(8), 1472–1488 (2008). https://doi.org/10.1016/j.disc.2007.07.073

20. Steiner, G.: On the k-path partition of graphs. Theor. Comput. Sci. **290**(3), 2147–2155 (2003). https://doi.org/10.1016/S0304-3975(02)00577-7

21. Yan, J.H., Chang, G.J., Hedetniemi, S.M., Hedetniemi, S.T.: k-path partitions in trees. Discret. Appl. Math. **78**(1–3), 227–233 (1997). https://doi.org/10.1016/S0166-218X(97)00012-7

On Proper Labellings of Graphs
with Minimum Label Sum

Julien Bensmail, Foivos Fioravantes$^{(\boxtimes)}$, and Nicolas Nisse

Université Côte d'Azur, CNRS, Inria, I3S, Sophia Antipolis, France
`foivos.fioravantes@inria.fr`

Abstract. The 1-2-3 Conjecture states that every nice graph G (without component isomorphic to K_2) admits a proper 3-labelling, i.e., a labelling of the edges with 1, 2, 3 such that no two adjacent vertices are incident to the same sum of labels. Another interpretation of this conjecture is that every nice graph G can be turned into a locally irregular multigraph M, i.e., with no two adjacent vertices of the same degree, by replacing each edge by at most three parallel edges. In other words, for every nice graph G, there should exist a locally irregular multigraph M with the same adjacencies and having few edges.

We study proper labellings of graphs with the extra requirement that the sum of assigned labels must be as small as possible. That is, given a graph G, we are looking for a locally irregular multigraph M^* with the fewest edges possible that can be obtained from G by replacing edges with parallel edges. This problem is quite different from the 1-2-3 Conjecture, as we prove that there is no k such that M^* can always be obtained from G by replacing each edge with at most k parallel edges.

We investigate several aspects of this problem. We prove that the problem of designing proper labellings with minimum label sum is $\mathcal{NP}$-hard in general, but solvable in polynomial time for graphs with bounded treewidth. We also conjecture that every nice connected graph G admits a proper labelling with label sum at most $\frac{3}{2}|E(G)| + \mathcal{O}(1)$, which we verify for several classes of graphs.

Keywords: Proper labelling · 1-2-3 Conjecture · Minimum label sum

1 Introduction

In this paper, we consider **proper labellings** of graphs, a notion related to the 1-2-3 Conjecture, with the extra constraint that the sum of assigned labels must be minimised. For any notation on graph theory not defined here, we refer the reader to [7]. For a graph G, a function $\ell : E(G) \mapsto \{1, \ldots, k\}$ is called a k-labelling of G. For any $v \in V(G)$, let $c_\ell(v) : V(G) \mapsto \mathbb{N}^*$ be the *colour* of v that is induced by ℓ, being the sum of labels assigned to the edges incident to v. That is, $c_\ell(v) = \sum_{u \in N(v)} \ell(vu)$ where $N(v) = \{u \in V(G) : uv \in E(G)\}$

Due to space limitation, several proofs have been omitted. They can be found in [3].

© Springer Nature Switzerland AG 2020
L. Gąsieniec et al. (Eds.): IWOCA 2020, LNCS 12126, pp. 56–68, 2020.
https://doi.org/10.1007/978-3-030-48966-3_5

is the neighbourhood of v. We say that ℓ is a *proper labelling* if the resulting colouring c_ℓ is a proper vertex-colouring of G, i.e., for every edge $uv \in E(G)$ we have $c_\ell(u) \neq c_\ell(v)$. Note that a graph admits a proper labelling if and only if it has no K_2 as a component [10]. Therefore, we here focus only on *nice graphs*, i.e., graphs without any component isomorphic to K_2. Given a nice graph G, let $\chi_\Sigma(G)$ be the smallest k such that G admits a proper k-labelling.

Maybe the most famous conjecture concerning proper labellings of graphs is the so-called **1-2-3 Conjecture**, introduced by Karoński, Łuczak and Thomason in 2004 [10]. This conjecture states that for every nice graph G, we have $\chi_\Sigma(G) \leq 3$. It is worth noting that there exist nice graphs, such as nice complete graphs [5], for which the upper bound is attained. Actually, given a graph G, deciding if $\chi_\Sigma(G) \leq 2$ holds is an $\mathcal{NP}$-complete problem [8]. The best currently known result towards the 1-2-3 Conjecture is that for any nice graph G, we have $\chi_\Sigma(G) \leq 5$ [9]. Another important result states that the conjecture is satisfied for nice 3-colourable graphs [10]. Quite recently, a characterisation of nice bipartite graphs G with $\chi_\Sigma(G) = 3$ was provided in [13]. Moreover, $\chi_\Sigma(G) \leq 4$ holds for every nice regular graph G [12] and $\chi_\Sigma(T) \leq 2$ holds for every nice tree T [5].

Our work takes place in a recent series of works dedicated to better understanding proper labellings by studying variations with additional requirements, such as minimising the number of distinct colours [1] or minimising the maximum colour [4] induced by a proper k-labelling. An additional motivation is the following [6]. Given a graph G and a proper labelling ℓ of G, by replacing every edge e by $\ell(e)$ parallel edges, we obtain a multigraph $M_{G,\ell}$ with the same adjacencies as G that is *locally irregular*, i.e., in which no two adjacent vertices have the same degree. In this setting, the 1-2-3 Conjecture states that, for every nice graph G, we can construct a corresponding $M_{G,\ell}$ by replacing each edge by at most three parallel edges, and thus construct such an $M_{G,\ell}$ with at most $3|E(G)|$ edges. One could argue however, that there might be cases in which it could be possible to obtain such a multigraph with fewer edges when being allowed to replace edges by more than three parallel edges. We study this through the following additional notions and definitions. Formally, for a labelling ℓ of a nice graph G, let $\sigma(\ell)$ be the sum of labels assigned to the edges of G by ℓ. That is, $\sigma(\ell) = \sum_{e \in E(G)} \ell(e)$. For any $k \geq 1$, let $\mathrm{mE}_k(G)$ be the minimum value of $\sigma(\ell)$ over all proper k-labellings ℓ of G. That is, $\mathrm{mE}_k(G) = \min\{\sigma(\ell) : \ell$ is a proper k-labelling of $G\}$. Let $\mathrm{mE}(G) = \min\{\mathrm{mE}_k(G) : k \geq \chi_\Sigma(G)\}$. Computing a proper labelling ℓ^* such that $\sigma(\ell^*) = \mathrm{mE}(G)$ is thus equivalent to finding a locally irregular multigraph M_{G,ℓ^*} with minimum number of edges.

Our Contributions. Section 2 starts by giving observations on labellings that are used to deduce the value of mE for nice complete bipartite graphs, complete graphs and cycles. We then exhibit an infinite family of graphs G showing that, for any fixed $k \geq 2$, the value $\mathrm{mE}_k(G)$ can be arbitrarily larger than $\mathrm{mE}_{k+1}(G)$, thereby establishing a fundamental property of our problem.

In Sect. 3, we study the complexity of computing the parameter $\mathrm{mE}_k(G)$ for some input integer k and nice graph G. We establish both positive and negative results. On the negative side we prove that determining $\mathrm{mE}_2(G)$ is $\mathcal{NP}$-complete, even when G is restricted to a planar bipartite graph. An important point is that this is contrasting with the complexity of determining whether $\chi_\Sigma(G) \le 2$ holds for a given bipartite graph G, which can be done in polynomial time [13]. On the positive side, we prove that determining $\mathrm{mE}_k(G)$ can be done in polynomial time whenever k is fixed and G is a graph with bounded treewidth.

Finally, Sect. 4 is dedicated to bounds on mE. Our guiding thread is a conjecture we raise, stating that, for any nice connected graph G, $\mathrm{mE}(G) \le \frac{3}{2}|E(G)| + \mathcal{O}(1)$. Towards this conjecture, we focus on the bipartite case. As support, we both provide infinite families of bipartite graphs G with "large" value of $\mathrm{mE}_2(G)$, and prove the conjecture for several classes of bipartite graphs.

2   First Insights into the Problem

In this section, we give first insights into the problem of determining the parameters $\mathrm{mE}(G)$ and $\mathrm{mE}_k(G)$ for a given graph G. We start off, in Sect. 2.1, by raising observations on labellings and by considering easy classes of graphs. For each G belonging to the classes we consider, we actually have $\mathrm{mE}_k(G) = \mathrm{mE}(G)$ for $k = \chi_\Sigma(G)$. Put differently, a larger label than $\chi_\Sigma(G)$ is not needed to achieve the smallest label sum. However, this behaviour is not systematic, as we exhibit, in Sect. 2.2, examples of trees T for which the smallest k such that $\mathrm{mE}_k(T) = \mathrm{mE}(T)$ is arbitrarily large.

2.1   Warm-Up Results

First off, note that in general, labellings have systematic properties that can be useful to establish bounds on mE and mE_k.

Observation 1. *Let ℓ be a k-labelling of a graph G. The following items hold:*

- $|E(G)| \le \sigma(\ell) \le k|E(G)|$.
- $\sum_{e \in E(G)} 2\ell(e) = 2\sigma(\ell) = \sum_{v \in V(G)} c_\ell(v)$.
- $\sum_{v \in V(G)} c_\ell(v)$ *must therefore be an even number.*

In particular, these observations allow to determine the value of mE for simple graph topologies, namely for complete bipartite graphs, complete graphs, and cycles. Due to lack of space, we only sketch the proof of the result about cycles.

Theorem 2. *Let $G = K_{n,m}$ be a complete bipartite graph with order $n + m > 2$.*

- *If $n \ne m$, then $\mathrm{mE}(G) = \mathrm{mE}_1(G) = |E(G)|$;*
- *otherwise, i.e., $n = m$, we have $\mathrm{mE}(G) = \mathrm{mE}_2(G) = |E(G)| + \sqrt{|E(G)|}$.*

Theorem 3. *Let K_n be a complete graph with order $n \ge 3$. Then:*

- *if $n = 3$, then $\mathrm{mE}(K_3) = \mathrm{mE}_3(K_3) = 6 = 2|E(K_3)|$;*
- *if $n \equiv 0$ or $1 \pmod 4$, then $\mathrm{mE}(K_n) = \mathrm{mE}_3(K_n) = \frac{3}{2}|E(K_n)|$;*
- *if $n \equiv 2$ or $3 \pmod 4$, then $\mathrm{mE}(K_n) = \mathrm{mE}_3(K_n) = \lceil \frac{3}{2}|E(K_n)| \rceil$.*

Theorem 4. *Let C_n be a cycle with length $n \geq 3$. Then:*

- *if $n \equiv 0 \pmod 4$, then $\mathrm{mE}(C_n) = \mathrm{mE}_2(C_n) = \frac{3}{2}|E(C_n)|$;*
- *if $n \equiv 1$ or $3 \pmod 4$, then $\mathrm{mE}(C_n) = \mathrm{mE}_3(C_n) = \lceil \frac{3}{2}|E(C_n)| \rceil + 1$;*
- *if $n \equiv 2 \pmod 4$, then $\mathrm{mE}(C_n) = \mathrm{mE}_3(C_n) = \frac{3}{2}|E(C_n)| + 3$.*

Sketch of Proof. The proof of the lower bounds follow mainly from the fact that, for any $l \leq k$, any proper k-labelling ℓ of C_n assigns label l to at most $\lfloor \frac{1}{2}|E(C_n)| \rfloor$ edges if n is odd and to at most $\frac{1}{2}|E(C_n)| - 1$ edges if $n \equiv 2 \pmod 4$. This claim is proved by considering the *conflict graph* that consists of one vertex per edge of C_n, with two vertices being adjacent when the corresponding edges of C_n cannot have the same label. We show that this conflict graph is either one cycle or two disjoint cycles (depending on the parity of n) and so the size of any of its independent sets (corresponding to a set of edges of C_n that can receive the same label) is bounded above as required.

The upper bounds on mE are proven by giving a proper labelling matching the lower bound. For instance, if $n \equiv 0 \pmod 4$, it is sufficient to alternate two consecutive edges labelled with 1, then two consecutive edges labelled with 2, and so on. When $n \equiv 1$ or $3 \pmod 4$, one single edge labelled with 3 is necessary and sufficient while two such edges are required and sufficient in the last case. ◊

2.2 Using Larger Labels can be Arbitrarily Better

In this section, we show that there is no absolute constant $k \in \mathbb{N}$ such that $\mathrm{mE}(G) = \mathrm{mE}_k(G)$ for all nice graphs. More precisely, for any integer k, we exhibit a tree T_k such that $\mathrm{mE}(T_k) = \mathrm{mE}_k(T_k) < \mathrm{mE}_{k-1}(T_k)$.

Let us first introduce the *auxiliary graph* $A(\alpha, \beta)$ (for $\alpha \geq 2$ and $\beta \geq 0$), which will serve as the building block for T_k. This auxiliary graph is a tree built recursively as follows. For any $\alpha^* \in \mathbb{N}$, define $A(\alpha^*, 0)$ as a leaf. For any $\beta > 0$, define $A(\alpha, \beta)$ as a tree of height β, rooted in a vertex r with α children. For each $1 \leq i \leq \alpha$, let c_i be the corresponding child of r; each c_i is the root of an $A(\alpha + i, \beta - 1)$ tree and thus $d(c_i) = \alpha + i + 1$ (since each c_i has $\alpha + i$ children of its own and an edge connecting it with his parent). Note that $d(c_i) \in D(\alpha) = [\alpha + 2, 2\alpha + 1]$ and that for $i \neq j$, we have $d(c_i) \neq d(c_j)$ (and thus all values of $D(a)$ are used exactly once). Finally, we say that $A(\alpha, \beta)$ is *represented* by r.

Let us also define the *pending auxiliary graph* that corresponds to $A(\alpha, \beta)$ as $P(\alpha, \beta) = (V, E)$, where $V = V(A(\alpha, \beta)) \cup \{v\}$ and $E = E(A(\alpha, \beta)) \cup \{vr\}$; in essence $P(\alpha, \beta)$ is $A(\alpha, \beta)$ with an extra vertex v connected to r. The vertex r is called the *representative* of $P(\alpha, \beta)$. The graph $P(\alpha, \beta)$ is said to be *pending* from v. Observe that $P(\alpha, \beta)$ is locally irregular and thus the labelling ℓ that assigns label 1 on every one of its edges is proper and $\mathrm{mE}(P(\alpha, \beta)) = |E|$.

Theorem 5. *For any $k \geq 2$, there is a graph T_k with $\mathrm{mE}_{k+1}(T_k) < \mathrm{mE}_k(T_k)$.*

Sketch of Proof. Let $k \geq 2$ and let us describe the construction of T_k. For $0 \leq j \leq k-1$, let $P(k+j, 2(k+1))$ be the graph pending from v_j that corresponds to an auxiliary graph $A(k + j, 2(k + 1))$ (represented by a vertex r_j) and let u, v be two adjacent vertices. The tree T_k is the graph that is produced by merging v with each one of the v_j. Observe that since r_j represents $A(k + j, 2(k + 1))$, each r_j has $d(r_j) = k + j + 1$ in T_k and that the height of T_k is $2(k+1) + 1$. Also observe that in T_k, since $N(v) = \{r_0, \ldots, r_{k-1}, u\}$, we have $d(v) = k + 1 = d(r_0)$.

Let ℓ be the $(k + 1)$-labelling of T_k that assigns label $k + 1$ to the edge uv and label 1 to the remaining edges of T_k. It easy to see that ℓ is a proper $(k + 1)$-labelling for T_k with $\sigma(\ell) = |E(T_k)| + k$.

Let ℓ' be any proper k-labelling of T_k. It suffices to show that $\sigma(\ell') > |E(T_k)| + k$. For any $w \in N(r_0) \setminus \{v\}$ and $y \in N(v) \setminus \{u, r_0\}$, since $d(v) = d(r_0) = k + 1$, at least one of the edges uv, r_0w or vy has to have a label different from 1 for ℓ' to be proper. Let us assume that $\ell(uv) \neq 1$ (the other cases being similar). Let $\ell'(uv) = l$ with $2 \leq l \leq k$ and assume that only this edge of T_k has a label different from 1. Then $c_{\ell'}(v) = k + l$ and $k + l \in [k + 2, 2k]$. Recall that for each $0 \leq j \leq k - 1$, r_j has $d(r_j) = k + j + 1$ and thus $d(r_j) \in [k + 1, 2k]$. Since uv is the only edge with a label different from 1, $c_{\ell'}(r_j) = d(r_j)$. It follows that there exists a $j \in [0, k - 1]$, such that $c_{\ell'}(r_j) = c_{\ell'}(v)$ leading to ℓ' not being proper. Thus, there must exist another edge $u'v'$ (with, say, u' being the parent of v') that is assigned a label different from 1 by ℓ'. Note that this edge $u'v'$ belongs to $P(q, 2(k + 1))$ (for some $q \in [k, 2k - 1]$) and either $v' = v$ or v' is the child of the representative v of $P(q, 2(k + 1))$. It can be shown that, for ℓ to be proper, at least one child of u' has one of its incident edges e assigned a label distinct from 1. This edge e belongs to some subtree $P(q, 2k)$ and e is incident to either the representative of this copy of $P(q, 2k)$ or to a child of this representative. Applying this argument recursively, it can be proved that, for ℓ to be proper, this copy of $P(q, 2k)$ must contain at least k edges with a label greater than 1. Overall, $mE(T_k) \geq |E(T_k)| + k + 1$.

Observe that the height of T_k can be freely controlled by changing the β value of the pending auxiliary graphs that form it. Furthermore, it follows from some of the arguments we have employed that $mE(T(\alpha, 2\beta)) < mE(T(\alpha, 2\beta'))$ for $\beta < \beta'$. Put simply, since the difference between $mE_{k+1}(T_k)$ and $mE_k(T_k)$ depends on the height of T_k and this can be an arbitrary number, the following holds:

Corollary 1. *For any $k \geq 2$, there exists a graph T_k such that $mE_{k+1}(T_k)$ is arbitrarily smaller than $mE_k(T_k)$.*

3 Complexity Aspects

This section is devoted to the complexity aspects of the problem of computing mE_k. On the negative side, we prove that the problem is $\mathcal{NP}$-complete in planar bipartite graphs. On the positive side, we prove that the problem can be solved in polynomial time for graphs with bounded treewidth, and that it is even FPT when parameterised by the treewidth plus the maximum degree.

3.1 $\mathcal{NP}$-hardness for Planar Bipartite Graphs

Let us first introduce the k-*gadget*, for $k \geq 11$, which will be useful for proving the main Theorem of this section. To build this gadget, start with $k - 1$ *stars*, each having a *center* denoted by s_i, $i \in [1, k - 1]$, such that $d(s_i) = k + 1$. For each star, pick an arbitrary edge $s_i y_i$ and identify all the y_i into a single vertex y, which is called the *representative* of the gadget. Finally add another vertex u, called the *root* of the gadget, which is connected to y. It is clear that $d(u) = 1$ and $d(y) = k$. Also, each k-gadget is a tree with $\mathcal{O}(k^2)$ edges. Let v be a vertex of a graph G, and H be a k-gadget. The operation of adding H to G and identifying the root u of H with v is called *attaching* H to v.

Theorem 6. *Let G be a nice planar bipartite graph, $k \geq 2$ and $q \in \mathbb{N}$. The problem of deciding if $\mathrm{mE}_k(G) \leq q$ is $\mathcal{NP}$-complete.*

Proof. The problem is clearly in $\mathcal{NP}$. We focus on showing it is also $\mathcal{NP}$-hard. The proof is done by reduction from PLANAR MONOTONE 1-IN-3 SAT, which was shown to be $\mathcal{NP}$-complete in [11]. In this problem, a 3CNF formula F is given as input, which has clauses with exactly three distinct variables all of which appear only positively. We say that a bipartite graph $G' = (V, C, E)$ *corresponds* to F if it is constructed in the following way: for each variable x_i of F we add a *variable vertex* v_i in V and for each clause C_j of F we add a *clause vertex* c_j in C. Then the edge $v_i c_j$ is added if variable x_i appears in clause C_j. In the PLANAR MONOTONE 1-IN-3 SAT problem, we also have that for any instance F the corresponding graph is planar. The question is whether there exists a 1-*in*-3 *truth assignment* of F; that is a truth assignment to the variables of F such that each clause has exactly one variable with the value *true*.

Let us prove the statement for $k = 2$. Let F be the 3CNF formula with c clauses that is given as input to the PLANAR MONOTONE 1-IN-3 SAT problem. Our goal is to construct a planar bipartite graph G such that F is 1-in-3 satisfiable if and only if $\mathrm{mE}_2(G) \leq |E(G)| + c$.

Start with $G' = (V, C, E)$ being the planar bipartite graph that corresponds to F, with V being the set of the variable vertices v_i, C being the set of the clause vertices c_j and $|C| = c$. In F, each clause has exactly three variables but there is no bound on how many times a variable appears in F. Thus for each $v_i \in V, d(v_i) \geq 1$ and for each $c_j \in C, d(c_j) = 3$. It follows that $|V| \leq 3c$.

Modify G' by adding the k-gadgets described earlier in the following way. For each variable vertex v_i of G, let d_i be the degree of v_i in G'. Let $d_{v,i} = (d_i - 1)(c + 1) + d_i$ and $d_c = 3(c + 1) + 3$. For each variable vertex v_i, for all $1 \leq j < d_i$, attach $c + 1$ copies of the $(d_{v,i} + j)$-gadget. Thus the degree of each v_i in G becomes equal to $d_{v,i}$. On each clause vertex c_j, attach $c + 1$ copies of the d_c-gadget, $c + 1$ copies of the $(d_c + 2)$-gadget and $c + 1$ copies of the $(d_c + 3)$-gadget. Thus the degree of each c_j in G becomes equal to d_c. Clearly, the construction of G is achieved in polynomial time. Observe also that since G' is planar and the attached gadgets are actually trees, G is also planar.

Claim. Let $G(V, C, E)$ be a bipartite graph and ℓ be any proper 2-labelling of G such that $\sigma(\ell) \leq |E(G)| + c$, for $c = |C|$. Let H be any p-gadget attached to G, where $p - 1 > c$. Let y be the representative of H. If at least one edge e of H incident to y is labelled 2, then at least two edges of H are labelled 2.

Let ℓ be a proper 2-labelling of G such that $\sigma(\ell) \leq |E(G)| + c$, i.e., there are at most c edges of G labelled 2 by ℓ. Observe that G contains p-gadgets for $p \in \{d_{v,i} + 1, d_{v,i} + 2, \ldots d_{v,i} + d_i - 1, d_c, d_c + 2, d_c + 3$ and $d_{v,i} - 1, d_c - 1 > c$. Thus the above claim holds for each gadget attached to G.

Claim. For any proper 2-labelling ℓ of G with $\sigma(\ell) \leq |E(G)| + c$, we have that:

- for each variable vertex $v_i \in V, c_\ell(v_i) \notin \{d_{v,i} + 1, d_{v,i} + 2, \ldots, d_{v,i} + d_i - 1\}$
- for each clause vertex $c_j \in C, c_\ell(c_j) \notin \{d_c, d_c + 2, d_c + 3\}$

Claim. Let ℓ be any proper 2-labelling of G with $\sigma(\ell) \leq |E(G)| + c$. Then all edges of the attached gadgets must be labelled 1.

Using the above Claims, it follows that the only possible colours induced by ℓ on the vertices of G' are in $\{d_{v,i}, d_{v,i} + 1, d_{v,i} + 2, \ldots, d_{v,i} + d_i - 1, d_{v,i} + d_i\}$ for each variable vertex $v_i \in V$, and in $\{d_c, d_c + 1, d_c + 2, d_c + 3\}$ for every clause vertex $c_j \in C$. Furthermore, for every variable vertex v_i, we have $c_\ell(v_i) \in \{d_{v,i}, d_{v,i} + d_i\}$, and observe that $c_\ell(v_i) = d_{v,i}$ if all edges of G' incident to v_i are labelled 1, while $c_\ell v_i = d_{v,i} + d_i$ if all edges of G' incident to v_i are labelled 2. For every clause vertex c_j, we have $c_\ell(c_j) = \{d_c + 1\}$, which corresponds to two edges of G' incident to c_j labelled 1 and only one edge labelled 2.

We are now ready to show the equivalence between finding a 1-in-3 truth assignment ϕ of F and finding a proper 2-labelling ℓ of G such that $\sigma(\ell) = \mathrm{mE}_2(G) \leq |E(G)| + c$. An edge $v_i c_j$ of G' labelled 2 (1, respectively) by ℓ corresponds to variable x_i bringing truth value *true* (*false*, respectively) to clause C_j by ϕ. Also, we know that in G', each variable vertex v_i is adjacent to $n \geq 1$ edges, all having the same label (either 1 or 2). Accordingly, the corresponding variable x_i brings, by ϕ, the same truth value to the n clauses of F that contain it. Finally, in G', each clause vertex c_j is adjacent to two edges labelled 1 and one labelled 2. This corresponds to the clause C_j being regarded as satisfied by ϕ only when it has exactly one true variable.   □

3.2   Polynomiality for Bounded-Treewidth Graphs

The following theorem is proved by a classical (while non trivial) dynamic programming algorithm on tree-decompositions. Due to lack of space, we only state our main theorem. The full description of the algorithm and of its proof can be found in [3].

Theorem 7. *Let $k \geq 2$ and* tw ≥ 1 *be two fixed integers. Given a nice graph G with $|V(G)| = n$ and an integer s, the problem of deciding whether* $\mathrm{mE}_k(G) \leq s$ *can be solved in polynomial time if G has treewidth at most* tw *(and in linear time if G is additionally of bounded maximum degree).*

Importantly, the above theorem provides a constructive polynomial-time algorithm to compute mE_k in the class of trees and in the class of odd multi-cacti (an important class in the context of the 1-2-3 Conjecture, that we detail below). Note however that k must be fixed and since, by Theorem 5, the smallest integer k such that $mE(T) = mE_k(T)$ for every tree T is not bounded, we leave open the question of the complexity of computing mE in the class of trees.

4 General Bounds

Recall that $mE(G) \leq \chi_\Sigma(G)|E(G)|$ and $\chi_\Sigma(G) \leq 5$ (see [9]) hold for every nice graph G. Thus $mE(G) \leq 5|E(G)|$ holds for every nice graph G, and even $mE(G) \leq 4|E(G)|$ holds when G is regular [12]. Moreover, for every graph satisfying the 1-2-3 Conjecture, even $mE(G) \leq 3|E(G)|$ holds. Throughout this section, we study how tight this bound is, in particular in the bipartite case.

4.1 Upper Bounds

Recall that bipartite graphs satisfy the 1-2-3 Conjecture [10]. For $i \in \{1, 2, 3\}$, let $\mathcal{B}_i$ be the set of bipartite graphs G with $\chi_\Sigma(G) = i$. In particular, $\mathcal{B}_1$ is the set of locally irregular bipartite graphs and the set $\mathcal{B}_3$ is that of the so-called *odd multi-cacti*, which are defined as follows [13]. The set $\mathcal{B}_3$ is exactly the set of graphs that can be obtained at any moment of the following procedure:

- Start from a cycle with length at least 6 congruent to 2 modulo 4 whose edges are properly coloured with red and green.
- Repeatedly consider a green edge uv, and join u and v by a path of length at least 5 congruent to 1 modulo 4 whose edges are properly coloured with red and green, where the edge incident to u and that incident to v are red.

Theorem 8. *Every nice bipartite graph G satisfies $mE(G) \leq mE_3(G) \leq 2|E(G)|$. Moreover, if $G \in \mathcal{B}_2$, then $mE(G) < 2|E(G)|$.*

Proof. The statement trivially holds for every $G \in \mathcal{B}_1$ since G is locally irregular and so $mE(G) = |E(G)|$. For every $G \in \mathcal{B}_2$ (so G is not locally irregular), if we had $mE_2(G) = 2|E(G)|$, then the only proper 2-labelling of G would be the one assigning label 2 to all edges, which can only be proper if G is locally irregular, a contradiction. Therefore, in any proper 2-labelling of G, there must be at least one edge assigned label 1, implying that $mE(G) < 2|E(G)|$.

Let us now assume $G \in \mathcal{B}_3$, i.e., G is an odd multi-cactus with bipartition (U, V) (both $|U|$ and $|V|$ are odd by construction). If G is a cycle with length at least 6 congruent to 2 modulo 4, then the result follows from Theorem 4. Thus, we may assume that the maximum degree $\Delta(G)$ of G is at least 3, i.e., some path attachments were made to build G starting from an original cycle.

Let us consider the last green edge xy to which a path $P = (x, v_1, \ldots, v_{4k}, y)$ was attached in the construction of G, where $k \geq 1$. Recall that $d(x) = d(y) \geq 3$ by construction. Consider $G' = G - \{v_1, v_2, v_3\}$. Assuming $v_1, v_3 \in U$ and $v_2 \in V$, the bipartition of G' is $(U', V') = (U \setminus \{v_1, v_3\}, V \setminus \{v_2\})$. This means that $|V'|$ is even. It is known that any bipartite graph with one part X of even size belongs to $\mathcal{B}_2$ and furthermore admits proper 2-labellings where all vertices of X have odd colour while all vertices of the other part Y have even colour [5]. Therefore, there is a proper 2-labelling ℓ' of G' such that all vertices of U' have even colour while all vertices of V' have odd colour. Since $x \in V'$, the colour $c_{\ell'}(x)$ is odd, and thus at least 3 since $d_{G'}(x) \geq 2$. Similarly, $v_4 \in V'$, so the colour $c_{\ell'}(v_4)$ is odd, and it is precisely 1 since $d_{G'}(v_4) = 1$.

We now extend ℓ' to a proper 3-labelling ℓ of G, by assigning label 1 to v_1v_2, label 2 to xv_1 and v_3v_4, and label 3 to v_2v_3. This way, note that $c_\ell(x)$ and $c_\ell(v_4)$ remain odd. Also, $c_\ell(v_1) = 3 < 5 \leq c_\ell(x)$, $c_\ell(v_3) = 5 > 3 = c_\ell(v_4)$ and $c_\ell(v_2) = 4 \notin \{c_\ell(v_1), c_\ell(v_3)\} = \{3, 5\}$. For these reasons, it should be clear that ℓ is indeed a proper 3-labelling of G. We additionally note that label 3 is actually assigned only once by ℓ, to v_2v_3. Furthermore, ℓ assigns label 1 at least once, e.g. to v_1v_2. From this, it follows that $\sigma(\ell) \leq 2|E(G)|$. □

Note that the upper bound in Theorem 8 is tight due to C_6 for which $\mathrm{mE}(C_6) = 12 = 2|E(C_6)|$ (recall Theorem 4). However this seems to be a pathological case due to the small size of C_6. For larger graphs, the next result shows that the upper bound can actually be improved.

Theorem 9. *Let G be a connected bipartite graph with bipartition (U, V) where $|U|$ is even. Then, we have $\mathrm{mE}_2(G) \leq |E(G)| + |V(G)| - 1$.*

Proof. Let U_e (U_o, respectively) be the set of vertices of U of even (odd, respectively) degree in G, and V_e (V_o, respectively) be the set of vertices of V of even (odd, respectively) degree in G. Note that either $|U_e|$ and $|V_o|$ must have the same parity, or $|U_o|$ and $|V_e|$ must have the same parity. This is because, otherwise, since $|U|$ is even and $|U| = |U_e| + |U_o|$, the sizes $|U_e|$ and $|U_o|$ must have the same parity, we would get that also $|V_e|$ and $|V_o|$ have the same parity. Then we would deduce that $\sum_{u \in U} d(u) \not\equiv \sum_{v \in V} d(v) \pmod{2}$, which is not possible.

Without loss of generality, we may assume that U_e and V_o have the same parity, thus that $|U_e| + |V_o|$ is even. Our aim now, is to design a 2-labelling of G that assigns label 2 on as few edges as possible, such that all vertices in U get an odd colour while all vertices in V get an even colour. Such a labelling will obviously be proper. To that aim, we proceed as follows. Let us start with assigning label 1 to all edges of G. This way, at this point the colour of every vertex is exactly its degree; so all vertices in U_o and V_e verify the desired colour property, while all vertices in U_e and V_o do not. To fix these vertices, we consider any spanning tree T of G. We now repeatedly apply the following fixing procedure: we consider any two vertices x and y of $U_e \cup V_o$ that remain to be fixed, and flip (i.e., turn the 1's into 2's, and *vice versa*) the labels of all edges on the unique path in T from x to y. This way, only the colours of x and y are altered modulo 2. Since $|U_e| + |V_o|$ is even, there is an even number of vertices to fix, and, by flipping labels along paths of T, we can fix the colour of all vertices in $U_e \cup V_o$. This results in a 2-labelling ℓ of G, with the desired properties, which is thus proper. Note now that ℓ assigns label 2 only to a subset of the edges of T. Since T has $|V(G)| - 1$ edges, the result follows. □

The arguments in the proof of Theorem 9 actually generalise to graphs with larger chromatic number. See [3] for the proof details.

Theorem 10. *Let G be a connected graph with chromatic number $k = \chi(G)$ at least 3. Then, we have $\mathrm{mE}(G) \leq \mathrm{mE}_{2\lfloor \frac{k}{2} \rfloor + 1}(G) \leq |E(G)| + 2 \lfloor \frac{k}{2} \rfloor |V(G)|$.*

4.2 General Conjecture and Refined Bounds for Bipartite Graphs

We are not aware of graphs for which all proper 3-labellings require more than a few edges labelled with 3. In general, it might actually be true that, for all nice graphs, there is a proper 3-labelling with a few 3's where the number of 1's is about the number of 2's. Also, we observed, during experimentation via computer programs, that only small graphs G seem to have their value of $\mathrm{mE}(G)$ close to $2|E(G)|$ (recall that K_3 and C_6 are such examples, by Theorems 3 and 4). This leads us to conjecture the following:

Conjecture 1. *There is an absolute constant $c \geq 1$ such that, for every nice connected graph G, we have $\mathrm{mE}(G) \leq \frac{3}{2}|E(G)| + c$.*

In the rest of this section, we investigate Conjecture 1 by giving a special focus to bipartite graphs. We exhibit several upper bounds for $\mathrm{mE}(G)$ in various subclasses of bipartite graphs. Each of these upper bounds support Conjecture 1. We also exhibit examples of graphs achieving these upper bounds.

Lower Bounds. We first show that it is not possible to lower $\mathrm{mE}(G)$ below the $\frac{3}{2}|E(G)|$ barrier for general graphs G. This is already illustrated by Theorem 4, which states that $\mathrm{mE}(C_n) = \frac{3}{2}|E(G)| + 3$ for every $n \equiv 2 \pmod 4$. Note that these cycles C_n are such that $\chi_\Sigma(C_n) = 3$. The lower bound even holds for bipartite graphs G with $\chi_\Sigma(G) = 2$. Indeed, there exist bipartite graphs for which label 2 must be assigned to at least half of the edges by any proper 2-labelling. This is a consequence of the following more general result.

Theorem 11. *There exist infinitely many bipartite graphs $G \in \mathcal{B}_2$ with various structure verifying $\mathrm{mE}_2(G) = \frac{3}{2}|E(G)|$. This remains true for trees.*

Sketch of Proof. Let G be any graph, and let H be a graph obtained from G by subdividing every edge e exactly n_e times, where $n_e = 4k_e + 3$ for some $k_e \geq 0$. Then $\chi_\Sigma(H) = 2$. Furthermore, $\mathrm{mE}_2(H) = \frac{3}{2}|E(H)|$.

Through our experimentation, we also managed to come up with the following class of bipartite graphs G for which $\mathrm{mE}_2(G)$ slightly exceeds $\frac{3}{2}|E(G)|$.

Theorem 12. *Let $x, y \geq 4$ be any two integers congruent to 0 modulo 4, and let H be the graph obtained by adding an edge joining any vertex of a cycle of length x and any vertex of a cycle of length y. Then, we have $\mathrm{mE}_2(H) = \lceil \frac{3}{2}|E(H)| \rceil$.*

Improved Upper Bounds. It is worth pointing out that a proper 2-labelling ℓ of a graph G where $\sigma(\ell)$ is about $\frac{3}{2}|E(G)|$ is actually a 2-labelling where the number of assigned 1's is about the same as the number of assigned 2's. Thus, Conjecture 1 relates to *equitable proper labellings* of graphs, introduced in [2], which are proper labellings where, for every two assigned labels i, j, the number of edges assigned label i differs by at most 1 from the number of edges assigned label j. Regarding Conjecture 1, observe that $mE_2(G) \leq \frac{3}{2}|E(G)| + 1$ holds for every graph G admitting an equitable proper 2-labelling.

The authors in [2] proved that nice forests admit equitable proper 2-labellings. This directly implies Theorem 13 below for trees with even size, while it does not for trees with odd size (as a 2-labelling where the number of assigned 2's is one more than the number of assigned 1's does not fulfill our claim), for which we need a dedicated proof. Recall that this result is optimal due to Theorem 11.

Theorem 13. *For every nice tree T, we have $mE_2(T) \leq \frac{3}{2}|E(T)|$.*

Sketch of Proof. The proof is by induction on the number k of branching vertices (i.e., vertices with degree at least 3) of T. Observe that, for a path $P = (v_1, \ldots, v_n)$ where $v_2, \ldots, v_{n-1}$ have degree 2, two inner vertices cannot be involved in a colour conflict by a 2-labelling assigning consecutive labels $1, 2, 2, 1, 1, \ldots$ (a path labelled in this fashion is called a 1-extension) or $2, 1, 1, 2, 2, \ldots$ (called a 2-extension) to the edges of P. Note also that 1-extensions and 2-extensions comply with equitability, as the numbers of 1's and 2's assigned to the edges of P differ by at most 1.

When $k = 0$, i.e., T is a path, the claim is proved by performing a 1-extension or a 2-extension from a degree-1 vertex to the other so that more 1's than 2's are assigned. For larger values of k, the claim is proved by rooting T at some degree-1 vertex r, considering a branching vertex v at largest distance from r, and removing all pendant paths attached to v, resulting in a tree T'. This tree T' can be assumed to be nice (as otherwise there would be a better choice for r), and it thus admits, by induction, a proper 2-labelling assigning more 1's than 2's. It can then be proved that this labelling can be extended, by performing 1-extensions and 2-extensions, to the paths attached to v, resulting in a proper 2-labelling of T where more 1's than 2's are assigned.

Towards Conjecture 1, refined bounds can be deduced in particular contexts. For instance, any graph G satisfies $|E(G)| + |V(G)| - 1 \leq \frac{3}{2}|E(G)|$ as soon as $|E(G)| \geq 2|V(G)| - 2$. As a consequence, Theorem 9 implies that a bipartite graph $G \in \mathcal{B}_2$ with a part of even size verifies $mE_2(G) \leq \frac{3}{2}|E(G)|$ as soon as G has minimum degree at least 4, or more generally when G is dense enough. The same holds for Hamiltonian bipartite graphs with a part of even size.

Lemma 1. *Let G be a Hamiltonian bipartite graph with bipartition (U, V) where $|U|$ is even. Then $mE(G) \leq mE_2(G) \leq \frac{3}{2}|E(G)|$.*

Proof. Just mimic the proof of Theorem 9, but repair pairs of defective vertices of G along a Hamiltonian cycle $C = (v_0, \ldots, v_{n-1}, v_0)$, matching each of them, say, with the next defective vertex in the ordering of C. If this fixing process turns more than half of the labels to 2, then, instead, repair pairs of vertices around C matching each of them with the previous defective vertex in the ordering (which is equivalent to flipping the labels along C). □

The same result holds when G is bipartite and cubic (in which case $\chi_\Sigma(G) = 2$ since $G \in \mathcal{B}_2$, by definition of odd multi-cacti), by a more general argument:

Lemma 2. *Let G be a regular graph with $\chi_\Sigma(G) = 2$. Then $\mathrm{mE}_2(G) \le \frac{3}{2}|E(G)|$.*

Proof. Let ℓ be a proper 2-labelling of G. Since G is regular, the edges labelled 1 by ℓ, and similarly the edges labelled 2, must induce a locally irregular subgraph of G. Then the 2-labelling ℓ' of G obtained by turning all 1's into 2's, and *vice versa*, is also proper. Now there is one of ℓ and ℓ' that assigns label 2 to at most half of the edges, and the conclusion follows. □

5 Conclusion

We have here studied the algorithmic complexity and bounds for the parameter mE. The main question we leave open is Conjecture 1 asking whether $\mathrm{mE}(G) \le \frac{3}{2}|E(G)| + \mathcal{O}(1)$ holds for every nice connected graph G. We think that the proof of Theorem 9 could be improved to prove the conjecture for bipartite graphs.

Regarding our algorithmic results in Sect. 3, note that they all deal, for a given graph G, with the parameter $\mathrm{mE}_k(G)$ (for some k), and not with the more general parameter $\mathrm{mE}(G)$. This is mainly because, as indicated by Theorem 5, in general there is no absolute constant that bounds, for all graphs G, the smallest k such that $\mathrm{mE}(G) = \mathrm{mE}_k(G)$. In particular, even for a graph G of bounded treewidth, although we can determine $\mathrm{mE}_k(G)$ in polynomial time for any fixed k (due to our algorithm in Theorem 7), running enough iterations of our algorithm to determine $\mathrm{mE}(G)$ is not feasible in polynomial time. Thus, the question of determining the complexity of $\mathrm{mE}(G)$ is left open, even when G is a tree.

References

1. Baudon, O., Bensmail, J., Hocquard, H., Senhaji, M., Sopena, E.: Edge weights and vertex colours: minimizing sum count. Discrete Appl. Math. **270**, 13–24 (2019)
2. Baudon, O., Pilśniak, M., Przybyło, J., Senhaji, M., Sopena, E., Woźniak, M.: Equitable neighbour-sum-distinguishing edge and total colourings. Discrete Appl. Math. **222**, 40–53 (2017)
3. Bensmail, J., Fioravantes, F., Nisse, N.: On proper labellings of graphs with minimum label sum. Research report (2020). https://hal.archives-ouvertes.fr/hal-02450521
4. Bensmail, J., Li, B., Li, B., Nisse, N.: On minimizing the maximum color for the 1-2-3 conjecture. Research report (2019). https://hal.archives-ouvertes.fr/hal-02330418
5. Chang, G., Lu, C., Wu, J., Yu, Q.: Vertex-coloring edge-weightings of graphs. Taiwanese J. Math. **15**, 1807–1813 (2011)

6. Chartrand, G., Erdős, P., Oellermann, O.: How to define an irregular graph. Coll. Math. J. **19**, 36–42 (1998)
7. Diestel, R.: Graph Theory. Graduate Texts in Mathematics, vol. 173, 4th edn. Springer, Heidelberg (2012)
8. Dudek, A., Wajc, D.: On the complexity of vertex-coloring edge-weightings. Discrete Math. Theoret. Comput. Sci. **13**, 45–50 (2011)
9. Kalkowski, M., Karoński, M., Pfender, F.: Vertex-coloring edge-weightings: towards the 1-2-3-conjecture. J. Comb. Theory Ser. B **100**(3), 347–349 (2010)
10. Karoński, M., Łuczak, T., Thomason, A.: Edge weights and vertex colours. J. Comb. Theory Ser. B **91**(1), 151–157 (2004)
11. Mulzer, W., Rote, G.: Minimum-weight triangulation is NP-hard. J. ACM **55**(2), 11:1–11:29 (2008)
12. Przybyło, J.: The 1-2-3 conjecture almost holds for regular graphs (2018)
13. Thomassen, C., Wu, Y., Zhang, C.Q.: The 3-flow conjecture, factors modulo k, and the 1-2-3-conjecture. J. Comb. Theory Ser. B **121**, 308–325 (2016)

Decremental Optimization of Dominating Sets Under the Reconfiguration Framework

Alexandre Blanché[1], Haruka Mizuta[2], Paul Ouvrard[1(✉)], and Akira Suzuki[2]

[1] Univ. Bordeaux, Bordeaux INP, CNRS, LaBRI, UMR5800, 33400 Talence, France
{alexandre.blanche,paul.ouvrard}@u-bordeaux.fr
[2] Graduate School of Information Sciences, Tohoku University, Aoba 6-6-05,
Aramaki-aza, Aoba-ku, Sendai, Miyagi 980-8579, Japan
haruka.mizuta.s4@dc.tohoku.ac.jp, a.suzuki@ecei.tohoku.ac.jp

Abstract. Given a dominating set, how much smaller a dominating set can we find through elementary operations? Here, we proceed by iterative vertex addition and removal while maintaining the property that the set forms a dominating set of bounded size. This can be seen as the optimization variant of the dominating set reconfiguration problem, where two dominating sets are given and the question is merely whether they can be reached from one another through elementary operations. We show that this problem is PSPACE-complete, even if the input graph is a bipartite graph, a split graph, or has bounded pathwidth. On the positive side, we give linear-time algorithms for cographs, trees and interval graphs. We also study the parameterized complexity of this problem. More precisely, we show that the problem is W[2]-hard when parameterized by the upper bound on the size of an intermediary dominating set. On the other hand, we give fixed-parameter algorithms with respect to the minimum size of a vertex cover, or $d + s$ where d is the degeneracy and s is the upper bound of the output solution.

Keywords: Combinatorial reconfiguration · Dominating set · Parameterized complexity

1 Introduction

Recently, *Combinatorial reconfiguration* [11] has been extensively studied in the field of theoretical computer science (See, e.g., surveys [10,18]). A reconfiguration problem is generally defined as follows: we are given two feasible solutions of

Partially supported by JSPS and MAEDI under the Japan-France Integrated Action Program (SAKURA). The first and third author is partially supported by ANR project GrR (ANR-18-CE40-0032). The second author is partially supported by JSPS KAK-ENHI Grant Number JP19J10042, Japan. The third author is partially supported by ANR project GraphEn (ANR-15-CE40-0009). The fourth author is partially supported by JST CREST Grant Number JPMJCR1402, and JSPS KAKENHI Grant Numbers JP17K12636 and JP18H04091, Japan. Full version available at https://arxiv.org/abs/1906.05163.

© Springer Nature Switzerland AG 2020
L. Gąsieniec et al. (Eds.): IWOCA 2020, LNCS 12126, pp. 69–82, 2020.
https://doi.org/10.1007/978-3-030-48966-3_6

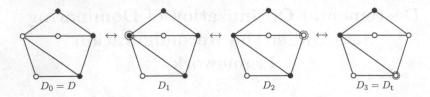

Fig. 1. Reconfiguration sequence between D_0 and D_3 via dominating sets D_1, D_2 with upper bound $k = 4$, where vertices contained in a dominating set are depicted by black circles, and added or removed vertices are surrounded by dotted circles.

a combinatorial search problem, and asked to determine whether we can transform one into the other via feasible solutions so that all intermediate solutions are obtained from the previous one by applying the specified reconfiguration rule. This framework is applied to several well-studied combinatorial search problems; for example, INDEPENDENT SET [3,9,13,14], VERTEX COVER [16,17], DOMINATING SET [8,15,17,19], and so on.

The DOMINATING SET RECONFIGURATION problem is one of the well-studied reconfiguration problems. For a graph $G = (V, E)$, a vertex subset $D \subseteq V$ is called a *dominating set* of G if D contains at least one vertex in the closed neighborhood of each vertex in V. Figure 1 illustrates four dominating sets of the same graph. Suppose that we are given two dominating sets D_0 and D_t of a graph whose cardinalities are at most a given upper bound k. Then the DOMINATING SET RECONFIGURATION problem asks to determine whether we can transform D_0 into D_t via dominating sets of cardinalities at most k such that all intermediate ones are obtained from the previous one by adding or removing exactly one vertex. Note that this reconfiguration rule, i.e. adding or removing exactly one vertex while keeping the cardinality constraint, is called *the token addition and removal* (TAR) rule. Figure 1 illustrates an example of transformation between two dominating sets D_0 and D_3 for an upper bound $k = 4$.

Combinatorial reconfiguration models "dynamic" transformations of systems, where we wish to transform the current configuration of a system into a more desirable one by a step-by-step transformation. In the current framework of combinatorial reconfiguration, we need to have in advance a target (a more desirable) configuration. However, it is sometimes hard to decide a target configuration, because there may exist exponentially many desirable configurations. Based on this situation, Ito *et al.* introduced the new framework of reconfiguration problems, called *optimization variant* [12]. In this variant, we are given a single solution as a current configuration, and asked for a more "desirable" solution reachable from the given one. This variant was introduced very recently, hence it has only been applied to INDEPENDENT SET RECONFIGURATION to the best of our knowledge. Therefore and since DOMINATING SET RECONFIGURATION is one of the well-studied reconfiguration problems as we already said, we focus on this problem and study it under this framework.

1.1   Our Problem

In this paper, we study the optimization variant of DOMINATING SET RECON-
FIGURATION, denoted by OPT-DSR. To avoid confusion, we call the original
DOMINATING SET RECONFIGURATION the *reachability variant*, and we denote
it by REACH-DSR. Suppose that we are given a graph G, two integers k, s,
and a dominating set D of G whose cardinality is at most k; we call k an *upper
bound* and s a *solution size*. Then OPT-DSR asks for a dominating set D_t satisfying the following two conditions: (a) the cardinality of D_t is at most s, and
(b) D_t can be transformed from D under the TAR rule with upper bound k. For
example, if we are given a dominating set D_0 in Fig. 1 and two integers $k = 4$
and $s = 2$, then one of the solutions is D_3, because D_3 can be transformed from
D_0 and $|D_3| \leq 2$ holds.

1.2   Related Results

Although OPT-DSR is being introduced in this paper, some results for
REACH-DSR relate to OPT-DSR in the sense that the techniques to show
the computational hardness or construct an algorithm will be used in our proof
for OPT-DSR. We thus list such results for REACH-DSR in the following.

There are several results for the polynomial-time solvability of REACH-
DSR. Haddadan *et al.* [8] showed that REACH-DSR under TAR rule is
PSPACE-complete for split graphs, for bipartite graphs, and for planar graphs,
while linear-time solvable for interval graphs, for cographs, and for forests.
REACH-DSR is also studied well from the viewpoint of fixed-parameter
(in)tractability. Mouawad *et al.* [17] showed that REACH-DSR under TAR
is W[2]-hard when parameterized by an upper bound k. As a positive result,
Lokshtanov *et al.* [15] gave a fixed-parameter algorithm with respect to $k + d$ for
graphs that exclude $K_{d,d}$ as a subgraph.

Ito *et al.* studied an optimization variant of INDEPENDENT SET RECON-
FIGURATION (denoted OPT-ISR) [12]. More precisely, they proved that this
problem is PSPACE-hard on bounded pathwidth, NP-hard on planar graphs,
while linear-time solvable on chordal graphs. They also gave an XP-algorithm
with respect to the solution size, and a fixed-parameter algorithm with respect
to both solution size and degeneracy.

1.3   Our Results

In this paper, we study OPT-DSR from the viewpoint of the polynomial-time
(in)tractability and fixed-parameter (in)tractability.

We first study the polynomial-time solvability of OPT-DSR with respect to
graph classes (See Fig. 2). Specifically, we show that the problem is PSPACE-
complete even for split graphs, for bipartite graphs, and for bounded pathwidth
graphs, and NP-hard for planar graphs with bounded maximum degree. On the
other hand, the problem is linear-time solvable for cographs, trees and interval
graphs. The inclusions of these graph classes are represented in Fig. 2.

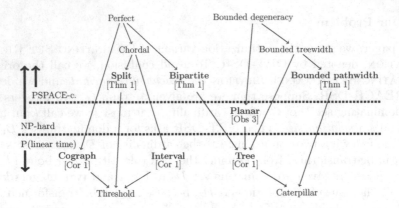

Fig. 2. Our results for polynomial-time solvability with respect to graph classes, where A → B means that the class A contains the class B.

We then study the fixed-parameter (in)tractability of OPT-DSR. We first focus on the following four graph parameters: the degeneracy d, the maximum degree Δ, the pathwidth pw, and the vertex cover number τ (that is the size of a minimum vertex cover). Figure 3(a) illustrates the relationship between these parameters, where A → B means that the parameter A is bounded by some function of B. This relation implies that if we have a result stating that OPT-DSR is fixed-parameter tractable for A then the tractability for B follows, while if we have a negative (i.e. intractability) result for B then it extends to A. From results for polynomial-time solvability, we show the PSPACE-completeness for fixed pw and NP-hardness for fixed Δ, and hence the problem is fixed-parameter intractable for each parameter pw, Δ and d under P $\neq$ PSPACE or P $\neq$ NP. As a positive result, we give an FPT algorithm for τ. We then consider two input parameters: the solution size s and the upper bound k. (See Fig. 3(c).) We show that OPT-DSR is W[2]-hard when parameterized by k. We note that we can assume without loss of generality that $s < k$ holds, as explained in Sect. 2. Therefore, it immediately implies W[2]-hardness for s. Most single parameters (except for τ) cause a negative (intractability) result. We thus finally consider combinations of one graph parameter and one input parameter. We give an FPT algorithm with respect to $s + d$. (See Fig. 3(b).) In the end, we can conclude from the discussion above that for any combination of a graph parameter $p \in \{d, \Delta, pw, \tau\}$ and an input parameter $q \in \{s, k\}$, OPT-DSR is fixed-parameter tractable when parameterized by $p + q$. Due to space limitations, proofs of statements marked with (*) have been omitted (see the arXiv version).

2 Preliminaries

For a graph G, we denote by $V(G)$ and $E(G)$ the vertex set of G and edge set of G, respectively. For a vertex $v \in V(G)$, we let $N_G(v) = \{w \mid vw \in E(G)\}$ and $N_G[v] = N_G(v) \cup \{v\}$; we call a vertex in $N_G(v)$ a *neighbor* of v in G. For a

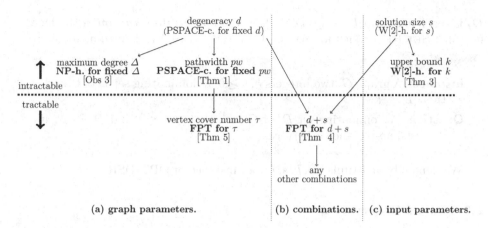

Fig. 3. Our results for fixed-parameter tractability, where A → B means that the parameter A is bounded on some function of B.

vertex subset $S \subseteq V(G)$, we let $N_G[S] = \bigcup_{v \in S} N_G[v]$. If there is no confusion, we sometimes omit G from the notation.

2.1 Optimization Variant of Dominating Set Reconfiguration

For a graph $G = (V, E)$, a vertex subset $D \subseteq V$ is a *dominating set* of G if $N[D] = V(G)$. For a dominating set D, we say that $u \in D$ *dominates* $v \in V$ if $v \in N[u]$ holds. We say that a vertex $v \in D$ has a *private neighbor* in D if there exists a vertex $u \in N[v]$ such that $N[u] \cap D = \{v\}$. In other words, the vertex u is dominated only by v in D. Note that the private neighbor of a vertex can be itself. A dominating set is (inclusion-wise) *minimal* if and only if each of its vertices has a private neighbor, and *minimum* if and only if the cardinality is minimum among all dominating sets. Notice that any minimum dominating set is minimal.

Let D and D' be two dominating sets of G. We say that D and D' are *adjacent* if $|D \Delta D'| = 1$, where $D \Delta D' = (D \setminus D') \cup (D' \setminus D)$ and we denote this by $D \leftrightarrow D'$. Let us now assume that D and D' are both of size at most k, for some given $k \geq 0$. Then, a *reconfiguration sequence* between D and D' under the TAR rule (or sometimes called a TAR-sequence) is a sequence $\langle D = D_0, D_1, \ldots, D_\ell = D' \rangle$ of dominating sets of G such that:

- for each $i \in \{0, 1, \ldots, \ell\}$, D_i is a dominating set of G such that $|D_i| \leq k$; and
- for each $i \in \{0, 1, \ldots, \ell - 1\}$, $D_i \leftrightarrow D_{i+1}$ holds.

Considering a reconfiguration sequence under the TAR rule, we sometimes write TAR(k) instead of TAR to emphasize the upper bound k on the size of a solution. We say that D' is *reachable from* D if there exists a reconfiguration sequence between D and D'; since a reconfiguration sequence is reversible, if D' is reachable from D, then D is also reachable from D'. We write $D \overset{k}{\leftrightsquigarrow} D'$ if D' (resp.

D) is reachable from D (resp. D'). Then, the *optimization variant* of the DOM-INATING SET RECONFIGURATION problem (OPT-DSR) is defined as follows:

OPT-DSR

Instance: A graph G, two integers $k, s \geq 0$, a dominating set D of G such that $|D| \leq k$.

Question: A dominating set D_t of G such that $|D_t| \leq s$ and $D \overset{k}{\leftrightsquigarrow} D_t$ if it exists, no-instance otherwise.

We denote by a 4-tuple (G, k, s, D) an instance of OPT-DSR.

2.2   Useful Observations

From the definition of OPT-DSR, we have the following observations.

Observation 1. *Let (G, k, s, D) be an instance of OPT-DSR. If k, s and $|D|$ violate the inequality $s < |D| \leq k$, then D is a solution of the instance.*

Proof. By the definition of D, we know $|D| \leq k$. Therefore if the inequality is violated, we have $|D| \leq s \leq k$ or $|D| \leq k \leq s$. In both cases, $|D| \leq s$ holds, and hence D is a solution.   □

It is observed that the condition in Observation 1 can be checked in linear time. Therefore, we sometimes assume without loss of generality that $s < |D| \leq k$ holds. Then, another observation follows.

Observation 2. *Let (G, k, s, D) be an instance of OPT-DSR such that $s < |D|$ holds. If D is minimal and $|D| = k$ holds, then the instance has no solution.*

Proof. Since $|D| = k$, we cannot add any vertex to D without exceeding the threshold k. Besides, since D is minimal, we cannot remove any vertex while maintaining the domination property. As a result, there is no dominating set D_t of size at most s reachable from D i.e. $D \overset{k}{\leftrightsquigarrow} D_t$ does not hold for any dominating set D_t such that $|D_t| \leq s$.   □

Again, the conditions in Observation 2 can be checked in linear time, and hence we can assume without loss of generality that D is not minimal or $|D| < k$ holds. Suppose that D is not minimal. Then we can always obtain a dominating set of size less than k by removing some vertex without private neighbor from D, that is, we have a dominating set D' with $D \overset{k}{\leftrightsquigarrow} D'$ and $|D'| < k$. Note that (G, k, s, D) has a solution if and only if (G, k, s, D') does. Therefore, it suffices to consider the case where $|D| < k$ holds. Combining it with Observation 1, we sometimes assume without loss of generality that $s < |D| < k$ holds.

Finally, we have the following observation which states that OPT-DSR is a generalization of the DOMINATING SET PROBLEM:

Observation 3. *Let $G = (V, E)$ be a graph and s be an integer. Then the instance $(G, |V|, s, V)$ of OPT-DSR is equivalent to finding a dominating set of G of size at most s.*

Proof. Let D_t be a dominating set of G of size at most s. Since we started from a dominating set containing all the vertices of G, it is sufficient to remove one by one each vertex in $V \setminus D_t$ to reach D_t.   □

Observation 3 implies that hardness results for the original DOMINATING SET problem extend to OPT-DSR. In particular, we get that OPT-DSR is NP-hard even for the case where the input graph has maximum degree 3, or is planar with maximum degree 4 [7]. However, we will show in Sect. 3.1 that this problem is actually PSPACE-complete.

3   Polynomial-Time (In)tractability

3.1   PSPACE-Completeness for Several Graph Classes

Theorem 1. OPT-DSR *is PSPACE-complete even when restricted to bounded pathwidth graphs, for split graphs, and for bipartite graphs.*

First, observe that OPT-DSR is in PSPACE. Indeed, when we are given a dominating set D_t as a solution for some instance of OPT-DSR, we can check in polynomial time whether it has size at most s or not. Furthermore, since REACH-DSR is in PSPACE, we can check in polynomial space whether it is reachable from the original dominating set D. Therefore, we can conclude that OPT-DSR is in PSPACE.

We now give three reductions to show the PSPACE-hardness for split graphs, bipartite graphs and bounded pathwidth graphs, respectively. These reductions are slight adaptations of the ones of PSPACE-hardness for REACH-DSR developed in [8]. We only give the hardness proof for split graphs (see the arXiv version for the two other proofs). To this end, we use a polynomial-time reduction from the optimization variant of VERTEX COVER RECONFIGURATION, denoted by OPT-VCR.

Given a graph $G = (V, E)$, a *vertex cover* is a subset of vertices that contains at least one endpoint of each edge in E. We now give the formal definition of OPT-VCR. Suppose that we are given a graph G, two integers $k, s \geq 0$, and a vertex cover C of G whose cardinality is at most k. Then OPT-VCR asks for a vertex cover C_t of size at most s reachable from C under the TAR(k) rule. This problem is known to be PSPACE-complete even for bounded pathwidth graphs[1] [12].

Lemma 1. OPT-DSR *is PSPACE-hard even for split graphs.*

---

[1] In [12], Ito *et al.* actually showed the PSPACE-completeness for the optimization variant of INDEPENDENT SET RECONFIGURATION. However, the result can easily be converted to OPT-VCR from the observation that any vertex cover of a graph is the complement of an independent set.

(a) Original graph G' with vertex cover $\{v_2, v_4\}$.   (b) Corresponding split graph G.

Fig. 4. Reduction for Lemma 1. Note that $\{v_2, v_4\}$ is a dominating set of G.

Proof. As we said, we give a polynomial-time reduction from OPT-VCR. More precisely, we extend the idea developed for the NP-hardness proof of DOMINATING SET problem on split graphs [2].

Let (G', k', s', C) be an instance of OPT-VCR with $V(G') = \{v_1, v_2, \ldots, v_n\}$ and $E(G') = \{e_1, e_2, \ldots, e_m\}$. We construct the corresponding split graph G as follows (see also Fig. 4). Let $V(G) = A \cup B$, where $A = V(G')$ and $B = \{w_1, w_2, \ldots, w_m\}$; the vertex $w_i \in B$ corresponds to the edge $e_i \in E(G')$. We join all pairs of vertices in A so that A forms a clique in G. In addition, for each edge $e_i = v_p v_q$ in $E(G')$, we join $w_i \in B$ with each of v_p and v_q. Let G be the resulting graph, and let $(G, k = k', s = s', D = C)$ be the corresponding instance of OPT-DSR (we will prove later that D is a dominating set of G). Clearly, this instance can be constructed in polynomial time. It remains to prove that (G', k', s', C) is a yes-instance if and only if (G, k, s, D) is a yes-instance.

We start by the only-if direction. Suppose that (G', k', s', C) is a yes-instance. Then, there exists a vertex cover C_t of size at most s' reachable from C under the TAR(k') rule. Since $k' = k$, $s = s'$ and both problems employ the same reconfiguration rule, it suffices to prove that any vertex cover of G' is a dominating set of G. Since $C \subseteq V(G') = A$ and A is a clique, all vertices in $A \setminus C$ are dominated by the vertices in C. Thus, consider a vertex $w_i \in B$, which corresponds to the edge $e_i = v_p v_q$ in $E(G')$. Then, since C is a vertex cover of G', at least one of v_p and v_q must be contained in C. This means that w_i is dominated by the endpoint v_p or v_q in G. Therefore, each vertex cover in the reconfiguration sequence between C and C_t is a dominating set of G (including $D = C$ and $D_t = C_t$) and thus, (G, k, s, D) is a yes-instance.

We now focus on the if direction. Suppose that (G, k, s, D) is a yes-instance. Then, there exists a dominating set D_t of G of size at most s reachable under the TAR(k) rule by a sequence $\mathcal{R} = \langle D_0, D_1, \ldots, D_t \rangle$, with $D = D_0$. Recall that $D = C$ and thus D is a vertex cover of G'. We want to produce a sequence of dominating sets that are subsets of A. To this end, we proceed by eliminating the vertices of B that appears in $\mathcal{R}$ one by one from the sequence. Let i be the smallest index such that $D_i \in \mathcal{R}$ contains a vertex $w \in B$ associated to the edge $v_a v_b \in E(G)$. Let $j \geq i$ be the largest index such that every dominating set $D_l \in \mathcal{R}$ ($i \leq l \leq j$) contains w. Now we show that D_{j+1} is reachable from D_{i-1} under TAR(k) rule without touching w, that is, there is a sequence where each dominating set in the sequence does not contain w. For every $D_l \in \mathcal{R}$ $(i \leq l \leq j)$ we instead consider the set $D'_l = (D_l \setminus w) \cup \{v_a\}$. Note that $v_a \in N_G(w)$, and $|D'_l| \leq |D_l| \leq k$. Observe that each D'_l is a dominating set since $N_G[w] \subseteq N_G[v_a]$. If $v_a \in D_{i-1}$, then $D_{i-1} = D'_i$. Otherwise, D'_i is obtainable from D_{i-1} in one step since we just replace the addition of w by the one of v_a. Moreover, due to the choice of j, $D_{j+1} = D_j \setminus \{w\}$. Hence, D_{j+1} contains a vertex in A adjacent to w. If this vertex is v_a, $D'_j = D_{j+1}$. Otherwise, $D_{j+1} = D'_j \setminus \{v_a\}$, which corresponds to a valid TAR move. Finally, since we ensure that each dominating set D'_l with $i \leq l \leq j$ contains v_a, we can ignore each move in the subsequence of $\mathcal{R}$ that touches v_a. Hence, either $D'_l = D'_{l+1}$ or $D'_l \leftrightarrow D'_{l+1}$ holds, for every $i \leq l < j$. By ignoring duplicates from the sequence $\langle D_{i-1}, D'_i, \ldots, D'_j, D_{j+1} \rangle$, we obtain a desired subsequence which does not touch w. Therefore, we can eliminate w in the subsequence $\langle D_{i-1}, D_i, \ldots, D_j, D_{j+1} \rangle$ of $\mathcal{R}$ by replacing it with the desired subsequence. Hence by repeating this process for each subsequence containing w we get a new sequence that does not touch w at all. We then repeat this process for every vertex of B that appears in $\mathcal{R}$ and we obtain a sequence $\mathcal{R}'$ where each dominating set is a subset of A. Finally, observe that any dominating set D of G such that $D \subseteq A = V(G')$ forms a vertex cover of G', because each vertex $w_i \in B$ is dominated by at least one vertex in $D \subseteq V(G')$. Therefore, (G', s', k', C) is a yes-instance. $\qquad\square$

Finally, the two following lemmas complete the proof of Theorem 1.

Lemma 2 (*).   OPT-DSR *is PSPACE-hard even for bounded pathwidth graphs.*

Lemma 3 (*).   OPT-DSR *is PSPACE-hard even for bipartite graphs.*

3.2   Linear-Time Algorithms

We now explain how OPT-DSR can be solved in linear time for several graph classes. To this end, we deal with the concept of a canonical dominating set. A dominating set D_c is *canonical* if D_c is a minimum dominating set which is reachable from any dominating set D under the $\mathsf{TAR}(|D|+1)$ rule. Then we have the following theorem.

Theorem 2. *Let $\mathcal{G}$ be a class of graphs such that any graph $G \in \mathcal{G}$ has a canonical dominating set and we can compute it in linear time. Then OPT-DSR can be solved in linear time on $\mathcal{G}$.*

Proof. Let (G, k, s, D) be an instance of OPT-DSR, where $G \in \mathcal{G}$. Recall that we can assume without loss of generality that $s < |D| < k$; we can check in linear time whether the inequality is satisfied or not, and if it is violated, then we know from Observation 1 and 2 that it is a trivial instance. Since $G \in \mathcal{G}$, G admits a canonical dominating set and we can compute in linear time an actual one. Let D_c be such a canonical dominating set. Then it follows from the definition that D_c is reachable from D under the $\mathsf{TAR}(k)$ rule since $k \geq |D| + 1$.

Since D_c is a minimum dominating set, we can output it if $|D_c| \leq s$ holds, and no-instance otherwise. All processes can be done in linear time, and hence the theorem follows.   □

Haddadan *et al.* showed in [8] that cographs, trees (actually, forests), and interval graphs admit a canonical dominating set. Their proofs are constructive, and hence we can find an actual canonical dominating set. It is observed that the constructions on cographs and trees can be done in linear time. The construction on interval graphs can also be done in linear time with a nontrivial adaptation by using an appropriate data structure. Therefore, we have the following linear-time solvability of OPT-DSR.

Corollary 1. *OPT-DSR can be solved in linear time on cographs, trees, and interval graphs.*

4   Fixed-Parameter (In)tractability

In this section, we study the fixed-parameter complexity of OPT-DSR with respect to several graph parameters. More precisely, we first show that OPT-DSR is W[2]-hard when parameterized by the upper bound k. To prove it, we use the idea of the reduction constructed by Mouawad *et al.* to show the W[2]-hardness of REACH-DSR [17].

Theorem 3 (*). *OPT-DSR is W[2]-hard when parameterized by the upper bound k.*

On the other hand, we give FPT algorithms with respect to the combination of the solution size s and the degeneracy d in Subsect. 4.1 and the vertex cover number τ in Subsect. 4.2.

4.1   FPT Algorithm for Degeneracy and Solution Size

The following is the main theorem in this subsection.

Theorem 4. *OPT-DSR is fixed-parameter tractable when parameterized by $d + s$, where d is the degeneracy and s the solution size.*

To prove the theorem, we give an FPT algorithm with respect to $d + s$. Note that our algorithm uses the idea of an FPT algorithm solving the reachability variant of DOMINATING SET RECONFIGURATION, developed by Lokshtanov *et al.* [15]. Their algorithm uses the concept of domination core; for a graph G, a *domination core* of G is a vertex subset $C \subseteq V(G)$ such that any vertex subset $D \subseteq V(G)$ is a dominating set of G if and only if $C \subseteq N_G[D]$ [6].

Suppose that we are given an instance (G, k, s, D) of OPT-DSR where G is a d-degenerate graph. By Observation 2, we can assume without loss of generality that $|D| < k$. We first check whether G has a dominating set of size at most s: this can be done in FPT$(d + s)$ time for d-degenerate graphs [1]. If G does not have it, then we can instantly conclude that this is a no-instance.

In the remainder of this subsection, we assume that G has a dominating set of size at most s. In this case, we kernelize the instance: we shrink G by removing some vertices while keeping the existence of a solution until the size of the graph only depends on d and s. To this end, we use the concept of domination core.

Lemma 4 (Lokshtanov et al. [15]). *If G is a d-degenerate graph and G has a dominating set of size at most s, then G has a domination core of size at most ds^d and we can find it in $FPT(d + s)$ time.*

Therefore, one can compute a domination core of G of size at most ds^d in $FPT(d + s)$ time by Lemma 4. In order to shrink G, we use the reduction rule **R1:** if there is a domination core C and two vertices $v_r, v_l \in V(G) \setminus C$ such that $N_G(v_r) \cap C \subseteq N_G(v_l) \cap C$, we remove v_r. We need to prove that **R1** is "safe", that is, we can remove v_r from G without changing the existence of a solution. However, if the input dominating set D contains v_r, we cannot do it immediately. Therefore, we first remove v_r from D.

Lemma 5 (*). *Let D be a dominating set such that both $|D| < k$ and $v_r \in D$ hold. Then there exists D' such that $v_r \notin D'$ and $D \overset{k}{\leadsto} D'$, and D' can be computed in linear time.*

We can now redefine D as a dominating set which does not contain v_r. We then consider removing v_r from G. Let $G' = G[V(G) \setminus \{v_r\}]$. The following lemma ensures that removing v_r keeps the existence of a solution.

Lemma 6 (*). *Let (G, k, s, D) be an instance where $v_r \notin D$. Then, (G, k, s, D) has a solution if and only if (G', k, s, D) has a solution.*

We exhaustively apply the reduction rule **R1** to shrink G. Let G_k and D_k be the resulting graph and dominating set, respectively. Then, any two vertices $u, v \in V(G_k) \setminus C$ satisfy $N_{G_k}(u) \cap C \neq N_{G_k}(v) \cap C$ (more precisely, $N_{G_k}(u) \cap C \nsubseteq N_{G_k}(v) \cap C$). Then the following lemma completes the proof of Theorem 4.

Lemma 7. *(G_k, k, s, D_k) can be solved in $FPT(d + s)$ time.*

Proof. We first show that the size of the vertex set of G_k is at most $f(d, s) = ds^d + 2^{ds^d}$. Since $|C| \leq ds^d$, it suffices to show that $|V(G_k) \setminus C| \leq 2^{ds^d}$ holds. Recall that any two vertices $u, v \in V(G_k) \setminus C$ satisfy $N_{G_k}(u) \cap C \neq N_{G_k}(v) \cap C$. Then since the number of combination of vertices in C is at most $2^{|C|} \leq 2^{ds^d}$, we have the desired upper bound $|V(G_k) \setminus C| \leq 2^{ds^d}$.

We now prove that (G_k, k, s, D_k) can be solved in $FPT(d + s)$ time. To this end, we construct an *auxiliary graph* G_A, where the vertex set of G_A is the set of all dominating sets of G_k, and any two nodes (that correspond to dominating sets of G_k) D and D' in G_A are adjacent if and only if $|D \triangle D'| = 1$ holds. Let $n = |V(G_k)|$ and $m = |E(G_k)|$. Then the number of candidate nodes in G_A (vertex subsets of G_k) is bounded by $O(2^n)$. For each candidate, we can check in

$O(n+m)$ time if it forms a dominating set. Thus we can construct the vertex set of G_A in $O(2^n(n+m))$ time. We then construct the edge set of G_A. There are at most $O(|V(G_A)|^2) = O(4^n)$ pairs of nodes in G_A. For each pair of nodes, we can check in $O(n)$ time if their corresponding dominating sets differ in exactly one vertex. Therefore we can construct the edge set of G_A in $O(4^n n)$ time, and hence the total time to construct G_A is $O(4^n n + 2^n(n+m))$ time. We finally search a solution by running a breadth-first search algorithm from D_k on G_A in $O(|V(G_A)| + |E(G_A)|) = O(4^n)$ time.

We can conclude that our algorithm runs in time $O(4^n n + 2^n(n+m))$ in total. Since $n \le f(d,s)$ and $m \le n^2 \le (f(d,s))^2$, this is an FPT time algorithm.   □

4.2   FPT Algorithm for Vertex Cover Number

Let (G, k, s, D) be an instance of OPT-DSR. As in the previous section, we may first assume by Observation 2 that $|D| < k$. We recall that $\tau(G)$ is the size of a minimum vertex cover of G. In order to lighten notations, we simply denote by τ the vertex cover number of the input graph. Then, we have the following:

Theorem 5. *OPT-DSR is fixed-parameter tractable when parameterized by τ.*

Observation 4 (*). *If G is d-degenerate, then $d \le \tau$.*

We are now able to get down to the proof of Theorem 5, by providing an algorithm that solves OPT-DSR and runs in time $\mathrm{FPT}(\tau)$. We first compute a minimum vertex cover $X \subseteq V(G)$ of G in time $\mathrm{FPT}(\tau)$ [4]. We partition the vertices of G into two components, the vertex cover X and the remaining vertices I. By definition of vertex cover, no edge can have both endpoints outside X, therefore I is an independent set. Note that if $s \le \tau$, then by Observation 4 we have $d + s \le 2\tau$, where d is the degeneracy of G. In this case we are able to use the algorithm of the last section, that runs in time $\mathrm{FPT}(d + s)$. We may therefore assume $\tau < s$. In that case, we have the following lemma:

Lemma 8 (*). *If $\tau < s$, then (G, k, s, D) is a yes-instance.*

It remains to discuss the complexity of this algorithm. As we already said, we first compute a minimum vertex cover X of G in time $\mathrm{FPT}(\tau)$. If $s \le \tau$, we run the FPT algorithm of Sect. 4.1. Otherwise, we first compute the set T and then run the subroutine which are both described in the proof of Lemma 8. The two rules used in this subroutine only apply to vertices that belong to the set I and whenever one is applied, exactly one vertex in I is removed (and none is added). Hence, they are applied at most $|I \cap D|$ times. Therefore, the subroutine runs in polynomial time and produces the desired dominating set D_t. As a result, this algorithm is FPT with respect to τ. This concludes the proof.

Concluding Remarks. In this paper, we showed that OPT-DSR is PSPACE-complete even if restricted to some graph classes. However, we only know that it is NP-hard for bounded maximum degree graphs or planar graphs, as an immediate corollary of Observation 3. Hence, it would be interesting to determine whether OPT-DSR is NP-complete or PSPACE-complete on these two graph classes. Note that the complexity on planar graphs remains open for OPT-ISR.

We also proved that OPT-DSR is W[2]-hard for parameter k but the question remains as to whether there exists an XP algorithm for upper bound k.

References

1. Alon, N., Gutner, S.: Linear time algorithms for finding a dominating set of fixed size in degenerated graphs. Algorithmica **54**(4), 544 (2008). https://doi.org/10.1007/s00453-008-9204-0
2. Bertossi, A.A.: Dominating sets for split and bipartite graphs. Inf. Process. Lett. **19**(1), 37–40 (1984)
3. Bonamy, M., Bousquet, N.: Token sliding on chordal graphs. In: Bodlaender, H.L., Woeginger, G.J. (eds.) WG 2017. LNCS, vol. 10520, pp. 127–139. Springer, Cham (2017). https://doi.org/10.1007/978-3-319-68705-6_10
4. Chen, J., Kanj, I.A., Xia, G.: Improved upper bounds for vertex cover. Theor. Comput. Sci. **411**(40), 3736–3756 (2010)
5. Downey, R.G., Fellows, M.R.: Parameterized Complexity. Springer, New York (1999). https://doi.org/10.1007/978-1-4612-0515-9
6. Drange, P., et al.: Kernelization and sparseness: the case of dominating set. In: 33rd Symposium on Theoretical Aspects of Computer Science (STACS 2016), pp. 31:1–31:14 (2016)
7. Garey, M.R., Johnson, D.S.: Computers and Intractability: A Guide to the Theory of NP-Completeness. Freeman, San Francisco (1979)
8. Haddadan, A., et al.: The complexity of dominating set reconfiguration. Theor. Comput. Sci. **651**, 37–49 (2016)
9. Hearn, R.A., Demaine, E.D.: PSPACE-completeness of sliding-block puzzles and other problems through the nondeterministic constraint logic model of computation. Theor. Comput. Sci. **343**(1–2), 72–96 (2005)
10. van den Heuvel, J.: The complexity of change. In: Surveys in Combinatorics 2013. London Mathematical Society Lecture Note Series, vol. 409, pp. 127–160. Cambridge University Press (2013)
11. Ito, T., et al.: On the complexity of reconfiguration problems. Theor. Comput. Sci. **412**(12–14), 1054–1065 (2011)
12. Ito, T., Mizuta, H., Nishimura, N., Suzuki, A.: Incremental optimization of independent sets under reachability constraints. In: Proceedings of the 25th International Computing and Combinatorics Conference (COCOON 2019), pp. 313–324 (2019)
13. Kamiński, M., Medvedev, P., Milanič, M.: Complexity of independent set reconfigurability problems. Theor. Comput. Sci. **439**, 9–15 (2012)
14. Lokshtanov, D., Mouawad, A.E.: The complexity of independent set reconfiguration on bipartite graphs. In: Proceedings of the 29th Annual ACM-SIAM Symposium on Discrete Algorithms (SODA 2018), pp. 7:1–7:19 (2019)
15. Lokshtanov, D., Mouawad, A.E., Panolan, F., Ramanujan, M.S., Saurabh, S.: Reconfiguration on sparse graphs. J. Comput. Syst. Sci. **95**, 122–131 (2018)
16. Mouawad, A.E., Nishimura, N., Raman, V.: Vertex cover reconfiguration and beyond. In: Ahn, H.-K., Shin, C.-S. (eds.) ISAAC 2014. LNCS, vol. 8889, pp. 452–463. Springer, Cham (2014). https://doi.org/10.1007/978-3-319-13075-0_36

17. Mouawad, A.E., Nishimura, N., Raman, V., Simjour, N., Suzuki, A.: On the parameterized complexity of reconfiguration problems. Algorithmica **78**(1), 274–297 (2016). https://doi.org/10.1007/s00453-016-0159-2
18. Nishimura, N.: Introduction to reconfiguration. Algorithms **11**(4), 52 (2018)
19. Suzuki, A., Mouawad, A.E., Nishimura, N.: Reconfiguration of dominating sets. J. Comb. Optim. **32**(4), 1182–1195 (2015). https://doi.org/10.1007/s10878-015-9947-x

On the Complexity of Stackelberg Matroid Pricing Problems

Toni Böhnlein[1]([⊠]) and Oliver Schaudt[2]

[1] Faculty of Engineering, Bar Ilan University, 52900 Ramat-Gan, Israel
toni.bohnlein@biu.ac.il
[2] Institut für Informatik, Universität zu Köln, Weyertal 80, 50321 Cologne, Germany
schaudto@uni-koeln.de

Abstract. In a Stackelberg pricing problem a distinguished player, the *leader*, chooses prices for a set of items, and one or several other players, the *followers*, seeks to buy a feasible subset of the items with minimal costs. The leader's goal is to maximize her revenue, which is determined by the sold items and their prices.

We are interested in cases where the followers' feasible subsets are given by a combinatorial optimization problem. For example, a pricing problem based on the shortest path problem was used by Labbé et al. [15] to model road-toll setting scenarios.

In this paper, we consider Stackelberg pricing problems that are based on matroids. The followers seek to buy a subset that is a basis. More specifically, we consider uniform, partition and laminar matroids.

We study the complexity of computing leader-optimal prices for a single and multiple followers. We show that optimal prices can be computed in polynomial time for all three matroids if there is one follower. In general, such pricing problems based on matroids are APX-hard (see [11]).

For multiple followers, we show that computing optimal prices for uniform matroids can be done in polynomial time. However, for partition and laminar matroids the pricing problem becomes NP-hard.

Keywords: Algorithmic pricing · Stackelberg games · Revenue maximization · Matroids

1 Introduction

We study pricing problems in a game-theoretic model known as Stackelberg games or *Stackelberg pricing problems*. In this model, one player, the *leader*, chooses prices for a number of items and one or several other players, the *followers*, are interested in buying subsets of the items. The followers buy subsets that minimize their expenses subject to some constraints while the leader's goal is to maximize her revenue, which is determined by the sold items and their prices. We are interested in the complexity of computing leader-optimal prices depending on the constraints of the followers.

© Springer Nature Switzerland AG 2020
L. Gąsieniec et al. (Eds.): IWOCA 2020, LNCS 12126, pp. 83–96, 2020.
https://doi.org/10.1007/978-3-030-48966-3_7

One line of research studies Stackelberg pricing problems where the followers' constraints are given by a combinatorial optimization problem. A well-motivated example of such a pricing problem was introduced by Labbé et al. [15] to compute optimal road-tolls: A road network is modeled by a graph where the edges have costs that have to be paid when traveling along an edge. A subset of the roads or edges belongs to the leader and she can charge a toll which increases their costs. Each follower is given by two nodes s and t in the graph and chooses a minimal-cost path connecting s and t. The leader gains revenue in the amount of paid tolls. When deciding on the tolls, the leader has to make to following consideration: On the one hand, low tolls might fail to produce maximum revenue. On the other hand, a large toll might cause the followers to avoid a road entirely resulting in zero revenue. Since the followers "buy" a shortest path, this variant is called *Stackelberg shortest path*.

Stackelberg minimum spanning tree was analyzed by Cardinal et al. and Bilò et al. [3,11,12]. The followers are again interested in subsets of a graph's edges, but the subsets have to form a spanning tree. This setting has applications, for example, when an internet service provider wants to connect hubs in a network. The leader charges additional costs for some of the edges and collects revenue if they are used by a follower (internet service provider). Moreover, *Stackelberg interval scheduling* models situations where the leader pays the follower to execute a set of jobs. The leader makes 'make or buy'-type decisions (see [6]).

Intuitively, the complexity of the pricing problem depends on the complexity of the followers' optimization problem. Stackelberg shortest path is hard to approximate within a factor of less than 2 (see Briest et al. [7]) and Stackelberg minimum spanning tree was shown to be APX-hard. However, Stackelberg interval scheduling is solvable in polynomial time.

Our Results. We study Stackelberg pricing problems that are based on matroids. A matroid is a family of subsets over a ground set that is subject to a set of constraints. The constraints are a bit technical and we spare them for the next section. As an example, for a matroid, we can think of the subsets of a graph's edges that are acyclic. This matroid is called the *graphic* matroid. The inclusion-wise maximal subsets of a matroid are its *bases*. If the elements of the ground set are associated with weights, then a minimal weight basis can be computed by the greedy algorithm. The minimal weight bases of a graphic matroid are the minimum weight spanning trees of the associated graph.

For a Stackelberg pricing problem based on a matroid, the ground set is partitioned into two blocks. One block contains items[1] that have *fixed-costs*. Think of these fixed-cost items as being offered by the leader's competitors. The second block contains the *priceable* items for which the leader chooses prices. Each follower comes equipped with a matroid over the ground set and is interested in buying a minimum weight basis. The weight of an item is either its fixed-cost or its price. Given prices, the decision which subset a follower buys can be computed by the greedy algorithm. If a follower buys a basis, the leader gains

---

[1] We call the element of the ground set items.

revenue for each contained priceable item in the amount of its price. The leader's goal is to maximize her revenue.

Finding a minimum weight basis of a matroid can be regarded as a rather simple problem since it can be done with the greedy algorithm. Therefore, it is surprising that the pricing problem based on the graphic matroid, Stackelberg minimum spanning tree, is APX-hard. To find cases that can be solved in polynomial time, we have to resort to even simpler matroids.

We study three different classes of matroids in two scenarios. For the simpler scenario, we assume that there is a *single* follower. Note that this scenario was studied in most of the literature on Stackelberg pricing problems so far. For the second scenario, there are *multiple* followers which implies a few questions regarding the availability of items and coordination between followers. We assume that the items are available in unlimited supply which makes coordination between the followers unnecessary. The leader sets one price for each item that is valid for all the followers.

The first class are *uniform* matroids. Here, followers are interested in buying a set of items that has a given size. Different followers may come with different sizes. There is no additional structure on the items and a follower buys a subset of his given size with minimum total weight. We show that leader-optimal prices can be computed in polynomial time for a single and the multiple followers. While the single follower scenario is quite simple, the multiple followers scenario requires a dynamic programming approach.

Second, we consider *partition* matroids which generalize uniform matroids. A follower is associated with a partition of the ground set into blocks. For each block, the follower buys a subset of a given size. Different followers may be associated with different partitions and sizes. We show that computing leader-optimal prices for a single follower can be done in polynomial time. For the multiple followers, this computational task is NP-hard.

Table 1. Summary of our results on matroid based Stackelberg pricing problems. The results on Stackelberg minimum spanning tree (MST) appear in [11].

	Single follower	Multiple followers
Uniform	poly-time	poly-time
Partition	poly-time	NP-hard
Laminar	poly-time	NP-hard
MST	APX-hard	APX-hard

The third class are *laminar* matroids which generalize partition matroids. A laminar matroid is based on a hierarchical family of subsets of the ground set, i.e., two subsets of the family are either disjoint or one is contained in the other. Such a hierarchical family is also called a laminar family. For each of the subsets of the laminar family, a follower has an upper bound on the number of items that he wants to buy from this subset. Finding leader-optimal prices has the same complexity for laminar matroids as it has for partition matroids.

Table 1 summarizes our results. The organization of the paper is as follows. In the next section, we give a more careful definition of matroids and Stackelberg pricing problems. In Sects. 3 and 4, we show how to solve Stackelberg uniform matroid with multiple followers and Stackelberg laminar matroid with one follower, respectively. Section 5 shows that Stackelberg partition matroid is hard with multiple followers. For the missing proofs, we refer to the full version of the paper. Finally, Sect. 6 discusses directions for future research.

Related Work. Additional literature includes surveys on Stackelberg shortest path by van Hoesel [20] and Labbé and Violin [16]. Roche et al. [18] present an algorithm with logarithmic approximation guarantee. The best lower bound is due to Briest et al. [7] showing approximation hardness within a factor of less than 2. This is an improvement over APX-hardness by Joret [13].

A Stackelberg shortest path tree game was studied by Bilo et al. [4] and Cabello [10]. Briest et al. [9] give a polynomial time algorithm for Stackelberg bipartite vertex cover game which was later improved by Baïou and Barahona [1].

Briest et al. [9] give a $log(k)$ approximation algorithm for Stackelberg pricing games where k is the number of items. Independently, a slightly more general result was obtained by Balcan et al. [2]. Their algorithms use a single price strategy which was studied in a more general setting by Böhnlein et al. [5].

Briest et al. [8] study Stackelberg pricing games where the follower is based on a NP-hard optimization problem and runs a known approximation algorithm.

2   Preliminaries

Stackelberg Pricing Problems. Let $E = E_f \dot\cup E_p$ be a finite set of items which consists of two blocks E_f and E_p. E_f contains the *fixed-cost* items and E_p contains the *priceable* items. Let

$$|E_f| = m \qquad \text{and} \qquad |E_p| = n.$$

The items in E_f have costs given by the function $c : E_f \to \mathbb{R}$.

We have one *leader* and ℓ *followers*, for an integer $\ell \geq 1$. The leader seeks to sell the items in E_p to the followers. But the followers can also buy items in E_f paying their costs c. The leader choose prices by specifying a price function $p : E_p \to \mathbb{R}$. From the followers' perspective, we do not distinguish between priceable and fixed-cost items. Hence, given a price function p, we compose a weight function $w : E \to \mathbb{R}$:

$$w(e) = \begin{cases} c(e), & e \in E_f, \\ p(e), & e \in E_p. \end{cases}$$

Each follower i is determined by a family of *feasible* subsets $\mathcal{S}_i \subseteq 2^E$ which contains the subsets that he is interested in buying. Given a price function p, the weight of a subset $S \in \mathcal{S}_i$ is defined as

$$w(S, p) = \sum_{e \in S} w(e).$$

The objective of the follower is to buy a feasible subset with minimum total weight which is $w_i^*(p) = \min_{S \in \mathcal{S}_i} w(S, p)$. A subset $S \subseteq E$ (if bought by a follower) yields *revenue* for the leader:

$$\mathrm{rev}(S) = \sum_{e \in S \cap E_p} p(e).$$

In case there are several feasible subsets of weight $w_i^*(p)$, we assume that followers are optimistic and buy a subset that yields maximum revenue for the leader. Hence, follower i buys the following feasible subset:

$$S_i^*(p) = \arg \max_{S \in \mathcal{S}_i} \{ \mathrm{rev}(S) : w(S, p) = w_i^*(p) \}. \tag{1}$$

The revenue from follower i is $\mathrm{rev}(S_i^*(p))$ and the leader's total revenue is

$$\mathrm{rev}(p) = \sum_{i=1}^{\ell} \mathrm{rev}(S_i^*(p)).$$

The leader's objective is to determine a price function p that maximizes $\mathrm{rev}(p)$.

A follower's decision is the solution to an optimization problem (given a price function). When the leader decides on the prices, she is aware of the fixed-cost items and their costs as well as the followers' objective functions and feasible subsets, i.e., we are in a full information setting. Moreover, we assume that each follower has a feasible subset that does not contain any priceable items; otherwise, the leader's revenue is unbounded. If there are multiple followers, we assume that items are available in unlimited supply.

STACKELBERG PRICING
Input: A ground set $E = E_f \cup E_p$, a cost function $c : E_f \to \mathbb{R}$, and ℓ followers given by families $\mathcal{S}_i \subseteq 2^E$, for $i \in [\ell]$.
Objective: Find prices $p : E_p \to \mathbb{R}$ maximizing $\mathrm{rev}(p)$.

Note that STACKELBERG PRICING captures the problems mentioned in the introduction.

Matroids. Given a ground set E, a family of subsets $\mathcal{S} \subset 2^E$ is a *matroid* if it satisfies the following conditions:

(M1) $\emptyset \in \mathcal{S}$.
(M2) If $X \subseteq Y \in \mathcal{S}$, then $X \in \mathcal{S}$.
(M3) If $X, Y \in \mathcal{S}$ and $|X| > |Y|$, there exists $x \in X \setminus Y$ such that $Y \cup x \in \mathcal{S}$.

Matroids are a well-studied combinatorial structure (cf. [17]). The *bases* of a matroid S are its inclusion-wise maximal elements. For example, the acyclic subsets of the edges of a graph G form a matroid. It is called the graphic matroid and its bases are the spanning forests of G.

Given weights $w : E \to \mathbb{R}$ on the ground sets, a minimum weight basis B can be computed using a greedy algorithm: To compute B, we (starting with $B = \emptyset$) consider the elements of E sorted by their weights in non-decreasing order and add an element e to B if $B \cup e \in S$.

Stackelberg Matroid is an instance of STACKELBERG PRICING where followers are given by a matroid and buy a minimum weight basis. Cardinal et al. [11] show that STACKELBERG PRICING based on the graphic matroid is APX-hard.

Theorem 2.1 (Cardinal et al. [11]). STACKELBERG MATROID *with one follower is APX-hard.*

Cardinal et al. observe that an optimal price function uses only values that appear as fixed-costs in c. Given an instance of STACKELBERG MATROID and a price function p, then $E = \{e_1, \ldots, e_{m+n}\}$ are the elements of E sorted non-decreasingly by their weights w. If $w(e_j) = w(e_i)$ where $e_j \in E_f$ and $e_i \in E_p$, then $i < j$. Hence, the optimistic follower computes his solution greedily based on this order. If p assigns a price that is not a fixed-cost, increasing this price to the next larger fixed-cost does not change the ordering but increases the leader's revenue. This observation also holds if there are multiple followers.

Lemma 2.1 (Cardinal et al. [11]). *There is an optimal price function that uses only values of the cost function c.*

We close this section with some more notation. Let $E_f = \{g_1, \ldots, g_m\}$ be the elements of E_f sorted non-decreasingly by their costs. Shorthand, we write $c(g_i) = c_i$ for the costs of $g_i \in E_f$. Similarly, given a price function p, let $E_p = \{h_1, \ldots, h_n\}$ be the elements of E_p sorted non-decreasingly by their price. Usually, the price function is clear from the context and we write $p(h_i) = p_i$.

3   Uniform Matroid

In light of Theorem 2.1, we consider an arguably simple class of matroids, namely *uniform* matroids. Given a ground set E and an integer $s \geq 1$, the uniform matroid S (of rank s) contains all subsets of E that have size at most s. Formally,

$$S = \{U \subseteq E : |U| \leq s\}.$$

The bases of the uniform matroid are the subsets of size exactly s.

Stackelberg Uniform Matroid with One Follower. Let the follower be based on a uniform matroid of rank s. Since E has no structure, the follower only cares about the weight of an element. Given a price function p, the follower buys the set $S^*(p) = \{e_1, \ldots, e_s\}$.

With the next lemma we analyze the conditions when the follower buys a fixed number $a \in \mathbb{N}$ of the priceable items, for $a \leq \min\{s, n\}$.

Lemma 3.1. *Let* $a \leq min\{s, n\}$. *Given a price function* p, $|S^*(p) \cap E_p| = a$ *if and only if* $p_a \leq c_{s-a+1}$ *and* $c_{s-a} < p_{a+1}$.

Proof. First, assume that $|S^*(p) \cap E_p| = a$. It follows that $|S^*(p) \cap E_f| = s - a$, $g_{s-a+1} \notin S^*(p)$, $g_{s-a} \in S^*(p)$, and that $h_a \in S^*(p)$. Hence, we must have that $p_a \leq c_{s-a+1}$ and $c_{s-a} < p_{a+1}$.

Now, assume that $p_a \leq c_{s-a+1}$ and $c_{s-a} < p_{a+1}$. It follows that $|S^*(p) \cap E_p| \geq a$ and that $|S^*(p) \cap E_f| \geq s - a$. Consequently, $|S^*(p) \cap E_p| = a$.   □

To gain maximum revenue when selling a items, the leader chooses the largest prices that satisfy the conditions of Lemma 3.1. It follows that an optimal price function assigns prices c_{s-a+1} for a many items while the prices of the remaining items must be larger. The revenue of such a price function is $a \cdot c_{s-a+1}$, and the maximum revenue rev* can be computed as follows:

$$\text{rev}^* = \max_{a \in [\min\{s, n\}]} a \cdot c_{s-a+1}.$$

Böhnlein et al. [5] showed that constant functions are optimal for STACKELBERG UNIFORM MATROID. The target values are the values of c.

Stackelberg Uniform Matroid with $\ell \geq 2$ Followers. Each follower i is determined by its ranks $s_i \in \mathbb{N}$. Without loss of generality, $s_\ell \leq \ldots \leq s_1$. For a given price function p, follower i buys the set $S_i^*(p) = \{e_1, \ldots, e_{s_i}\}$. It follows that $S_\ell^*(p) \subseteq \ldots \subseteq S_1^*(p)$ and that an item $e \in S_i^*(p)$ is bought by i many followers.

Observation 1. *If* $|S_i^*(p) \cap E_f| = a$, *then the leader sells a items at least i times. Moreover,* $|S_i^*(p) \cap E_f| = a$ *if and only if* $p_a \leq c_{s_i-a+1}$ *and* $c_{s_i-a} < p_{a+1}$.

The second part of Observation 1 follows from Lemma 3.1.

To convince ourselves that an optimal price function does not have to be constant when there are more than one followers, we consider a small example. Assume that there are 4 items with fixed-costs $c_1 = 3$ and $c_{2/3/4} = 5$ as well as 4 priceable items. We have two followers of rank 1 and 4, respectively. Verify that the constant price functions with values 3 and 5 each yield a revenue of 15. But the price function that assigns prices $(3, 5, 5, 5)$ yields a revenue of 16.

From the small example, we get the intuition that an optimal price function p can have several steps (assuming a non-decreasing ordering of the function's values). To determine these steps or equivalently the step-lengths, we use dynamic programming. We construct an algorithm based on solving elementary cases, in which the leader sells a many priceable items $i+1$ times and $a+b$ many priceable items i times. The set of all price functions that satisfy these conditions is

$$\mathcal{P}^i_{a,b} = \{p : E_p \to \mathbb{R} : |S^*_{i+1} \cap E_p| = a \text{ and } |S^*_i \cap E_p| = a + b\},$$

for $a, b \in \mathbb{N}$. With the next lemma we characterize optimal price functions under these conditions.

Lemma 3.2. Let $i \in [\ell]$ and $a, b \in \mathbb{N}$ such that $a + b \leq n$, $a \leq s_{i+1}$ and $b \leq s_i - s_{i+1}$. A price function $p \in \mathcal{P}^i_{a,b}$ is optimal, if

$$p_{a+1} = \ldots = p_{a+b} = c_{s_i-(a+b)+1}.$$

Proof. Let $p \in \mathcal{P}^i_{a,b}$ be an optimal price function. With Observation 1 we have that $p_{a+1} \leq \ldots \leq p_{a+b} \leq c_{s_i-(a+b)+1}$. Suppose towards a contradiction that $p_{a+1} < c_{s_i-(a+b)+1}$. But this implies that p_{a+1} can be increased to $c_{s_i-(a+b)+1}$ without changing the followers' decision and increasing the leader's revenue, contradicting that p was optimal. □

For $i \leq \ell$ and $a \leq s_{i+1}$, we compute recursively the maximum revenue that the items $h_{a+1}, \ldots, h_n$ can yield under the conditions of $\mathcal{P}^i_{a,b}$.

Definition 3.1. Let $i \leq \ell$ and $a, b \in \mathbb{N}$ such that $a \leq min\{s_{i+1}, n\}$ and $b \leq min\{s_i - s_{i+1}, n - i\}$. The maximum revenue that items $h_{a+1}, \ldots, h_n$ can yield under a price function $p \in \mathcal{P}^i_{a,b}$ is

$$\lambda^i_{a,b} = max_{p \in \mathcal{P}^i_{a,b}} \sum_{k=1}^i rev(S^*_k(p) \setminus S^*_{i+1}(p)).$$

Note that $\mathcal{P}^i_{a,b}$ can be empty, and for a sound definition max is replaced by sup. For a practical algorithm, we define $\lambda^i_{a,b} = -\infty$ in these cases. Moreover, we set $S^*_{\ell+1} = \emptyset$ and $s_{\ell+1} = 0$. The main technical feat of this section is the next lemma which derives a recursive formula for $\lambda^i_{a,b}$.

Lemma 3.3. $\lambda^i_{a,b} = max_{\theta \in [min\{n-(a+b), s_{i-1}-s_i\}]} \lambda^{i-1}_{a+b,\theta} + b \cdot i \cdot c_{s_i-(a+b)+1}$

Following Lemma 3.2, the base cases for the dynamic program $(i = 1)$ can be determined as follows: $\lambda^1_{a,b} = b \cdot c_{s_1-(a+b)+1}$. Lemma 3.3 allows us to compute the values of $\lambda^i_{a,b}$ (for $i \geq 2$).

Lemma 3.4. *The leader's maximum revenue can be computed as follows:*

$$max_p \; rev(p) = max_{b \leq min\{s_\ell, n\}} \lambda^\ell_{0,b}.$$

The lemmas above imply an algorithm to compute the leader's optimal revenue for STACKELBERG UNIFORM MATROID when there are multiple followers. The running time of the algorithm is of order $\mathcal{O}(\ell \cdot n^3)$. Hence, the main result of this section is as follows.

Theorem 3.1. STACKELBERG UNIFORM MATROID *with multiple followers can be solved in polynomial time.*

4 Laminar Matroid

We continue with positive results showing that STACKELBERG MATROID based on laminar matroids can be solved in polynomial time if there is only one follower.

Let E be a ground set. Then, $\mathcal{F} \subseteq 2^E$ is a *laminar family* if for all $U, V \in \mathcal{F}$, either $U \subseteq V$, $V \subseteq U$ or $U \cap V = \emptyset$. Each element of $\mathcal{F}$ has a *capacity* $\varphi : \mathcal{F} \to \mathbb{N}$. The *laminar* matroid $\mathcal{S}$ based on $\mathcal{F}$ and φ is defined as follows

$$\mathcal{S} = \{W \subseteq E : |W \cap U| \leq \varphi(U) \text{ for all } U \in \mathcal{F}\}.$$

Without loss of generality, we assume that $E \in \mathcal{F}$. In case $E \notin \mathcal{F}$, we add E to $\mathcal{F}$ and set $\varphi(E) = |E|$ without changing the matroid.

Let $N, M \in \mathcal{F}$ such that $N \subset M$. If there is no subset $T \in \mathcal{F}$ such that $N \subset T \subset M$, N is a *direct* subset of M. Let $D(M)$ be the set of M's direct subsets. If $D(M) = \emptyset$, M is *minimal*. Observe that laminar families have a hierarchical structure. We can associate a laminar family with a rooted tree where we identify the nodes of the tree with the elements of the laminar family. Set E is the root. The children of a node are its direct subsets and the minimal subsets are leafs.

We say that $\mathcal{F} = \{M_1, \ldots, M_s\}$ is a *topological* order of $\mathcal{F}$ if for $M_i \subseteq M_j$ it follows that $i > j$. For $i \in [s]$, let $\mathcal{F}_i = \{M_j \in \mathcal{F} : j \geq i\}$. Observe that $\mathcal{F}_i$ is again a laminar family on E. Moreover, we define $\mathcal{S}_i$ to be the laminar matroid based on $\mathcal{F}_i$ and $\varphi|_{\mathcal{F}_i}$. It holds that $\mathcal{S} \subseteq \mathcal{S}_i$.

Stackelberg Laminar Matroid with One Follower. We are given a ground set $E = E_f \cup E_p$. The follower is determined by a laminar matroid $\mathcal{S}$ based on a laminar family $\mathcal{F} = \{M_1, \ldots M_s\}$ with a topological order.

To solve the pricing problem, we use a dynamic program whose structure is based on the tree structure of $\mathcal{F}$. First, we compute the optimal revenue for the minimal elements for several *configurations*. For an inner node of the tree, the optimal revenue is computed based on the configurations of its direct subsets.

Let $\mathcal{P} = \{c(e) : e \in E_f\}$ be the set of the fixed-costs and $\mathcal{P}_\infty = \mathcal{P} \cup \{-\infty, \infty\}$. For a set M that contains priceable items, we define $\mathcal{P}[M] = \{p : M \cap E_p \to \mathcal{P}\}$ to be the set of all functions that map a priceable item of M to a value of $\mathcal{P}$. According to Lemma 2.1 only the values in $\mathcal{P}$ are relevant prices.

Definition 4.1. *Let $M_i \in \mathcal{F}$, $x \leq \varphi(M_i)$ and $Q^-, Q^+ \in \mathcal{P}_\infty$. Then $\Theta^{M_i}_{x, Q^-, Q^+}$ is the set of all pairs $(p, S) \in \mathcal{P}[M_i] \times 2^{M_i}$ such that*

T1 $S \subseteq M_i$ and $|S| = x$.

T2 $\max_{e \in S} w(e) = Q^-$.

T3 $\min\{w(e) : e \in M_i \setminus S \text{ and } S \cup e \in \mathcal{S}_i\} = Q^+$.

T4 $\nexists\, e \in S, e' \in M_i \setminus S$ such that $(S \setminus e) \cup e' \in \mathcal{S}_i$ and $w(e) < w(e')$.

For a pair $(p, S) \in \Theta^{M_i}_{x, Q^-, Q^+}$, S is a minimal weight subset of M_i under price function p. Set S has size x and satisfies the capacity constraint of $\varphi(M_i)$. Q^-

is the weight of maximum weight element of S. Adding an element of $M_i \setminus S$ to S such that the capacity constraints are still met increases the weight of S by at least Q^+. Based on $\Theta^{M_i}_{x,Q^-,Q^+}$ we define $\lambda^{M_i}_{x,Q^-,Q^+}$ as follows:

Definition 4.2. *Let* $M_i \in \mathcal{F}$, $x \leq \varphi(M_i)$ *and* $Q^-, Q^+ \in \mathcal{P}_\infty$. *Then*

$$\lambda^{M_i}_{x,Q^-,Q^+} = \max \left\{ \sum_{e \in S \cap E_p} p(e) : (p, S) \in \Theta^{M_i}_{x,Q^-,Q^+} \right\}.$$

Note that if $Q^+ < Q^-$, $\Theta^{M_i}_{x,Q^-,Q^+} = \emptyset$, and for a sound definition max is replaced by sup. For a practical algorithm, we set $\lambda^{M_i}_{x,Q^-,Q^+} = -\infty$ for these cases.

First, we show that if we indeed know all the values of $\lambda^{M_i}_{x,Q^-,Q^+}$, we can compute the optimal revenue. Note that each basis of a matroid has the same size and that we can compute this size by computing a basis for any price function.

Lemma 4.1. *Let* b *be the size of a basis of* S. *The maximum revenue* rev^* *of the leader can be computed as follows:*

$$rev^* = \max_{Q^-, Q^+ \in \mathcal{P}_\infty} \lambda^E_{b,Q^-,Q^+}.$$

Our dynamic program computes all values $\lambda^{M_i}_{x,Q^-,Q^+}$ in reverse order of the topological order. First, we derive a direct formula for the minimal elements of $\mathcal{F}$. Second, we derive a recursive formula for the non-minimal elements of $\mathcal{F}$.

Minimal elements. Let $M_i \in \mathcal{F}$ be minimal, $x \leq \varphi(M_i)$ and $p \in \mathcal{P}[M_i]$. We define the set $S^{M_i}_{x,p}$ to contain the x items of M_i with the smallest weight under price function p. The priceable items are preferred in this selection.

Definition 4.3. *Let* $M_i \in \mathcal{F}$ *be minimal,* $x \leq \varphi(M_i)$ *and* $p \in \mathcal{P}[M_i]$. *Moreover, let* $Q^-, Q^+ \in \mathcal{P}_\infty$ *such that* $Q^- < Q^+$. *Then,* $J^{M_i}_{x,Q^-,Q^+}$ *is the set of indices* $j \in [min\{x, |M_i \cap E_p|\}]$ *such that there is a* $p \in \mathcal{P}[M_i]$ *where*

J1 $|S^{M_i}_{x,p} \cap E_p| = j$.
J2 $\max_{e \in S^{M_i}_{x,p}} w(e) = Q^-$.
J3 $\min\{w(e) : e \in M_i \setminus S^{M_i}_{x,p} \text{ and } S^{M_i}_{x,p} \cup e \in \mathcal{S}_i\} = Q^+$.

The set $J^{M_i}_{x,Q^-,Q^+}$ contains the numbers of possible priceable items among the x minimum weight items in M_i if their maximum weight item has weight Q^- and the item in M_i with the $x + 1$ largest weight has weight Q^+. The set $J^{M_i}_{x,Q^-,Q^+}$ is essential in determining $\lambda^{M_i}_{x,Q^-,Q^+}$ for the minimal sets M_i.

Lemma 4.2. *Let* $M_i \in \mathcal{F}$ *be minimal,* $x \leq \varphi(M_i)$ *and* $p \in \mathcal{P}[M_i]$. *Moreover, let* $Q^-, Q^+ \in \mathcal{P}_\infty$ *such that* $Q^- < Q^+$. *Then,*

$$\lambda^{M_i}_{x,Q^-,Q^+} = \begin{cases} \max J^{M_i}_{x,Q^-,Q^+} \cdot Q^-, & J^{M_i}_{x,Q^-,Q^+} \neq \emptyset, \\ -\infty, & \text{otherwise.} \end{cases}$$

We show how to compute $J^{M_i}_{x,Q^-,Q^+}$ only in the full version of the paper. Basically, this can be done by simple routines that inspect the fixed-cost items in M_i.

Non-minimal elements. We show how to compute $\lambda^{M_i}_{x,Q^-,Q^+}$ if M_i is not minimal.

Definition 4.4. *Let* $M_i \in \mathcal{F}$ *such that* $D(M) = \{N_1, \ldots, N_s\}$, $x \leq \varphi(M_i)$ *and* $Q^-, Q^+ \in \mathcal{P}_\infty$. *Then,* $\Lambda^{M_i}_{x,Q^-,Q^+}$ *is the set of all tuples* $(\bar{x}, \bar{Q}^-, \bar{Q}^+) \in [\varphi(M_i)]^s \times \mathcal{P}^s_\infty \times \mathcal{P}^s_\infty$ *such that*

L1 $\sum_{j=1}^s \bar{x}_j = x$

L2 $Q^- = \max_{j \in [s]} \bar{Q}^-_j$

L3 $Q^+ = \begin{cases} \min_{j \in [s]} \bar{Q}^+_j, & x < \varphi(M_i), \\ \infty, & x = \varphi(M_i). \end{cases}$

L4 $\max_{j \in [s]} \bar{Q}^-_j \leq \min_{j \in [s]} \bar{Q}^+_j$

Lemma 4.3. *Let* $M_i \in \mathcal{F}$ *such that* $D(M) = \{N_1, \ldots, N_s\}$, $x \leq \varphi(M_i)$ *and* $Q^-, Q^+ \in \mathcal{P}_\infty$ *where* $Q^- \leq Q^+$. *Then,*

$$\lambda^{M_i}_{x,Q^-,Q^+} = \max \left\{ \sum_{j=1}^s \lambda^{N_i}_{\bar{x}_j, \bar{Q}^-_j, \bar{Q}^+_j} : (\bar{x}, \bar{Q}^-, \bar{Q}^+) \in \Lambda^{M_i}_{x,Q^-,Q^+} \right\}.$$

With Lemma 4.3, we are able to compute all values of $\lambda^{M_i}_{x,Q^-,Q^+}$. First, we compute the values $\lambda^{M_i}_{x,Q^-,Q^+}$ for the minimal elements of $\mathcal{F}$. The running time of this step is in $\mathcal{O}(|\mathcal{F}|(m+n)^2 n^2)$. Applying the recursive formula of Lemma 4.3 involves checking all the configurations for the direct subsets. Observe that the number of direct subsets in a laminar family can be bounded by 2: Assume we have a laminar family $\mathcal{F}$ where an element M has more than 2 direct subsets $D(M) = \{N_1, N_2, \ldots\}$. In this case, we add $N_1 \cup N_2$ to $\mathcal{F}$ and set $\varphi(N_1 \cup N_2) = \varphi(N_1) + \varphi(N_2)$. Observe that $\mathcal{F}$ remains a laminar family. And as such it holds that $|\mathcal{F}| \leq 2|E|$ (cf. [19]). For a family where the direct subsets of an element are at most 2, the running time of applying the recursion takes times of order $\mathcal{O}((m+n)^2 \cdot n^4)$. Finally, we can compute the maximum revenue according to Lemma 4.1.

Theorem 4.1. STACKELBERG LAMINAR MATROID *with one follower can be solved in polynomial time.*

It follows that STACKELBERG PARTITION MATROID with one follower can be solved in polynomial time since laminar matroids generalize partition matroids.

5   Partition Matroid

For a ground set E, let $\mathcal{A}$ be a partition of E into blocks $A_1, \ldots, A_s$. Moreover, there are *capacities* $\varphi : \mathcal{A} \to \mathbb{N}$ associated with each block of $\mathcal{A}$. The *partition matroid* $\mathcal{S}$ with respect to $\mathcal{A}$ and φ contains a subsets of E if its intersection with each block is at most the block's capacity. Formally,

$$\mathcal{S} = \{S \subseteq E : |A_i \cap S| \leq \varphi(A_i) \text{ for each } i \in [s]\}.$$

In this section, we show that STACKELBERG PARTITION MATROID is computationally hard to solve if there are multiple followers. An instance on a ground set E with $\ell \geq 2$ followers is given by partition matroids $\mathcal{S}_1, \ldots \mathcal{S}_\ell$ (each with a possibly different partition and capacities).

Theorem 5.1. STACKELBERG PARTITION MATROID *with ℓ followers is NP-hard, for $\ell \geq 2$.*

It follows that STACKELBERG LAMINAR MATROID with multiple followers is also NP-hard since partition matroids are a special case of laminar matroids.

Proof (incomplete). We consider the decision variant of STACKELBERG PARTITION MATROID where we need to decide if the leader can make more revenue than a given threshold. Our reduction is from the HITTING SET problem which is known to be NP-complete (cf. [14]). Here, we are given a set T, a value $t \in \mathbb{N}$ and a family of subsets $\mathcal{U} = \{U_1, \ldots U_r\}$ of T. The question is if we can find a *hitting set* $H \subset T$ such that $H \cap U \neq \emptyset$ for all $U \in \mathcal{U}$ and $|H| \leq t$.

Given an instance of HITTING SET, we construct an instance of STACKELBERG PARTITION MATROID as follows:

- We have $E_f = \{g_1, g_2\}$ with costs $c(g_1) = 1$ and $c(g_2) = 2$.
- The set of priceable items is $E_p = T = \{h_1, \ldots, h_n\}$.

In total, there are $2 \cdot r + n$ followers.

- For each $i \in [r]$ there are two identical followers $F_{i,1}$ and $F_{i,2}$ with blocks $X = U_i \cup g_1$ and $Y = E \setminus X$ where the capacities are $\varphi(X) = 1$ and $\varphi(Y) = 0$.
- For each $h \in T$ there is one follower $\bar{F}_h$ with blocks $X = \{h, g_2\}$ and $Y = E \setminus X$ where the capacities are $\varphi(X) = 1$ and $\varphi(Y) = 0$.

The idea of the construction is that the leader chooses a hitting set H by setting $p(h) = 1$ if $h \in H$ and $p(h) = 2$ if $h \notin H$. We claim that H is indeed a hitting set of size at most t if p yields revenue at least $2r - t + 2n$. The leader receives a revenue of 1 from each of the two identical follower if for each U_i there exists an priceable item h with $p(h) = 1$. To gain revenue 2 from a follower $\bar{F}_h$, the leader has to set $p(h) = 2$.

To complete the proof, we show (in the full version of the paper) that our STACKELBERG PARTITION MATROID instance admits revenue of at least $2r - t + 2n$ if and only if the HITTING SET instance has a hitting set of size at most t.   □

The decision version of STACKELBERG PARTITION MATROID is NP-complete since we can compute the leader's revenue for a given price function and compare it to a threshold. Moreover, our reduction covers several special cases. Note that we used only two different fixed-cost values and that the partition of each follower contains only 2 blocks.

6 Conclusion and Future Work

We make progress towards the more general question of determining the complexity of a Stackelberg pricing problem depending on the complexity of the underlying (follower) optimization problem. With the uniform matroid we identified a case that is solvable in polynomial time if there are more than one followers. We are not aware of another Stackelberg pricing problems of this kind.

A direction for further research is to consider the multiple followers scenario where the items are available in limited supply. Intuitively, pricing problems become harder in this setting (cf. [2]). Several models on how the followers coordinate themselves can be considered. For example, there might be a fixed order in which followers buy a subset of the available item.

References

1. Baïou, M., Barahona, F.: Stackelberg bipartite vertex cover and the preflow algorithm. Algorithmica **74**(3), 1174–1183 (2016)
2. Balcan, M.-F., Blum, A., Mansour, Y.: Item pricing for revenue maximization. In: Proceedings of the 9th ACM Conference on Electronic Commerce, pp. 50–59 (2008)
3. Bilò, D., Gualà, L., Leucci, S., Proietti, G.: Specializations and generalizations of the stackelberg minimum spanning tree game. Theoret. Comput. Sci. **562**, 643–657 (2015)
4. Bilò, D., Gualà, L., Proietti, G., Widmayer, P.: Computational aspects of a 2-player stackelberg shortest paths tree game. In: Papadimitriou, C., Zhang, S. (eds.) WINE 2008. LNCS, vol. 5385, pp. 251–262. Springer, Heidelberg (2008). https://doi.org/10.1007/978-3-540-92185-1_32
5. Toni Böhnlein, S.K., Schaudt, O.: Revenue maximization in stackelberg pricing games: beyond the combinatorial setting. In: 44th International Colloquium on Automata, Languages, and Programming (ICALP 2017) (2017)
6. Böhnlein, T., Schaudt, O., Schauer, J.: Stackelberg packing games. In: Friggstad, Z., Sack, J.-R., Salavatipour, M.R. (eds.) WADS 2019. LNCS, vol. 11646, pp. 239–253. Springer, Cham (2019). https://doi.org/10.1007/978-3-030-24766-9_18
7. Briest, P., Chalermsook, P., Khanna, S., Laekhanukit, B., Nanongkai, D.: Improved hardness of approximation for stackelberg shortest-path pricing. In: Saberi, A. (ed.) WINE 2010. LNCS, vol. 6484, pp. 444–454. Springer, Heidelberg (2010). https://doi.org/10.1007/978-3-642-17572-5_37
8. Briest, P., Gualà, L., Hoefer, M., Ventre, C.: On stackelberg pricing with computationally bounded customers. Networks **60**(1), 31–44 (2012)
9. Briest, P., Hoefer, M., Krysta, P.: Stackelberg network pricing games. Algorithmica **62**(3–4), 733–753 (2012)
10. Cabello, S.: Stackelberg shortest path tree game, revisited. arXiv preprint arXiv:1207.2317 (2012)
11. Cardinal, J., et al.: The stackelberg minimum spanning tree game. Algorithmica **59**(2), 129–144 (2011)
12. Cardinal, J., Demaine, E.D., Fiorini, S., Joret, G., Newman, I., Weimann, O.: The stackelberg minimum spanning tree game on planar and bounded-treewidth graphs. J. Comb. Optim. **25**(1), 19–46 (2013)

13. Joret, G.: Stackelberg network pricing is hard to approximate. Networks **57**(2), 117–120 (2011)
14. Karp, R.M.: Reducibility among combinatorial problems. In: Miller, R.E., Thatcher, J.W., Bohlinger, J.D. (eds.) Complexity of Computer Computations. The IBM Research Symposia Series, pp. 85–103. Springer, Boston (1972). https://doi.org/10.1007/978-1-4684-2001-2_9
15. Labbé, M., Marcotte, P., Savard, G.: A bilevel model of taxation and its application to optimal highway pricing. Manag. Sci. **44**(12–part–1), 1608–1622 (1998)
16. Labbé, M., Violin, A.: Bilevel programming and price setting problems. Ann. Oper. Res. **240**(1), 141–169 (2016)
17. Oxley, J.G.: Matroid Theory, vol. 3. Oxford University Press, USA (2006)
18. Roch, S., Savard, G., Marcotte, P.: An approximation algorithm for stackelberg network pricing. Netw. Int. J. **46**(1), 57–67 (2005)
19. Schrijver, A.: Combinatorial Optimization: Polyhedra and Efficiency, vol. 24. Springer Science & Business Media (2003)
20. Van Hoesel, S.: An overview of stackelberg pricing in networks. Eur. J. Oper. Res. **189**(3), 1393–1402 (2008)

Nonexistence Certificates for Ovals in a Projective Plane of Order Ten

Curtis Bright[1,2]([✉]), Kevin K. H. Cheung[2], Brett Stevens[2], Ilias Kotsireas[3], and Vijay Ganesh[1]

[1] Department of Electrical and Computer Engineering, University of Waterloo, Waterloo, Canada
cbright@uwaterloo.ca
[2] School of Mathematics and Statistics, Carleton University, Ottawa, Canada
[3] Department of Physics and Computer Science, Wilfrid Laurier University, Waterloo, Canada
https://cs.uwaterloo.ca/~cbright/

Abstract. In 1983, a computer search was performed for ovals in a projective plane of order ten. The search was exhaustive and negative, implying that such ovals do not exist. However, no nonexistence certificates were produced by this search, and to the best of our knowledge the search has never been independently verified. In this paper, we rerun the search for ovals in a projective plane of order ten and produce a collection of nonexistence certificates that, when taken together, imply that such ovals do not exist. Our search program uses the cube-and-conquer paradigm from the field of satisfiability (SAT) checking, coupled with a programmatic SAT solver and the nauty symbolic computation library for removing symmetries from the search.

Keywords: Combinatorial search · Satisfiability checking · Symbolic computation

1 Introduction

Projective geometry—a generalization of the familiar Euclidean geometry where parallel lines do not exist—has been extensively studied since the 1600s. A special case of projective geometry occurs when only a finite number of points exist. A two-dimensional projective geometry with a finite number of points is known as a *finite projective plane*.

Despite a huge amount of study some basic questions about finite projective planes are still open—for example, how many points can a finite projective plane contain? It is well-known [14] that a finite projective plane must contain $n^2 + n + 1$ points for some integer n (known as the plane's *order*) and finite projective planes can be explicitly constructed in all orders that are prime powers. The order six case is excluded by a theoretical result of Bruck and Ryser [8] making ten the first uncertain order.

L. Gąsieniec et al. (Eds.): IWOCA 2020, LNCS 12126, pp. 97–111, 2020.
https://doi.org/10.1007/978-3-030-48966-3_8

In the 1970s and 1980s, a significant amount of mathematical ingenuity and computer searches successfully eliminated the possibility of a projective plane of order ten [27]. Today, this remains one of the most prominent achievements of computational combinatorial classification [23]. The search was made feasible due to results of MacWilliams, Sloane, and Thompson [32] concerning the error-correcting code generated by a hypothetical projective plane of order ten. They showed that the weight distribution of this code depends on just two unknown parameters. One of these parameters is the number of ovals that exist in the projective plane of order ten—here an *oval* being a set of twelve points, no three of which are collinear.

In 1983, Lam, Thiel, Swiercz, and McKay [30] showed the nonexistence of ovals in a projective plane of order ten via a computer search. The search space is of a significant size and required about 4,400 h of computation time on the super-mini computer VAX 11/780 (clock speed 5 MHz) to search exhaustively. Because of the nature of the search, Lam et al. specifically encouraged an independent verification:

> Since the existence of ovals is an important question, we hope that someone will do an independent search to verify the result.

Despite this hope, there has been little published work independently verifying the search for ovals or their subsequent searches [28,29] that culminated in the proof that projective planes of order ten do not exist. In his 2011 master's thesis, Roy [37] performed a verification of the nonexistence of a projective plane of order ten using about 35,000 h on a cluster of desktop machines. However, he did not specifically run a search for the ovals as it was nonessential to his ultimate goal. To the best of our knowledge, there has been no published work specifically replicating the search for ovals.

In this paper, we report our results on verifying the nonexistence of ovals in a projective plane of order ten. Our method relies on a satisfiability (SAT) solver and produces certificates that a third party can use to verify that our search completed successfully. In total, our search used about 1,850 core hours on the supercomputer Graham at the University of Waterloo (clock speed 2.1 GHz) and produced SAT proofs that when compressed use about 3 terabytes of storage.

In addition to a using a SAT solver our method also takes advantage of the nauty symbolic computation library [34] to reduce the size of the search space by eliminating redundant symmetries. We present the necessary background on projective geometry, satisfiability checking, and symbolic computation in Sect. 2, describe our SAT encoding in Sect. 3, give details on our implementation and results in Sect. 4, and finally discuss future work in Sect. 5.

2   Preliminaries

The main background necessary to understand our results are some familiarity with projective geometry (see Sect. 2.1), satisfiability checking (see Sect. 2.2), and symbolic computation (see Sect. 2.3).

2.1   Projective Geometry

A *finite projective plane* of order n is a collection of $n^2 + n + 1$ points and $n^2 + n + 1$ lines and an incidence relationship between points and lines where any two points are incident with a unique line and any two lines are incident with a unique point. Furthermore, every line is incident with $n + 1$ points and every point is incident with $n + 1$ lines.

An *oval* of a projective plane of even order n is a set of $n + 2$ points (or $n + 1$ points when n is odd) with no three points collinear (incident with the same line). It can be shown that it is not possible to find a larger set of points with no three points collinear [15], but no characterization of ovals in general projective planes is known. In particular, prior to the search of Lam et al. [30] it was not known if a projective plane of order ten could contain ovals or not.

From a computational perspective, a convenient way of representing a finite projective plane of order n is by a square $\{0, 1\}$ incidence matrix whose (i, j)th entry contains a 1 exactly when the ith line is incident to the jth point. We say that two $\{0, 1\}$-vectors *intersect* when they share a 1 in the same location and the *weight* of a $\{0, 1\}$-vector is the number of nonzero entries it contains. In this framework, a projective plane of order n is a $\{0, 1\}$-matrix with $n^2 + n + 1$ rows that each have weight $n + 1$ and pairwise intersect exactly once (and similarly for the columns). In other words, a $\{0, 1\}$-matrix A with $n^2 + n + 1$ rows represents a projective plane exactly when it satisfies $AA^T = A^T A = nI + J$ where I denotes the identity matrix and J denotes the matrix consisting of all 1s. Two projective planes that are identical up to row or column permutations are called *isomorphic* and we call a submatrix of a projective plane a *partial projective plane*.

Suppose that A is a projective plane of order ten that contains an oval. Without loss of generality we assume that the first twelve points of the plane consist of an oval. By definition, each pair of points in the oval must define a unique line, and therefore there are $\binom{12}{2} = 66$ lines incident to the oval. Without loss of generality, we assume these lines are ordered in lexicographically increasing order. In other words, the first 66 rows of A have the form

$$B = \begin{bmatrix} \begin{array}{c} 110000000000 \\ 101000000000 \\ \vdots \\ 000000000011 \end{array} & \Big| & B' \end{bmatrix}.$$

The first twelve columns contain two 1s on each row, so B' must contain nine 1s in each row. Furthermore, by definition of a projective plane each column in B' must intersect each of the first twelve columns. Each 1 in B' induces an intersection with two of the first twelve columns, so each column in B' contains exactly six 1s.

Without loss of generality, we assume the columns of B' are sorted in lexicographic order. This implies the first nine columns of B' will be incident with the first line (the line through the first and second points). As noted by [26], this also means the ith column of B' (for $1 \le i \le 9$) will be incident with the line through the third and $(3 + i)$th points. We call the nine columns of B' that are

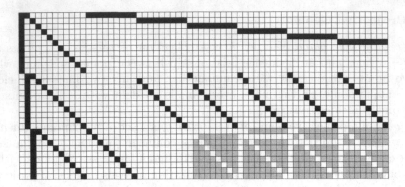

Fig. 1. The upper-left 30×66 submatrix of B under the assumption that the rows are lexicographically ordered and the columns outside the oval are lexicographically ordered. Black entries denote 1s, white entries denote 0s, and gray entries are unknown.

incident with the ith row the ith *block*. In general, all blocks' columns may be ordered similarly to those in the first block [26], and this fixes the first two 1s in each column of B'. Figure 1 contains a visual depiction of the first 30 rows of B up to the sixth block.

Some entries of B' are still undetermined (shown as gray in Fig. 1). At this stage, it is still uncertain if they can be completed in a consistent way to make B' a partial projective plane—since the above description assumes that an oval exists in A. Thus, a proof that there is no way of completing the unknown entries of B' in a consistent way would also imply the nonexistence of ovals in a projective plane of order ten.

The *symmetry group* of a matrix is the group of row and column permutations that fix the entries of the matrix. For example, consider the symmetry group S of the first twelve columns of B. Each row of this submatrix is completely specified by the two columns incident to it, so any column permutation completely specifies a row permutation that undoes the permutation. It follows that S is isomorphic to S_{12}, the symmetric group on twelve elements.

The group S acts on the entries of B' as follows: Given a permutation $\varphi \in S$ the row permutations from φ are applied to the entries of B', then column permutations are applied to reorder its columns in lexicographic order. The result $\varphi(B')$ is a partial projective plane that is isomorphic to B'. To avoid duplication of work, any search for B' should ideally avoid exploring parts of the search space that are isomorphic under S. Exploiting this leads to a huge reduction in the size of the search space, since S contains about 479 million permutations.

2.2 Satisfiability Checking

Given a formula of Boolean logic, satisfiability (SAT) checking is to determine whether or not the formula is *satisfiable*—that is, is there a way of assigning true and false to its variables that results in the whole formula becoming true? A *SAT solver* is a program that performs SAT checking on a given formula. Modern SAT solvers require their input to be given in conjunctive normal form or CNF: if x is a Boolean variable then x and $\neg x$ are known as *literals*, expressions of the form $l_1 \lor \cdots \lor l_n$ for literals l_i are known as *clauses*, and expressions of the form $c_1 \land \cdots \land c_m$ for clauses c_i are in *CNF*. The literal x is satisfied when x is assigned true, $\neg x$ is satisfied when x is assigned false, $l_i \lor \cdots \lor l_n$ is satisfied when at least one l_i is satisfied, and $c_1 \land \cdots \land c_m$ is satisfied when every c_i is satisfied.

In order to reduce the search for ovals in a projective plane of order ten to a SAT problem we use the incidence structure described in Sect. 2.1 that was based on the assumption that ovals exist. The SAT instance will have a solution when there is a completion of the unknown entries of the matrix B to a partial projective plane—so showing the instance has no solution implies that no ovals exist.

SAT solvers are effective as combinatorial search tools—for example, they were used to resolve the first open case of the Erdős discrepancy conjecture [25]. The *cube-and-conquer* SAT solving paradigm has been particularly effective at solving very large combinatorial search problems [21]. First developed by Heule, Kullmann, Wieringa, and Biere for computing van der Waerden numbers [22], the cube-and-conquer method has since been used to resolve the Boolean Pythagorean triples problem [20] and determine the value of the fifth Schur number [16].

A *cube* is a formula of the form $l_1 \land \cdots \land l_n$ where l_i are literals. In the cube-and-conquer paradigm a SAT instance is split into a number of distinct subinstances specified by cubes. Each subinstance contains a single cube and the cube is assumed to be true for the purposes of solving the subinstance. The cubes are typically generated by running a "cubing solver" on the SAT instance which attempts to find a set of cubes which split the instance into subinstances of approximately equal difficulty. After the cubes have been generated a "conquering solver" solves the subinstances either in sequence or in parallel. Ideally, the literals in each cube are added to the solver as incremental assumptions. In this case, after each cube is solved the assumptions are removed and the literals from the next cube are added without restarting the SAT solver.

2.3 Symbolic Computation and SAT+CAS

Symbolic computation is a branch computer science devoted to manipulating and simplifying mathematical expressions. Many computer algebra systems (CASs) are available today that contain extensive symbolic computation functionality from a huge number of mathematical domains. However, although CASs contain many sophisticated algorithms, they have not typically been optimized to perform searches in the way that SAT solvers have [1,4].

For problems that need *both* mathematical sophistication and finely-tuned search it can be useful to combine computer algebra and SAT solvers [10]. Recently SAT+CAS methods have been used in a number of various problems—for example, they have been used to verify the correctness of Boolean arithmetic circuits [24], improve the best known result in the Hadwiger–Nelson plane-colouring problem [17], find many new algorithms for multiplying 3×3 matrices [19], and improve the best known result in the Ruskey–Savage hypercube conjecture [40].

In addition to a SAT solver we use the nauty symbolic computation library [34] in order to show the nonexistence of ovals in a projective plane of order ten. We call nauty from within the callback function of a "programmatic" SAT solver. A solver is called *programmatic* if it allows learning clauses on-the-fly through a piece of code supplied to the SAT solver. A programmatic SAT solver will run the supplied code from time to time as it is performing its search. The code will examine the current assignment to the variables and test whether the current assignment may be discarded (possibly using knowledge queried from a CAS). If the assignment can be discarded a clause is added to the SAT instance on-the-fly that blocks the current assignment (and ideally other similar assignments). Programmatic SAT solvers were introduced by Ganesh et al. [11] in order to solve an RNA folding problem. They have since been used to search for various combinatorial objects such as Williamson matrices [5], best matrices [7], and complex Golay sequences [6].

3   Satisfiability Encoding

We now describe the encoding that we use to search for ovals in a projective plane of order ten. As described in Sect. 2.1, we may assume a number of entries of this projective plane have been fixed in advance, including all entries in the first twelve columns and all entries in the first 21 rows (see Fig. 1). Specifying these entries removes a substantial amount of symmetry from the search space, however, as described in Sect. 2.1, the remaining search space is still symmetric under the action of the group S generated by permuting the twelve points of the oval. In Sect. 3.1, we give our basic encoding without removing symmetries from S. In Sect. 3.2, we provide a programmatic SAT method of removing symmetries from the group S.

3.1   Basic SAT Encoding

Following Sect. 2.1, let B be the first 66 rows of the incidence matrix of a partial projective plane of order ten whose first twelve points form an oval. As previously outlined, up to isomorphism some points of B can be assumed in advance, but most points remain unspecified. For each unspecified point we define a Boolean variable $b_{i,j}$ that will be true exactly when the ith line is incident to the jth point, i.e., the (i,j)th entry of B is 1.

We now give properties that necessarily hold in B as Boolean constraints in conjunctive normal form. In particular, we encode the two facts that (1) columns of B intersect at most once and (2) each column of B intersects a column in the oval at least once. A similar encoding has been previously used to verify MacWilliams et al.'s result that vectors of weight 15 do not exist in the rowspace of any projective plane of order ten [3].

Columns Intersect at Most Once. Let i and j be arbitrary column indices of B, so $i, j \in \{1, \ldots, 111\}$. By definition of a projective plane these columns cannot intersect twice, so we know there do not exist rows k and l mutually incident to columns i and j. In Boolean logic we write this constraint as

$$\bigwedge_{1 \leq k < l \leq 66} (\neg b_{k,i} \vee \neg b_{k,j} \vee \neg b_{l,i} \vee \neg b_{l,j}).$$

Each Column Intersects a Column in the Oval. Let i be an arbitrary column in the oval, i.e., $i \in \{1, \ldots, 12\}$ and let j be an arbitrary column not in the oval, i.e., $j \in \{13, \ldots, 111\}$. By definition of a projective plane columns i and j must intersect somewhere in the plane. Since all rows incident with column i occur in the first 66 rows, the intersection of columns i and j must occur in B. In Boolean logic we write this constraint as $\bigvee_{k : B[k,i]=1} b_{k,j}$.

Abbreviated Constraints. For the first set of constraints there are $\binom{111}{2} \cdot \binom{66}{2} \approx 13$ million clauses of the first form and $12 \cdot 99 = 1188$ clauses of the second form. After removing variables whose values are already fixed there are 2696 undetermined variables in these clauses.

SAT solvers often perform better if the number of constraints can be significantly decreased. In our case, we found that it was only necessary to consider a submatrix of B before reaching a contradiction. In particular, our primary searches only used the variables in the blocks 2 to 6 (columns 22 to 66). The first block was skipped since its columns did not intersect the columns of any other block in the known entries (see Fig. 1). This increases the efficiency of the search because contradictions are generally easier to derive from two already-intersecting columns. Using columns 22 to 66 meant there were $\binom{57}{2} \cdot \binom{66}{2} \approx 3.4$ million clauses of the first form, $12 \cdot 45 = 540$ clauses of the second form, and 1199 unknown variables.

Known Row Intersections. We included one further set of constraints that, while not strictly necessary, improved the performance of the SAT solver by enforcing row intersections that must occur. In particular, note that rows 2–6 must intersect rows 22–66 in B and all 1s in row $i \in \{2, \ldots, 6\}$ outside the oval occur in the columns $B_i := \{4 + 9i, \ldots, 12 + 9i\}$. Thus, we also included clauses of the form $\bigvee_{k \in B_i} b_{j,k}$ for rows $i \in \{2, \ldots, 6\}$ and $j \in \{22, \ldots, 66\}$ that do not intersect in the oval.

3.2 Symmetry Breaking

The encoding described in Sect. 3.1 could in theory be used to show there is no way of completing the unknown entries of B subject to the given constraints. However, as discussed in Sect. 2.1 the search space is symmetric under the action of relabelling the twelve points of the oval (while appropriately reordering the rows and remaining columns to preserve our lexicographic presentation of the search space). Since this is an enormous group of symmetries it is worthwhile developing a method that will reduce or "break" these symmetries. Using an "orderly generation" algorithm [36,38] is one way to avoid generating isomorphic partial solutions at each stage of the search. Our approach is similar, though it will only avoid isomorphic partial solutions violating a property that we can show the entries of B satisfy (up to isomorphism).

Mathon [33] provided a characterization of an oval in a projective plane of order ten in terms of K_{12}, the complete graph on vertices $\{1, \ldots, 12\}$. Note that a *1-factor* of a graph is a perfect matching of its edges and a *1-factorization* of a graph is a decomposition of its edges into 1-factors. If rows denote edges and columns denote points then the first twelve columns of B are precisely the incidence matrix of K_{12}. Every column of B outside the oval contains six 1s on rows that will not be adjacent (as edges of K_{12}). Therefore, each column of B outside the oval forms a 1-factor of K_{12}.

Furthermore, consider the set of columns in the first block of B. These rows are all incident to the row through points 1 and 2. The other five 1s in each column must each occur on distinct rows and will cover the remaining $\binom{10}{2} = 45$ rows through the points $\{3, \ldots, 12\}$. Therefore, the first block of B forms a 1-factorization of $K_{12} \setminus \{1, 2\} \cong K_{10}$ and in general the ith block forms a 1-factorization of $K_{12} \setminus \{1, i+1\}$. Gelling [12] determined that there are exactly 396 nonisomorphic 1-factorizations of K_{10} and we assume that each has been given a distinct label in the set $\{1, \ldots, 396\}$.

Note that the symmetry group S generated by permuting the columns of the oval acts transitively on the set of blocks: there is a permutation in S that will send any block to any other block. Suppose we tag each completed block of B with the label (as described above) of the 1-factorization that it is isomorphic with. We may assume that block 2 of B has the minimal label amongst the blocks of B—if it didn't, we could send the block with minimal label to block 2 by an appropriate permutation of S. Our symmetry breaking method will enforce the condition that block 2 has the minimal label amongst the other blocks of B for which we are searching (blocks 2–6). However, it is not very easy to concisely express this constraint as clauses in Boolean logic. Therefore, we make use of the programmatic SAT paradigm in order to enforce this constraint on-the-fly.

Programmatic Symmetry Breaking. A programmatic SAT solver is compiled with a "callback" function that often examines the solver's current assignment (the mapping from variables to truth values). When the callback function determines the current state should be discarded it will add clauses to the SAT instance that block the current assignment.

If all the variables in the ith block have been assigned and p is one of these variables then we let $B_i \models p$ denote that variable p has been assigned true. Suppose all the variables in block 2 and block $i \in \{3, \ldots, 6\}$ have been assigned. If the label of block 2 is strictly larger than the label of block i we want to block this configuration from the search space. In such a case we want to add the Boolean constraint

$$\bigwedge_{B_2 \models p} p \rightarrow \left(\neg \bigwedge_{B_i \models p} p\right) \quad \text{or equivalently} \quad \bigvee_{B_2 \models p} \neg p \vee \bigvee_{B_i \models p} \neg p$$

which says that the ith block cannot be assigned the way it currently is while the second block is assigned the way it currently is.

4   Implementation and Results

Our SAT encoding is implemented as a part of the MathCheck project; our scripts are open source and available online at uwaterloo.ca/mathcheck. The search proceeds in three main parts: First, we verify the result of Gelling [12] that there are exactly 396 nonisomorphic 1-factorizations of K_{10}. Second, we generate 396 separate SAT instances (one for each nonisomorphic way of filling in block 2 of B). The cube-and-conquer method is used in parallel to solve each SAT instance. A cubing solver generates a set of cubes from each SAT instance and a programmatic conquering solver is used to show that (up to the symmetry breaking method of Sect. 3.2) there are 58 ways of completing the blocks 2–6. Finally, we generate a new SAT instance for each of the 58 solutions and verify that there are no consistent ways of extending these completions to block 7. Additionally, to increase the confidence that the SAT instances were successfully solved the SAT solvers produced DRUP (delete reverse unit propagation) certificates [18] which were subsequently verified. A flowchart of these steps is available in this paper's appendix at uwaterloo.ca/mathcheck.

4.1   Generating the SAT Instances

The SAT instances are generated by a Python script that writes the clauses described in Sect. 3.1 to a file in DIMACS (Discrete Mathematics and Theoretical Computer Science) CNF format. The script accepts as a parameter the columns to include in the SAT instance and by default uses the columns in blocks 2–6 (those used in our primary search).

4.2   Generating the Nonisomorphic 1-Factorizations

The 396 nonisomorphic 1-factorizations of K_{10} as reported by Gelling [12] can be quickly generated using a straightforward search, but we used a SAT approach as that was convenient for our purpose. The SAT instance only uses the variables in block 2, the columns of this block corresponding with 1-factors of $K_{12} \setminus \{1,3\}$. As noted by Gelling, up to isomorphism the entries of the first 1-factor can be completely assumed. By the lexicographic ordering assumption the first 1-factor includes the edge $(2,4)$ and by permuting the columns $\{5,\ldots,12\}$ of the oval we can assume the first 1-factor contains the edges $(5,6)$, $(7,8)$, $(9,10)$, and $(11,12)$. Gelling also noted that after fixing the first 1-factor there are exactly two ways (up to isomorphism) of fixing the second 1-factor and the union of the first two 1-factors either form a 4-cycle and a 6-cycle or a 10-cycle.

The entries that can be fixed are given to the SAT solver as unit clauses and a programmatic implementation finds all nonisomorphic 1-factorizations. Whenever a completion of block 2 is found, the program Traces from the nauty graph isomorphism library determines if the completion is new or isomorphic to a previously found completion. (The graph provided to Traces is the *incidence graph* representation [13] of the first 12 columns of B and the columns of block 2.) A new completion is recorded for later use and a clause that blocks the completion (i.e., $\bigvee_{B_2 \models p} \neg p$) is added to the SAT instance until all possible completions have been examined. A programmatic implementation of MapleSAT [31] confirms the result of Gelling that 396 nonisomorphic 1-factorizations of K_{10} exist in about 8 s.

4.3 Solving the SAT Instances: Cubing

We now generate 396 distinct SAT instances, one for each of the 396 nonisomorphic ways of filling in block 2. Variables from blocks 2–6 are used in each SAT instance, with the variables in block 2 completely determined by the specific nonisomorphic 1-factorization chosen in each case. We simplify these instances with the preprocessor of the SAT solver Lingeling [2] which produces proofs of simplification without renaming variables. After simplification, these instances each contained 912 unknown variables and on average contained 22,883 clauses. Simplifying all 396 SAT instances requires about 15 min in total.

Next, we apply the cubing solver March_cu [22] on each of the 396 individual SAT instances. The conquering solver (see Sect. 4.4) typically performs better when the variables in the cubes are not split across blocks. Thus, we modified March_cu so that it only produces cubes using variables occurring in the same block as the first variable in the cubes. We controlled the cubing cutoff using the -n parameter of March_cu which stops cubing once the number of free variables falls below the given bound. Each block contains 228 unknown variables and we stop cubing once the subproblems specified by each cube contain at least 228 fewer free variables than the original instance. On average, March_cu produced about 180,000 cubes per SAT instance and spent about 17.5 total hours in this step.

4.4 Solving the SAT Instances: Conquering

The majority of the search work was done by the conquering solver. A programmatic version of MapleSAT [31] was used to complete this step. Each of the 396 SAT instances along with the cubes previously computed for each instance were given to separate instances of MapleSAT and solved in parallel. The literals in each cube were specified as incremental assumptions [35], so that it was not necessary to restart the SAT solver after solving each cube.

The programmatic encoding from Sect. 3.2 was used to ignore any completions of blocks 3–6 whose label was strictly smaller than the label of block 2 (which was fixed in each SAT instance). The label of each block completion can be computed by calling nauty on the incidence graph representation of the block (as described in Sect. 4.2). However, these incidence graphs contain 87 vertices (from 66 rows and 21 columns) and we found there was significant overhead from calling nauty in this way.

Our final implementation makes use of a simpler check based on Gelling's observation that up to isomorphism each pair of columns in a block are of two types (either a 4-cycle and 6-cycle or a 10-cycle). Given a complete block, we check all $\binom{9}{2} = 36$ pairs of columns and generate the *cycle pattern* for each block up to isomorphism—for example, one cycle pattern consists of the case when all pairs of columns form 10-cycles. In general, a cycle pattern graph on 9 vertices is constructed where two vertices are adjacent exactly when their associated two columns form a 10-cycle. Using nauty we determined that the 396 distinct block types gave rise to 359 distinct cycle patterns and in our programmatic implementation we used the cycle pattern as a proxy for determining the block label. In most cases the cycle pattern could be used to uniquely identify the block label, but otherwise the block label was assigned to the largest possible label consistent with the given cycle pattern (i.e., the most pessimistic choice in terms of symmetry breaking).

Following [29], block labels were chosen for the blocks by sorting the blocks in ascending order by the size of their stabilizer groups. Additionally, blocks with identical cycle patterns were given adjacent labels when possible in order to minimize the impact of the above pessimistic choice.

In total, this step required about 1,832 core hours on a cluster of Intel E5-2683 CPUs running at 2.1 GHz. The search produced 58 valid completions of the blocks 2–6 (see uwaterloo.ca/mathcheck for one explicit completion). Whenever a valid completion B was found, a clause $\bigvee_{B \models p} \neg p$ was programmatically added to the SAT instance. The added clause blocked the completion from occurring again later in the search.

Finally, for each of the 58 completions of blocks 2–6 a SAT instance was generated that included the constraints from blocks 2–7 and a cube specifying the completion (i.e., $\bigwedge_{B \models p} p$). It was found that none of the completions of blocks 2–6 could be extended to block 7 and this final step required less than a second.

4.5 Certificate Verification

The runs from the solvers produced DRUP proofs totalling about 33 terabytes. These were verified using the proof verification tool DRAT-trim [39] which was also used to trim and compress the proofs. These optimized proofs were archived using 7z data compression and produced archives totalling about 3 terabytes. These archives are available from the authors by request.

In order for the proofs to be verified by DRAT-trim the clauses which were programmatically generated during the solver's run also need to be provided to DRAT-trim. One way of doing this is to add the programmatic clauses directly into the CNF file provided to DRAT-trim. However, this method was found to suffer from very poor performance because this significantly increased the size of the initial active clause database tracked by DRAT-trim.

To get around this we modified DRAT-trim to support the addition of "trusted" clauses midway through the proof. Normally, each step of a proof consists of either an *addition* or *deletion* to the active clause database. In the case of an addition, DRAT-trim verifies that the added clause is a logical consequence[1] of the current set of active clauses. In our proofs we have a third kind of step, a *trusted addition* that adds the clause into the current set of active clauses without checking its provability. The justification for these clauses relies on our symmetry breaking method and not on a property easily checkable in Boolean logic, so the symmetry breaking clauses were not verified by DRAT-trim. However, if you believe in the correctness of our SAT encoding, generation scripts, DRAT-trim, and the trusted additions (whose correctness relies on our symmetry breaking method and a call to nauty) then you must believe in our certificates.

These proofs were checked using a system configured to limit each core to at most 4 GB of memory. In order to meet this limit it was necessary to ensure that each proof did not grow too large. To do this, March_cu was used generate a second "toplevel" set of cubes that partitioned the ith SAT instance into $398 - i$ subinstances. (As the label increased fewer subinstances were used because the instances became easier due to symmetry breaking.) Each of the subinstances were solved and had their proofs verified separately (each using at most 4 GB of memory and 10 min of computing time).

5   Conclusions and Future Work

In this paper we have completed an independent search showing the nonexistence of ovals in a projective plane of order ten. This was accomplished using a reduction to the Boolean satisfiability problem along with a SAT solver to show the resulting SAT instances are unsatisfiable. However, in order to make the amount of computation feasible it was necessary to use a symmetry breaking method. We used a "programmatic" SAT solver coupled with the symbolic computation library nauty [34] in order to learn symmetry breaking clauses on-the-fly during the search.

Our implementation uses the SAT+CAS interface as developed by the Math-Check project [40]. We are currently working on using MathCheck to verify more of the searches that were necessary in order to show the nonexistence of projective planes of order ten [3]. To date, we have verified the searches of MacWilliams et al. [32], Carter [9], and Lam et al. [29] that show that the rowspace of a projective plane of order ten does not contain vectors of weight 15 or 16. A consequence of these searches is that the weight enumerator of the error-correcting code generated by a projective plane of order ten can be specified exactly [32].

---

[1] DRAT-trim also supports a more general kind of provability termed "resolution asymmetric tautology" that we did not use in our proofs.

In particular, the rowspace of a projective plane of order ten must contain exactly 24,675 vectors of weight 19. The search for such vectors is the only case that remains in order to provide a complete SAT-based independent verification of the nonexistence of projective planes of order ten. We are currently exploring the feasibility of this and believe MathCheck will be useful in this case as well. The same basic encoding can be used but it seems necessary to tailor the symmetry breaking method and the structure of the search. This will be the subject of future research.

Acknowledgements. The authors would like to thank the reviewers and the coordinators at Springer Nature whose work improved the quality and correctness of this publication.

References

1. Ábrahám, E.: Building bridges between symbolic computation and satisfiability checking. In: Proceedings of the 2015 ACM on International Symposium on Symbolic and Algebraic Computation, pp. 1–6. ACM (2015). https://doi.org/10.1145/2755996.2756636

2. Biere, A.: CaDiCaL, Lingeling, Plingeling, Treengeling and YalSAT entering the SAT competition. In: Proceedings of SAT Competition 2018: Solver and Benchmark Descriptions (2018). http://fmv.jku.at/lingeling

3. Bright, C., Cheung, K., Stevens, B., Roy, D., Kotsireas, I., Ganesh, V.: A nonexistence certificate for projective planes of order ten with weight 15 codewords. Applicable Algebra in Engineering, Communication and Computing (2020). https://doi.org/10.1007/s00200-020-00426-y

4. Bright, C., Kotsireas, I., Ganesh, V.: SAT solvers and computer algebra systems: a powerful combination for mathematics. In: Proceedings of the 29th Annual International Conference on Computer Science and Software Engineering, pp. 323–328. IBM Corporation (2019). https://dl.acm.org/doi/10.5555/3370272.3370309

5. Bright, C., Kotsireas, I., Ganesh, V.: Applying computer algebra systems with SAT solvers to the Williamson conjecture. J. Symb. Comput. **100**, 187–209 (2020). https://doi.org/10.1016/j.jsc.2019.07.024

6. Bright, C., Kotsireas, I., Heinle, A., Ganesh, V.: Complex golay pairs up to length 28: a search via computer algebra and programmatic SAT. J. Symb. Comput. (2019). https://doi.org/10.1016/j.jsc.2019.10.013

7. Bright, C., Ðoković, D.Ž., Kotsireas, I., Ganesh, V.: The SAT+CAS method for combinatorial search with applications to best matrices. Ann. Math. Artif. Intell. **87**(4), 321–342 (2019). https://doi.org/10.1007/s10472-019-09681-3

8. Bruck, R.H., Ryser, H.J.: The nonexistence of certain finite projective planes. Can. J. Math. **1**(1), 88–93 (1949). https://doi.org/10.4153/CJM-1949-009-2

9. Carter, J.L.: On the existence of a projective plane of order ten. Ph.D. thesis, University of California, Berkeley (1974). https://hdl.handle.net/2027/uc1.c3475138

10. Davenport, J.H., England, M., Griggio, A., Sturm, T., Tinelli, C.: Symbolic computation and satisfiability checking. J. Symb. Comput. **100**, 1–10 (2020). https://doi.org/10.1016/j.jsc.2019.07.017

11. Ganesh, V., O'Donnell, C.W., Soos, M., Devadas, S., Rinard, M.C., Solar-Lezama, A.: Lynx: a programmatic SAT solver for the RNA-folding problem. In: Cimatti, A.,

Sebastiani, R. (eds.) SAT 2012. LNCS, vol. 7317, pp. 143–156. Springer, Heidelberg (2012). https://doi.org/10.1007/978-3-642-31612-8_12

12. Gelling, E.N.: On 1-factorizations of the complete graph and the relationship to round robin schedules. Master's thesis, University of Victoria (1973). http://hdl.handle.net/1828/7341

13. Godsil, C., Royle, G.F.: Algebraic Graph Theory, vol. 207. Springer, New York (2013). https://doi.org/10.1007/978-1-4613-0163-9

14. Hall Jr., M.: Finite projective planes. Am. Math. Mon. **62**(7P2), 18–24 (1955). https://doi.org/10.2307/2308176

15. Hall Jr., M.: Configurations in a plane of order ten. In: Annals of Discrete Mathematics, vol. 6, pp. 157–174. Elsevier (1980). https://doi.org/10.1016/S0167-5060(08)70701-5

16. Heule, M.J.H.: Schur number five. In: Proceedings of the Thirty-Second AAAI Conference on Artificial Intelligence, pp. 6598–6606. AAAI Press (2018). https://www.aaai.org/ocs/index.php/AAAI/AAAI18/paper/view/16952

17. Heule, M.J.H.: Trimming graphs using clausal proof optimization. In: Schiex, T., de Givry, S. (eds.) CP 2019. LNCS, vol. 11802, pp. 251–267. Springer, Cham (2019). https://doi.org/10.1007/978-3-030-30048-7_15

18. Heule, M.J.H., Hunt Jr., W.A., Wetzler, N.: Trimming while checking clausal proofs. In: 2013 Formal Methods in Computer-Aided Design, pp. 181–188. IEEE (2013). https://doi.org/10.1109/FMCAD.2013.6679408

19. Heule, M.J.H., Kauers, M., Seidl, M.: New ways to multiply 3×3-matrices. arXiv preprint arXiv:1905.10192 (2019). https://arxiv.org/abs/1905.10192

20. Heule, M.J.H., Kullmann, O., Marek, V.W.: Solving and verifying the boolean pythagorean triples problem via cube-and-conquer. In: Creignou, N., Le Berre, D. (eds.) SAT 2016. LNCS, vol. 9710, pp. 228–245. Springer, Cham (2016). https://doi.org/10.1007/978-3-319-40970-2_15

21. Heule, M.J.H., Kullmann, O., Marek, V.W.: Solving very hard problems: cube-and-conquer, a hybrid SAT solving method. In: Proceedings of the Twenty-Sixth International Joint Conference on Artificial Intelligence, IJCAI 2017, pp. 4864–4868 (2017). https://doi.org/10.24963/ijcai.2017/683

22. Heule, M.J.H., Kullmann, O., Wieringa, S., Biere, A.: Cube and conquer: guiding CDCL SAT solvers by lookaheads. In: Eder, K., Lourenço, J., Shehory, O. (eds.) HVC 2011. LNCS, vol. 7261, pp. 50–65. Springer, Heidelberg (2012). https://doi.org/10.1007/978-3-642-34188-5_8

23. Kaski, P., Östergård, P.R.J.: Classification Algorithms for Codes and Designs. Springer, Heidelberg (2006). https://doi.org/10.1007/3-540-28991-7

24. Kaufmann, D., Biere, A., Kauers, M.: Verifying large multipliers by combining SAT and computer algebra. In: 2019 Formal Methods in Computer Aided Design (FMCAD), pp. 28–36. IEEE (2019). https://doi.org/10.23919/FMCAD.2019.8894250

25. Konev, B., Lisitsa, A.: Computer-aided proof of Erdős discrepancy properties. Artif. Intell. **224**, 103–118 (2015). https://doi.org/10.1016/j.artint.2015.03.004

26. Lam, C., Thiel, L., Swiercz, S.: A feasibility study of a search for ovals in a projective plane of order 10. In: Billington, E.J., Oates-Williams, S., Street, A.P. (eds.) Combinatorial Mathematics IX. LNM, vol. 952, pp. 349–352. Springer, Heidelberg (1982). https://doi.org/10.1007/BFb0061988

27. Lam, C.W.H.: The search for a finite projective plane of order 10. Am. Math. Mon. **98**(4), 305–318 (1991). https://doi.org/10.1080/00029890.1991.12000759

28. Lam, C.W.H., Thiel, L., Swiercz, S.: The non-existence of finite projective planes of order 10. Can. J. Math. **41**(6), 1117–1123 (1989). https://doi.org/10.4153/CJM-1989-049-4

29. Lam, C.W.H., Thiel, L., Swiercz, S.: The nonexistence of code words of weight 16 in a projective plane of order 10. J. Comb. Theory Ser. A **42**(2), 207–214 (1986). https://doi.org/10.1016/0097-3165(86)90091-9

30. Lam, C.W.H., Thiel, L., Swiercz, S., McKay, J.: The nonexistence of ovals in a projective plane of order 10. Discret. Math. **45**(2–3), 319–321 (1983). https://doi.org/10.1016/0012-365X(83)90049-3

31. Liang, J.H., Govind V.K., H., Poupart, P., Czarnecki, K., Ganesh, V.: An empirical study of branching heuristics through the lens of global learning rate. In: Gaspers, S., Walsh, T. (eds.) SAT 2017. LNCS, vol. 10491, pp. 119–135. Springer, Cham (2017). https://doi.org/10.1007/978-3-319-66263-3_8

32. MacWilliams, F.J., Sloane, N.J.A., Thompson, J.G.: On the existence of a projective plane of order 10. J. Comb. Theory Ser. A **14**(1), 66–78 (1973). https://doi.org/10.1016/0097-3165(73)90064-2

33. Mathon, R.: The partial geometries pg(5, 7, 3). Congr. Numer. **31**, 129–139 (1981)

34. McKay, B.D., Piperno, A.: Practical graph isomorphism, II. J. Symb. Comput. **60**, 94–112 (2014). https://doi.org/10.1016/j.jsc.2013.09.003

35. Nadel, A., Ryvchin, V.: Efficient SAT solving under assumptions. In: Cimatti, A., Sebastiani, R. (eds.) SAT 2012. LNCS, vol. 7317, pp. 242–255. Springer, Heidelberg (2012). https://doi.org/10.1007/978-3-642-31612-8_19

36. Read, R.C.: Every one a winner or how to avoid isomorphism search when cataloguing combinatorial configurations. In: Annals of Discrete Mathematics, vol. 2, pp. 107–120. Elsevier (1978). https://doi.org/10.1016/S0167-5060(08)70325-X

37. Roy, D.J.: Confirmation of the non-existence of a projective plane of order 10. Master's thesis, Carleton University (2011). https://doi.org/10.22215/etd/2011-09202

38. Royle, G.F.: An orderly algorithm and some applications in finite geometry. Discret. Math. **185**(1–3), 105–115 (1998). https://doi.org/10.1016/S0012-365X(97)00167-2

39. Wetzler, N., Heule, M.J.H., Hunt Jr., W.A.: DRAT-trim: efficient checking and trimming using expressive clausal proofs. In: Sinz, C., Egly, U. (eds.) SAT 2014. LNCS, vol. 8561, pp. 422–429. Springer, Cham (2014). https://doi.org/10.1007/978-3-319-09284-3_31

40. Zulkoski, E., Bright, C., Heinle, A., Kotsireas, I., Czarnecki, K., Ganesh, V.: Combining SAT solvers with computer algebra systems to verify combinatorial conjectures. J. Autom. Reason. **58**(3), 313–339 (2016). https://doi.org/10.1007/s10817-016-9396-y

Edge-Disjoint Branchings in Temporal Graphs

Victor Campos[3], Raul Lopes[3], Andrea Marino[1], and Ana Silva[2(✉)]

[1] Dipartimento di Sistemi, Informatica, Applicazioni,
Università degli Studi di Firenze, Florence, Italy
`andrea.marino@unifi.it`
[2] Departamento de Matemática, Universidade Federal do Ceará,
Fortaleza, CE, Brazil
`anasilva@mat.ufc.br`
[3] Departamento de Computação, Universidade Federal do Ceará,
Fortaleza, CE, Brazil
`{campos,raul.lopes}@lia.ufc.br`

Abstract. A temporal digraph $\mathcal{G}$ is a triple (G, γ, λ) where G is a digraph, γ is a function on $V(G)$ that tells us the time stamps when a vertex is active, and λ is a function on $E(G)$ that tells for each $uv \in E(G)$ when u and v are linked. Given a static digraph G, and a subset $R \subseteq V(G)$, a spanning branching with root R is a subdigraph of G that has exactly one path from R to each $v \in V(G)$. In this paper, we consider the temporal version of Edmonds' classical result about the problem of finding k edge-disjoint spanning branchings respectively rooted at given $R_1, \cdots, R_k$. We introduce and investigate different definitions of spanning branchings, and of edge-disjointness in the context of temporal graphs. A branching $\mathcal{B}$ is vertex-spanning if the root is able to reach each vertex v of G at some time where v is active, while it is temporal-spanning if v can be reached from the root at every time where v is active. On the other hand, two branchings $\mathcal{B}_1$ and $\mathcal{B}_2$ are edge-disjoint if they do not use the same edge of G, and are temporal-edge-disjoint if they can use the same edge of G but at different times. This lead us to four definitions of disjoint spanning branchings and we prove that, unlike the static case, only one of these can be computed in polynomial time, namely the temporal-edge-disjoint temporal-spanning branchings problem, while the other versions are NP-complete, even under very strict assumptions.

1 Introduction

In this paper, we refer to digraphs in the classical sense as *static digraphs*. A *temporal digraph* is a digraph that exists and changes in a time interval $\mathcal{T}$. That

Partially supported by FUNCAP/CNPq/Brazil, Project PRONEM PNE-0112-00061.01.00/16, CNPq Universal 401519/2016-3/ Produtividade 304576/2017-4, MIUR under PRIN Project n. 20174LF3T8 AHeAD (Efficient Algorithms for HArnessing Networked Data).

L. Gąsieniec et al. (Eds.): IWOCA 2020, LNCS 12126, pp. 112–125, 2020.
https://doi.org/10.1007/978-3-030-48966-3_9

is, given a static digraph G, a temporal digraph $\mathcal{G}$ with *base static digraph* G and lifetime $\mathcal{T}$ changes as follows: at each *time stamp* $t \in \mathcal{T}$, only a subdigraph of G is *active*, and edges might have a delay, leaving a vertex at some time stamp but arriving only later. If a vertex $v \in V(G)$ is active at every $t \in \mathcal{T}$, we say that v is *permanent*.

In this paper we deal with *disjoint spanning branchings* in temporal digraphs, which are well-understood structures in digraphs. Given a digraph G, and a subset $R \subseteq V(G)$, we say that $H \subseteq G$ is a *spanning branching* of G with root R if $V(H) = V(G)$, and H contains exactly one path between some $r \in R$ and u, for each $u \in V(G)$. Given subsets $R_1, \cdots, R_k$, a classical result by Edmonds [9] gives a necessary and sufficient condition for the existence of k edge-disjoint branchings with roots $R_1, \cdots, R_k$, respectively. His result also gives a polynomial algorithm that constructs these branchings.

When translating concepts to temporal graphs, it is often the case that theorems coming from graph theory, in the classical sense, can hold or not depending on the adopted definition. Indeed, in [14] the authors give an example where Edmonds' result on branchings does not hold on the temporal context. However, as we will see later, their concept is just one of many possible definitions, and in fact there is even one case where polynomiality holds.

Another example of such behavior is the validity of Menger's Theorem. It has been shown that the edge version of Menger's Theorem holds [3], even if one considers weights on the edges [2]. However, the vertex version of Menger's Theorem holds or not, depending on how one interprets what a cut should be. If a cut is understood as a subset of $V(G)$, then Menger's Theorem does not hold [3,14]; and if it is understood as a subset of the appearances of vertices in time (alternatively, a cut can be seen as deactivating vertices at some time stamps), then Menger's Theorem holds [18].

Our Contribution. Given a temporal digraph $\mathcal{G}$ with base digraph G, and subsets of *vertices in time* $R_1, \cdots, R_k$, i.e. sets of pairs (u, t) where u is a vertex of G and t a time stamp, here we investigate the many variations of finding (pairwise) disjoint spanning branchings with roots $R_1, \cdots, R_k$. Spanning can mean that one wants to pass by at least one appearance of each $u \in V(G)$ (called *vertex spanning*), or by all appearances of each $u \in V(G)$ (called *temporal spanning*). Similarly, edge-disjoint can have different interpretations, as it can refer to edges of G or to the appearances of these edges in $\mathcal{G}$. We say that two branchings are *edge-disjoint* if they do not share any edge of G, and that they are *temporal-edge-disjoint* (or *t-edge-disjoint* for short) if they do not share any appearance of an edge of G in $\mathcal{G}$. We found that the only case in which this problem is polynomial (as its static counterpart) is when we want t-edge-disjoint temporal-spanning branchings. We also found that if vertices are permanent (this is the more popular case where vertices are always active), the problem is polynomial for temporal-spanning branchings and NP-complete otherwise. Our results are summarized in Table 1 and detailed in the following main theorem. A digraph G is a *in-star* if there exists $u \in V(G)$ such that all the edges in G are incoming edges to u.

Table 1. Our results. Vertices are permanent if they are always active.

	NOT PERMANENT VERTICES		PERMANENT VERTICES	
	EDGE-DISJOINT	T-EDGE-DISJOINT	EDGE-DISJOINT	T-EDGE-DISJOINT
TEMPORAL-SPANNING	Poly	NP-c	Poly	Poly
VERTEX-SPANNING	NP-c	NP-c	NP-c	NP-c

Theorem 1. *Let $\mathcal{G}$ be a temporal digraph with base digraph G, and consider subsets of vertices in time, $R_1, \cdots, R_k$. The problem of finding k branchings rooted at $R_1, \cdots, R_k$ is:*

1. *Polynomial for t-edge-disjoint temporal-spanning,*
2. *NP-complete for edge-disjoint temporal-spanning even if G is a in-star, and each snapshot has constant size, or if $\mathcal{G}$ has lifetime 3. And if vertices are permanent or $\mathcal{G}$ has lifetime 2, then edge-disjoint temporal-spanning becomes polynomial.*
3. *NP-complete for edge-disjoint vertex-spanning even if G is a DAG, the lifetime of $\mathcal{G}$ is 2, and vertices are permanent.*
4. *NP-complete for t-edge-disjoint vertex-spanning even if G is a DAG, the lifetime of $\mathcal{G}$ is 2, and vertices are permanent.*

As said before, Edmonds' condition is the characterization behind the polynomial algorithm for finding k edge disjoint spanning branchings in digraphs. Because of our NP-completeness results, it is worth remarking that, unless P=NP, any such characterization for the NP-complete cases in temporal digraphs should be checkable in superpolynomial time, unlike the one provided by Edmonds.

Finally, our reductions further imply that, in the case of edge-disjoint temporal-spanning, even if the base digraph G is a in-star, the problem cannot be solved by an algorithm running in time $O^*(2^{o(\mathcal{T})})$ unless ETH fails, where $\mathcal{T}$ is the lifetime of $\mathcal{G}$. Moreover, in the vertex-spanning variations, the problem also cannot be solved in $O^*(2^{o(n+m)})$ under the same assumption, where n and m are respectively the number of nodes and edges of the base digraph of $\mathcal{G}$.

Related Work: While it is easy to imagine a variety of graph problems that can profit from considering changes in time, it is hard to pin-point when the study of temporal graphs and similar structures began. Nevertheless, in the last decade or so, it has attracted a lot of attention from the community, with a considerable number of papers being published in the field (we refer the reader to the surveys [15,19]). We mention that temporal graphs (or other very similar structures) appear in the literature under a number of names, such as dynamic networks [4], time-varying graphs [8], evolving networks [5], and link streams [15]. Also, many works consider a temporal graph $\mathcal{G}$ as having vertices that are always active, and edges have the same starting and ending time [2,6,14,18,20]. While models where edges that have a delay are more common [8,25], models where nodes can be inactive have already been considered in [8,15].

A path in temporal graphs is generally understood as a sequence of edges respecting time, i.e. the arrival time in each vertex of the path must be lower than the departing time of the next edge taken. In this context, a number of metrics can be related to a path, such as earliest arrival time, latest departure time, minimum number of temporal edges, and minimum traveling time [25]. When vertices can be inactive, we have to further ensure that, when waiting for the next edge on a certain vertex, it must remain active in the waiting period [8]. In this scenario, the definitions of reachability and connectivity change accordingly, and it is natural to ask how well-known structures and results from graph theory in the classical sense change taking into account the temporal constraint.

Temporal definitions of trees [6,15] and (minimum) spanning trees [13], which are related to our definition of branching, have been proposed and investigated, and usually consist of ensuring that the root-to-node path in the tree is a valid temporal path. Analogously, temporal cuts from a vertex s to t aim to break any temporal path from s to t and can be related to extending the max-flow min-cut Theorem to temporal graphs [2]. And as we have already mentioned, different conclusions have been made about a temporal version of Menger's Theorem depending on the adopted translation in terms of temporal graphs [3,14,18].

Edmonds' Theorem on disjoint branchings is a classical theorem in graph theory, with many distinct existing proofs (e.g. Lovász [16], Tarjan [24], and Fulkerson and Harding [12]), and has many interesting consequences on digraph theory (e.g., one can derive Menger's Theorem from it, characterize arc-connectivity [22], characterize branching cover [11], ensure integer decomposition of the polytope of branchings of size k [17], etc). As far as we know, the only other time that Edmonds' Theorem has been investigated on the temporal context has been in [14], where the authors give an example where the theorem does not hold. The definition used by them falls into our category of edge-disjoint vertex-spanning branchings, which we prove to be NP-complete even under very strict constraints.

Structure of the Paper. The paper is organized as follows. In Sect. 2, we formalize the definitions of spanning branchings and disjointness, also showing that having multiple roots in each of the k branchings is computationally equivalent to having a single root for all of the k branchings. In Sect. 3, we present the results about temporal-spanning branchings. In Sect. 4 we present our results concerning vertex-spanning branchings. Finally, in Sect. 5, we draw our conclusions and make some final remarks. The proofs of the results marked with '($\star$)' can be found online in [7].

2   The Temporal Disjoint Branchings Problems

This section is devoted to formally define the several concepts of temporal graphs and disjoint branchings we introduce in this paper. A temporal digraph $\mathcal{G}$ is a triple (G, γ, λ) where G is a digraph and γ and λ are functions on $V(G)$ and $E(G)$, respectively, that tell us when the vertices and the edges appear. More formally, for each $v \in V(G)$ we have $\gamma(v) \subseteq \mathbb{N}$, and for each edge $e \in E(G)$ we have $\lambda(e) \subseteq \mathbb{N} \times \mathbb{N}$. Also, if $(t, t') \in \lambda(uv)$, then $t \leq t'$, $t \in \gamma(u)$ and $t' \in \gamma(v)$. Here, we consider only finite temporal digraphs, i.e., $\mathcal{T} = \max \bigcup_{v \in V(G)} \gamma(v)$ is defined and is called the *lifetime of* $\mathcal{G}$. We call G the *base digraph of* $\mathcal{G}$. In what follows, unless said otherwise, we work on general digraphs, i.e., directions, loops and multiple edges are allowed.

In particular, if $\mathcal{T}$ is the lifetime of $\mathcal{G} = (G, \gamma, \lambda)$, $\gamma(v) = [\mathcal{T}]$ for each $v \in V(G)$, and $t = t'$ for every $(t, t') \in \lambda(E(G))$, then the above definition corresponds to the definition of temporal graph given in [14] and many other works. The above definition also generalizes the definition of stream graph given in [15], and of time-varying graphs given in [1].

The *vertices* and *edges* of $\mathcal{G}$ are the vertices and edges of G. We say that a vertex v *is active* at time t if $t \in \gamma(v)$, and that v *is active* from t_1 to t_2 if v is active for every time t with $t_1 \leq t \leq t_2$. Also, if v is active throughout the lifetime of $\mathcal{G}$, then we say that v is *permanent*. The set V_T of *temporal vertices* is the set $\{(v, t) \mid v \in V(G) \ and \ t \in \gamma(v)\}$, and the set E_T of *temporal edges* is the set $\{(u, t)(v, t') \mid e = uv \in E(G) \ and \ (t, t') \in \lambda(e)\}$. Observe that a temporal digraph $\mathcal{G} = (G, \gamma, \lambda)$ can be also seen as a pair of digraphs (G, G_T) where $G_T = (V_T, E_T)$. This is similar to what has been proposed in [1] and [2]. We call the digraph G_T the *(γ, λ)-digraph of* $\mathcal{G}$.

Since in our more general case, also vertices appear and disappear, the definition of *walk* must take into account that it is possible to wait only on vertices which are active, as formally defined next. Given temporal vertices $s_1, s_k \in V_T$, an s_1, s_k-*temporal walk* in (G, G_T) is a sequence of temporal vertices and temporal edges, $(s_1, \ldots, s_k)$, that either goes through a temporal edge, or stays on different copies of the same vertex of G. More formally: if s_i is a temporal edge, then s_{i-1} and s_{i+1} are temporal vertices and s_i goes from s_{i-1} to s_{i+1}; and if s_i and s_{i+1} are temporal vertices, then $s_i = (v, t)$ and $s_{i+1} = (v, t+1)$ for some vertex v and some time t. If such a walk exists, we say that s_1 *reaches* s_k.

A temporal digraph $\mathcal{B} = (G', \gamma', \lambda')$ such that $G' \subseteq G$, $\gamma' \subseteq \gamma$ and $\lambda' \subseteq \lambda$ is called a *temporal subdigraph of* $\mathcal{G}$.[1] Let $R \subseteq V_T$; a temporal subdigraph $\mathcal{B}$ of $\mathcal{G}$ is a *temporal-spanning branching* of $\mathcal{G}$ with root R if $\mathcal{B}$ has a unique temporal walk from R to every vertex in V_T, i.e. for any $(u, i) \in V_T$ there is exactly one temporal walk in $\mathcal{B}$ starting at some vertex $r \in R$ and arriving at (u, i). And $\mathcal{B}$ is a *vertex-spanning branching* of $\mathcal{G}$ with root R if $\mathcal{B}$ has exactly one temporal walk from R to some vertex in $\{(u, i) \in V_T\}$ for every $u \in V(G)$.

Given two branchings $\mathcal{B}_1 = (G_1, \gamma_1, \lambda_1)$ and $\mathcal{B}_2 = (G_2, \gamma_2, \lambda_2)$ rooted at R_1, R_2, respectively, either both temporal-spanning or both vertex-spanning, we say that $\mathcal{B}_1$ and $\mathcal{B}_2$ are *temporal-edge-disjoint* (or t-edge-disjoint for short) if they have no common temporal edges; more formally, if $\lambda_1(e) \cap \lambda_2(e) = \emptyset$ for

---

[1] Here, a function is seen as a set of ordered pairs, and the containment relation is the usual one for sets.

every $e \in E(G)$. And we say that $\mathcal{B}_1$ and $\mathcal{B}_2$ are *edge-disjoint* if there is no edge $uv \in E(G)$ that has copies in both $\mathcal{B}_1$ and $\mathcal{B}_2$; more formally, $E(G_1) \cap E(G_2) = \emptyset$.

Problem 1 (k X-disjoint Y-spanning Branching). Let $X \in \{edge, t\text{-}edge\}$, $Y \in \{temporal, vertex\}$, and k be a fixed positive integer. Given a temporal digraph $\mathcal{G}$, and subsets of temporal vertices $R_1, \ldots, R_k \subseteq V_T$, find k X-disjoint Y-spanning branchings $\mathcal{B}_1, \ldots, \mathcal{B}_k$ respectively with roots $R_1, \ldots, R_k$.

We introduce the following restriction of Problem 1, which corresponds to finding branchings that have a single root (also called out-arborescence).

Problem 2 (k Single Source X-disjoint Y-spanning Branching). Let $X \in \{edge, t\text{-}edge\}$, $Y \in \{temporal, vertex\}$, and k be a fixed positive integer. Given a temporal digraph $\mathcal{G}$, and a temporal vertex $r \in V_T$, find k X-disjoint Y-spanning branchings $\mathcal{B}_1, \ldots, \mathcal{B}_k$ each one with root r.

Lemma 1. *Problem 1 is computationally equivalent to Problem 2.*

Proof. Problem 2 is clearly a restriction of Problem 1. In the following we provide the reduction in the opposite direction, from the problem where each branching has a subset of V_T as roots to the problem where each branching has a single same root. For this, for each $i \in [k]$ add a new vertex r_i to G adjacent to every $u \in V(G)$ such that $(u, t) \in R_i$, for some $t \in [T]$. Then, make $\gamma(r_i) = \{0\}$, and for each $(u, t) \in R_i$, add $(0, t)$ to $\lambda(r_i u)$ (which is the same as adding the temporal edge $(r_i, 0)(u, t)$ to $\mathcal{G}$). Moreover, add a vertex r and make it adjacent to $\{r_1, \cdots, r_k\}$; also make $\gamma(r) = \{0\}$ and $\lambda(r r_i) = \{(0, 0)\}$ (which is the same as adding temporal edges $(r, 0)(r_i, 0)$ for every $i \in [k]$).

One can see that k vertex-spanning (resp. temporal-spanning) branchings rooted at r give k vertex-spanning (resp. temporal-spanning) branchings rooted at $R_1, \cdots, R_k$, and vice-versa. The edge-disjointness, both for t-edge or edge-disjoint versions, clearly are not altered by adding the new temporal edges.   □

The next easy proposition tells us that if finding k disjoint spanning branchings is hard, for some fixed k, then so is finding $k + 1$ of them.

Proposition 1. *Let $X \in \{edge, t\text{-}edge\}$, $Y \in \{temporal, vertex\}$ and k be a fixed positive integer. If Problem k X-disjoint Y-spanning Branching is NP-complete, then the same holds for Problem $k + 1$ X-disjoint Y-spanning Branching.*

Proof. To reduce from k to $k + 1$, it suffices to add $R_{k+1} = V_T$ as entry. Surely the $(k+1)$-th branching has no temporal edges, which means that the other ones form a solution to the initial problem.   □

3   Temporal-Spanning Branchings

This section is devoted to study Problem 1 in the case where Y is temporal, i.e. we aim to find k X-disjoint temporal-spanning branchings, with $X \in \{edge, t\text{-}edge\}$. We will hence prove Item 1 and Item 2 of Theorem 1 respectively in Sect. 3.1 and in Sect. 3.2.

3.1 T-Edge-Disjoint Temporal-Spanning Branchings

Let $\mathcal{G} = (G, \gamma, \lambda)$, and let V_T, E_T be its set of temporal vertices and edges, respectively. Also, let $R_1, \cdots, R_k \subseteq V_T$, and $H = (V_T, E_T \cup E')$, where E' contains k copies of the edge $(u,t)(u,t+1)$ whenever $\{(u,t),(u,t+1)\} \subseteq V_T$. We prove that $\mathcal{G}$ has the desired branchings iff H has k edge-disjoint spanning branchings with roots $R_1, \cdots, R_k$. Then, Item 1 of Theorem 1 follows by Edmonds' result [9].

Lemma 2. *Let $\mathcal{G} = (G, \gamma, \lambda)$ be a temporal digraph, $R_1, \cdots, R_k \subseteq V_T$, and H be constructed as above. Then, $\mathcal{G}$ has k t-edge-disjoint temporal-spanning branchings rooted at $R_1, \cdots, R_k$ iff H has k edge-disjoint spanning branchings rooted at $R_1, \cdots, R_k$.*

Proof. Let $\mathcal{B}_1, \cdots, \mathcal{B}_k$ be t-edge-disjoint temporal-spanning branchings rooted at $R_1, \cdots, R_k$, respectively. For each $\mathcal{B}_i$, let B_i be a spanning subgraph of H initially containing the temporal edges of $\mathcal{B}_i$; then for each $(u,t) \in V(B_i)$, if the only walk in $\mathcal{B}_i$ from R_i to (u,t) contains $(u,t)(u,t+1)$ as a subsequence, then add an unused copy of $(u,t)(u,t+1) \in$ to B_i. Because this walk is unique and cannot pass twice from time stamp t to time stamp $t+1$, we get that at most k copies are needed, and, hence, the produced branchings are edge-disjoint. The converse can be easily proved by deleting the edges in E' from the solution to obtain the temporal subgraphs. □

3.2 Edge-Disjoint Temporal-Spanning Branchings

In this section, we prove Item 2 of Theorem 1. For this, we first prove that the problem is NP-complete, and then that it is polynomial when each vertex is active for a consecutive set of time stamps. This includes the popular case where vertices are assumed to be permanent, as well as the case where $T = 2$.

Theorem 2 and Theorem 3 below detail our NP-completeness results. In the next proof, we make a reduction from the k-WEAK DISJOINT PATHS problem (k-WDP), where we are given a digraph G and a set I of k pairs of vertices $\{(s_1, t_1), \ldots, (s_k, t_k)\}$ (called the *requests*) of $V(G)$ and the goal is to find a collection of pairwise edge-disjoint paths $\{P_1, \ldots, P_k\}$ such that P_i is a path from s_i to t_i in G, for $i \in \{1, \ldots, k\}$. The k-WDP problem is NP-complete for $k = 2$ [10] and W[1]-hard with parameter k in DAGs [23].

Theorem 2. *Let $k \geq 2$ be a fixed integer, $\mathcal{G} = (G, \gamma, \lambda)$ be a temporal digraph, and $R_1, \ldots, R_k \subseteq V_T$. Deciding whether $\mathcal{G}$ has k edge-disjoint temporal-spanning branchings rooted at $R_1, \cdots, R_k$ is NP-complete even if $\mathcal{G}$ has lifetime 3.*

Proof. Let (G, I) be an instance of 2-WDP with $I = \{(s_1, t_1), (s_2, t_2)\}$, and define $W = \{s_1, t_1, s_2, t_2\}$. Assume that s_1, s_2 are sources, t_1, t_2 are sinks, and all vertices in W are distinct. We construct the temporal graph $\mathcal{G} = (G, \gamma, \lambda)$ with subsets R_1, R_2 such that $\mathcal{G}$ has 2 edge-disjoint temporal-spanning branchings rooted at R_1, R_2 if and only if (G, I) is a "yes" instance of 2-WDP. The NP-completeness for higher values of k follows from Proposition 1.

In the constructed temporal graph, there are no temporal edges of the type $(u,t)(v,t')$ with $t \neq t'$. For this reason, it is easier to describe our temporal graph by describing, for each timestamp, what are the vertices and edges that are active. These are called *snapshots* and consist of subgraphs of G formed at each timestamp.

We let the first snapshot of $\mathcal{G}$ initially consist of $G - \{s_2, t_2\}$, and the third snapshot initially consist of $G - \{s_1, t_1\}$. Then, we add a new vertex x to snapshot 1, and add the edges: $\{xv \mid v \in V(G) \setminus \{s_2, t_2\}\} \cup \{t_1 v \mid v \in (V(G) \cup \{x\}) \setminus \{s_1, s_2, t_2\}\}$. Similarly, we add a new vertex y to snapshot 3, and add the edges: $\{yv \mid v \in V(G) \setminus \{s_1, t_1\}\} \cup \{t_2 v \mid v \in (V(G) \cup \{y\}) \setminus \{s_2, s_1, t_1\}\}$. Observe Fig. 1.

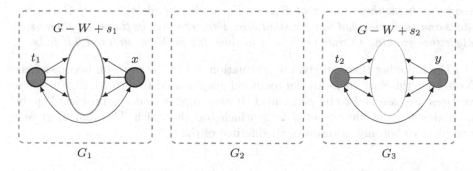

Fig. 1. Temporal graph constructed from an instance (G, I) of 2-WDP, where $I = \{(s_1, t_1), (s_2, t_2)\}$ and $W = \{s_1, t_1, s_2, t_2\}$. Edges arriving in t_1 and t_2 originally from G are omitted.

Define $R_1 = \{(s_1, 1), (y, 3)\}$ and $R_2 = \{(s_2, 3), (x, 1)\}$. Now, we prove that (G, I) is a "yes" instance of 2-WDP if and only if $\mathcal{G}$ contains two edge-disjoint temporal-spanning branchings rooted at R_1 and R_2, respectively. Notice that snapshot 2 of $\mathcal{G}$ is empty, thus each path in G can be represented by either a temporal path on snapshot 1 or a temporal path on snapshot 2.

First, let P_1 and P_2 be two edge-disjoint paths from s_1 to t_1 and from s_2 to t_2 in G, respectively. Let T_1 be initially the copy of P_1 in snapshot 1, and T_2 be initially the copy of P_2 in snapshot 3. Note that the vertices not spanned by T_1 are all the copies of $v \notin V(P_1)$ in snapshot 1, together with all the vertices in snapshot 3, and vertices $\{(x, 1), (y, 3)\}$. To span snapshot 3, add to T_1 all edges between $(y, 3)$ and $(v, 3)$, for every $v \in V(G) \setminus \{s_1, t_1\}$. To span the remainder of snapshot 1, add all edges between $(t_1, 1)$ and $(v, 1)$, for every $v \in V(G) \setminus (V(P_1) \cup \{s_2, t_2\})$, and the edge from $(t_1, 1)$ to $(x, 1)$. A similar argument can be applied to span every temporal vertex also with T_2. Because P_1 and P_2 are edge-disjoint, we get that T_1 and T_2 could only intersect in the added edges, which does not occur because all edges added to T_1 are incident to t_1 and y, all edges added to T_2 are incident to t_2 and x, and there is no intersection between these.

Now, let T_1 and T_2 be edge-disjoint temporal-spanning branchings in $\mathcal{G}$ with roots R_1, R_2. Denote snapshot 1 by G_1. Since t_1 appears only in G_1, and the only root of R_1 in G_1 is $(s_1, 1)$, we get that in T_1 there exists a path of G_1 going from $(s_1, 1)$ to $(t_1, 1)$. Because the only incoming edge to $(x, 1)$ is $(t_1, 1)(x, 1)$, we get that $(x, 1)$ cannot be an internal vertex in this path, and hence it corresponds to a path in G, P_1. Applying a similar argument, we get a path P_2 from s_2 to t_2 in G taken from T_2, and since T_1 and T_2 are edge-disjoint, so are P_1 and P_2. □

The next result concludes the proof of Item 2 of Theorem 1.

Theorem 3 ($\star$). *Let $k \geq 2$ be a fixed integer, $\mathcal{G} = (G, \gamma, \lambda)$ be a temporal digraph, and $R_1, \ldots, R_k \subseteq V_T$. Deciding whether $\mathcal{G}$ has k edge-disjoint temporal-spanning branchings rooted at $R_1, \cdots, R_k$ is NP-complete, even if G is a in-star, and each snapshot has constant size. Furthermore, in this case, there is no algorithm running in time $O^*(2^{o(\mathcal{T})})$ to solve the problem, unless ETH fails.*

The following theorem gives us a situation where the problem becomes easy. Note that this case includes the temporal graphs used in [2, 6, 14, 18, 20], where vertices are assumed to be permanent. It also implies that the problem is polynomial when the lifetime of $\mathcal{G}$ is 2, which together with Theorem 2, gives a complete dichotomy in terms of the lifetime of $\mathcal{G}$.

Theorem 4. *Let $\mathcal{G} = (G, \gamma, \lambda)$ be a temporal digraph with temporal vertices V_T, and let $R_1, \cdots, R_k \subseteq V_T$. If for every $v \in V(G)$, $\gamma(v)$ is exactly one interval of consecutive integers, then finding k edge-disjoint temporal-spanning branchings rooted at $R_1, \cdots, R_k$ can be done in polynomial time.*

Proof. Let $\mathcal{T}$ be the lifetime of $\mathcal{G}$. We first construct digraphs $G_0, \cdots, G_{\mathcal{T}}$ and subsets $R_1^j, \cdots, R_k^j$ for each $j \in \{0, \cdots, \mathcal{T}\}$, then we prove that $\mathcal{G}$ has the desired branchings if and only if G_j has k edge-disjoint branchings rooted at $R_1^j, \ldots, R_k^j$ for each $j \in \{0, \cdots, \mathcal{T}\}$, which can be checked in polynomial time, applying Edmonds' result [9].

First, let $G_0 = (V_0, E_0)$ be the digraph in time stamp 0, i.e., $V_0 = \{u \in V(G) \mid 0 \in \gamma(u)\}$ and $E_0 = \{e \in E(G) \mid (0, 0) \in \gamma(e)\}$. Also, for every $i \in [k]$, let R_i^0 be the roots at time stamp 0, i.e., the set $\{u \in V(G) \mid (u, 0) \in R_i\}$. Now, for each $j \in [\mathcal{T}]$, let $G_j = (V_j, E_j)$ be the digraph containing the edges arriving at time stamp j together with their endpoints; more formally, $E_j = \{e \in E(G) \mid (t, j) \in \lambda(e)$, for some $t\}$ and $V_j = \{u \in V(G) \mid (u, j) \in V_T$ or $uv \in E_j$, for some $v\}$. Also, for each $i \in [k]$, let R_i^j be the set of roots at time stamp j together with vertices still active from the previous time stamp, i.e., $R_i^j = \{u \in V(G) \mid (u, j) \in R_i\} \cup \{u \in V(G) \mid \{j-1, j\} \subseteq \gamma(u)\}$.

Now, let $\mathcal{B}_1, \cdots, \mathcal{B}_k$ be edge-disjoint temporal-spanning branchings rooted at $R_1, \cdots, R_k$; denote by $E_T(\mathcal{B}_i)$ the set of temporal edges of $\mathcal{B}_i$. Consider $j \in \{0, \cdots, \mathcal{T}\}$, and for each $i \in [k]$, let B_i^j be the set of edges of $\mathcal{B}_i$ that have a copy ending at time stamp j, i.e., $B_i^j = \{uv \in E(G) \mid$

$(u, h)(v, j) \in E_T(\mathcal{B}_i)$ for some h}. Because $\mathcal{B}_1, \cdots, \mathcal{B}_k$ are edge-disjoint, we get that $B_1^j, \cdots, B_k^j$ are also disjoint. It remains to prove that each B_i^j is the edge set of a spanning branching of G_j rooted at R_i^j. So, consider any $i \in [k]$. Because $\mathcal{B}_i$ is a temporal-spanning branching of $\mathcal{G}$, we know that each $u \in V(G)$ is either the head of some edge in B_i^j, in which case u is spanned by B_i^j, or u is a root in B_i^j. We prove that in the latter case we get that $u \in R_i^j$. Because u is not the head of any edge in B_i^j, this means that either $(u, j) \in R_i$ or (u, j) is spanned by $\mathcal{B}_i$ just by waiting, i.e., $\{j-1, j\} \subseteq \gamma(u)$. In both cases, we get that $u \in R_i^j$, as we wanted to prove.

Now, for each $j \in \{0, \cdots, T\}$, let $B_1^j, \ldots, B_k^j$ be the edge sets of k edge-disjoint spanning branchings of G_j. First, we prove that if $uv \in B_i^j$, then $v \in R_{i'}^{j'}$ for every $i' \in [k]$ and every $j' \in \{j+1, \cdots, T\} \cap \gamma(v)$; hence if $B_i = \bigcup_{j=0}^{T} B_i^j$, then we get that $B_1, \cdots, B_k$ are disjoint (these will be used later to construct the desired temporal branchings). So let $j' \in \{j+1, \cdots, k\} \cap \gamma(v)$ and observe that if $uv \in E(G_j)$ then $j \in \gamma(v)$. Because $\gamma(v)$ is an interval of consecutive integers and $j < j' \in \gamma(v)$, we get that $j' - 1 \in \gamma(v)$, which implies that $v \in R_{i'}^{j'}$ for every $i' \in [k]$, as we wanted to show. Now, for each $i \in [k]$, let $\mathcal{B}_i = (G, \gamma, \lambda^i)$ be a spanning temporal subdigraph of $\mathcal{G}$ having as temporal edges the temporal copies of each $e \in B_i$, i.e., $\lambda^i(e) = \lambda(e)$ if $e \in B_i$, and $\lambda^i(e) = \emptyset$ otherwise. Because $B_1, \cdots, B_k$ are disjoint, it follows that $\mathcal{B}_1, \cdots, \mathcal{B}_k$ are edge-disjoint, so it remains to prove that each $\mathcal{B}_i$ is a temporal-spanning branching rooted at R_i. Let $u \in V(G)$, and recall that $\gamma(u)$ is an interval of consecutive integers; denote by s_u the minimum value in $\gamma(u)$. Note that we just need to prove that if $(u, s_u) \notin R_i$, then there exists a temporal edge in $\mathcal{B}_i$ arriving in (u, s_u); this is because the other copies can be spanned simply by waiting in the interval $\gamma(u)$. Since $(u, s_u) \notin R_i$ and $s_u - 1 \notin \gamma(u)$, we get that $u \notin R_i^{s_u}$. So, let $vu \in B_i^{s_u}$ (it exists since $B_i^{s_u}$ is the edge set of a spanning branching of G_{s_u}), and recall that $\lambda^i(vu) = \lambda(vu)$. We know that $vu \in E(G_{s_u})$ only if $(v, j)(u, s_u)$ is a temporal edge of $\mathcal{G}$ for some $j \leq s_u$ (i.e. $(j, s_u) \in \lambda(vu)$). This means that there is a temporal edge arriving in (u, s_u) in $\mathcal{B}_i$, completing the proof.   □

4 Vertex-Spanning Branchings

In this section, we provide an NP-completeness proof to prove both Item 3 and Item 4 of Theorem 1. We make a reduction from NAE-SAT, which consists of, given a CNF formula ϕ such that each clause contains exactly 3 literals, deciding whether there is a truth assignment to ϕ such that each clause has at least one true and one false literal. This is problem is NP-complete [21], and in fact it is a well known standard procedure to make a reduction from 3-SAT to it that produces a formula of size linear on the size of the original 3-SAT formula. Therefore, applying ETH we get that NAE-SAT also cannot be solved in time $O(2^{o(n+m)})$ where n, m are the number of variables and clauses of an input, respectively.

Let ϕ be a CNF formula on variables $\{x_1, \ldots, x_n\}$ and clauses $\{c_1, \ldots, c_m\}$. A variable gadget related to x_i is formed by the set of vertices

$$V_i = \{x_i, F_i, T_i, a_i\}$$

and the set of edges

$$E_i = \{x_i T_i, x_i F_i, T_i a_i, F_i a_i\}.$$

Now, consider a clause $c_i = \{\ell_{i_1}, \ell_{i_2}, \ell_{i_3}\}$, and for each $i \in [3]$ let x_{i_j} be the variable related to literal ℓ_{i_j}. For each $i \in [3]$, if x_{i_j} appears positively in c_i, then add edge $T_{i_j} c_i$ to the clause gadget related to c_i; otherwise, add edge $F_{i_j} c_i$. See Fig. 2 for the digraph related to $\phi = (x_1 \vee x_2 \vee x_3) \wedge (\overline{x}_2 \vee x_3 \vee \overline{x}_4)$.

Denote by C_i the set of vertices in the clause gadget of c_i, and by E_i', the set of edges. Now, let G_ϕ be the digraph formed by the union of all variable and clause gadgets, i.e., $V(G) = \bigcup_{i=1}^n V_i \cup \bigcup_{i=1}^m C_i$ and $E(G) = \bigcup_{i=1}^n E_i \cup \bigcup_{i=1}^m E_i'$. Finally, add to G_ϕ two new vertices, g, r, and add edges $\{gx_i, rx_i\}$ for every $i \in \{1, \cdots, n\}$.

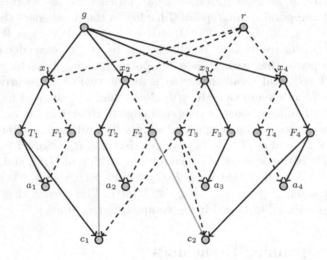

Fig. 2. Snapshot 1 related to formula $\phi = (x_1 \vee x_2 \vee x_3) \wedge (\overline{x}_2 \vee x_3 \vee \overline{x}_4)$, and branchings related to the assignment (T, T, F, F) to (x_1, x_2, x_3, x_4).

Finally, let G' be the graph having $A \cup \{g, r\}$ as vertex set, where $A = \{T_i, F_i \mid i \in [n]\}$, and having every edge going from $\{g, r\}$ to A. Let $\mathcal{G}$ be the temporal digraph with lifetime 2, having G_ϕ as first snapshot and G' as second snapshot (therefore, the basic digraph of $\mathcal{G}$ is given by $(V, E(G_\phi) \cup E(G'))$, where $V = V(G_\phi) \supset V(G')$).

Theorem 5. *For each $k \geq 2$, given a temporal digraph $\mathcal{G} = (G, \gamma, \lambda)$ with lifetime $\mathcal{T}$, and set of temporal vertices V_T, and subsets $R_1, \cdots, R_k \subseteq V_T$, it is*

NP-*complete to decide whether* $\mathcal{G}$ *has* k *(t-edge-disjoint or edge-disjoint) vertex-spanning branchings rooted at* $R_1, \cdots, R_k$, *even if* $\mathcal{T} = 2$ *and* G *is a DAG. Furthermore, letting* $n = |V(G)|$ *and* $m = |E(G)|$, *no algorithm running in time* $O^*(2^{o(n+m)})$ *can exist for the problem, unless ETH fails.*

Proof. The second part follows easily since the reduction is linear. We prove the theorem for $k = 2$, and NP-completeness for bigger values of k follows by Lemma 1. Let ϕ be an instance of NAE-SAT, and let $\mathcal{G}$ be the temporal digraph constructed as before; denote by G the base digraph. We prove that ϕ is a "yes" instance if and only if $\mathcal{G}$ has k edge-disjoint vertex-spanning branchings rooted at $\{(g, 1), (r, 1)\}$ (we will see that the branchings are also t-edge disjoint).

First, suppose that ϕ is a "yes" instance of NAE-SAT. We construct a solid and a dotted branching that satisfy our conditions. For each true variable x_i, add to the solid branching the following edges of snapshot 1: $\{gx_i, x_iT_i, T_ia_i\}$, together with edge T_ic_j for each clause c_j containing x_i that is not reached by the solid branching yet; also add to the dotted branching edges $\{rx_i, x_iF_i, F_ia_i\}$, , together with edge F_ic_j for each clause c_j containing $\overline{x}_i$ that is not reached by the dotted branching yet. Do something similar to the false variables, but switching the branchings. Figure 2 gives the branchings related to the assignment (T, T, F, F) to (x_1, x_2, x_3, x_4), respectively.

Observe that every $u \in V(G)$ is spanned by both branchings, with the exception of vertices in $B = \{(T_i, 2), (F_i, 2) \mid i \in [n]\}$. However, these can easily be spanned in the second snapshot since $\{(g, 2), (r, 2)\}$ is complete to B.

Now, let $\mathcal{B}_1, \mathcal{B}_2$ be two edge-disjoint vertex-spanning branchings. Because each a_i can only be reached at the first snapshot, it is reached by exactly two paths from $\{(g, 1), (r, 1)\}$, one of them going through $(x_i, 1)(T_i, 1)$ and the other through $(x_i, 1)(F_i, 1)$. We then put x_i as true if and only if $(x_i, 1)(T_i, 1)$ is in branching $\mathcal{B}_1$. Now, consider clause $c_i = (\ell_{i_1} \vee \ell_{i_2} \vee \ell_{i_3})$. One can verify that, because c_i is spanned by $\mathcal{B}_1$ and $\mathcal{B}_2$, we get that at least one of the edges in E_i' is in $\mathcal{B}_1$, and at least one in $\mathcal{B}_2$, which implies that at least one of $\ell_{i_1}, \ell_{i_2}, \ell_{i_3}$ is true, and at least one is false, as desired. $\qquad\square$

5   Conclusions and Open Problems

In this paper we have investigated the temporal version of Edmonds' classical result about the problem of finding k edge-disjoint spanning branchings rooted at given $R_1, \cdots, R_k$. We have introduced different definitions of spanning branchings, and of edge-disjointness in temporal digraphs. We have proved that, unlike the static case, only one of the these can be computed in polynomial time, namely the temporal-edge-disjoint temporal-spanning branchings problem, while the other versions are NP-complete under very strict constraints. Given a temporal digraph $\mathcal{G} = (G, \gamma, \lambda)$, in the particular case of edge-disjoint temporal-spanning, we give separate NP-complete results for fixed lifetime, and for when G is a in-star. A good question then might be whether there exists a polynomial algorithm for fixed lifetime and treewidth (a in-star has treewidth 1).

Another interesting question is whether the problem remains hard for fixed life-time when the base digraph is a DAG. Also, as we have provided computational lower bounds under ETH in Theorem 3 and in Theorem 5, we wonder whether there exist algorithms matching these lower bounds.

References

1. Casteigts, A., Flocchini, P., Quattrociocchi, W., Santoro, N.: Time-varying graphs and dynamic networks. In: Frey, H., Li, X., Ruehrup, S. (eds.) ADHOC-NOW 2011. LNCS, vol. 6811, pp. 346–359. Springer, Heidelberg (2011). https://doi.org/10.1007/978-3-642-22450-8_27
2. Akrida, E.C., Czyzowicz, J., Gasieniec, L., Kuszner, L., Spirakis, P.G.: Temporal flows in temporal networks. J. Comput. Syst. Sci. **103**, 46–60 (2019)
3. Berman, K.A.: Vulnerability of scheduled networks and a generalization of Menger's theorem. Networks **28**, 125–134 (1996)
4. Bhadra, S., Ferreira, A.: Complexity of connected components in evolving graphs and the computation of multicast trees in dynamic networks. In: Pierre, S., Barbeau, M., Kranakis, E. (eds.) ADHOC-NOW 2003. LNCS, vol. 2865, pp. 259–270. Springer, Heidelberg (2003). https://doi.org/10.1007/978-3-540-39611-6_23
5. Borgnat, P., Fleury, E., Guillaume, J.L., Magnien, C., Robardet, C., Scherrer, A.: Evolving networks. In: Mining Massive Data Sets for Security, pp. 198–203 (2007)
6. Bui-Xuan, B.-M., Ferreira, A., Jarry, A.: Computing shortest, fastest, and foremost journeys in dynamic networks. Int. J. Found. Comput. Sci. **14**(2), 267–285 (2003)
7. Campos, V., Lopes, R., Marino, A., Silva, A.: Edge-disjoint branchings in temporal graphs. arXiv e-prints, page arXiv:2002.12694 (2020)
8. Casteigts, A., Flocchini, P., Quattrociocchi, W., Santoro, N.: Time-varying graphs and dynamic networks. IJPEDS **27**(5), 387–408 (2012)
9. Edmonds, J.: Edge-disjoint branchings. Combinatorial algorithms (1973)
10. Fortune, S., Hopcroft, J., Wyllie, J.: The directed subgraph homeomorphism problem. Theor. Comput. Sci. **10**(2), 111–121 (1980)
11. Frank, A.: Covering branchings. Acta Scientiarum Mathematicarum (Szeged) **41**, 77–81 (1979)
12. Fulkerson, D.R., Harding, G.: On edge-disjoint branchings. Networks **6**(2), 97–104 (1976)
13. Huang, S., Fu, A.W.C., Liu, R.: Minimum spanning trees in temporal graphs. In: Proceedings of the 2015 ACM SIGMOD International Conference on Management of Data (SIGMOD 2015), pp. 419–430. ACM, New York (2015)
14. Kempe, D., Kleinberg, J., Kumar, A.: Connectivity and inference problems for temporal networks. In STOC 2000: Proceedings of the Thirty-Second Annual ACM Symposium on Theory of Computing (2000)
15. Latapy, M., Viard, T., Magnien, C.: Stream graphs and link streams for the modeling of interactions over time. Soc. Netw. Anal. Min. **8**(1), 1–29 (2018). https://doi.org/10.1007/s13278-018-0537-7
16. Lovász, L.: On two minimax theorems in graph. J. Comb. Theory, Ser. B **21**(2), 96–103 (1976)
17. McDiarmid, C.: Integral decomposition in polyhedra. Math. Program. **25**(2), 183–198 (1983)
18. Mertzios, G.B., Michail, O., Spirakis, P.G.: Temporal network optimization subject to connectivity constraints. Algorithmica **81**, 1416–1449 (2019)

19. Michail, O.: An introduction to temporal graphs: an algorithmic perspective. Internet Math. **12**(4), 239–280 (2016)

20. Santoro, N., Quattrociocchi, W., Flocchini, P., Casteigts, A., Amblard, F.: Time-varying graphs and social network analysis: temporal indicators and metrics. CoRR, abs/1102.0629 (2011)

21. Schaefer, T.J.: The complexity of satisfiability problems. In: Proceedings of the 10th Annual ACM Symposium on Theory of Computing, 1–3 May 1978, San Diego, California, USA, pp. 216–226 (1978)

22. Shiloach, Y.: Edge-disjoint branching in directed multigraphs. Inf. Process. Lett. **8**(1), 24–27 (1979)

23. Slivkins, A.: Parameterized tractability of edge-disjoint paths on directed acyclic graphs. SIAM J. Discrete Math. **24**(1), 146–157 (2010)

24. Robert Endre Tarjan: A good algorithm for edge-disjoint branching. Inf. Process. Lett. **3**(2), 51–53 (1974)

25. Huanhuan, W., Cheng, J., Huang, S., Ke, Y., Yi, L., Yanyan, X.: Path problems in temporal graphs. PVLDB **7**(9), 721–732 (2014)

Optimal In-place Algorithms for Basic Graph Problems

Sankardeep Chakraborty[1], Kunihiko Sadakane[2(✉)], and Srinivasa Rao Satti[3]

[1] National Institute of Informatics, Tokyo, Japan
sankar@nii.ac.jp
[2] The University of Tokyo, Tokyo, Japan
sada@mist.i.u-tokyo.ac.jp
[3] Seoul National University, Seoul, South Korea
ssrao@cse.snu.ac.kr

Abstract. We present linear time *in-place* algorithms for several fundamental graph problems including the well-known graph search methods (like depth-first search, breadth-first search, maximum cardinality search), connectivity problems (like biconnectivity, 2-edge connectivity), decomposition problem (like chain decomposition) among various others, improving the running time (by polynomial multiplicative factor) of the recent results of Chakraborty et al. [ESA, 2018] who designed $O(n^3 \lg n)$ time in-place algorithms for some of the above mentioned problems. The running times of all our algorithms are essentially optimal as they run in linear time. One of the main ideas behind obtaining these algorithms is the detection and careful exploitation of sortedness present in the input representation for any graph without loss of generality. This observation alone is powerful enough to design some basic linear time in-place algorithms, but more non-trivial graph problems require extra techniques which, we believe, may find other applications while designing in-place algorithms for different graph problems in future.

1   Introduction

Inspired by the rapid growth of humongous data set ("big data phenomenon"), *space efficient algorithms* are becoming increasingly more crucial than ever before. The dire need of such algorithms is also propelled by the pervasive usage of small specialized handheld devices and embedded systems which come equipped with tiny memory. To design such algorithms, a vast array of computational models have already been proposed in the literature. In what follows, we briefly mention a few of them in the order they are historically developed.

In the *read-only memory* model (henceforth ROM) where the input is read-only, output is write only, and a limited sized random access read/write work space is available, researchers have designed space efficient algorithms

The full version of this paper appears as [15]. The work of the first author is supported by JSPS KAKENHI Grants Number 18H05291.

L. Gąsieniec et al. (Eds.): IWOCA 2020, LNCS 12126, pp. 126–139, 2020.
https://doi.org/10.1007/978-3-030-48966-3_10

for selection and sorting [18,26,33,39,40], problems in computational geometry [2,4,6,17,22], and graphs [3,5,10,12–14,25] among various others. In the *in-place* model, it is assumed that the input elements are given in an array, and the algorithm may use the input array as working space, hence the algorithm is allowed to modify the array during its execution. However, at any point during the execution, all the input elements should be present in the array (maybe in a permuted order), and the output maybe put in the same array or sent to an output stream. The extra space usage during the entire execution of the algorithm is limited to $O(\lg n)$ bits only. A prime example of an in-place algorithm is the classic heap-sort. Other than in-place sorting [32], searching [30,37] and selection [36], many in-place algorithms were designed in areas such as computational geometry [8] and string algorithms [31]. A very recent addition to this long list is the in-place algorithms for the graph problems [11]. Other than these, researchers have also designed space efficient algorithms in *(semi)-streaming* models [1,29,39] and recently introduced *restore* [19] and *catalytic-space* [9] models.

Previous Work on Space Efficient Graph Algorithms. Inspired by the pervasive practical applications of the fundamental graph algorithms, recently there has been a surge of interest in improving the space complexity of graph algorithms without paying too much penalty in the running time. Thus the goal is to design space-efficient yet reasonably time-efficient graph algorithms on the ROM. Generally most of the standard implementations of classical graph algorithms take linear or near-linear running time and use $O(n \lg n)$ (or sometimes $O(m \lg n)$ for graphs with n vertices and m edges) bits. A recent series of papers [3,5,13,16,25] with this point of view showed such results for a vast array of basic graph problems, namely depth-first search (henceforth DFS), breadth-first search (henceforth BFS), minimum spanning tree (henceforth MST), (strong) connectivity, topological sorting, recognizing chordal graphs, biconnectivity, *st*-numbering, shortest path and many others.

Even though these results are still both time and space efficient, they still require $\Theta(n)$ bits for most of important graph algorithms, and this is a major concern in places with severe space constraints. In order to break this inherent space bound barrier and still obtain reasonable time efficiency, Chakraborty et al. [11] initiated a systematic study of designing efficient *in-place* (i.e., using $O(\lg n)$ bits of extra space other than the input space) algorithms for graph problems by defining a new framework which is a slight relaxation of the ROM. Using this framework they designed in-place DFS, BFS, MST, reachability algorithms taking time $O(n^3 \lg n)$. Despite being optimal in space usage, observe that these results still leave a polynomial gap in the running time from the optimal value. In this work, we essentially obtain the best of the both worlds by closing this gap. More specifically, we show how one can design optimal in-place algorithms i.e., $O(m+n)$ time and using $O(\lg n)$ bits of extra space, for several of these (and a lot more) basic graph algorithms in this work. Recently Kammer et al. [34] also considered a similar model where they showed efficient in-place algorithms for DFS, unordered-BFS (will be defined shortly) only.

In-place Model for Graph Algorithms and Input Representations.
Before explaining our in-place algorithms and stating main results, in this section
we first describe the input graph representation. Note that, as in the case of
the standard in-place model, we need to ensure that the graph (adjacency)
structure must remain intact throughout the entire execution of the algorithm.
Let $G = (V, E)$ be the input graph with $n = |V|$, $m = |E|$, and as usual let
$V = \{1, 2, \cdots, n\}$ denote the vertex set of G. We assume that the input graph
is given in the standard adjacency array format, and throughout this paper, we
refer to this array as Z. More specifically, it is an array having size $(n + m + 1)$
($(n+2m+1)$ resp.) words for directed (undirected resp.) graphs where $Z[1]$ stores
the number of vertices in G, the next n entries (which we refer to as the *offsets*
part of Z) store n pointers (one per vertex) pointing to the location in Z of the
last neighbor for each vertex, and finally the last m ($2m$ for undirected graphs)
entries are reserved for the edges of G. At this point, we should emphasize a
small, yet important, technical detail. The Z array can be thought of as a single
bit array as follows. For a directed graph G, the array Z is a concatenation of
$Z[1]$ of length $\lceil \lg n \rceil$ bits, $Z[2] \ldots Z[n+1]$ of length $\lceil \lg m \rceil$ bits each[1], and finally
$Z[n+2] \ldots Z[n+m+1]$ of length $\lceil \lg n \rceil$ bits each. For undirected graphs, only the
second part changes to size $\lceil \lg m \rceil + 1$ bits (instead of $\lceil \lg m \rceil$) each. Thus, if we
just remember the boundaries, we know exactly how many bits we need to read
in order to extract useful information from the relevant parts of Z. For the sake
of simplicity, we drop the ceiling notations from now on. Moreover, throughout
this paper, it should be clear from the context the word size depending on which
part of Z we are currently working on. See Fig. 1 for an example. Note that this
representation implicitly captures the degree information for every vertex in G.
Given this format, we say an algorithm $\mathcal{A}$ is an *in-place* algorithm if $\mathcal{A}$ (a) may
modify any part of Z during its execution, (b) retains all the initial elements
of Z (in any order) when it finishes execution; and (c) uses just $O(\lg n)$ bits of
extra space. Our goal is to design such algorithms in this paper for a vast array
of fundamental graph problems.

In this paper we assume the standard word RAM model of computation. We
count space in terms of number of *extra* bits used by the algorithm other than the
input, and this quantity is referred as "extra space" and "space" interchangeably
throughout the paper.

Graph Terminology and Notations. In general we will assume the knowledge
of basic graph theoretic terminology as given in [23] and basic graph algorithms
as given in [21]. Still here we collect all the necessary graph theoretic definitions
that will be used throughout the paper for quick reference and making the paper
self-contained. For BFS traversal that we study here, there are two versions
studied in the literature. In the *ordered* BFS (sometimes also known as queue
BFS [16]), vertices are extracted from the queue in the first in first out (FIFO)
order, whereas in the *unordered* BFS [5], vertices can be taken out from the

---

[1] Note that it is enough to store the offset values starting from 0, since we can add
$n + 1$ to the offset value to find the corresponding location in Z; hence the offset
values can be stored using $\lceil \lg m \rceil$ bits.

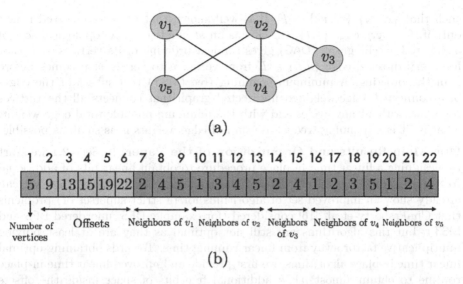

Fig. 1. (a) An undirected graph G with 5 vertices and 8 edges. (b) The standard adjacency array representation of G. To avoid cluttering the diagram, we drop the superscript v from the vertex labels while referring to them as neighbors.

queue in any order as long as no elements are extracted from a higher level of the BFS tree before finishing all the vertices from a lower level of the tree. In this paper, by a BFS/DFS traversal of the input graph G, as in [3,5,13,16] we refer to reporting the vertices of G in the BFS/DFS ordering, i.e., in the order in which the vertices are visited for the first time. Tarjan et al. [45] defined another method called maximum cardinality search (MCS) and used this to give a recognition algorithms for chordal graphs. MCS works as follows: assuming that every vertex is unnumbered at the beginning, at each iteration of the execution of MCS, an unnumbered vertex that is adjacent to the largest number of numbered vertices is chosen (breaking the ties arbitrarily), and is numbered with the next available label. Thus, the output of the MCS algorithm is a numbering of the vertices from 1 to n.

A cut vertex in an undirected graph G is a vertex v that when removed (along with its incident edges) from a graph creates more components than previously in the graph. A (connected) graph with at least three vertices is biconnected if and only if it has no cut vertex. Similarly in an undirected graph G, a bridge is an edge that when removed, creates more components than previously in the graph. A connected graph with at least two vertices is 2-edge-connected if and only if it has no bridge. Given a biconnected graph G, and two distinguished vertices s and t in V such that $s \neq t$, st-numbering is a numbering of the vertices of the graph so that s gets the smallest number, t gets the largest and every other vertex is adjacent both to a lower-numbered and to a higher-numbered vertex i.e., a numbering $s = v_1, v_2, \cdots, v_n = t$ of the vertices of G is called an st-numbering, if for all vertices $v_j, 1 < j < n$, there exist $1 \leq i < j < k \leq n$

such that $\{v_i, v_j\}, \{v_j, v_k\} \in E$. It is well-known that G is biconnected if and only if, for every edge $\{s, t\} \in E$, it has an st-numbering. A topological sort of a directed acyclic graph (DAG) gives a linear ordering of its vertices such that for every directed edge $(u, v) \in E$ from vertex u to vertex v, u comes before v in the ordering. A minimum spanning tree (MST) is a subset of the edges of a connected, edge-weighted undirected graph that connects all the vertices together, without any cycles and with the minimum possible total edge weight. That is, it is a spanning tree whose sum of edge weights is as small as possible.

Our Main Results and Organization of the Paper. In Sect. 2.1 we start by designing a linear time in-place procedure to obtain linear bits of additional free space inside the offsets part of the adjacency array. Using this, we can already show an improved set of algorithms for (a strict superset of) problems that Chakraborty et al. [16] considered (for example, DFS, unordered BFS and MST), but this algorithms are still not optimal as they are at least polylog multiplicative factor away from linear running time. Towards obtaining optimal linear time in-place algorithms, we first provide an improved linear time in-place routine to obtain almost $n \lg n$ additional free bits of space inside the offsets part, which is what we use crucially along with other additional ideas to show the following main result of this paper in Sect. 2.2.

Theorem 1. *Using linear time in the in-place model, one can*

1. *traverse the vertices of any graph in (un)ordered* BFS *and* DFS *manner,*
2. *recognize bipartite graphs, and compute connected components,*
3. *report the vertices of a* DAG *in topologically sorted order,*
4. *obtain a maximum cardinality search ordering of any graph,*
5. *output an st-numbering of given biconnected graph, given two vertices s and t,*
6. *perform a chain decomposition of any undirected graph, and*
7. *determine whether any given undirected graph G is biconnected (and/or 2-edge connected resp.) and if not, we can also compute and report all the cut vertices (bridges resp.) of G.*

Also, given an undirected edge-weighted (where weights are bounded by some polynomial in n) graph G, we can find a minimum spanning tree (MST) of G in $O(m \lg n)$ time in-place.

Techniques. All the results of our paper stem from the following very simple yet absolutely crucial observation: *numbers in sorted order have less entropy than in any arbitrary order.* More specifically, assuming we have n numbers from a universe of size m, when these numbers are in any arbitrary order their binary entropy is $n \lg m$ but when they are in sorted order, binary entropy becomes $n \lg m - \Theta(n \lg n)$. This clearly indicates that we can exploit the sorted structure assumption to gain some additional space. Now, note that, without loss of any generality, by construction, the *offsets* part of the adjacency array Z for any given graph G is sorted. Thus, we can use the above mentioned idea in the offsets part of Z to gain some free space which is what we use finally to design our optimal in-place graph algorithms. Towards this, we also have to handle several other key technical issues which we describe in respective sections in detail.

2   Exploiting Input Redundancy to Create Working Space

In this section, we describe how one can exploit the redundancy in the input representation to save almost $n \lg n$ bits, which can then be used as part of the working space for a graph algorithm.

2.1   Saving Linear Bits and Its Applications

As a warm up, we start by showing how we can squeeze in linear sized free bits inside the offsets part of Z while still being able to access any element inside the offsets part in $O(1)$ time, as well as returning to the original configuration of the offsets part of Z before freeing linear bits. Towards this, we first reprove the following lemma, which is essentially same as [34, Lemma 5]. See the full version [15] for a proof.

Lemma 1. *Given a sorted list of n integers from the universe $[0, m-1]$, it can be represented either simply as an array $A[1...n]$ with the integers in sorted order or as an array of n integers, such that for some fixed constant $c > 1$, the last cn bits of this array are all zero. Moreover, there exists an in-place $O(n)$ time algorithm for switching between both these formats.*

The above lemma alone is powerful enough to help us design in-place algorithms (albeit with sub-optimal time complexity as we will see shortly) for a variety of fundamental graph algorithms. In the full version [15] of this paper, we describe how to obtain efficient in-place algorithms for a variety of graph algorithms, using the above lemma. The main idea is to simulate the corresponding ROM algorithms in the in-place model. Next, we further improve the running times to optimal, by providing an improved version of Lemma 1.

2.2   Saving $n \lg n - 2n$ Bits

In what follows, we show how one can improve Lemma 1 so that almost $n \lg n$ bits become free to be used, and using this we will design optimal in-place algorithms for the above mentioned graph problems. Our main result can be described as follows:

Theorem 2. *Given a sorted list of n integers from the universe $[0, m-1]$, it can be represented either simply as an array $A[1...n]$ with the integers in sorted order or as an array of n integers, such that the last $n \lg n - 2n$ bits of this array are all zero. Moreover, there exists an in-place $O(n)$ time algorithm for switching between both these formats.*

Proof. One can easily obtain the space bound mentioned in the second representation by applying the Elias-Fano encoding [24,28] on the array A. But to implement this encoding in-place, we apply this encoding in two steps.

We first split the array A into two subarrays of size $n/2$ each (assume, for simplicity, that n is even) - call them A_1 and A_2. One can replace the most significant $\lg n$ bits of each of the elements in A_1 by a bit vector, say B, length

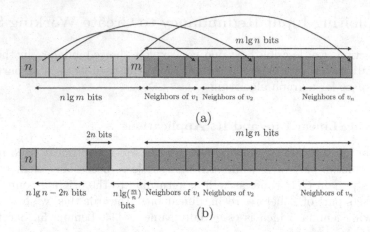

Fig. 2. (a) General adjacency array structure Z of a given input directed graph. (b) Configuration of Z after freeing $n \lg n - 2n$ bits in the offsets part of Z.

$n + n/2$, using the Elias-Fano encoding. To store B (of length $3n/2$), we first replace the most significant 3 bits of each of the elements in A_2 by storing 8 positions into the array A_2 (using Lemma 1, with $c = 3$). We store the bit vector B inside the most-significant 3 bits of every element of A_2, and compact the remaining (least-significant $\lg m - \lg n$) bits of every element in A_1 into a consecutive chunk of $(n/2) \lg(m/n)$ bits in A_1, so that the first $(n/2) \lg n$ bits of A_1 is free (i.e., filled with all zeros). We now copy the bit vector B into this free space, and restore the 3 most significant bits of all the elements of A_2. We now replace the most-significant $\lg n$ bits of each element in A_2 by a bit vector C of length $3n/2$, and store it inside free space in A_1 (here, we assume that $3n \leq (n/2) \lg n$), and compact the remaining (least-significant $\lg m - \lg n$) bits into a consecutive chunk of $(n/2) \lg(m/n)$ bits in A_2. Finally, we copy all the lower order bits (of total length $n \lg(m/n)$ bits) into a single chunk, and also merge the two bit vectors of length $3n/2$ each into a single bit vector of length $2n$. Thus the array A is replaced by a total of $n \lg(m/n) + 2n$ bits, giving a free space of $n \lg n - 2n$ bits. These steps can be essentially performed in reverse order to restore the original representation from the second representation. To support the operation of accessing the i-th element of A in $O(1)$ time, we can store an additional $o(n)$-bit auxiliary structure that support the *rank* and *select* operations [20,38] on the $2n$ bit sequence, which can then be used to access the most-significant $\lg n$ bits of any element in $O(1)$ time. The remaining $\lg m - \lg n$ bits can be simply read from the array of values stored in the second representation. See Fig. 2 for a visual description of the final outcome of application of this theorem.

3   Optimal In-place Graph Algorithms

In this section, we show how one can use Theorem 2 for solving the graph problems mentioned before. Before giving specific details, we would like to sketch the general pattern for designing optimal in-place algorithms for some of these graph problems. Given the adjacency array representation (as in Z) of the input graph G, we now first apply Theorem 2 on the offsets part of Z to make $n \lg n - 2n$ bits free. Now the classical linear time algorithms [21,27,42–45] for these problems typically take $cn \lg n + dn$ bits where both the constants c and d are at most 2. Hence, our idea is to run these algorithms as it is but in some constant number of phases. More specifically, we store only, say $n/3$ vertices, explicitly at any point of time during the execution of these algorithms, and when these vertices are taken care of by the respective algorithms, we refresh the data structures by initiating it with a new set of $n/3$ vertices and proceed again till we exhaust all the vertices, thus, the entire algorithm would finish in three phases ultimately. Now the exact details of refreshing the data structure with a new set of vertices and start the algorithm again where it left off depends on specific problems. This idea would work for most of the algorithms that we discuss in this paper except a few important ones. More specifically, a few of the algorithms for those graph problems are two (or more) pass algorithms, i.e., in the first pass it computes some function which is what used in the second pass to solve the problem finally, for example, chain decomposition, biconnectivity etc. For these kinds of algorithms, it seems hard to make them work using the previously described constant phase algorithmic idea. Thus, we handle them differently by first proving some related lemmata which might be of independent interest, and then use these lemmata to design in-place algorithms for these graph problems. We discuss these after giving proofs for the algorithms which we can handle in constant phases only. In what follows we provide the proofs of linear time in-place algorithms for DFS and its applications, especially chain decomposition, biconnectivity, 2-edge connectivity, and also develop/prove the necessary ideas for these algorithms. The missing proofs of Theorem 1 cen be found in the full version [15].

The classical implementation of DFS (see for example, Cormen et al. [21]) uses three colors and a stack to traverse the whole graph. More specifically, every vertex v is white initially while it has not been discovered yet, becomes grey when DFS discovers v for the first time and pushes on the stack, and is colored black when it is finished i.e., all its neighbors have been explored completely, and it leaves the stack. The algorithm maintains a color array C of length $O(n)$ bits that stores the color of each vertex at any point in the algorithm, along with a stack (which could grow to $O(n \lg n)$ bits) for storing all the grey vertices at any point during the execution. Our idea is to run essentially the same DFS algorithm but we limit the stack size so that it contains at most $n/2$ latest grey vertices all the time. More specifically, whenever the stack grows to have more than $n/2$ vertices, we delete the bottom most vertex from the stack so that above invariant is always maintained along with storing the last such vertex to be deleted in order to enforce the invariant. At some point during the execution of the algorithm, when we arrive at a vertex v such that none of v's neighbors are white, then we color the vertex v as black, and we pop it from the stack. If the stack is still non-empty, then the parent of v (in the DFS tree) would be at the top of the stack, and we continue the DFS from this vertex. On the other hand, if the stack becomes empty after removing v, we need to reconstruct it to the state such that it holds the last $n/2$ grey vertices after all the pops done so far. We refer to this phase of the algorithm as reconstruction step. For this, using ideas from [3, 25], we basically repeat the same algorithm but with one twist which also enables us now to skip some of the vertices during this reconstruction phase. In detail, we again start with an empty stack, insert the root s first and scan its adjacency list from the first entry to skip all the black vertices and insert into the stack the leftmost grey vertex. Then the repeat the same for this newly inserted vertex into the stack until we reconstruct the last $n/2$ grey vertices. As we have stored the last vertex to be deleted for maintaining the invariant true, we know when to stop this reconstruction procedure. It is not hard to see that this procedure correctly reconstructs the latest set of grey vertices in the stack. We continue this process until all the vertices become black. Moreover, this algorithm runs in $O(m+n)$ time as it involves two phases each taking linear time in the worst case, and uses at most $(n \lg n)/2 + n \lg 3$ bits which fits in our budget of free space in the offsets part of the adjacency array. This completes the description of the linear time in-place DFS algorithm.

Before providing the algorithms for other problems, we need a few additional ideas which we will describe next. In the following theorem, we are interested in dynamically maintaining the degree sequence of all vertices that belong to a spanning subgraph of the original graph. More specifically, given a graph $G = (V, E)$, we want to run some algorithm on G for constructing a sparse spanning subgraph $G' = (V, E')$ (which is a spanning subgraph of G i.e., $E' \subseteq E$ and $|E'| = O(V)$) of G, and we are interested in dynamically maintaining the degree of all the vertices v in G' i.e., degree of a vertex v in G' is defined as the number of neighbors u such that the edge (v, u) belongs to G'. Thus, degree of a vertex v in G' may not be same as degree of v in G. Also note that, by the notion of dynamic, we mean that the algorithm starts with an empty graph and gradually add edges to it before finally culminating with a sparse spanning subgraph, thus during the execution of this algorithm degrees of the individual vertices are changing, and it is this dynamically changing degrees that we want to efficiently maintain. We refer to this as the *dynamic maintenance of degree sequence* phase. Towards this goal, we prove the following general theorem.

Theorem 3. *Given a graph G with n vertices and m edges, let G' be a spanning subgraph of G with m' edges, and also let $d' = m'/n$ be the average degree of G'. Then, we can construct the dynamically created degree sequence for the vertices of G' in $O(m + n)$ time using $O(n(\lg d' + \lg \lg n))$ bits of construction space. Moreover, the final degree sequence can be stored using $O(n \lg d')$ bits such that degree of any vertex can be returned in $O(1)$ time.*

Proof. We divide the vertices into $n/\lg n$ groups of $\lg n$ vertices each. For each group, we allocate a block of $\lg n(\lg d' + \lg \lg n)$ ($\leq \lg^2 n$) bits initially (uniformly for all the vertices in the block), to store their degrees. We also maintain another parallel bit vector for each block that simply stores the delimiters for each vertex's degree (i.e., a 1 bit to indicate the last bit corresponding to each vertex's degree, and 0 everywhere else). To access the degree of the i-th vertex in a block, we first find the positions of the $i - 1$-th and the i-th 1 bits in the corresponding delimiter sequence, and read the bits between these two positions in the block. To perform this efficiently during the construction, we maintain an auxiliary structure that supports *select* operation in $O(1)$ time [20,38]. At any point, the representation of each block and delimiter sequence consists of an integral number of words, and these representations are maintained as a collection of "extendible arrays" using the structure of [41, Lemma 1].

At any time, a vertex has some number of bits allocated to store its degree. If the degree of the vertex can be updated in-place, then we first access the position where the degree of the vertex is stored, using the select data structure stored for the corresponding delimiter sequence, and update the degree of the node stored within the block. Otherwise, we first note that at least $\lg n$ increments have been performed to some vertex within the block (since each vertex has a 'slack' of $\lg \lg n$ bits at the beginning of the latest re-construction of the block). Now, we spend $O(\lg n)$ time to re-construct the block (and also the corresponding delimiter sequence with its select structure) so that the degree of each vertex v in the block is stored $\lceil \lg d_v \rceil + \lg \lg n$ bits, where d_v is the current degree of v. This $\lg n$ construction time can be amortized over the $\lg n$ increments performed on the block before its re-construction, incurring an $O(1)$ amortized cost per increment. Once we construct the degree sequence for the entire subgraph G', we can scan all the blocks, and compact the degree sequence so that it occupies $O(n \lg d')$ bits. The space usage during the construction is bounded by $O(n(\lg d' + \lg \lg n))$ bits of space. Note that, the above task can be performed while executing the linear time DFS algorithm described before, and this completes the proof.

Corollary 1. *When G' is the DFS tree of G, then we can store the dynamically created degree sequence of G', whose size is bounded by $2n$ bits, by running a linear time DFS procedure while using $O(n \lg \lg n)$ bits of space during construction such that the degree of any vertex in G' can be accessed in $O(1)$ time.*

For the following discussion, assume that we are working with connected undirected graphs only, and given this, now we are going to describe the *setting up parent* phase. More specifically, while performing DFS, suppose we visit the vertex u for the first time from the vertex v (hence v becomes the parent of u in the DFS tree), at that point we perform one or more swaps in the portion of the adjacency array Z where the neighbors of u are located so that the vertex v becomes the first neighbor of u now. If the initial configuration of Z already satisfies this property in u's neighborhood, we don't need to do anything else. We repeat this procedure for every vertex $v \in V$ so that when DFS ends, the first neighbor of every vertex v (except the root vertex) is its parent in the DFS tree. Note that we can perform this step of setting up parent in the first location of every neighborhood list of every vertex alongside performing the linear time DFS algorithm of Theorem 1. Thus, we obtain the following.

Lemma 2. *There exists a linear time in-place algorithm for performing the setting up parent procedure for every vertex of G.*

Note that, by choosing appropriate parameters, we can actually perform the *dynamic maintenance of degree sequence* and the *setting up parent* phase together while running the linear time in-place DFS algorithm of Theorem 1 in any graph G. More specifically, suppose we choose to run the linear time in-place DFS algorithm of Theorem 1 coupled with the setting up parent procedure (to implement Lemma 2) by storing $n/2$ vertices (thus taking $n \lg n/2$ bits) in the free space of the offsets part of Z, thus, leaving roughly $(n \lg n/2 - 2n)$ bits of space still free, which can be used to construct and store the degree sequence of all the vertices in the DFS tree (to implement Corollary 1) while running the same linear time in-place DFS algorithm of Theorem 1. By degree of a vertex v in the DFS tree T, we mean the number of children v has in T, and it is this number that gets stored using the algorithm of Corollary 1. Hence, at the end of this linear time in-place procedure, we have the following invariant: (a) the first neighbor of every vertex (except the root) is its parent in the DFS tree, and (b) the offsets part of Z contains the degree sequence of every vertex v in the DFS tree, and this occupies at most $2n$ bits.

Armed with the above algorithm, we are going to explain next the *implicit representation of the search tree* phase. The goal of this phase is to rearrange the neighbors of any vertex v in such a way that the first neighbor of v becomes its parent in the DFS tree (except for the root vertex), followed by all of v's children in the DFS tree (if any) one by one, finally all the non-child neighbors. Thanks to the setting up parent phase, we can implement the implicit representation the search tree phase in linear time overall by doing a reverse search. More specifically, for every non-root vertex v, we start by scanning v's list from the second neighbor onward (as first neighbor is its parent), and for each one of them, say u, we go to the first location of u's neighbor list to check if v is u's parent if so, we move u in v's list closer to v's parent (i.e., towards the beginning of v's list) by swapping, and repeat this procedure for all the neighbors of v's so that at the end all the children of v are clustered together followed by v's parent. The root vertex can be handled similarly, but we need to start the scanning procedure from the first neighbor itself as it doesn't have any parent. Hence, we spend time proportional to its degree at every vertex, and obtain the following.

Lemma 3. *There exists a linear time in-place algorithm for implicitly representing the search tree of G.*

Thus, from now on we can assume that the neighbor list of every vertex is represented in the search tree format implicitly. We choose to call it so as, note that, given in this format, it is very convenient to answer the following queries for any given vertex v in the DFS tree T: (a) return the parent of v in T in $O(1)$ time, (b) return the number of children v has in T in $O(1)$ time (from the dynamically maintained degree sequence), and finally, (c) enumerate all the children of v one by one optimally in time proportional to its number of children. Not only this, observe that we can still perform the DFS traversal of G optimally in linear-time using essentially the same algorithm of Theorem 1 given this representation. We can even slightly optimize this DFS algorithm by stop scanning the neighbor list of any vertex v as soon as we encounter its last child u in the DFS tree (can be derived from the dynamically maintained degree sequence) as neighbors after u will not be of significance in performing the DFS traversal of G. Hence, we obtain the following.

Lemma 4. *There exists a linear time in-place algorithm for performing the* DFS *traversal of a given graph G using the implicit search tree representation of G.*

Topological Sorting. One of the standard algorithms for computing topological sort [21] works by simply reporting the vertices of a DFS traversal of a given directed acyclic graph in reverse order. We can easily implement this in-place in linear time by running our DFS algorithm in two phases. More specifically, in the first phase, we run the DFS algorithm completely to generate/store the last $n/2$ vertices in the DFS traversal order, and then report them in reverse order. This is followed by running the DFS algorithm one more time but stopping just when we obtain the other $n/2$ vertices, then we reverse the order of this vertices and report. This completes the description of generating the vertices in topologically sorted order of an input directed acyclic graph in-place in linear time.

In the full version [15] of this paper, we describe linear-time in-place algorithms for all the remaining graph algorithms mentioned in Theorem 1, namely, ordered BFS, MCS, st-numbering, MST, chain decomposition, checking biconnectivity and/or 2-edge connectivity, and finding cut vertices and bridges.

4   Conclusions

In this paper, we designed linear time in-place algorithms for a variety of graph problems. As a consequence, many interesting and contrasting observations follow. For example, for *directed st-reachability*, the most space efficient polynomial time algorithm [7] in ROM uses $n/2^{\Theta(\sqrt{\lg n})}$ bits. In sharp contrast, we obtain optimal linear time using logarithmic extra space algorithms for this problem as a simple corollary of both BFS and DFS. Thus, in terms of workspace this is exponentially better than the best known polynomial time algorithm [7] in ROM. This provided us with one of the main motivations for designing algorithms in the *in-place* model. A somewhat incomparable result obtained by Buhrman et al. [9,35] where they gave an algorithm for *directed st-reachability* on catalytic Turing machines in space $O(\lg n)$ with catalytic space $O(n^2 \lg n)$ and time $O(n^9)$. Finally, we conclude by mentioning that we barely scratched the surface of designing in-place graph algorithms with plenty of more to be studied in this model in future. For example, can we design linear time in-place algorithms for testing planarity of a graph? Can we compute the max-flow/min-cut in-place? Can we compute shortest paths between any two vertices of a given graph in-place? We leave these problems as our future directions of study.

References

1. Alon, N., Matias, Y., Szegedy, M.: The space complexity of approximating the frequency moments. J. Comput. Syst. Sci. **58**(1), 137–147 (1999)
2. Asano, T., et al.: Reprint of: memory-constrained algorithms for simple polygons. Comput. Geom. **47**(3), 469–479 (2014)

3. Asano, T., et al.: Depth-first search using $O(n)$ bits. In: Ahn, H.-K., Shin, C.-S. (eds.) ISAAC 2014. LNCS, vol. 8889, pp. 553–564. Springer, Cham (2014). https://doi.org/10.1007/978-3-319-13075-0_44

4. Asano, T., Mulzer, W., Rote, G., Wang, Y.: Constant-work-space algorithms for geometric problems. JoCG **2**(1), 46–68 (2011)

5. Banerjee, N., Chakraborty, S., Raman, V., Satti, S.R.: Space efficient linear time algorithms for BFS, DFS and applications. Theory Comput. Syst. **62**(8), 1736–1762 (2018)

6. Barba, L., Korman, M., Langerman, S., Sadakane, K., Silveira, R.I.: Space-time trade-offs for stack-based algorithms. Algorithmica **72**(4), 1097–1129 (2015)

7. Barnes, G., Buss, J., Ruzzo, W., Schieber, B.: A sublinear space, polynomial time algorithm for directed s-t connectivity. SIAM J. Comput. **27**(5), 1273–1282 (1998)

8. Brönnimann, H., Chan, T.M., Chen, E.Y.: Towards in-place geometric algorithms and data structures. In: SOCG, pp. 239–246 (2004)

9. Buhrman, H., Cleve, R., Koucký, M., Loff, B., Speelman, F.: Computing with a full memory: catalytic space. In: STOC, pp. 857–866 (2014)

10. Chakraborty, S.: Space efficient graph algorithms. Ph.D. thesis, The Institute of Mathematical Sciences, HBNI, India (2018)

11. Chakraborty, S., Mukherjee, A., Raman, V., Satti, S.R.: A framework for in-place graph algorithms. In: ESA, pp. 13:1–13:16 (2018)

12. Chakraborty, S., Mukherjee, A., Satti, S.R.: Space efficient algorithms for breadth-depth search. In: Gąsieniec, L.A., Jansson, J., Levcopoulos, C. (eds.) FCT 2019. LNCS, vol. 11651, pp. 201–212. Springer, Cham (2019). https://doi.org/10.1007/978-3-030-25027-0_14

13. Chakraborty, S., Raman, V., Satti, S.R.: Biconnectivity, st-numbering and other applications of DFS using O(n) bits. J. Comput. Syst. Sci. **90**, 63–79 (2017)

14. Chakraborty, S., Sadakane, K.: Indexing graph search trees and applications. In: 44th MFCS. LIPIcs, vol. 138, pp. 67:1–67:14. Schloss Dagstuhl - Leibniz-Zentrum für Informatik (2019)

15. Chakraborty, S., Sadakane, K., Satti, S.R.: Optimal in-place algorithms for basic graph problems. CoRR, abs/1907.09280 (2019)

16. Chakraborty, S., Satti, S.R.: Space-efficient algorithms for maximum cardinality search, its applications, and variants of BFS. J. Comb. Optim. **37**(2), 465–481 (2018)

17. Chan, T.M., Chen, E.Y.: Multi-pass geometric algorithms. Discret. Comput. Geom. **37**(1), 79–102 (2007)

18. Chan, T.M., Munro, J.I., Raman, V.: Faster, space-efficient selection algorithms in read-only memory for integers. In: Cai, L., Cheng, S.-W., Lam, T.-W. (eds.) ISAAC 2013. LNCS, vol. 8283, pp. 405–412. Springer, Heidelberg (2013). https://doi.org/10.1007/978-3-642-45030-3_38

19. Chan, T.M., Munro, J.I., Raman, V.: Selection and sorting in the "restore" model. ACM Trans. Algorithms **14**(2), 11:1–11:18 (2018)

20. Clark, D.R.: Compact pat trees. Ph.D. thesis. University of Waterloo, Canada (1996)

21. Cormen, T.H., Leiserson, C.E., Rivest, R.L., Stein, C.: Introduction to Algorithms, 3rd edn. MIT Press, Cambridge (2009)

22. Darwish, O., Elmasry, A.: Optimal time-space tradeoff for the 2D convex-hull problem. In: Schulz, A.S., Wagner, D. (eds.) ESA 2014. LNCS, vol. 8737, pp. 284–295. Springer, Heidelberg (2014). https://doi.org/10.1007/978-3-662-44777-2_24

23. Diestel, R.: Graph Theory, 4th edn (2012)

24. Elias, P.: Efficient storage and retrieval by content and address of static files. J. ACM **21**(2), 246–260 (1974)
25. Elmasry, A., Hagerup, T., Kammer, F.: Space-efficient basic graph algorithms. In: 32nd STACS, pp. 288–301 (2015)
26. Elmasry, A., Juhl, D.D., Katajainen, J., Satti, S.R.: Selection from read-only memory with limited workspace. Theor. Comput. Sci. **554**, 64–73 (2014)
27. Even, S., Tarjan, R.E.: Computing an st-numbering. Theor. Comput. Sci. **2**(3), 339–344 (1976)
28. Fano, R.M.: On the number of bits required to implement an associative memory. Memorandum 61, Computer Structures Group, MIT, Cambridge (1971)
29. Feigenbaum, J., Kannan, S., McGregor, A., Suri, S., Zhang, J.: On graph problems in a semi-streaming model. Theor. Comput. Sci. **348**(2–3), 207–216 (2005)
30. Franceschini, G., Munro, J.I.: Implicit dictionaries with $O(1)$ modifications per update and fast search. In: SODA, pp. 404–413 (2006)
31. Franceschini, G., Muthukrishnan, S.: In-place suffix sorting. In: Arge, L., Cachin, C., Jurdziński, T., Tarlecki, A. (eds.) ICALP 2007. LNCS, vol. 4596, pp. 533–545. Springer, Heidelberg (2007). https://doi.org/10.1007/978-3-540-73420-8_47
32. Franceschini, G., Muthukrishnan, S., Pătraşcu, M.: Radix sorting with no extra space. In: Arge, L., Hoffmann, M., Welzl, E. (eds.) ESA 2007. LNCS, vol. 4698, pp. 194–205. Springer, Heidelberg (2007). https://doi.org/10.1007/978-3-540-75520-3_19
33. Frederickson, G.N.: Upper bounds for time-space trade-offs in sorting and selection. J. Comput. Syst. Sci. **34**(1), 19–26 (1987)
34. Kammer, F., Sajenko, A.: Linear-time in-place DFS and BFS on the word RAM. In: Heggernes, P. (ed.) CIAC 2019. LNCS, vol. 11485, pp. 286–298. Springer, Cham (2019). https://doi.org/10.1007/978-3-030-17402-6_24
35. Koucký, M.: Catalytic computation. Bull. EATCS **118** (2016). http://eatcs.org/beatcs/index.php/beatcs/article/view/400
36. Lai, T.W., Wood, D.: Implicit selection. In: Karlsson, R., Lingas, A. (eds.) SWAT 1988. LNCS, vol. 318, pp. 14–23. Springer, Heidelberg (1988). https://doi.org/10.1007/3-540-19487-8_2
37. Munro, J.I.: An implicit data structure supporting insertion, deletion, and search in $O(\log^2 n)$ time. J. Comput. Syst. Sci. **33**(1), 66–74 (1986)
38. Munro, J.I.: Tables. In: FSTTCS, pp. 37–42 (1996)
39. Munro, J.I., Paterson, M.: Selection and sorting with limited storage. Theor. Comput. Sci. **12**, 315–323 (1980)
40. Pagter, J., Rauhe, T.: Optimal time-space trade-offs for sorting. In: FOCS, pp. 264–268 (1998)
41. Raman, R., Rao, S.S.: Succinct dynamic dictionaries and trees. In: Baeten, J.C.M., Lenstra, J.K., Parrow, J., Woeginger, G.J. (eds.) ICALP 2003. LNCS, vol. 2719, pp. 357–368. Springer, Heidelberg (2003). https://doi.org/10.1007/3-540-45061-0_30
42. Schmidt, J.M.: A simple test on 2-vertex- and 2-edge-connectivity. Inf. Process. Lett. **113**(7), 241–244 (2013)
43. Tarjan, R.E.: Depth-first search and linear graph algorithms. SIAM J. Comput. **1**(2), 146–160 (1972)
44. Tarjan, R.E.: A note on finding the bridges of a graph. Inf. Process. Lett. **2**(6), 160–161 (1974)
45. Tarjan, R.E., Yannakakis, M.: Simple linear-time algorithms to test chordality of graphs, test acyclicity of hypergraphs, and selectively reduce acyclic hypergraphs. SIAM J. Comput. **13**(3), 566–579 (1984)

Further Results on Online Node- and Edge-Deletion Problems with Advice

Li-Hsuan Chen[1], Ling-Ju Hung[2], Henri Lotze[3], and Peter Rossmanith[3(✉)]

[1] Kenkone Medical Co., Taipei, Taiwan
[2] National Taipei University of Business, Taipei, Taiwan
[3] RWTH Aachen University, Aachen, Germany

Abstract. In online edge- and node-deletion problems the input arrives node by node and an algorithm has to delete nodes or edges in order to keep the input graph in a given graph class at all times. We consider graph classes that can be characterized by forbidden sets of induced subgraphs and analyze the advice complexity of getting an optimal solution. We give almost tight lower and upper bounds for the DELAYED H-NODE DELETION PROBLEM, where there is one forbidden induced subgraph that may or may not be disconnected and tight bounds on the DELAYED $\mathcal{F}$-NODE DELETION PROBLEM, where we have an arbitrary number of forbidden connected graphs. For the latter result we present an algorithm that computes the advice complexity directly from $\mathcal{F}$. For the DELAYED H-NODE DELETION PROBLEM the advice complexity is basically an easy function of the size of the biggest component in H.

Keywords: Online algorithm · Advice complexity · Node deletion · Edge deletion · Delayed decision model · Graph modification

1 Introduction

Many classical online problems can be formulated as follows: Given an instance $I = \{x_1, \ldots, x_n\}$ as a series of elements ordered from x_1 to x_n, an algorithm receives them iteratively in this order, having to decide whether to include x_i into its solution at the point it receives it. It can base this decision only on the previously revealed $x_1, \ldots, x_{i-1}$ and must neither remove x_i from its solution later nor include any of the previously revealed elements into its solution. A way to measure the performance of such an online algorithm is the *competitive ratio*, which compares how much worse it performs compared to an optimal offline algorithm [4]. An algorithm is c-competitive if the competitive ratio of the algorithm is bounded by a constant c.

In most classical online problems such as the k-SERVER PROBLEM, the PAGING PROBLEM or the KNAPSACK PROBLEM as well as most other online problems, receiving the next x_i of an instance coincides with an algorithm having to process this request. This makes a lot of sense in the previously mentioned

© Springer Nature Switzerland AG 2020
L. Gąsieniec et al. (Eds.): IWOCA 2020, LNCS 12126, pp. 140–153, 2020.
https://doi.org/10.1007/978-3-030-48966-3_11

problems, but arguably less sense when there is no "need to act" after an item of the instance is presented, which may regularly happen in the instances of the problem that we study in this paper: Informally, the requests are single nodes of a graph that are iteratively revealed and our task is to keep the graph induced by these nodes free of a set $\mathcal{F}$ of forbidden induced subgraphs by deleting nodes or edges. Obviously, there are sets and instances in which an arbitrary number of nodes can be revealed before any forbidden induced substructure is revealed. The offline variant of this problem was shown to be NP-Hard by Yannakakis [17].

In this work, we use a modified version, which we call the delayed decision model, which was already used in [16] and which is similar to the preemptive model used by Komm et al. [13]. We consider an instance $I = \{x_1, \ldots, x_n\}$ of an online minimization problem for which a solution $S \subseteq I$ has to satisfy some condition C. Again, an algorithm ALG has to decide whether to include any element into its solution S. We denote the intermediate solution of an algorithm on an instance I at the revelation of element x_i – before the decision on whether to include it in S – by $S_i^I(ALG)$. While in the classical definition, an algorithm has to decide on whether to include an element into its solution at the point of revelation, the algorithm may now wait until the condition C is violated by $S_i^I(ALG)$. It may then include any of the previously revealed elements into its solution, but is still unable to revert any of its previous selections.

A selection of online problems that do not admit any algorithm with a constantly bounded competitive ratio, such as the MINIMUM VERTEX COVER PROBLEM and in particular general node and edge deletion problems are constantly competitive with delayed decision.

A simple example is the online MINIMUM VERTEX COVER PROBLEM. The input I is a series of induced subgraphs $G[\{v_1\}], G[\{v_1, v_2\}], \ldots, G[\{v_1, \ldots, v_n\}]$ for which C states that $S_i^I(ALG)$ *is a vertex cover on* $G[\{v_1, \ldots, v_i\}]$. In this setting, an algorithm has to include nodes into its current solution only once an edge is revealed that is not covered yet. While the MINIMUM VERTEX COVER PROBLEM is competitive in the maximum degree Δ of an input graph in the classical online setting [5], a competitive ratio of 2 can be proven for the delayed decision setting: The upper bound is given by always taking both nodes of an uncovered edge into the solution (this is the classical 2-approximation algorithm). The lower bound can be achieved by presenting an edge $\{v_i, v_j\}$ and adding another edge to either v_i or v_j, depending on which node is not taken into the solution by a deterministic online algorithm. If both nodes are taken into the solution then no additional edge is introduced. This gadget can be repeated and forces a deterministic algorithm to take two nodes into the vertex cover where one suffices.

We denote by H a finite graph and by $\mathcal{F}$ a finite set of finite graphs. For a problem Π we denote the optimal solution size on an input I by $opt_\Pi(I)$.

The competitive ratio is a standard method to analyze online algorithms and a relatively new alternative is the *advice complexity* introduced by Dobrev, Královič, and Pardubská [7], revised by Hromkovič, Královič and Královič [11] and refined by Böckenhauer et al. [2]. The advice complexity measures the amount of information about the future that is necessary to solve an online problem optimally or with a given competitive ratio. There is an oracle called "advisor" that knows the whole input instance and gives the online algorithm "advice" in the form of a binary string that can be read from a special advice tape. Many problems have been successfully analyzed in this model including the k-SERVER PROBLEM [8], the KNAPSACK PROBLEM [3], JOB-SHOP SCHEDUL-ING [1] and many more. One criticism on the advice model is that in the real world such a powerful advisor usually cannot exist. However, the new research area of *learning-augmented algorithms* uses an AI-algorithm to guide classical algorithms to solve optimization problems and they are closely related to the advice complexity [14,15]. A strong application of advice complexity are the lower bounds it provides: For example, the online knapsack problem can be solved with a competitive ratio of two by a randomized algorithm. It has been shown that this competitive ratio cannot be improved with $o(\log n)$ advice bits.

We base our work on the definitions of advice complexity from [12] and [2], with a variation due to the modified online model we are working on: The length of the advice string is often measured as a function in the input length n, which usually almost coincides with the number of decisions an online algorithm has to make during its run. In the delayed decision model, the number of decisions may be smaller than n by a significant amount and we can measure the advice as $f(opt_\Pi(I))$, i.e., a function of the size of the optimum solution. This usually does not work in classical online algorithms.

Tight results for the advice complexity of the DELAYED CONNECTED $\mathcal{F}$-NODE DELETION PROBLEM and of the DELAYED CONNECTED H-EDGE DELETION PROBLEM were shown in [16]. We show upper and lower bounds for the general DELAYED H-NODE DELETION PROBLEM and a tight bound for the DELAYED CONNECTED $\mathcal{F}$-EDGE DELETION PROBLEM. We leave open the exact advice complexity for the general DELAYED $\mathcal{F}$-NODE DELETION PROBLEM and DELAYED $\mathcal{F}$-EDGE DELETION PROBLEM, for which we can only provide lower bounds. Some proofs can be found only in the full version of this paper.

2   The $\mathcal{F}$-Node Deletion Problem and $\mathcal{F}$-Edge Deletion Problem Without Advice

For a graph $G = (V, E)$ we write $|G|$ to denote $|V(G)|$ and $||G||$ to denote $|E(G)|$. We use the symbol $\unlhd$ to denote an induced subgraph relation, i.e. $A \unlhd B$ iff A is an induced subgraph of B. We write $\mathcal{G}$ to denote the set of all graphs.

We write $G - U$ for $G[V(G) - U]$ and $G - u$ for $G - \{u\}$ and also use $G - E$ similarly for an edge set E. For graphs H and G we write $H \unlhd_\varphi G$ if there exists an isomorphism φ such that $\varphi(H) \unlhd G$. We call a set of graphs *unordered* if the members are pairwise maximal according to the induced subgraph relation $\unlhd$. It is easy to see that every DELAYED $\mathcal{F}$-NODE DELETION PROBLEM can be reduced to one with an unordered $\mathcal{F}$. A graph G is called $\mathcal{F}$-free if there is no $H_i \unlhd_\varphi G$ for any $H_i \in \mathcal{F}$.

Definition 1. Let $\mathcal{F}$ be an unordered set of graphs. Let I be a sequence of growing induced subgraphs $G[\{v_1\}], \ldots, G[\{v_1, \ldots, v_n\}]$. The $\mathcal{F}$-NODE DELETION PROBLEM is to delete a minimum size set of nodes S from G such that $G - S$ is $\mathcal{F}$-free. We call $S_i^I \subseteq \{v_1, \ldots, v_i\}$ an (intermediate) solution for the $\mathcal{F}$-NODE DELETION PROBLEM on $G[\{v_1, \ldots, v_i\}]$ if $G[\{v_1, \ldots, v_i\}] - S_i^I$ is $\mathcal{F}$-free.

The DELAYED $\mathcal{F}$-NODE DELETION PROBLEM is defined accordingly, with the condition C stating *The Graph* $G[\{v_1, \ldots, v_i\} - S_i^I(ALG)]$ *is $\mathcal{F}$-free* for all $i \in \{1, \ldots, n\}$ and some algorithm ALG. $\mathcal{F}$-EDGE DELETION and DELAYED $\mathcal{F}$-EDGE DELETION are defined accordingly, with the solution being a set of edges. The graph is always revealed as a sequence of nodes. We will denote the DELAYED $\mathcal{F}$-NODE DELETION PROBLEM for $\mathcal{F} = \{H\}$ as the DELAYED H-NODE DELETION PROBLEM.

Lemma 1. *There is at least one $\mathcal{F}$ for which the $\mathcal{F}$-NODE DELETION PROBLEM is not c-competitive for any constant c.*

Lemma 1 is not surprising. It generalizes that VERTEX COVER admits no constantly bounded competitive ratio [5].

Lemma 2. *There is at least one $\mathcal{F}$ for which the $\mathcal{F}$-EDGE DELETION PROBLEM is not c-competitive for any constant c.*

Lemma 3. *The DELAYED $\mathcal{F}$-NODE DELETION PROBLEM is k-competitive for $k = \max_{H \in \mathcal{F}}\{|H|\}$. The DELAYED $\mathcal{F}$-EDGE DELETION PROBLEM is k-competitive for $k = \max_{H \in \mathcal{F}}\{\|H\|\}$.*

Proof. Whenever an algorithm finds an induced H, it deletes all of its nodes, resp. edges. $\qquad\square$

3   The Delayed H-Node Deletion Problem with Advice

If $\mathcal{F}$ consists of connected subgraphs, tight results have already been proven in [16]. The advice complexity is exactly $opt_{\mathcal{F}}(G) \log(|H|) + O(1)$ for a biggest graph $H \in \mathcal{F}$. The problem becomes harder when the graphs in $\mathcal{F}$ are disconnected and was left as an open question. We answer it partially by determining the advice complexity for the DELAYED H-NODE DELETION PROBLEM, where H can be disconnected.

Definition 2. Let $C_G = \{C_1, C_2, \ldots, C_j\}$ denote the set of components of G.

If a forbidden graph H is disconnected, it may contain multiple copies of the same component, e.g., three disjoint triangles among other components. If we were only to delete triangles, we would thus have to delete all but two copies to make the graph of an instance H-free. We introduce some notation to determine the number and the actual copies of a *type* of component.

Definition 3. Given a graph G. For a connected graph C we define the packing $p_C(G)$ of C in G as the set of sets of pairwise node-disjoint copies of C in G and the packing number of C in G, $\nu_C(G)$, as $\max_{H \in p_C(G)}(|H|)$.

In other words, $\nu_C(G)$ is the maximal number of C's that can be packed node-disjointly into G.

We use the multiplicity of components in H in a lower bound that forces any algorithm to leave specific components such as the two specific triangles in our small example. To punish a wrong selection, we use a *redundancy construction* that maps a component C into a C' such that $C \trianglelefteq C'$ and even $C \trianglelefteq C' - \{v\}$ for every v holds, while C' does not contain two disjoint copies of C.

Definition 4. We call the graph H' a redundancy construction of a connected graph H with $|H| > 1$ if there exists an isomorphism $\varphi_1 \colon \mathcal{G} \to \mathcal{G}$ such that for every isomorphism $\varphi_2 \colon \mathcal{G} \to \mathcal{G}$ the following holds:

- $\varphi_1(H) \trianglelefteq H' - v$ for all $v \in V(H')$
- $\varphi_1(H) \ntrianglelefteq H' - V(\varphi_2(H))$ if $V(\varphi_2(H)) \subseteq V(H')$

To show that such a redundancy construction actually exists, we use the following transformation.

Definition 5. Given a connected graph $H = (V, E)$ with $V = \{v_1, \ldots, v_n\}$, $n > 1$, in some order and some $k \in [2, n]$ s.t. $(v_1, v_k) \in E(H)$. H' is then constructed in the following way: $V(H') = V(H) \cup \{v_i' \mid v_i \in V(H), i \geq 2\}$ and $E(H') = E(H) \cup \{(v_i', v_j') \mid (v_i, v_j) \in E(H), v_i', v_j' \in V(H')\} \cup \{(v_1, v_i') \mid (v_1, v_i) \in E(H)\} \cup \{(v_k, v_j') \mid (v_1, v_j) \in E(H)\}$

Intuitively, we create a copy of H except for a single node v_1. The copied neighbors of v_1 are then connected with v_1. Lastly, some copied node is chosen and connected with the original neighbors of v_1.

Example 1. A graph H and its redundancy construction H':

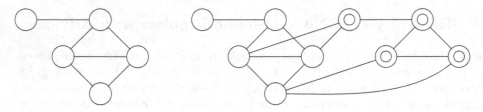

Lemma 4. *The transformation in Definition 5 is a redundancy construction.*

We denote an optimal solution of the DELAYED H-NODE DELETION PROBLEM on a graph G by $sol_H(G)$.

3.1 Lower Bound

Theorem 1. *Let H be a graph. Let C_{max} be a component of H of maximum size. Any online algorithm optimally solving the* DELAYED H-NODE DELETION *PROBLEM uses at least* $opt_H(G) \cdot \log |V(C_{max})| + (\nu_{C_{max}}(H) - 1) \cdot \log(opt_H(G))$ *many advice bits on input G.*

Proof. Let $C_H = \{C_1, \ldots, C_j\}$ and $|V(C_1)| \leq \ldots \leq |V(C_j)|$. The adversary first presents $k \geq \max\{\nu_{C_i}(H) \mid C_i \in C_H\}$ disjoint copies of each $C_i \in C_H$ in an iterative way such that in each iteration one copy of each C_i is revealed node by node. If an algorithm deleted any nodes before an H is completed, the adversary would simply stop and the algorithm would not be optimal.

As soon as G is no longer H-free, any algorithm has to delete some node(s). For a $C_i \in C_H$ it can either delete all C_i except for $\nu_{C_i}(H) - 1$ occurrences and optionally some additional node(s). Obviously, deleting an additional node is not optimal, as the adversary would simply stop presenting nodes.

The following strategy will force an optimal online algorithm always to delete copies of C_{max}. After all k copies of all $C_i \in C_H$ are presented, additionally $\max_{C_i \in C_H}\{\nu_{C_i}(H)\} - \nu_{C_{max}}(H) + 1$ copies of each $C_i \in C_H \setminus C_{max}$ are presented. Deleting all C_{max} except for $\nu_{C_{max}}(H) - 1$ occurrences will thus only need $k - \nu_{C_{max}}(H) + 1$ deletions, while deleting any other component will need at least $k - \max_{C_i \in C_H}\{\nu_{C_i}(H)\} + 1 + \max_{C_i \in C_H}\{\nu_{C_i}(H)\} - \nu_{C_{max}}(H) + 1 = k - \nu_{C_{max}}(H) + 2$ deletions. Thus, it is always optimal for any algorithm to focus on C_{max} for deletion.

After all components have been revealed - and some deletion(s) had to be made - a redundancy construction such as the one from Definition 5 is used in order to *repair* an arbitrary set of $\nu_{C_{max}}(H) - 1$ copies of C_{max}. Every optimal algorithm will leave exactly $\nu_{C_{max}}(H) - 1$ copies of C_{max} after G is completely revealed. There are $\binom{opt_H(G) + \nu_{C_{max}}(H) - 1}{opt_H(G)}$ many different ways to distribute the affected components onto all components and an algorithm without advice cannot distinguish them. In particular, each of these instances is part of a different, unique optimal solution, which deletes a node from all but the $\nu_{C_{max}}(H) - 1$ subgraphs. If an algorithm has chosen to delete a node from a component that is affected by the redundancy construction, this component is now *repaired* and demands an additional deletion. By definition, applying the redundancy construction does not result in additional disjoint copies of C_{max}. Thus, it is still optimal to focus on C_{max} for deletion.

Finally, for every component that is not affected by a redundancy construction, the adversary glues a copy of C_{max} to one of its nodes as defined in [16]. It has $|V(C_{max})|$ ways to do so for each copy of C_{max}. Intuitively, the glueing operation joins two graphs by identifying a single node from both and connecting them by joining these two nodes into one.

We now measure how much advice an algorithm needs at least. First of all, it should be easy to see that the adversary is able to present $|V(C_{max})|^{opt_H(G)}$ many different instances regarding the deletion of nodes for the copies of C_{max} not selected for the redundancy construction.

Assuming $\nu_{C_{max}}(H) > 1$, any algorithm needs to determine the correct subset of $opt_H(G)$ components out of $k - 1$ presented ones to delete one node from. As the adversary has $\binom{opt_H(G) + \nu_{C_{max}}(H) - 1}{opt_H(G)}$ different ways to distribute these redundancies and since every single of these instances has a different unique optimal solution, any correct algorithm has to get advice on the complete distribution

---

Algorithm 1. Upper Bound: DELAYED H-NODE DELETION PROBLEM

1: Input: Online graph G with $V(G) = \{v_1, \ldots, v_n\}$, H
2: Advisor computes $C_{min\nu} \in \arg\min_{C \in C_H} \{\nu_C(G) - \nu_C(H)\}$
3: Advisor computes $C_{min} \in \arg\min_{C \in C_{min\nu}} \{|V(C)|\}$
4: Advisor computes L, the list of labels marked for keeping
5: Read advice: C_{min} ▷ Which $C_{min} \in C_H$ to delete
6: Read advice: List L of numbers in range $[1, opt_H(G) + O(1)]$
7: $k \leftarrow 1$
8: Define $l : \mathcal{G} \rightarrow \mathbf{N}$, $l(G) = 0$
9: Define $labeled : \mathcal{G} \rightarrow \{0,1\}$, $labeled(G) = 1$ iff $l(G) \neq 0$, otherwise $labeled(G) = 0$
10: **for all** $i \in \{1, \ldots, n\}$ **do**
11:     $G_i \leftarrow G[v_1, \ldots, v_i]$ ▷ Reveal next node
12:     **if** $\nu_{C_{min}}(G_i) \geq \nu_{C_{min}}(H)$ **then**
13:         $W \leftarrow \arg\max_{P \in p_{C_{min}}(G_i)} |P|$ ▷ Biggest Packings
14:         $\mathcal{H} \leftarrow \arg\max_{P \in W} \sum_{g \in P} labeled(g)$ ▷ Most labels
15:         Select $P \in \mathcal{H}$ ▷ Arbitrary set
16:         **for all** $C \in P$ **do**   ▷ Label everything unlabeled
17:             **if** $l(C) = 0$ **then** $l(C) \leftarrow k$; $k \leftarrow k + 1$
18:         $S \leftarrow \{C \in P \mid l(C) \notin L\}$ ▷ Select everything not marked for keeping
19:         **for all** $C \in S$ **do**
20:             Read advice: Which $v \in V(C)$ to delete
21:             Delete v out of G_i

---

in the size of at least $\log \left({opt_H(G) + \nu_{C_{max}}(H) - 1 \atop opt_H(G)} \right) \geq (\nu_{C_{max}}(H) - 1) \cdot \log(opt_H(G))$ advice bits.   □

3.2 Upper Bound

For simplicity of writing down the algorithm, we will assume in this section that we are only ever presented graphs which induce at least one forbidden subgraph H. Our algorithm can be easily transformed into one that only starts to read any advice once the first forbidden subgraph is completely revealed. For an instance with an online graph G with $V(G) = \{v_1, \ldots, v_n\}$ and a forbidden subgraph H, the advisor first computes the advice the algorithm is going to read during its run. It first identifies the set of components $C_{min\nu}$ which each require the fewest node deletions in G to make the graph H-free. Of these possible components, the advisor chooses the component with the fewest nodes which an optimal offline algorithm would choose, named C_{min} from here on. Finally, the advisor computes a list L of labels which will coincide with labels given by the algorithm to copies of C_{min} which are not to be deleted in an optimal solution. As there are at most $\nu_{C_{min}}(H) \cdot opt_H(G)$ node-disjoint copies of H in G and as Lemma 7 states that our algorithm uses at most $opt_H(G) + O(1)$ labels, we can limit the range of possible labels by $[1, opt_H(G) + O(1)]$. Finally, a number of advice bits is written for every deletion that the algorithm will make which encode the concrete node out of a copy of C_{min} is optimal to delete.

The algorithm starts by reading from the advice tape which component C_{min} to focus on for deletion and the list L, using self-delimiting encoding.

Whenever the next node x_i of the instance is revealed which fulfills $\nu_{C_{min}}(G_i) \geq \nu_{C_{min}}(H)$, i.e. that there are at least as many node-disjoint copies of C_{min} in the current graph as in H, the algorithm will delete nodes from the graph as described in the following, otherwise the algorithm simply waits for the next node to be revealed.

To identify which node(s) of G_i are to be deleted, the algorithm first identifies all biggest sets of node-disjoint copies of C_{min}. Of them it identifies a set P of which the most components have already received a label. Then all previously unlabeled copies of $C_{min} \in P$ receive a new unique label. The algorithm now looks at the label list L given by the advisor. Every copy of $C_{min} \in P$ whose label is not in L is now marked for deletion. The algorithm reads advice which concrete node out of every copy of C_{min} is optimal to delete.

Lemma 5. *Algorithm 1 is correct.*

Lemma 6. *Algorithm 1 is optimal.*

Definition 6. Given graphs G, H and a labeling function $l\colon \mathcal{G} \to \mathbf{N}$. We call a family $\mathcal{C}$ of induced subgraphs of G a *configuration*, if every element of $\mathcal{C}$ is isomorphic to H, $l(C) \neq 0$ for each $C \in \mathcal{C}$ and $V(C_1) \cap V(C_2) = \emptyset$ for all $C_1, C_2 \in \mathcal{C}$, $C_1 \neq C_2$. The *size* of a configuration is the number of induced subgraphs it contains.

Informally speaking, a configuration is a set of disjoint induced subgraphs of G that already have a label.

Lemma 7. *Given an online graph G, a forbidden graph H, as well as a subgraph $C \in C_H$ of which there may be at most $k = \nu_C(H) - 1$ disjoint copies present in G. Algorithm 1 assigns no more than $opt_H(G) + O(1)$ labels to G if the advisor assigns $C_{min} = C$ as specified in line 5.*

Theorem 2. *Let H be a graph. Let $C_{min\nu} = arg\,min_{C \in C_H}\{\nu_C(G) - \nu_C(H)\}$ and $C_{min} = arg\,min_{C \in C_{min\nu}}\{|V(C)|\}$. The* DELAYED H-NODE DELETION PROBLEM *can be solved optimally using at most $opt_H(G) \cdot \log|V(C_{min})| + O(\log opt_H(G))$ many advice bits on input G.*

Proof. We count the number of advice bits used by Algorithm 1. We know by Lemma 5 and 6 that it is correct and optimal. The advice in line 5 is of constant size. As L only contains the labels for components which are not to be deleted and we limited the number of them by a constant in Lemma 7, only $O(\log opt_H(G))$ advice, using self-delimiting encoding, is needed in line 6.

Finally, the algorithm reads advice on which node of each copy of C_{min} that is part of $sol_H(G)$ to delete in line 21. This can be done using $opt_H(G) \cdot \log|V(C_{min})|$ advice bits in total.   □

4   The Delayed Connected $\mathcal{F}$-Edge Deletion Problem

Let $(d_1,\ldots,d_k) \in \mathbf{N}^k$. Let $m(n)$ be the solution to the recurrence relation

$$m(n) = \begin{cases} \sum_{i=1}^{k} m(n - d_i) & \text{if } n \geq \max\{d_1,\ldots,d_k\} \\ c_n & \text{otherwise} \end{cases}$$

where $c_n \geq 0$ and some $c_i > 0$ for $0 \leq i < \max\{d_1,\ldots,d_k\}$. Let $\beta(d_1,\ldots,d_k) = \inf_\tau\{\tau \mid m(n) = O(\tau^n)\}$. Note that β does not depend on the c_i's.

If $S = \{D_1,\ldots,D_k\}$ is a set of sets, then we define $\beta(S) = \beta(|D_1|,\ldots,|D_k|)$.

A homogeneous linear recurrence relation with constant coefficients usually has a solution of the form $\Theta(n^{k-1}\tau^n)$ if τ is the dominant singularity of the characteristic polynomial with multiplicity k [9]. However, here the coefficients of the characteristic polynomial are real numbers and there is exactly one sign change. By Descartes' rule of signs there is exactly one positive real root and therefore its multiplicity has to be one [6,10]. Therefore $m(n) = \Theta(\beta(S)^n)$.

Definition 7. Let $\mathcal{F}$ be a set of forbidden connected induced subgraphs and $H \in \mathcal{F}$. Let $S \subseteq 2^{E(H)}$.

1. A set $D \subseteq E(H)$ is H-optimal for a graph G if $H \trianglelefteq G$ and $G - D$ is $\mathcal{F}$-free and $opt_{\mathcal{F}}(G) = |D|$.
2. A set $D \subseteq E(H)$ is H-good for a graph G if $H \trianglelefteq G$ and D is a non-empty subset of some $\bar{D} \subseteq E(G)$ where $opt_{\mathcal{F}}(G) = |\bar{D}|$ and $G - \bar{D}$ is $\mathcal{F}$-free.
3. S is H-sound if $H - D$ is $\mathcal{F}$-free for every $D \in S$.
4. S is H-sufficient if for every connected graph G with $H \trianglelefteq G$ there is a $D \in S$ such that D is H-good for G.
5. S is H-minimal if for every $D \in S$, there is a graph G such that D is H-good for G, but every $D' \in S$, $D' \neq D$ is not.

Lemma 8. Let $\mathcal{F} = \{H_1,\ldots,H_k\}$ be a set of connected graphs, G a graph and $D \subseteq E(H_i)$ that is H_i-good for G. Then there is a subgraph $G' \subseteq G$ such that D is H_i-optimal for G'.

4.1   Upper Bound

Theorem 3. Let $\mathcal{F} = \{H_1,\ldots,H_k\}$ be a set of connected graphs and let S_i be H_i-sound and H_i-sufficient for all $i \in \{1,\ldots,k\}$. Then there is an $m \in \mathbf{R}$ and an algorithm that solves the DELAYED CONNECTED $\mathcal{F}$-EDGE DELETION PROBLEM for every graph G with $m \cdot opt_{\mathcal{F}}(G) + O(1)$ many advice bits where $2^m \leq \beta(S_i)$ for all $i \in \{1,\ldots,k\}$.

Proof. The algorithm receives $opt_{\mathcal{F}}(G) \cdot \log(\max_i\{\beta(S_i)\}) + O(1)$ many advice bits and then a graph G as a sequence of growing induced subgraphs. The algorithm interprets the advice as a number that can be between 0 and $O((\max_i\{\beta(S_i)\})^{opt_{\mathcal{F}}(G)})$.

The algorithm will delete in total exactly $opt_{\mathcal{F}}(G)$ edges. We analyze the total number of different advice strings the algorithm might use when deleting $opt_{\mathcal{F}}(G)$ edges.

When the algorithm receives a new node and its incident edges to form the next graph G it proceeds as follows: While G is not $\mathcal{F}$-free, choose some $H_i \in \mathcal{F}$ for which $H_i \trianglelefteq_{\varphi} G$. The advisor chooses one $D \in S_i$ for which $\varphi(D)$ is $\varphi(H_i)$-good for the graph at hand and puts it in the advice.

The advice strings are therefore partitioned into $|S_i|$ subsets, one for each $D \in S_i$. After deleting $\varphi(D)$ the algorithm proceeds on the graph $G - \varphi(D)$, where $opt_{\mathcal{F}}(G)$ is now by $|D|$ smaller. If $m(opt_{\mathcal{F}}(G))$ is the total number of advice strings we get the recurrence $m(opt_{\mathcal{F}}(G)) = \max_i(\sum_{D \in S_i} m(opt_{\mathcal{F}}(G) - |D|))$ if $opt_F(G)$ is at least as big as every $D \in S_i$. Standard techniques show that $m(opt_{\mathcal{F}}(G)) = O(\max\{\beta(S_1), \ldots, \beta(S_k)\}^{opt_{\mathcal{F}}(G)})$. $\qquad\square$

4.2   Lower Bound

Let $\mathcal{F} = \{H_1, \ldots, H_k\}$ be a set of connected graphs. We fix some correct algorithm A for the DELAYED CONNECTED $\mathcal{F}$-EDGE DELETION PROBLEM.

We define the sets $S_i = S_i(A)$ for $i = 1, \ldots, k$ as follows: $D \in S_i$ if and only if there is some input sequence $G_1, G_2, \ldots, G_t$ such that algorithm A deletes the edge set D' from G_t. Moreover, there is a set X and an isomorphism φ such that $G[X] \cong H_i$, $\varphi \colon V(H) \to X$, and $\varphi(D) = D' \cap E(G[X])$. Informally speaking, the edge sets in S_i are those that are deleted from some isomorphic copy of H_i by algorithm A in some scenario.

We will need the following technical lemma. It states that we can find a matching with special properties in every connected bipartite graph. The matching should have the following properties. Let U' be the partners in the matching on top and V' on the bottom.

The first property is $N(U') = V$, i.e., every node in V is connected to at least one node in U'. The second property states that we have an *induced* matching, i.e., that the graph induced by $U' \cup V'$ *is* a matching. The third property concerns the vertices in V': If $v \in V'$ then $N(v)$ contains several vertices from U, but exactly one node in U', i.e., its partner in the matching. We require that this partner is the *smallest* one in $N(v)$.

Lemma 9. *Let $G = (U, V, E)$ be a bipartite graph where $U = \{u_1, \ldots, u_k\}$. Let $\leq$ be a preorder on U such that $u_1 \leq \cdots \leq u_k$. Moreover, assume that $V \subseteq N(U)$, i.e., every node in V is connected to some node in U. Then there is a $U' \subseteq U$ and $V' \subseteq V$ such that*

1. *$N(U') = V$,*
2. *$G[U' \cup V']$ is a matching,*
3. *$\min N(v) \in U'$ for every $v \in V'$.*

Lemma 10. *Let $\mathcal{F} = \{H_1, \ldots, H_k\}$ be a set of connected graphs and S_i be H_i-sound and H_i-sufficient for all $i \in \{1, \ldots, k\}$. Then there are $S_i' \subseteq S_i$ such that S_i' is H_i-sound, H_i-sufficient and H_i-minimal and moreover:*

For every $D' \in S_i'$ there is a graph G with $H_i \trianglelefteq G$ such that D' is H_i-good for G and for every $D \in S_i \setminus S_i'$ that is also H_i-good for G, it holds that $|D| \geq |D'|$.

Theorem 4. *Let $\mathcal{F} = \{H_1, \ldots, H_k\}$ be a set of connected graphs and assume that there is an algorithm A that can solve the* DELAYED CONNECTED $\mathcal{F}$-EDGE DELETION PROBLEM *for all inputs G with at most $m \cdot opt_{\mathcal{F}}(G) + O(1)$ advice for some $m \in \mathbf{R}$. Then there exist S_i' that are H_i-sound, H_i-sufficient, and H_i-minimal and $\beta(S_i') \leq 2^m$ for every $i \in \{1, \ldots, k\}$.*

Proof. By Lemma 10 there is an $S_i' = \{D_1, \ldots, D_r\} \subseteq S_i$ that is H_i-sound, H_i-sufficient, and H_i-minimal. It additionally has the property that for every $D' \in S_i'$ there is a graph G with $H_i \trianglelefteq G$ such that D' is H_i-good for G and for every $D \in S_i \setminus S_i'$ that is also H_i-good for G, it holds that $|D| \geq |D'|$.

Let $l \in \mathbf{N}$. The adversary prepares $\Theta(\beta(S')^l)$ many instances by repeating the following procedure until the size of the optimum solution for the presented graph exceeds $l - \max\{|D_1|, \ldots, |D_r|\}$.

1. The adversary presents a disjoint copy of H_i.
2. Then the adversary computes an induced supergraph G_j of H_i for which D_j is H_i-good, but all $D_{j'} \in S_i'$ with $j' \neq j$ are not H_i-good, for all $1 \leq j \leq r$. The existence of the graph G_j is guaranteed by the H_i-minimality of S_i'. In particular there is a $\bar{D}_j \supseteq D_j$ such that $\bar{D}_j$ is H_i-optimal for G_j. Let $D_j' = \bar{D}_j - D_j$. Let $G_j' = G_j - D_j'$. It is easy to see that D_j is H_i-optimal for G_j'.

We show that no other $D_{j'} \in S_i'$ is H_i-good for G_j'. Assume otherwise. If $D_{j'}$ is H_i-good for G_j' then there must be a $\bar{D}_{j'} \supseteq D_{j'}$ that is H_i-optimal for G_j'. Then $G_j - D_{j'} - ((\bar{D}_{j'} - D_{j'}) \cup D_j')$ is $\mathcal{F}$-free. This implies that $D_{j'}$ is H_i-good for G_j contradicting the H_i-minimality of S_i'. Next the adversary transforms the H_i into one of the r possible G_j's and presents the new vertices. Then $opt_{\mathcal{F}}(G_j') = |D_j|$. Hence, the optimal solution size increases by $|D_j|$.

In each round the input graph grows and the optimal solution size grows by $|D_j|$. As soon as that size exceeds $l - \max\{|D_1|, \ldots, |D_r|\}$ the adversary keeps presenting disjoint copies of H_i without turning them into bigger connected graphs until the size reaches exactly l. The number $N(l)$ of different instances is given by the following recurrence:

$$N(l) = \begin{cases} \sum_{j=1}^r N(l - |D_j|) & \text{if } l \geq \max\{|D_1|, \ldots, |D_r|\} \\ 1 & \text{otherwise} \end{cases}$$

It is easy to see that $N(l) = \Theta(\beta(S_i')^l)$. The algorithm has to react differently on all of these instances: When the algorithm sees a new H_i to be turned into one of $G_1', \ldots, G_r'$, it deletes different edge sets for each of the r possibilities.

The adversary constructed an instance that consists of a sequence of disjoint graphs $G_{i_1}', \ldots, G_{i_t}'$ from the set $\{G_1', \ldots, G_r'\}$ of which the total size is at least $\sum_{j=1}^t opt_{\mathcal{F}}(G_{i_j}) - \max\{|D_1|, \ldots, |D_r|\}$ and $O(1)$ many copies of H_i. If G is the whole constructed instance we have $opt_{\mathcal{F}}(G) = l + O(1)$ because $opt_F(H_i)_{\mathcal{F}} = O(1)$. Together with $N(l) = \Theta(\beta(S'_i)^l)$ this means that Algorithm A uses at least $\log N(l) = l \cdot \log \beta(S'_i) + O(1) = opt_{\mathcal{F}}(G) \log \beta(S'_i) + O(1)$ advice bits. Assume Algorithm A uses at most $m \cdot opt_{\mathcal{F}}(G) + O(1)$ advice bits on every graph G as stated in the precondition above. Then m cannot be smaller than $\log \beta(S'_i)$ for every $i \in \{1, \ldots, k\}$ because $opt_{\mathcal{F}}(G)$ can be become arbitrarily big. $\square$

Lemma 11. *Let $\mathcal{F}$ be a set of connected forbidden graphs, $H \in \mathcal{F}$, and $S \subseteq 2^{E(H)}$. There is an algorithm that can decide whether S is H-sufficient.*

Proof. It is sufficient to verify for all connected graphs G with $H \trianglelefteq G$ that some $D \in S$ is H-good for G, i.e., there is an optimal solution for G that contains D. By Lemma 8 we can restrict our search to all such G's that have an optimal solution that is a subset of $E(H)$. There are infinitely many graphs G to check. To overcome this we define the *unfolding* of G, written $\Upsilon(G)$, as the set of the following graphs: Remember that $H \trianglelefteq G$. If there is some $H' \in \mathcal{F}$ with $H' \trianglelefteq_{\varphi} G$ then $G[E(H) \cup E(\varphi(H'))] \in \Upsilon(G)$ (for every possible φ). If, however, $\Upsilon(G)$ contains two graphs G' and G'' that are isomorphic via an isomorphism that is the identity on $V(H)$, then only the lexicographically smaller one is retained.

This means that the unfolding of G contains all induced subgraphs that consist of H and one other copy of some forbidden induced subgraph from $\mathcal{F}$ that must overlap with H in some way (because we assumed that G has an optimal solution that consists solely of edges from H). Here is a small example: Let $\mathcal{F} = \{\square, \triangle\}$, $H = \{\square\}$, $G =$ ⬡⊳∘. Then $\Upsilon(G) = \{\square, \lhd\!\square, \square\!\rhd, \boxminus\}$.

It is easy to see that deleting some $D \subseteq E(H)$ from G makes it $\mathcal{F}$-free iff deleting the same D from all graphs $G' \in \Upsilon(G)$ makes all these G' $\mathcal{F}$-free. Hence, there is an optimal solution for G that is a subset of $E(H)$ iff there is such a subset that is "optimal" for $\Upsilon(G)$ (i.e., deletion of no smaller edge set can make all graphs in $\Upsilon(G)$ $\mathcal{F}$-free).

There are only finitely possibilities for $\Upsilon(G)$ and we can enumerate all of them. Let us say this enumeration is $\Upsilon_1, \ldots, \Upsilon_t$. For each Υ_i we first find out, whether there is a G with $\Upsilon(G) = \Upsilon_i$. We can do this by enumerating all graphs G up to a size that does not exceed the sum of the sizes of all graphs in Υ_i and computing $\Upsilon(G)$ for them. If indeed $\Upsilon(G) = \Upsilon_i$ then we test whether S is H-good for G. Iff these tests pass for all i then S is indeed H-sufficient. $\square$

Theorem 5. *Let $\mathcal{F} = \{H_1, \ldots, H_k\}$ be connected graphs. The advice complexity for* DELAYED CONNECTED $\mathcal{F}$-EDGE DELETION *is $m \cdot opt_{\mathcal{F}}(G) + O(1)$ where $m = \max_{i \in \{1, \ldots, k\}} \min\{\log \beta(S) \mid S \subseteq 2^{E(H)}, S$ is H_i-sound and H_i-sufficient$\}$. There is an algorithm that can compute m from $\mathcal{F}$. More specifically, there is an algorithm that gets $\mathcal{F}$ and $l \in \mathbf{N}$ as the input and returns the tth bit of the binary representation of m.*

Proof. "$\leq$" by Theorem 3. "$\geq$" by Theorem 4. An algorithm can enumerate all possible $S \subseteq E(H)$ and then test if S is H_i-sound and H_i-sufficient (by Lemma 11). Then $\beta(S)$ is computed by finding the only real root of the characteristic polynomial of the corresponding recurrence relations [9]. $\square$

Acknowledgement. We like to thank Ratislav Královič for helping significantly to simplify the proof of Lemma 7.

References

1. Böckenhauer, H., Komm, D., Královic, R., Královic, R., Mömke, T.: On the advice complexity of online problems. In: Algorithms and Computation, 20th International Symposium, ISAAC 2009, Honolulu, Hawaii, USA, 16–18 December 2009, Proceedings, pp. 331–340 (2009). https://doi.org/10.1007/978-3-642-10631-6_35
2. Böckenhauer, H., Komm, D., Královic, R., Královic, R., Mömke, T.: Online algorithms with advice: the tape model. Inf. Comput. **254**, 59–83 (2017). https://doi.org/10.1016/j.ic.2017.03.001
3. Böckenhauer, H., Komm, D., Královic, R., Rossmanith, P.: The online knapsack problem: advice and randomization. Theoret. Comput. Sci. **527**, 61–72 (2014). https://doi.org/10.1016/j.tcs.2014.01.027
4. Borodin, A., El-Yaniv, R.: Online Computation and Competitive Analysis. Cambridge University Press, Cambridge (1998)
5. Demange, M., Paschos, V.T.: On-line vertex-covering. Theoret. Comput. Sci. **332**(1–3), 83–108 (2005). https://doi.org/10.1016/j.tcs.2004.08.015
6. Descartes, R.: Discours de la methode pour bien conduire sa raison, et chercher la verité dans les sciences. Plus la Dioptriqve. Les Meteores. Et la Geometrie. - Qui sont des essais de cete Methode. De l'Imprimerie de Ian Maire (1637)
7. Dobrev, S., Královic, R., Pardubská, D.: Measuring the problem-relevant information in input. ITA **43**(3), 585–613 (2009). https://doi.org/10.1051/ita/2009012
8. Emek, Y., Fraigniaud, P., Korman, A., Rosén, A.: Online computation with advice. Theoret. Comput. Sci. **412**(24), 2642–2656 (2011). https://doi.org/10.1016/j.tcs.2010.08.007
9. Greene, D.H., Knuth, D.E.: Mathematics for the Analysis of Algorithms, 3rd edn. Birkhäuser, Boston (1990)
10. Henrici, P.: Applied and Computational Complex Analysis, vol. 1. Wiley, New York (1988)
11. Hromkovič, J., Královič, R., Královič, R.: Information complexity of online problems. In: Hliněný, P., Kučera, A. (eds.) MFCS 2010. LNCS, vol. 6281, pp. 24–36. Springer, Heidelberg (2010). https://doi.org/10.1007/978-3-642-15155-2_3
12. Komm, D.: An Introduction to Online Computation - Determinism, Randomization. Advice Texts in Theoretical Computer Science. An EATCS Series. Springer, Cham (2016). https://doi.org/10.1007/978-3-319-42749-2
13. Komm, D., Královic, R., Královic, R., Kudahl, C.: Advice complexity of the online induced subgraph problem. In: Faliszewski, P., Muscholl, A., Niedermeier, R. (eds.) 41st International Symposium on Mathematical Foundations of Computer Science, MFCS 2016, 22–26 August 2016 - Kraków, Poland, LIPIcs, vol. 58, pp. 59:1–59:13. Schloss Dagstuhl - Leibniz-Zentrum für Informatik (2016). https://doi.org/10.4230/LIPIcs.MFCS.2016.59
14. Lykouris, T., Vassilvitskii, S.: Competitive caching with machine learned advice. In: Proceedings of the 35th International Conference on Machine Learning, ICML 2018, Stockholmsmässan, Stockholm, Sweden, 10–15 July 2018, pp. 3302–3311 (2018)
15. Purohit, M., Svitkina, Z., Kumar, R.: Improving online algorithms via ML predictions. In: Advances in Neural Information Processing Systems, vol. 31, pp. 9684–9693 (2018)

16. Rossmanith, P.: On the advice complexity of online edge- and node-deletion problems. In: Adventures Between Lower Bounds and Higher Altitudes - Essays Dedicated to Juraj Hromkovič on the Occasion of His 60th Birthday, pp. 449–462 (2018). https://doi.org/10.1007/978-3-319-98355-4_26
17. Yannakakis, M.: Node- and edge-deletion np-complete problems. In: Lipton, R.J., Burkhard, W.A., Savitch, W.J., Friedman, E.P., Aho, A.V. (eds.) Proceedings of the 10th Annual ACM Symposium on Theory of Computing, 1–3 May 1978, San Diego, California, USA, pp. 253–264. ACM (1978). https://doi.org/10.1145/800133.804355

Fair Packing of Independent Sets

Nina Chiarelli[1], Matjaž Krnc[2], Martin Milanič[1], Ulrich Pferschy[3], Nevena Pivač[1], and Joachim Schauer[3,4(✉)]

[1] FAMNIT and IAM, University of Primorska, Koper, Slovenia
nina.chiarelli@famnit.upr.si, martin.milanic@upr.si,
nevena.pivac@iam.upr.si
[2] FAMNIT, University of Primorska, Koper, Slovenia
matjaz.krnc@upr.si
[3] University of Graz, Graz, Austria
{ulrich.pferschy,joachim.schauer}@uni-graz.at
[4] FH JOANNEUM, Kapfenberg, Austria

Abstract. In this work we add a graph theoretical perspective to a classical problem of fairly allocating indivisible items to several agents. Agents have different profit valuations of items and we allow an incompatibility relation between pairs of items described in terms of a conflict graph. Hence, every feasible allocation of items to the agents corresponds to a partial coloring, that is, a collection of pairwise disjoint independent sets. The sum of profits of vertices/items assigned to one color/agent should be optimized in a maxi-min sense. We derive complexity and algorithmic results for this problem, which is a generalization of the classical PARTITION and INDEPENDENT SET problems. In particular, we show that the problem is strongly NP-complete in the classes of bipartite graphs and their line graphs, and solvable in pseudo-polynomial time in the classes of cocomparability graphs and biconvex bipartite graphs.

Keywords: Fair division · Conflict graph · Partial coloring

1 Introduction

Allocating resources to several agents in a satisfactory way is a classical problem in combinatorial optimization. In particular, interesting questions arise if agents have different valuations of resources or if additional constraints are imposed for a feasible allocation. In this work we study the fair allocation of n indivisible goods or items to a set of k agents. Each agent has its own additive utility function over the set of items. The goal is to assign every item to exactly one of the agents such that the minimal utility over all agents is as large as possible. Related problems of fair allocation are frequently studied in Computational Social Choice, see, e.g., [9]. In the area of Combinatorial Optimization a similar problem is well-known as the *Santa Claus* problem (see [5]), which can be also seen as weight partitioning as well as a scheduling problem.

In this paper we look at the problem from a graph theoretical perspective and add a major new aspect to the problem. We allow an incompatibility relation

© Springer Nature Switzerland AG 2020
L. Gąsieniec et al. (Eds.): IWOCA 2020, LNCS 12126, pp. 154–165, 2020.
https://doi.org/10.1007/978-3-030-48966-3_12

between pairs of items, meaning that incompatible items should not be allocated to the same agent. This can reflect the fact that items rule out their joint usage or simply the fact that certain items are identical (or from a similar type) and it does not make sense for one agent to receive more than one of these items. We will represent such a relation by a *conflict graph* where vertices correspond to items and edges express incompatibilities. Now, every feasible allocation to one agent must be an independent set in the conflict graph. This means that the overall solution can also be expressed as a *partial k-coloring* of the conflict graph G, but in addition every vertex/item has a profit value for every color/agent and the sum of profits of vertices/items assigned to one color/agent should be optimized in a maxi-min sense.

We believe that this problem combines aspects of independent sets, graph coloring, and weight partitioning in an interesting way, offering new perspectives to look at these classical combinatorial optimization problems.

Disjunctive constraints represented by conflict graphs were considered for a wide variety of combinatorial optimization problems. We just mention the knapsack problem [20,21], bin packing [18], scheduling (e.g., [8,13]) and problems on graphs (e.g., [11]).

For a formal definition of our problem we consider a set V of items with cardinality $|V| = n$ and k profit functions $p_1, \ldots, p_k : V \to \mathbb{Z}_+$. The *satisfaction level* of an ordered k-partition $(X_1, \ldots, X_k)$ of V (with respect to $p_1, \ldots, p_k$) is defined as the minimum of the resulting profits $p_j(X_j) := \sum_{v \in X_j} p_j(v)$, where $j \in \{1, \ldots, k\}$. The classical fair division problem can be stated as follows.

FAIR k-DIVISION OF INDIVISIBLE GOODS

Input: A set V of n items, k profit functions $p_1, \ldots, p_k : V \to \mathbb{Z}_+$.

Task: Compute an ordered k-partition of V with maximum satisfaction level.

For the special case, where all k profit functions are identical, i.e., $p_1 = p_2 = \ldots = p_k$, the problem can also be represented in a scheduling setting. There are k identical machines and n jobs, which have to be assigned to the machines by a k-partitioning. The goal is to maximize the minimal completion time (corresponding to the satisfaction level) over all k machines. It was pointed out in [12] that this problem is weakly NP-hard even for $k = 2$ machines. Indeed, it is easy to see that an algorithm deciding the above scheduling problem for two machines would also decide the classical PARTITION problem: given n integers $a_1, \ldots, a_n$, can they be partitioned into two subsets with equal sums? For $k \geq 3$, one can simply add jobs of length one half of the sum of weights in the instance of PARTITION. If k is not fixed, but part of the input, the same scheduling problem is strongly NP-hard as mentioned in [4]. In fact, an instance of the strongly NP-complete 3-PARTITION problem with $3m$ elements and target bound B could be decided by any algorithm for the scheduling problem with $n = 3m$ jobs, $k = m$ machines and a desired minimal completion time equal to B. We conclude for later reference.

Observation 1. FAIR k-DIVISION OF INDIVISIBLE GOODS, *even with k identical profit functions, is weakly NP-hard for any constant $k \geq 2$ and strongly NP-hard for k being part of the input.*

Note that the problem is still only weakly NP-hard for constant k even for arbitrary profit functions, since we can construct a pseudo-polynomial algorithm solving the problem with a k-dimensional dynamic programming array.

The first elaborate treatment of FAIR k-DIVISION OF INDIVISIBLE GOODS was given in [7], where two approximation algorithms with bounded (but not constant) approximation ratio were given. They also mention that the problem cannot be approximated by a factor better than $1/2$ (under P $\neq$ NP). In [14] further approximation results were derived. In 2006 Bansal and Sviridenko [5] coined the term *Santa Claus* problem, which corresponds to the variant of the above problem when k is not fixed but part of the input. Since then a huge number of approximation results have appeared on this problem of allocating indivisible goods exploring different concepts of objective functions and various approximation measures.

A different specialization is assumed in the widely studied *Restricted Max-Min Fair Allocation* problem. This is a special case of FAIR k-DIVISION OF INDIVISIBLE GOODS where every item $v_i \in V$ has a fixed valuation $p(v_i)$ and every kid either likes or ignores item v_i, i.e., the profit function $p_j(v_i) \in \{0, p(v_i)\}$. A fairly recent overview of approximation results both for this restricted setting as well as for the general case of the Santa Claus problem can be found in [3].

In this paper we study a generalization of FAIR k-DIVISION OF INDIVISIBLE GOODS, where a *conflict graph* $G = (V, E)$ on the set V of items to be divided is introduced. An edge $\{i, j\} \in E$ means that items i and j should not be assigned to the same subset of the partition. The conflict graph immediately gives rise to (partial) colorings of the graph which were studied by Berge [6] and de Werra [22].

Definition 1. *A partial k-coloring of a graph G is a sequence $(X_1, \ldots, X_k)$ of pairwise disjoint independent sets in G.*

Combining the profit structure with the notion of coloring we define for the k profit functions $p_1, \ldots, p_k : V \to \mathbb{Z}_+$ and for each partial k-coloring $c = (X_1, \ldots, X_k)$ a k-tuple $(p_1(X_1), \ldots, p_k(X_k))$, called the *profit profile* of c. The minimum profit of a profile, i.e., $\min_{j=1}^{k}\{p_j(X_j)\}$, is the *satisfaction level* of c. Now we can define the problem considered in this paper:

FAIR k-DIVISION UNDER CONFLICTS

Input: A graph $G = (V, E)$, k profit functions $p_1, \ldots, p_k : V \to \mathbb{Z}_+$.
Task: Compute a partial k-coloring of G with maximum satisfaction level.

In the hardness reductions of this paper we will frequently use the decision version of this problem: for a given $q \in \mathbb{Z}_+$, does there exists a partial k-coloring of G with satisfaction level at least q?

Note that an optimal partial k-coloring $(X_1, \ldots, X_k)$ does not necessarily select all vertices from V. Furthermore, note also that for $k = 1$, the problem coincides with the weighted independent set problem. In particular, since the case of unit weights and $k = 1$ generalizes the independent set problem, we obtain the following result.

Observation 2. FAIR 1-DIVISION UNDER CONFLICTS *is strongly NP-hard.*

Thus, the addition of the conflict structure gives rise to a much more complicated problem, since FAIR k-DIVISION OF INDIVISIBLE GOODS (which arises naturally as a special case for an edgeless conflict graph G) is trivial for $k = 1$ and only weakly NP-hard for $k \geq 2$ (see Observation 1).

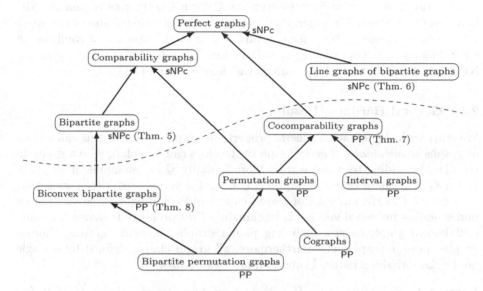

Fig. 1. Relationships between various graph classes and the complexity of the FAIR k-DIVISION UNDER CONFLICTS problem. The arrow from a class $\mathcal{G}_1$ to a class $\mathcal{G}_2$ means that every graph in $\mathcal{G}_1$ is also in $\mathcal{G}_2$. Label 'PP' means that the problem is solvable in pseudo-polynomial time for each fixed k in the given class, label 'sNPc' means that the problem is strongly NP-complete for all fixed $k \geq 2$. For all graph classes in the figure, the problem is solvable in strongly polynomial time for $k = 1$, as it coincides with the weighted independent set problem.

In this contribution we first introduce a general concept of extendable graph families and show that for every such graph class $\mathcal{G}$ in which INDEPENDENT SET is NP-complete, the decision version of our FAIR k-DIVISION UNDER CONFLICTS is strongly NP-complete when the conflict graphs are in $\mathcal{G}$ (Sect. 2.1). By a similar reasoning we can also reach a strong inapproximability result for our problem. For bipartite conflict graphs as well as their line graphs FAIR k-DIVISION UNDER CONFLICTS can be shown to be strongly NP-hard (Sect. 2.2) although the corresponding INDEPENDENT SET problem is polynomial-time solvable. On the other hand, for the relevant special case of biconvex bipartite graphs (cf. [16,17]), FAIR k-DIVISION UNDER CONFLICTS can be solved by a pseudo-polynomial time algorithm. This result is based on an insightful pseudo-polynomial algorithm for the problem on a cocomparability conflict graph (Sect. 3). See Fig. 1 for a summary of these results. Many proofs had to be omitted for lack of space. They can be found in the extended version of this paper posted on http://arxiv.org/abs/2003.11313.

2   Hardness Results

Observation 2 shows that FAIR k-DIVISION UNDER CONFLICTS is strongly NP-hard even for $k = 1$ for general graphs, while Observation 1 shows the weak NP-hardness of the problem for constant $k \geq 2$ in the absence of conflicts. In what follows, we show that FAIR k-DIVISION UNDER CONFLICTS is strongly NP-hard also for all $k \geq 2$, for various well-known graph classes.

2.1   General Hardness Results

We start with the following general property of graph classes. Let us call a class of graphs $\mathcal{G}$ *sustainable* if every graph in the class can be enlarged to a graph in the class by adding to it one vertex. More formally, $\mathcal{G}$ is sustainable if for every graph $G \in \mathcal{G}$ there exists a graph $G' \in \mathcal{G}$ and a vertex $v \in V(G')$ such that $G' - v = G$. Clearly, any class of graphs closed under adding isolated vertices, or under adding universal vertices is sustainable. This property is shared by many well known graph classes, including planar graphs, bipartite graphs, chordal graphs, perfect graphs, etc. Furthermore, all graph classes defined by a single nontrivial forbidden induced subgraph are sustainable.

Lemma 1. *For every graph H with at least two vertices, the class of H-free graphs is sustainable.*

For an example of a non-sustainable graph class $\mathcal{G}$ closed under vertex deletion, consider the family of all cycles and their induced subgraphs. Then every cycle is in $\mathcal{G}$ but cannot be extended to a larger graph in $\mathcal{G}$. The importance of sustainable graph classes for FAIR k-DIVISION UNDER CONFLICTS is evident from the following theorem.

Theorem 3. *Let $\mathcal{G}$ be a sustainable class of graphs for which the decision version of FAIR k-DIVISION UNDER CONFLICTS is (strongly) NP-complete. Then, for every $\ell \geq k$, the decision version of FAIR ℓ-DIVISION UNDER CONFLICTS with conflict graphs from $\mathcal{G}$ is (strongly) NP-complete.*

Since the INDEPENDENT SET problem is a special case of the FAIR 1-DIVISION UNDER CONFLICTS, Theorem 3 immediately implies the following.

Corollary 1. *Let $\mathcal{G}$ be a sustainable class of graphs for which INDEPENDENT SET is NP-complete. Then, for every $k \geq 1$, the decision version of FAIR k-DIVISION UNDER CONFLICTS with conflict graphs from $\mathcal{G}$ is strongly NP-complete.*

It is known (see, e.g., [2]) that for every graph H that has a component that is not a path or a subdivision of the claw, INDEPENDENT SET is NP-complete on H-free graphs. Thus, for every such graph H, Lemma 1 and Corollary 1 imply that for every $k \geq 1$, FAIR k-DIVISION UNDER CONFLICTS (decision version) with H-free conflict graphs is strongly NP-complete. By using a similar argument, we even get a strong inapproximability result for general graphs.

Theorem 4. *For every $k \geq 1$ and every $\varepsilon > 0$, it is NP-hard to approximate* FAIR k-DIVISION UNDER CONFLICTS *within a factor of $|V(G)|^{1-\varepsilon}$, even for unit profit functions.*

2.2 Bipartite Graphs and Their Line Graphs

In this section we show that for all $k \geq 2$, FAIR k-DIVISION UNDER CONFLICTS is NP-hard in two classes of graphs where the INDEPENDENT SET problem is solvable in polynomial time: the class of bipartite graphs and the class of line graphs of bipartite graphs. Recall that for a given graph G, its line graph has a vertex for each edge of G, with two distinct vertices adjacent in the line graph if and only if the corresponding edges share an endpoint in G.

The proof for bipartite graphs shows strong NP-hardness even for the case when all the profit functions are equal.

Theorem 5. *For each integer $k \geq 2$, the decision version of* FAIR k-DIVISION UNDER CONFLICTS *is strongly NP-complete in the class of bipartite graphs.*

Proof. We use a reduction from the decision version of the CLIQUE problem: Given a graph G and an integer ℓ, does G contain a clique of size ℓ? Consider an instance (G, ℓ) of CLIQUE such that $2 \leq \ell < n := |V(G)|$. We define an instance of FAIR k-DIVISION UNDER CONFLICTS (decision version) consisting of a bipartite conflict graph G', profit functions $p_1, \ldots, p_k$, and a lower bound q on the required satisfaction level. The graph $G' = (A \cup B, E')$ has a vertex for each vertex of the graph G as well as for each edge of G and k new vertices $x_1, \ldots, x_k$. It is defined as follows:

$$A = V(G) \cup \{x_1\}, \quad B = E(G) \cup \{x_i \mid 2 \leq i \leq k\},$$
$$E' = \{ve \mid v \in V(G) \text{ is an endpoint of } e \in E(G)\} \cup \{vx_i \mid v \in V(G), 2 \leq i \leq k\}.$$

The lower bound q on the satisfaction level is defined by setting $q = n^4 + \binom{\ell}{2}n + (n - \ell)$. For ease of notation we set $N_1 = n^4$ and we furthermore introduce a second integer N_2 such that $q = N_2 + \left(m - \binom{\ell}{2}\right)n$, where $m = |E(G)|$. (Note that $N_2 \geq n^3$.) With this, the profit functions $p_i : V(G') \to \mathbb{Z}_+$, for all $i \in \{1, \ldots, k\}$, are defined as

$$p_i(v) = \begin{cases} 1; & \text{if } v \in V(G); \\ n; & \text{if } v \in E(G); \\ N_1; & \text{if } v = x_1; \\ N_2; & \text{if } v = x_2; \\ q; & \text{if } v = x_j \text{ for some } j \in \{3, \ldots, k\}. \end{cases}$$

Note that all the profits introduced as well as the number of vertices and edges of G' are polynomial in n. To complete the proof, we show that G has a clique of size ℓ if and only if G' has a partial k-coloring with satisfaction level at least q. First assume that G has a clique C of size ℓ. We construct a partial k-coloring $c = (X_1, \ldots, X_k)$ of G' by setting

$$
\begin{aligned}
X_1 &= \{x_1\} \cup \{e \in E(G) \mid e \subseteq C\} \cup (V(G) \setminus C), \\
X_2 &= \{x_2\} \cup (E(G) \setminus X_1), \\
X_j &= \{x_j\} \text{ for } 3 \leq j \leq k.
\end{aligned}
$$

Observe that the partial k-coloring c gives rise to the corresponding profit profile with all entries equal to q, which establishes one of the two implications.

Suppose now that there exists a partial k-coloring $c = (X_1, \ldots, X_k)$ of G' for which the profit profile has all entries $\geq q$. Since for each $i \in \{1, \ldots, k\}$, the total profit of the set $V(G) \cup E(G)$ is only $mn + n < n^4$, the partial coloring c must use exactly one of the k vertices $x_1, \ldots, x_k$ in each color class. We may assume without loss of generality that $x_i \in X_i$ for all $i \in \{1, \ldots, k\}$. Let U be the set of uncolored vertices in G' w.r.t. the partial coloring c. Since for each of the profit functions p_i, the difference between the overall sum of the profits of vertices of G' and $k \cdot q$ is equal to ℓ, we clearly have $\sum_{v \in U} p_i(v) \leq \ell < n$, which implies that $U \subseteq V(G)$. Next, observe that every vertex of $E(G)$ belongs to either X_1 or to X_2, since otherwise we would have $p_1(X_1) + p_2(X_2) < 2q$, contrary to the assumption that the satisfaction level of c is at least q.

Consider the sets $W = X_1 \cap V(G)$ and $F = X_1 \cap E(G)$. Then $X_1 = \{x_1\} \cup W \cup F$ and, since $\sum_{v \in X_1} p_1(v) \geq q = N_1 + \binom{\ell}{2} n + (n - \ell)$, it follows that X_1 contains exactly $\binom{\ell}{2}$ vertices from $E(G)$ (if $|F| > \binom{\ell}{2}$, then $p_2(X_2) < q$) and at least $n - \ell$ vertices from $V(G)$. Let C denote the set of all vertices of G' with a neighbor in F. By the construction of G' and since $|F| = \binom{\ell}{2}$, it follows that C is of cardinality at least ℓ. Furthermore, since X_1 is independent, we have $C \cap W = \emptyset$. Consequently, $n = |V(G)| \geq |C| + |W| \geq \ell + (n - \ell) = n$, hence equalities must hold throughout. In particular, C is a clique of size ℓ in G. □

Theorem 6. *For each integer $k \geq 2$, the decision version of* FAIR k-DIVISION UNDER CONFLICTS *is strongly NP-complete in the class of line graphs of bipartite graphs.*

The proof is based on a reduction from the following problem, shown to be NP-complete by Pálvölgi (see [19]): Given a bipartite graph G and an integer q, does G contain a perfect matching and a disjoint matching of size q?

3   Pseudo-Polynomial Algorithms for Special Graph Classes

As shown in Theorem 5, for each $k \geq 2$, FAIR k-DIVISION UNDER CONFLICTS is strongly NP-complete in the class of bipartite graphs. This rules out the existence of a pseudo-polynomial time algorithm for the problem in the class of bipartite graphs, unless $P = NP$. In this section we show that for every k there is a pseudo-polynomial time algorithm for the FAIR k-DIVISION UNDER CONFLICTS in a subclass of bipartite graphs, the class of *biconvex bipartite graphs*. The algorithm reduces the problem to the class of bipartite permutation graphs. To solve the problem in the class of bipartite permutation graphs, we develop a solution in a more general class of graphs, the class of cocomparability graphs. A graph $G = (V, E)$ is a *comparability graph* if it has a transitive orientation, that is, if each of the edges $\{u, v\}$ of G can be replaced by exactly one of the ordered pairs (u, v) and (v, u) so that the resulting set A of directed edges is transitive (that is, for every three vertices $x, y, z \in V$, if $(x, y) \in A$ and $(y, z) \in A$, then $(x, z) \in A$). A graph G is a *cocomparability graph* if its complement is a comparability graph. Comparability graphs and cocomparability graphs are well-known subclasses of perfect graphs. The class of cocomparability graphs is a common generalization of the classes of interval graphs, permutation graphs, and trapezoid graphs (see, e.g., [10, 15]).

Since every bipartite graph is a comparability graph, Theorem 5 implies that for each $k \geq 2$, FAIR k-DIVISION UNDER CONFLICTS is strongly NP-complete in the class of comparability graphs. For cocomparability graphs, we prove that the problem is solvable in pseudo-polynomial time. The key result in this direction is the following lemma, which will also be used in our proof of Theorem 8.

Lemma 2. *For every $k \geq 1$, given a cocomparability graph $G = (V, E)$ and k profit functions $p_1, \ldots, p_k : V \to \mathbb{Z}_+$, the set of all profit profiles of partial k-colorings of G can be computed in time $\mathcal{O}(n^{k+2}(Q + 1)^k)$, where $Q = \max_{1 \leq j \leq k} p_j(V)$.*

The proof is based on a directed acyclic graph representing a transitive orientation of the complement of G.

Lemma 2 implies the following.

Theorem 7. *For every $k \geq 1$, FAIR k-DIVISION UNDER CONFLICTS is solvable in time $\mathcal{O}(n^{k+2}(Q + 1)^k)$ for cocomparability conflict graphs G, where $Q = \max_{1 \leq j < k} p_j(V(G))$.*

Recall from Theorem 5 that FAIR k-DIVISION UNDER CONFLICTS is strongly NP-hard for bipartite conflict graphs. Thus, we consider in the following the more restricted case of *biconvex* bipartite conflict graphs. Recall that a bipartite graph $G = (A \cup B, E)$ is biconvex if it has a *biconvex ordering*, that is, an ordering of A and B such that for every vertex $a \in A$ (resp. $b \in B$) the neighborhood $N(a)$ (resp. $N(b)$) is a consecutive interval in the ordering of B (resp. ordering of A).

It is known that a connected biconvex bipartite graph G can always be ordered in such a way that the first and last vertices on one side have a special structure. Fix a biconvex ordering of G, say $A = (a_1, \ldots, a_s)$ and $B = (b_1, \ldots, b_t)$. Define a_L (resp. a_R) as the vertex in $N(b_1)$ (resp. $N(b_t)$) whose neighborhood is not properly contained in any other neighborhood set (see [1, Def. 8]). In case of ties, a_L is the smallest such index (and a_R the largest). We always assume that $a_L \leq a_R$, otherwise the ordering in A could be mirrored. Under these assumptions, the neighborhoods of vertices appearing in the ordering before a_L and after a_R are nested.

Lemma 3 (Abbas and Stewart [1]). *Let $G = (A \cup B, E)$ be a connected biconvex graph. Then there exists a biconvex ordering of the vertices of G such that:*

i For all a_i, a_j with $a_1 \leq a_i < a_j \leq a_L$ there is $N(a_i) \subseteq N(a_j)$.
ii For all a_i, a_j with $a_R \leq a_i < a_j \leq a_s$ there is $N(a_j) \subseteq N(a_i)$.
iii The subgraph G' of G induced by vertex set $\{a_L, \ldots, a_R\} \cup B$ is a bipartite permutation graph.

Property (iii) can be put in context with Theorem 7. Indeed, it is known that permutation graphs are a subclass of cocomparability graphs (see, e.g., [10]). This gives rise to the following result that FAIR k-DIVISION UNDER CONFLICTS on biconvex graphs is indeed easier (from the complexity point of view) than on general bipartite graphs. It should be pointed out that the contribution of Theorem 8 is the identification of the complexity status of the problem, but not a practically relevant algorithm, since the pseudo-polynomial running time will be prohibitive in practice. The high-level idea of the algorithm is illustrated in Algorithm 1.

Algorithm 1. Algorithmic Idea for a Connected Biconvex Graph G

---

apply Lemma 3 for getting the cocomparability graph G' and vertices a_L, a_R
let $A_L := \{a_1, \ldots, a_{L-1}\}$ and $A_R := \{a_{R+1}, \ldots, a_s\}$
for all $j \in \{1, \ldots, k\}$ **do**
  guess $\overline{a}_j \in A_L$ with largest index (resp. smallest index $\underline{a}_j \in A_R$) included in X_j
end for
each such guess can be represented by a $2k$-tuple $\sigma = (\overline{a}_1, \ldots, \overline{a}_k, \underline{a}_1, \ldots, \underline{a}_k)$
for each guess σ **do**
  for all $j \in \{1, \ldots, k\}$ **do**
    exclude all vertices v of the neighborhood $N(\overline{a}_j) \subseteq B$ (and $N(\underline{a}_j) \subseteq B$)
    from insertion into X_j by setting their profit $p_j(v) := 0$
  end for
  apply Lemma 2 to the cocomparability graph G' and the modified profit functions
  to obtain the set Π_σ of all profit profiles $(q_1, \ldots, q_k)$ of partial k-colorings of G'
  with respect to the modified profits
  increase each profit profile by setting $q_j := q_j + p_j(\overline{a}_j) + p_j(\underline{a}_j)$
  augment these profiles with vertices from A_L and A_R
end for
choose the best solution over all guesses σ

Theorem 8. *For every $k \geq 1$, FAIR k-DIVISION UNDER CONFLICTS is solvable in time $\mathcal{O}(n^{3k+2}(Q+1)^k)$ for connected biconvex bipartite conflict graphs G, where $Q = \max_{1 \leq j \leq k} p_j(V(G))$.*

Proof. Assuming at first that G is connected, Lemma 3 is applied for obtaining from G the cocomparability graph G'. However, we have to consider also the vertex sets $A_L := \{a_1, \ldots, a_{L-1}\}$ and $A_R := \{a_{R+1}, \ldots, a_s\}$. This is done by considering assignments of vertices in $A_L \cup A_R$ to the k subsets of a partial k-coloring of G in an efficient way as follows.

For every $j \in \{1, \ldots, k\}$, we guess, by going through all possibilities, the largest index vertex $\bar{a}_j \in A_L$ (resp. smallest index $\underline{a}_j \in A_R$) inserted in X_j. One can add an artificial vertex a_0 (resp. a_{s+1}) to represent the case that no vertex from A_L (resp. A_R) is inserted in X_j. Thus, every guess is represented by a $2k$-tuple $\sigma = (\bar{a}_1, \ldots, \bar{a}_k, \underline{a}_1, \ldots, \underline{a}_k)$. The total number of such guesses (i.e., iterations) is bounded by $(n+1)^k$ for each of A_L and A_R, i.e., $\mathcal{O}(n^{2k})$ selections to be considered in total.

For each such guess σ we perform the following computations. For every $j \in \{1, \ldots, k\}$ the vertices in the neighborhood $N(\bar{a}_j) \subseteq B$ (and $N(\underline{a}_j) \subseteq B$) of the chosen index must be excluded from insertion into the corresponding set X_j. This can be easily realized by setting to 0 the profits p_j of all vertices in $N(\bar{a}_j)$ (resp. $N(\underline{a}_j)$). With these slight modifications of the profits we can apply Lemma 2 for the cocomparability graph G' and the modified profit functions p_j^σ to obtain the set Π_σ of all (pseudo-polynomially many) profit profiles $(q_1, \ldots, q_k)$ of partial k-colorings of G' with respect to p^σ. Every entry q_j of a profit profile in Π_σ is increased by $p_j(\bar{a}_j) + p_j(\underline{a}_j)$, to account for inclusion of the vertices selected by the guess σ.

In every guess there are the two vertices $\bar{a}_j$ and $\underline{a}_j$ permanently assigned to X_j for every j and their neighborhoods $N(\bar{a}_j)$ and $N(\underline{a}_j)$ are excluded from X_j. Now it follows from properties (i) and (ii) of Lemma 3 that for each vertex $a' \in A_L$ with $a' < \bar{a}_j$ (resp. $a' \in A_R$ with $a' > \underline{a}_j$) the neighborhood $N(a')$ is a subset of $N(\bar{a}_j)$ (resp. $N(\underline{a}_j)$). Thus, these vertices a' could also be inserted in X_j without any violation of the conflict structure. Therefore, we can start from the set Π_σ of profit profiles computed for (G', p^σ) and consider iteratively (in arbitrary order) the addition of a vertex $a' \in A_L$ to one of the color classes X_j, as is usually done in dynamic programming. Each a' is considered as an addition to every profit profile $(q_1, \ldots, q_k) \in \Pi_\sigma$ and for every index j with $a' < \bar{a}_j$ yielding new profit profiles $(q_1, \ldots, q_{j-1}, q_j + p_j(a'), q_{j+1}, \ldots, q_k)$ to be added to Π_σ. An analogous procedure is performed for all vertices $a' \in A_R$ where the addition is restricted to indices j with $a' > \underline{a}_j$.

For every guess σ, the running time is dominated by the effort of computing the $\mathcal{O}((Q+1)^k)$ profit profiles of (G', p^σ) according to Lemma 2, since adding any of the $\mathcal{O}(n)$ vertices a' requires only k operations for each profit profile.

In this way, we construct the set Π_σ of all profit profiles of partial k-colorings of G for each guess σ. It remains to identify the optimal solution in the set $\Pi := \bigcup_\sigma \Pi_\sigma$ similarly as in the proof of Theorem 7. Going over all $\mathcal{O}(n^{2k})$ guesses σ, the total running time can be given from Lemma 2 as $\mathcal{O}(n^{3k+2}(Q+1)^k)$. $\square$

4   Conclusions

In this paper we introduced the FAIR k-DIVISION UNDER CONFLICTS and studied it from a computational complexity point of view, with respect to various restrictions on the conflict graph. In particular, we could show that the problem is strongly NP-hard on general bipartite conflict graphs, but it can be solved in pseudo-polynomial time on biconvex bipartite graphs. The latter also contains the class of bipartite permutation graphs. There are other graph classes sandwiched between the two classes of our results, for which the complexity of FAIR k-DIVISION UNDER CONFLICTS is still open. In particular, we can derive open problems from the following sequence of inclusions: biconvex bipartite $\subseteq$ convex bipartite $\subseteq$ interval bigraph $\subseteq$ chordal bipartite $\subseteq$ bipartite. We believe that a result for convex bipartite graphs should be the next attempt. Outside this chain of inclusions, we pose the complexity of the problem for planar bipartite conflict graphs as another interesting open question.

Beside the results given in this work we can also derive pseudo-polynomial algorithms for FAIR k-DIVISION UNDER CONFLICTS if the conflict graph is chordal or if its treewidth or clique-width is bounded. These results will be described in a future publication.

Acknowledgements. The work of this paper was done in the framework of a bilateral project between University of Graz and University of Primorska, financed by the OeAD (SI 22/2018) and the Slovenian Research Agency (BI-AT/18-19-005). The authors acknowledge partial support of the Slovenian Research Agency (I0-0035, research programs P1-0285, P1-0383, P1-0297, research projects J1-9110, N1-0102, J1-1692, J1-9187, and a Young Researchers Grant) and the European Commission for funding the InnoRenew CoE project (Grant Agreement #739574) under the Horizon2020 Widespread-Teaming program and the Republic of Slovenia (Investment funding of the Republic of Slovenia and the European Union of the European Regional Development Fund) and by the Field of Excellence "COLIBRI" at the University of Graz and by the Federal Ministry for Digital and Economic Affairs of the Republic of Austria through the COIN project FIT4BA.

References

1. Abbas, N., Stewart, L.K.: Biconvex graphs: ordering and algorithms. Discret. Appl. Math. **103**(1–3), 1–19 (2000)
2. Alekseev, V.E.: The effect of local constraints on the complexity of determination of the graph independence number. In: Combinatorial-Algebraic Methods in Applied Mathematics, pp. 3–13. Gorky University Press (1982). (in Russian)
3. Annamalai, C., Kalaitzis, C., Svensson, O.: Combinatorial algorithm for restricted max-min fair allocation. ACM Trans. Algorithms **13**(3), 1–28 (2017)
4. Azar, Y., Epstein, L.: On-line machine covering. J. Sched. **1**, 67–77 (1998)
5. Bansal, N., Sviridenko, M.: The Santa Claus problem. In: STOC 2006: Proceedings of the 38th Annual ACM Symposium on Theory of Computing, pp. 31–40 (2006)
6. Berge, C.: Minimax relations for the partial q-colorings of a graph. Discret. Math. **74**(1–2), 3–14 (1989)

7. Bezakova, I., Dani, V.: Allocating indivisible goods. ACM SIGecom Exchanges **5**(3), 11–18 (2005)
8. Bodlaender, H.L., Jansen, K.: On the complexity of scheduling incompatible jobs with unit-times. In: Borzyszkowski, A.M., Sokołowski, S. (eds.) MFCS 1993. LNCS, vol. 711, pp. 291–300. Springer, Heidelberg (1993). https://doi.org/10.1007/3-540-57182-5_21
9. Bouveret, S., Chevaleyre, Y., Maudet, N.: Fair allocation of indivisible goods. In: Brandt, F., Conitzer, V., Endriss, U., Lang, J., Procaccia, A.D. (eds.) Handbook of Computational Social Choice, pp. 284–310. Cambridge University Press, Cambridge (2016)
10. Brandstädt, A., Le, V.B., Spinrad, J.P.: Graph Classes: A Survey. SIAM Monographs on Discrete Mathematics and Applications. Society for Industrial and Applied Mathematics (SIAM) (1999)
11. Darmann, A., Pferschy, U., Schauer, J., Woeginger, G.: Paths, trees and matchings under disjunctive constraints. Discret. Appl. Math. **159**, 1726–1735 (2011)
12. Deuermeyer, B.L., Friesen, D.K., Langston, M.A.: Scheduling to maximize the minimum processor finish time in a multiprocessor system. SIAM J. Algebraic Discret. Methods **3**(2), 190–196 (1982)
13. Even, G., Halldórsson, M.M., Kaplan, L., Ron, D.: Scheduling with conflicts: online and offline algorithms. J. Sched. **12**(2), 199–224 (2009)
14. Golovin, D.: Max-min fair allocation of indivisible goods. Technical report, CMU-CS-05-144, Carnegie Mellon University (2005)
15. Golumbic, M.C.: Algorithmic Graph Theory and Perfect Graphs. Annals of Discrete Mathematics, vol. 57. Elsevier, Amsterdam (2004)
16. Khodamoradi, K., Krishnamurti, R., Rafiey, A., Stamoulis, G.: PTAS for ordered instances of resource allocation problems. In: Proceedings of the 33rd International Conference on Foundations of Software Technology and Theoretical Computer Science, FSTTCS 2013. LIPICS, vol. 24, pp. 461–473 (2013)
17. Mastrolilli, M., Stamoulis, G.: Restricted max-min fair allocations with inclusion-free intervals. In: Gudmundsson, J., Mestre, J., Viglas, T. (eds.) COCOON 2012. LNCS, vol. 7434, pp. 98–108. Springer, Heidelberg (2012). https://doi.org/10.1007/978-3-642-32241-9_9
18. Muritiba, A., Iori, M., Malaguti, E., Toth, P.: Algorithms for the bin packing problem with conflicts. INFORMS J. Comput. **22**(3), 401–415 (2010)
19. Pálvölgi, D.: Partitioning to three matchings of given size is NP-complete for bipartite graphs. Acta Univ. Sapientiae Informatica **6**(2), 206–209 (2014)
20. Pferschy, U., Schauer, J.: The knapsack problem with conflict graphs. J. Graph Algorithms Appl. **13**(2), 233–249 (2009)
21. Pferschy, U., Schauer, J.: Approximation of knapsack problems with conflict and forcing graphs. J. Comb. Optim. **33**(4), 1300–1323 (2017)
22. de Werra, D.: Packing independent sets and transversals. In: Combinatorics and Graph Theory, Banach Center Publications, vol. 25, pp. 233–240. PWN, Warsaw (1989)

Polynomial Time Algorithms for Tracking Path Problems

Pratibha Choudhary[(⊠)]

Indian Institute of Technology Jodhpur, Jodhpur, India
pratibhac247@gmail.com

Abstract. Given a graph G, and terminal vertices s and t, the TRACK-ING PATHS problem asks to compute a minimum number of vertices to be marked as trackers, such that the sequence of trackers encountered in each s-t path is unique. TRACKING PATHS is NP-hard in both directed and undirected graphs in general. In this paper we give a collection of polynomial time algorithms for some restricted versions of TRACKING PATHS. We prove that TRACKING PATHS is polynomial time solvable for chordal graphs and tournament graphs. We prove that TRACKING PATHS is NP-hard in graphs with bounded maximum degree $\delta \geq 6$, and give a $2(\delta + 1)$-approximate algorithm for the same. We also analyze the version of tracking s-t paths where paths are tracked using edges instead of vertices, and we give a polynomial time algorithm for the same. Finally we give a polynomial algorithm which, given an undirected graph G, a tracking set $T \subseteq V(G)$, and a sequence of trackers π, returns the unique s-t path in G that corresponds to π, if one exists.

Keywords: Graphs · Paths · Chordal graphs · Tournaments · Approximation · Bounded degree graphs · Tracking paths

1 Introduction

Tracking moving objects in networks has been studied extensively due to applications in surveillance and monitoring. Specific cases include secure system surveillance, habitat monitoring, vehicle tracking, and other similar scenarios. Object tracking in networks also finds applications in analyzing disease spreading patterns, information dissemination patterns on social media, and data packet flow in large networks like the world wide web. Tracking has been largely studied in the fields of machine learning, artificial intelligence, networking systems among other fields.

The problem of tracking paths in a network was first graphically modeled by Banik et al. in [2]. Let $G = (V, E)$ be an undirected graph without any self loops or parallel edges and suppose that G has a unique entry vertex (source) s and a unique exit vertex (destination) t. A simple path from s to t is called

This work was done while the author was visiting The Institute of Mathematical Sciences, Chennai, India.

L. Gąsieniec et al. (Eds.): IWOCA 2020, LNCS 12126, pp. 166–179, 2020.
https://doi.org/10.1007/978-3-030-48966-3_13

an s-t path. The problem requires finding a set of vertices $T \subseteq V$, such that for any two distinct s-t paths, say P_1 and P_2, in G, the sequence of vertices in $T \cap V(P_1)$ as encountered in P_1 is different from the sequence of vertices in $T \cap V(P_2)$ as encountered in P_2. Here T is called a *tracking set* for the graph G, and the vertices in T are referred to as *trackers*. Banik et al. [2] proved that the problem of finding a minimum-cardinality tracking set to track *shortest* s-t paths (TRACKING SHORTEST PATHS problem) is NP-hard and APX-hard. Later, the problem of tracking all s-t paths (TRACKING PATHS) in an undirected graph was studied in [4,7,10]. TRACKING PATHS is formally defined as follows.

TRACKING PATHS (G, s, t)

Input: An undirected graph $G = (V, E)$ with terminal vertices s and t.

Question: Find a minimum cardinality tracking set T for G.

TRACKING PATHS was proven to be NP-complete in [4]. Here, the authors studied the parameterized version of TRACKING PATHS, which asks if there exists a tracking set of size at most k, and showed it to be fixed-parameter tractable (FPT) by giving a polynomial kernel. Specifically, it was proven that an instance of TRACKING PATHS can be reduced to an equivalent instance of size $\mathcal{O}(k^7)$ in polynomial time, where k is the desired size of the tracking set. In [7], the authors improved this kernel to $\mathcal{O}(k^2)$, and gave an $\mathcal{O}(k)$ kernel for planar graphs. In [10], Eppstein et al. proved that TRACKING PATHS is NP-complete for planar graphs and gave a 4-approximation algorithm for this setting. Here, the authors also proved that TRACKING PATHS can be solved in linear time for graphs of bounded clique width, when the clique decomposition is given in advance.

TRACKING SHORTEST PATHS was also studied in [3] and [5]. In [3], Banik et al. studied TRACKING SHORTEST PATHS and proved the problem to be fixed-parameter tractable. In [5], Bilò et al. prove that TRACKING SHORTEST PATHS is NP-hard for cubic planar graphs in case of multiple source-destination pairs, and give an FPT algorithm parameterized by the number of vertices equidistant from the source s.

In this paper we study TRACKING PATHS for chordal graphs, tournament graphs, and degree bounded graphs. A *chordal* graph is a graph in which each cycle of length greater than three has a chord (an edge between non-adjacent vertices of the cycle). A *tournament* is a directed graph in which there exists a directed edge between each pair of vertices. So far all the work done on TRACKING PATHS has been focused on tracking s-t paths (or shortest s-t paths) using vertices. In this paper, we also study tracking s-t paths using edges. We also give a path reconstruction algorithm that finds the unique s-t path corresponding to the given sequence of trackers, if one exists. Chordal graphs find applications in computational biology, computer vision and artificial intelligence [9,13,16,18]. Tournament graphs are used in voting theory and social choice theory to graphically depict pairwise relationships between entities in a community [17,19]. Tournament graphs are particularly used to study the Condorcet voting model, where a preference is indicated between each pair of contestants [11].

Our Results and Methods. In this paper we give polynomial time results for some variants of the TRACKING PATHS problem. We prove that TRACKING PATHS is polynomial time solvable for chordal graphs and tournaments. The key idea in proofs for chordal and tournament graphs is that if two s-t paths differ in only one vertex, than that vertex necessarily needs to be marked as a tracker. Next we prove that TRACKING PATHS is NP-hard for graphs with maximum degree δ ($\delta \geq 6$). We also give a $2(\delta + 1)$-approximation algorithm for graphs with maximum degree δ. Here the idea is to ensure that sufficient vertices are marked as trackers in each cycle. This derives from the fact that each cycle in a graph necessarily needs a tracker [4]. In order to give a complete solution for tracking paths in a graph, we also give an algorithm that reconstructs the required s-t path given a sequence of trackers and a constant size tracking set for the input graph. This uses the fact that by the definition of a tracking set, each maximal sequence of trackers in a tracking set should correspond to at most one s-t path in a graph. The reconstruction algorithm uses the disjoint path algorithms for undirected graphs [15] and tournaments [8] to construct the required s-t path.

Towards the end of the paper we analyze the problem of tracking s-t paths in an undirected graph using edges rather than vertices. We prove that even while using edges, each cycle in the graph needs at least one edge to be marked as a tracker. Further, a minimum feedback edge set (set of edges whose removal makes a graph acyclic) is also a minimum tracking edge set.

2   Notations and Definitions

Throughout the paper, while analyzing tracking paths using vertices in a graph, we assume graphs to be simple i.e. there are no self loops and multi-edges. When considering tracking set for a graph $G = (V, E)$, we assume that the given graph is an s-t graph, i.e. the graph contains a unique source $s \in V$ and a unique destination $t \in V$ (both s and t are known), and we aim to find a tracking set that can distinguish between all simple paths between s and t. Here s and t are also referred as the terminal vertices. If $a, b \in V$, then unless otherwise stated, $\{a, b\}$ represents the set of vertices a and b, and (a, b) represents an edge between a and b. For a vertex $v \in V$, *neighborhood* of v is denoted by $N(v) = \{x \mid (x, v) \in E\}$. We use $deg(v) = |N(v)|$ to denote degree of vertex v. For a subgraph G', $V(G')$ represents the vertex set of G' and $E(G')$ represents those edges whose both endpoints belong to $V(G')$. For a vertex $v \in V$ and a subgraph G', $N_{G'}(v) = N(v) \cap V(G')$. For a subset of vertices $V' \subseteq V$ we use $N(V')$ to denote $\bigcup_{v \in V'} N(v)$. With slight abuse of notation we use $N(G')$ to denote $N(V(G'))$. For a graph G and a set of vertices $S \subseteq V(G)$, $G - S$ denotes the subgraph induced by the vertex set $V(G) \setminus V(S)$. If S is a singleton, we may use $G - x$ to denote $G - S$, where $S = \{x\}$. A *chord* in a cycle is an edge between two vertices of the cycle, such that the edge itself not part of the cycle.

In an undirected graph, a feedback vertex set (FVS) is a set of vertices whose removal makes the graph acyclic and feedback edge set (FES) is the set of edges whose removal makes the graph acyclic. An edge-weighted graph is a graph with real valued weights assigned to each of its edges. Let P_1 be a path between vertices a and b, and P_2 be a path between vertices b and c, such that $V(P_1) \cap V(P_2) = \{b\}$. By $P_1 \cdot P_2$, we denote the path between a and c, formed by concatenating paths P_1 and P_2 at b. Two paths P_1 and P_2 are said to be *vertex disjoint* if their vertex sets do not intersect except possibly at the end points, i.e. $V(P_1) \cap V(P_2) \subseteq \{a, b\}$, where a and b are the starting and end points of the paths. By distance we mean length of the shortest path, i.e. the number of edges in that path. For a sequence of vertices π, by $V(\pi)$ we mean the set of vertices in the sequence π. If there exists a path P such that (a, b) is an edge that lies at one end point of P, then $P - (a, b)$ denotes the subpath of P obtained after removing the edge (a, b). Graphs which have maximum degree of vertices as three are known as *cubic* graphs. By a *bounded degree graph*, we mean a graph whose vertices have a maximum degree of d, where d is some constant.

3 Preliminary Analysis

In this section, we give some basic claims which are used for proving results in subsequent sections. We start by first recalling a reduction rule from [4] that ensures that each vertex and edge in the input graph participates in an s-t path.

Reduction Rule 1. *[4] In a graph G, if there exists a vertex or an edge that does not participate in any s-t path in G, then delete it.*

It is known that Reduction Rule 1 is safe and can be applied in quadratic time on undirected graphs [4]. In the rest of the paper, by *reduced graph* we mean a graph that is preprocessed using Reduction Rule 1. Let G' be a subgraph of graph G, and $u, v \in V(G')$. If there exists a path in G from s to u, say P_{su}, and another path from v to t, say P_{vt}, such that $V(P_{su}) \cap V(P_{vt}) = \emptyset$, $V(P_{su}) \cap V(G') = \{u\}$ and $V(P_{vt}) \cap V(G') = \{v\}$, then u is a *local source* for G' and v is a *local destination* for G'. Next we recall the following lemma from [4], which is used to define some commonly used terms in this paper.

Lemma 1. *In a reduced graph G, any subgraph G' consisting of at least one edge, contains a local source and local destination.*

Now we state the *tracking set condition*, which is useful for validation of a tracking set [4].

Tracking Set Condition:
For a graph $G = (V, E)$, with terminal vertices $s, t \in V$, a set of vertices $T \subseteq V$, is said to satisfy the **tracking set condition** if there does not exist a pair of vertices $u, v \in V$, such that the following holds:

- there exist two distinct paths, say P_1 and P_2, between u and v in $(G \setminus (T \cup \{s, t\})) \cup \{u, v\}$, and
- there exists a path from s to u, say P_{su}, and a path from v to t, say P_{vt}, in $(G \setminus (V(P_1) \cup V(P_2))) \cup \{u, v\}$, and $V(P_{su}) \cap V(P_{vt}) = \emptyset$, i.e. P_{su} and P_{vt} are mutually vertex disjoint, and also vertex disjoint from P_1 and P_2.

It is known that for a reduced graph G, a set of vertices $T \subseteq V(G)$ is a tracking set if and only if T satisfies the *tracking set condition*. We use this fact, to prove the following lemma.

Lemma 2. ⊛[1] *In a graph G, if $T \subseteq V(G)$ is not a tracking set for G, then there exist two s-t paths with the same sequence of trackers, and they form a cycle C in G, such that C has a local source a and a local destination b, and $T \cap (V(C) \setminus \{a, b\}) = \emptyset$.*

4   Tracking Paths in Chordal Graphs and Tournaments

In this section, we consider polynomial time algorithms for solving TRACKING PATHS for chordal graphs and tournaments.

We start by giving a polynomial time algorithm for finding a tracking set for undirected chordal graphs. Recall that chordal graphs are those graphs in which each cycle of length greater than three has a chord. Many problems that are known to be NP-hard on general graphs are polynomial time solvable for chordal graphs e.g. chromatic number, feedback vertex set, independent set [14].

In undirected graphs, a tracking set is also a feedback vertex set [4]. However, a tracking set can be arbitrarily larger in size compared to a feedback vertex set. This holds true for chordal graphs as well.

Algorithm 1: Finding Tracking Set for a Chordal Graph.

  Input: Chordal graph $G = (V, E)$ and vertices $s, t \in V$.
  Output: Tracking Set $T \subseteq V$ for G.

1 Initialize $T = \emptyset$; Apply Reduction Rule 1;
2 **foreach** $e = (a, b) \in E$ **do**
3   **foreach** $x \in (N(a) \cap N(b)) \setminus T$ **do**
4   **if** $\exists$ an s-t path P in $G - x$ such that $e \in E(P)$ **then**
5              $T = T \cup \{x\}$;
6   **end**
7   **end**
8 **end**
9 Return T;

Algorithm 1 gives a procedure to compute a minimum tracking set for a chordal graph G. We prove its correctness in the following lemma.

Lemma 3. *Algorithm 1 gives an optimum tracking set for a chordal graph.*

---

[1] Proofs of Lemmas marked with ⊛ can be found in the full version of the paper [6].

Proof. Algorithm 1 first ensures that each vertex and edge in the input graph G participates in an s-t path. Next for each edge $e = (a, b) \in E$, if there exists a vertex $x \in (N(a) \cap N(b)) \setminus T$, we check if there exists an s-t path in $G - x$ that contains the edge e. Let P such a path in $G - x$. Now consider the path P' that can be obtained by replacing the edge e in P by the path $(a, x) \cdot (b, x)$ along with the vertex x. Observe that the vertex sets of P and P' differ only in vertex x. Hence, x necessarily belongs to a tracking set for G.

Now we prove that Algorithm 1 indeed returns an optimal tracking set T for G. Suppose not. Then T is not a tracking set for G. Due to Lemma 2, there exists two s-t paths, say P_1, P_2 and they form a cycle C in G, such that C has a local source u and a local destination v, and $V(C) \setminus \{u, v\}$ does not contain any trackers. See Fig. 1.

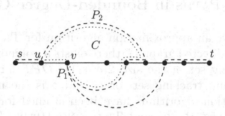

Fig. 1. Indistinguishable s-t paths in a graph form a cycle (marked in dotted lines)

Path P_1 is marked in solid lines, while path P_2 is marked in dashed lines. Since P_1 and P_2 contain the same sequence of trackers, no vertex in $V(C) \setminus \{u, v\}$ can be a tracker. Since we consider graphs without any parallel edges, there exists at least one vertex in $V(C) \setminus \{u, v\}$.

First, consider the case where C is a triangle. Due to Algorithm 1, the vertex in $V(C) \setminus \{u, v\}$ would have been marked as a tracker. This contradicts the assumption that no vertex in $V(C) \setminus \{u, v\}$ is marked as a tracker.

Next, consider the case when C is not a triangle (C contains four or more vertices). Since G is a chordal graph, C contains a chord. Consider the following two cases based on whether a chord is incident on the vertex u or not. Let w and x be two vertices such that $w, x \in N(u) \cap V(C)$ and $(w, x) \in E$. Without loss of generality, let $x \in V(P_1)$. Observe that the edge (u, x) in path P_1 can be replaced by the concatenated path $(u, w) \cdot (w, x)$, to obtain a new path that differs from P_1 only at the vertex w. Hence w must have been marked as a tracker by Algorithm 1. Next, consider the case when a chord in C is incident on u. Let $a \in N(u) \cap V(C)$ and $b \in N(a) \cap V(C)$, such that $a \neq b \neq u$, and $(b, u) \in E$. Note that there exists an s-t path containing the edge (b, u), where (b, u) can be replaced with the path $(b, a) \cdot (a, u)$, to obtain a new path that differs only at the vertex a. Hence a must have been marked as a tracker by Algorithm 1. Both the above cases contradict the assumption that no vertex in $V(C) \setminus \{u, v\}$ is a tracker. Hence Algorithm 1 gives an optimum tracking set for a chordal graph. □

Lemma 4. ⊛ *Algorithm 1 runs in time $\mathcal{O}(m.n^3)$.*

From Lemma 3 and Lemma 4, we have the following theorem.

Theorem 1. TRACKING PATHS *can be solved in polynomial time in chordal graphs.*

A similar technique can be used to prove that TRACKING PATHS is polynomial time solvable for tournament graphs. However, the analysis of tournament graphs is slightly more involved and due to space constraint it is deferred to the full version of the paper [6].

5   Approximation Algorithm and NP-Hardness of TRACKING PATHS in Bounded-Degree Graphs

In this section, we give an approximation algorithm for TRACKING PATHS. We show that given an undirected graph G, there exists a polynomial time algorithm that returns a tracking set of the size $2(\delta + 1) \cdot OPT$ for G, where OPT is the size of an optimum tracking set for G and δ is the maximum degree of graph G. Approximation algorithms have been studied for restricted versions of TRACKING SHORTEST PATHS and TRACKING PATHS. Banik et al. gave a 2-approximate algorithm for TRACKING SHORTEST PATHS in planar graphs in [2]. Eppstein et al. gave a 4-approximate algorithm for TRACKING PATHS in planar graphs in [10]. Bilò et al. gave an $\tilde{O}(\sqrt{n})$-approximate algorithm for TRACKING SHORTEST PATHS in case of multiple source-destination pairs in [5]. Next we show that TRACKING PATHS for bounded degree graphs is polynomial time reducible from VERTEX COVER for bounded degree graphs.

Lemma 5. *Given an undirected graph G with maximum degree d, there exists an s-t graph G' with maximum degree $2d$, such that G has a vertex cover of size k if and only if G' has a tracking set for all s-t paths, of size $k + |E|^2 + 3|E| - 2$.*

Proof. Let G be and undirected graph with maximum degree d. For reference, let G be the graph in Fig. 2.

We create the graph G' as follows. For each vertex $a \in V(G)$, we introduce a vertex v_a in $V(G')$, and we refer to this set of newly introduced vertices in G' as V_v. For each edge $i \in E(G)$, we introduce two vertices v_i, v_i' in $E(G')$, and we call the set of vertices v_i as V_e, and the set of vertices v_i' as V_e'. The adjacencies between V_v and V_e, V_e' are introduced as follows. If an edge i is incident on vertices a, b in G, then we add edges between the corresponding vertices $v_i, v_i' \in V_e, V_e'$ and the vertices $v_a, v_b \in V_v$ in G'. Next, we add the source and destination vertices s and t in G'. We then create a triangular grid Tg_1 between s and the vertices in V_e, and another triangular grid between the vertices in V_e' and t. See Fig. 3. The vertices of V_v are marked with blank boxes, while the ones from $V_e \cup V_e'$ are marked with solid boxes. The circled vertices form a tracking set. Observe that the maximum degree of vertices in Tg_1 and

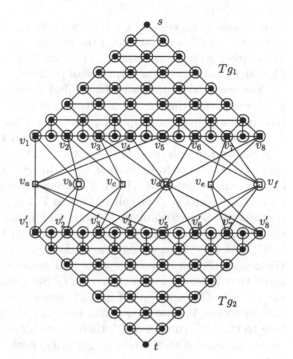

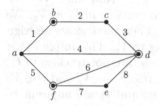

Fig. 2. Depiction of an undirected graph G with maximum
degree d

Fig. 3. Depiction of graph G' mentioned in
Lemma 5

Tg_2, including the vertices in $V_e \cup V'_e$, is 6. The maximum degree of vertices in
V_v is at most $2d$.

Now we prove that there exists a vertex cover of size k in G if and only if
there exists a tracking set in G' of size $k + |E|^2 + 3|E| - 2$. First consider the case
when G has a vertex cover V_c of size k. We now prove that there exists a tracking
set of size $k + |E|^2 + 3|E| - 2$ in G'. We mark the vertices in G' corresponding
to V_c as trackers. In addition we mark all the vertices in Tg_1 and Tg_2 (except s
and t) as trackers. Now the size of tracking set T in G' is $k + |E|^2 + 3|E| - 2$.
We claim that T is a valid tracking set for G'. Suppose not. Then there exists
two distinct s-t paths, say P_1, P_2 in G', such that the sequence of trackers in
P_1 is same as that in P_2. Observe that two distinct subpaths (subpaths of some
s-t paths) contained in Tg_1 (Tg_2) cannot have the same sequence of trackers
from $Tg_1 - \{s\}$ ($Tg_2 - \{t\}$). Since all vertices in Tg_1, Tg_2 are marked as trackers,
this implies that P_1, P_2 contain the same sequence of vertices from Tg_1 and Tg_2,
and they necessarily differ in vertices from V_v. Let $x, y \in V_v$ be the vertices that
distinguish P_1 and P_2, and $x \in V(P_1)$ and $y \in V(P_2)$. Since P_1 and P_2 can not
differ in their vertex set from Tg_1 and Tg_2, the vertex preceding x, y has to be
common in both P_1, P_2. Without loss of generality, we assume that the z is the
vertex preceding x, y, and $z \in V(Tg_1)$. This implies that $z \in V_e$. Note that z

corresponds to an edge in G. Since we marked the vertices corresponding to V_c as trackers in G', at least one of the neighbors of z in V_v is necessarily a tracker. Thus either x or y is necessarily a tracker. This contradicts the assumption that P_1 and P_2 have the same sequence of trackers.

Now we consider the case when G' has a tracking set T of size $k+|E|^2+3|E|-2$. We claim that there exists a vertex cover of size k in G. Suppose not. Consider the triangular grid subgraphs Tg_1 and Tg_2. Observe that for each edge (a, b) in Tg_1, there exists a vertex $c \in N(a) \cap N(b)$, and there exists an s-t path, say P_1, that passed through (a, b) in $G - c$, such that we can replace edge (a, b) in P_1 by edges $(a, c),(c, b)$ to form another s-t path, say P_2. Observe that P_1 and P_2 differ in only one vertex i.e. c. Hence c is necessarily a tracker. The same holds true for each edge in Tg_2. Thus all vertices in $V(Tg_1) \cup V(Tg_2) \setminus \{s, t\}$ are necessarily trackers and hence belong to T. Since $|V(Tg_1) \cup V(Tg_2) \setminus \{s, t\}| = |E|^2+3|E|-2$, the remaining k trackers in T are vertices from V_v. Let V_t be the set of vertices in V_v that have been marked as trackers, i.e. $V_t = V_v \cap T$. Note that $|V_t| = k$. We denote the set of vertices in G that correspond to vertices in V_t as V_c. We claim that V_c forms a vertex cover for G. Suppose not. Then there exists an edge, say (a, b) in G, such none of its end points a, b belong to V_c. This implies that the vertices in V_v that correspond to a and b, say v_a, v_B, are not trackers in G'. Due to the construction of G', there exists a pair of vertices $v_i \in V_e$ and $v'_i \in V'_e$ (v_i, v'_i correspond to the edge (a, b) in G) such that v_a and v_b are adjacent to both v_i and v'_i.

Observe that for each pair of vertices v_i, v'_i, where $v_i \in V_e$ and $v'_i \in V'_e$, there exists two vertices in V_v (the vertices in $V(G)$ that correspond to the endpoints of the edge i in G) that are adjacent to both v_i and v'_i. Thus for each pair of vertices v_i, v'_i, there exists two paths between them passing through two distinct vertices in V_v. Further, there exists a path from s to v_i that is completely contained in Tg_1, and there exists a path from v'_i to t that is completely contained in Tg_2. Thus at least one of the vertices from V_v that are adjacent to v_i, v'_i, must necessarily be a tracker. This contradicts the fact that neither v_a nor v_b is a tracker in G'. This completes the proof.   □

Since VERTEX COVER is known to be NP-hard for graphs with maximum degree d ($d \geq 3$) [12], due to Lemma 5 we have the following corollary.

Corollary 1. TRACKING PATHS *is* NP-*hard for graphs with maximum degree* $\delta \geq 6$.

Algorithm 2 gives a procedure to find a $2(\delta + 1)$-approximate tracking set for undirected graphs with maximum degree δ. We prove its correctness in the following lemma.

Lemma 6. ⊛ *Algorithm 2 gives a* $2(\delta + 1)$-*approximate tracking set for an undirected graph.*

Lemma 7. ⊛ *Algorithm 2 runs in time* $\mathcal{O}(n^2)$.

From Lemma 6 and Lemma 7, we have the following theorem.

Algorithm 2: Finding a $2(\delta + 1)$-approximate tracking set for undirected graphs with maximum degree δ.

Input: Undirected graph $G = (V, E)$ such that $deg(x) \leq \delta$, $\forall x \in V$, a pair of vertices $s, t \in V$.

Output: Tracking Set $T \subseteq V$ for G.

1 Apply Reduction Rule 1;
2 Find a 2-approximate feedback vertex set S for G;
3 Set $T = S$;
4 **foreach** $v \in S$ **do**
5 | **foreach** $x \in N(v)$ **do**
6 | | $T = T \cup \{x\}$;
7 | **end**
8 **end**
9 Return T;

Theorem 2. *For an undirected graph G on n vertices such that the maximum degree of vertices in G is δ, there exists an $\mathcal{O}(n^2)$ algorithm that finds a $2(\delta + 1)$-approximate tracking set for G.*

The approximation ratio for our algorithm can be improved slightly by using the improved approximation bounds known for FVS in bounded degree graphs [1].

6   Reconstructing Paths Using Trackers

In real-world applications, it might be required to identify the s-t path which corresponds to a given sequence of trackers. Banik et al. [2] gave a polynomial time algorithm to reconstruct the shortest s-t path corresponding to a subset of trackers, given a tracking set for shortest s-t paths. Here we give an algorithm which, given a graph G, a constant size tracking set T, and a sequence of trackers π, returns the unique s-t path in G that corresponds to π, if one exists. Our algorithm works for both undirected graphs as well as tournaments.

Theorem 3. *Given a graph G, a tracking set T of constant size k for G, and a sequence of trackers π, the unique s-t path in G corresponding to π, if exists, can be found in polynomial time.*

Proof. Let $V(\pi)$ denote the vertices in the sequence π. Without loss of generality, let $|V(\pi)| = k$ and $\pi = (v_1, v_2, \ldots, v_k)$. Let P be the unique s-t path in G that corresponds to π. Let S be the set of pairs of vertices formed by consecutive vertices in π, preceding and ending with s and t respectively, i.e. $S = \{\{s, v_1\}, \{v_1, v_2\}, \ldots, \{v_k, t\}\}$. Since π is the sequence of trackers in P, $V(P)$ does not contain any trackers from T, other than those in π. In order to find P, we need to find the vertex disjoint paths between each pair of vertices (v_i, v_{i+1}) in S, where $v_0 = s$ and $v_{k+1} = t$. We create a copy v'_i for each vertex v_i in π, and introduce and edge between v_i' and each vertex in $N(v_i)$ in the graph G. Let $S' = \{\{s, v_1\}, \{v_1', v_2\}, \{v_2', v_3\} \ldots, \{v_{k-1}, v_k\}, \{v_k', t\}\}$ and $V(S')$ be the set of all vertices in S'. Consider the graph $G' = G - (T \setminus V(S'))$. If G is an undirected graph, then using the algorithm for disjoint paths in undirected graphs from [15], find the vertex disjoint paths between the pairs of vertices in S', in the graph G'. If G is a tournament graph, then using the algorithm for disjoint paths in tournaments from [8], find the vertex disjoint paths between the pairs of vertices in S', in the graph G'. Since disjoint path problem can be solved in polynomial time for undirected graphs and tournaments [8,15], we can perform this step in polynomial time. Observe that the sequence of these vertex disjoint paths will form an s-t path in G', which will also be an s-t path in G. Note that if the paths between pairs of vertices in S' are not vertex disjoint, it is a violation of the tracking set condition, as there are two vertex disjoint paths between a pair of vertices that have disjoint paths to s and t themselves. Next we prove that the path found will be a unique s-t path. If not, then there exists two s-t paths in G, containing the sequence of trackers π. This contradicts the assumption that T is a tracking set for G. Since T is assumed to be a tracking set for G, if vertex disjoint paths are not found between all pair of vertices in the S', then P does not exist. In this case the algorithm returns NO.   □

7   Tracking Edge Set for Undirected Graphs

In this section we study the problem of identifying s-t paths in an undirected edge-weighted graph using the edges of the graph. For a graph G, we define a *tracking edge set* as the set of edges whose intersection with each s-t path results in a unique sequence of edges. Here we allow parallel edges in the input graph. We formally define the problem of tracking paths using edges as follows.

TRACKING PATHS USING EDGES (G, s, t)
Input: An undirected edge-weighted graph $G = (V, E)$ with terminal vertices s and t.
Question: Find a minimum weight tracking edge set $T \subseteq E$ for G.

We start by first applying Reduction Rule 1, which ensures that each vertex and edge in the graph participates in some s-t path. Next we prove that each cycle in the reduced graph needs an edge as a tracker.

Lemma 8. ⊛ *For a reduced graph $G = (V, E)$, if $T \subseteq E$ is a tracking edge set, then each cycle in G contains an edge e such that $e \in T$.*

Note that the *tracking set condition* mentioned in Sect. 3 holds for a tracking edge set as well if we consider trackers as edges instead of vertices. Further, by using the arguments similar to those in Lemma 2, the following lemma for tracking using edges (instead of vertices) can be derived. Details are skipped to avoid repetition.

Lemma 9. *In a graph G, if $T \subseteq E(G)$ is not a tracking set for G, then there exist two s-t paths with the same sequence of trackers, and they form a cycle C in G, such that C has a local source a and a local destination b, and $T \cap (E(C) \setminus \{a, b\}) = \emptyset$.*

Next we prove that a feedback edge set (FES) is a tracking edge set for a reduced graph. An FES is a set of edges whose removal makes the graph acyclic.

Lemma 10. *For a reduced graph G, a feedback edge set F is also a tracking edge set for G.*

Proof. Consider graph $G = (V, E)$ reduced by Reduction Rule 1, and an FES $F \subseteq E$ for G. We claim that $T = F$ is a tracking edge set for G. Suppose not. Then there exists two s-t paths, say P_1 and P_2, in G, such that the sequence of tracking edges in both these paths is the same. Due to Lemma 9, the graph induced by P_1 and P_2 contains at least one cycle, say C, such that $C \cap T = \emptyset$. However, since T is an FES for G, it must necessarily contain an edge, say e, from the cycle C marked as a tracking edge. Observe that e can belong to either P_1 and P_2, but not both of them. This contradicts the assumption that P_1 and P_2 contain the same sequence of tracking edges. $\square$

Although finding a minimum FVS is an NP-hard problem, an FES can be found in polynomial time. We now prove that TRACKING PATHS USING EDGES can be solved in polynomial time.

Theorem 4. *For an undirected edge-weighted graph G on n vertices, TRACKING PATHS USING EDGES can be solved in $\mathcal{O}(n^2)$ time.*

Proof. Let G be an undirected edge-weighted graph on n vertices. From Lemma 10 it is known that an FES is a tracking edge set for G. In order to find a minimum weighted tracking edge set for G, we first find a maximum weight spanning tree T for G using Prim's algorithm or Kruskal's algorithm in $\mathcal{O}(n^2)$ time. Now the edges in $G - T$ comprise a minimum weight FES, which is also a minimum weight tracking edge set for G. $\square$

A path reconstruction algorithm similar to the one mentioned in Sect. 6 can be obtained by considering a sequence of tracking edges, and finding vertex disjoint paths between their endpoints in the graph obtained after removal of remaining tracking edges from the tracking edge set for that graph.

8 Conclusion

In this paper, we give polynomial time results for some variants of the TRACKING PATHS problem. Specifically, we solve TRACKING PATHS for chordal graphs and tournaments, along with giving an approximation algorithm for degree bounded graphs. We also analyze the problem TRACKING PATHS USING EDGES, and prove it to be polynomial time solvable. A constructive algorithm has also been given that helps identify an *s-t* path, given the unique sequence of trackers it contains. Future scope of this work lies in improving the running times of these algorithms and identifying more graph classes where TRACKING PATHS may be easily solvable. Open problems include finding approximation algorithms for other NP-hard variants of the problem for both undirected and directed graphs.

Acknowledgement. We thank Prof. Venkatesh Raman for the insightful discussions and suggestions.

References

1. Bafna, V., Berman, P., Fujito, T.: Constant ratio approximations of the weighted feedback vertex set problem for undirected graphs. In: Staples, J., Eades, P., Katoh, N., Moffat, A. (eds.) ISAAC 1995. LNCS, vol. 1004, pp. 142–151. Springer, Heidelberg (1995). https://doi.org/10.1007/BFb0015417
2. Banik, A., Katz, M.J., Packer, E., Simakov, M.: Tracking paths. In: Fotakis, D., Pagourtzis, A., Paschos, V.T. (eds.) CIAC 2017. LNCS, vol. 10236, pp. 67–79. Springer, Cham (2017). https://doi.org/10.1007/978-3-319-57586-5_7
3. Banik, A., Choudhary, P.: Fixed-parameter tractable algorithms for tracking set problems. In: Panda, B.S., Goswami, P.P. (eds.) CALDAM 2018. LNCS, vol. 10743, pp. 93–104. Springer, Cham (2018). https://doi.org/10.1007/978-3-319-74180-2_8
4. Banik, A., Choudhary, P., Lokshtanov, D., Raman, V., Saurabh, S.: A polynomial sized kernel for tracking paths problem. Algorithmica **82**(1), 41–63 (2020)
5. Bilò, D., Gualà, L., Leucci, S., Proietti, G.: Tracking routes in communication networks. In: Censor-Hillel, K., Flammini, M. (eds.) SIROCCO 2019. LNCS, vol. 11639, pp. 81–93. Springer, Cham (2019). https://doi.org/10.1007/978-3-030-24922-9_6
6. Choudhary, P.: Polynomial time algorithms for tracking path problems. CoRR abs/2002.07799 (2020). https://arxiv.org/abs/2002.07799
7. Choudhary, P., Raman, V.: Improved kernels for tracking path problems. CoRR abs/2001.03161 (2020). http://arxiv.org/abs/2001.03161
8. Chudnovsky, M., Scott, A., Seymour, P.: Disjoint paths in tournaments. Adv. Math. **270**, 582–597 (2015)
9. Duraisamy, K., Dempsey, K., Ali, H., Bhowmick, S.: A noise reducing sampling approach for uncovering critical properties in large scale biological networks. In: 2011 International Conference on High Performance Computing Simulation, pp. 721–728, July 2011. https://doi.org/10.1109/HPCSim.2011.5999898
10. Eppstein, D., Goodrich, M.T., Liu, J.A., Matias, P.: Tracking paths in planar graphs. In: 30th International Symposium on Algorithms and Computation, ISAAC 2019, 8–11 December 2019, Shanghai University of Finance and Economics, Shanghai, China, pp. 54:1–54:17 (2019)
11. Fisher, D.C., Ryan, J.: Tournament games and condorcet voting. Linear Algebra Appl. **217**, 87–100 (1995). Proceedings of a Conference on Graphs and Matrices in Honor of John Maybee
12. Garey, M., Johnson, D., Stockmeyer, L.: Some simplified np-complete graph problems. Theoret. Comput. Sci. **1**(3), 237–267 (1976)
13. Geman, D.: Random fields and inverse problems in imaging. In: Hennequin, P.-L. (ed.) École d'Été de Probabilités de Saint-Flour XVIII - 1988. LNM, vol. 1427, pp. 115–193. Springer, Heidelberg (1990). https://doi.org/10.1007/BFb0103042

14. Golumbic, M.C.: Perfect graphs (chapter 3). In: Golumbic, M.C. (ed.) Algorithmic Graph Theory and Perfect Graphs, pp. 51–80. Academic Press (1980)
15. Kawarabayashi, K., Kobayashi, Y., Reed, B.: The disjoint paths problem in quadratic time. J. Comb. Theory Ser. B **102**(2), 424–435 (2012)
16. Lauritzen, S.L., Spiegelhalter, D.J.: Local computations with probabilities on graphical structures and their application to expert systems. J. Roy. Stat. Soc.: Ser. B (Methodol.) **50**(2), 157–224 (1988)
17. McGarvey, D.C.: A theorem on the construction of voting paradoxes. Econometrica **21**(4), 608–610 (1953)
18. Rabiner, L.R.: A tutorial on hidden Markov models and selected applications in speech recognition. Proc. IEEE **77**(2), 257–286 (1989). https://doi.org/10.1109/5.18626
19. Stearns, R.: The voting problem. Am. Math. Mon. **66**(9), 761–763 (1959)

New Bounds for Maximizing Revenue in Online Dial-a-Ride

Ananya Christman[1]([✉])(iD), Christine Chung[2](iD), Nicholas Jaczko[1], Tianzhi Li[1], Scott Westvold[1], Xinyue Xu[1], and David Yuen[3](iD)

[1] Middlebury College, Middlebury, VT 05753, USA
achristman@middlebury.edu
[2] Connecticut College, New London, CT 06320, USA
cchung@conncoll.edu
[3] 92-1507 Punawainui St., Kapolei, HI 96707, USA

Abstract. In the Online-Dial-a-Ride Problem (OLDARP) a server travels to serve requests for rides. We consider a variant where each request specifies a source, destination, release time, and revenue that is earned for serving the request. The goal is to maximize the total revenue earned within a given time limit. We prove that no non-preemptive deterministic online algorithm for OLDARP can be guaranteed to earn more than half the revenue earned by OPT. We then investigate the SEGMENTED BEST PATH (SBP) algorithm of [8] for the general case of weighted graphs. The previously-established lower and upper bounds for the competitive ratio of SBP are 4 and 6, respectively, under reasonable assumptions about the input instance. We eliminate the gap by proving that the competitive ratio is 5 (under the same assumptions). We also prove that when revenues are uniform, SBP has competitive ratio 4. Finally, we provide a competitive analysis of SBP on complete bipartite graphs.

1 Introduction

In the On-Line Dial-a-Ride Problem (OLDARP), a server travels through a graph to serve requests for rides. Each request specifies a *source*, which is the pick-up (or start) location of the ride, a *destination*, which is the delivery (or end) location, and the *release time* of the request, which is the earliest time the request may be served. Requests arrive over time; specifically, each arrives at its release time and the server must decide whether to serve the request and at what time, with the goal of meeting some optimality criterion. The server has a *capacity* that specifies the maximum number of requests it can serve at any time. Common optimality criteria include minimizing the total travel time (i.e. makespan) to satisfy all requests, minimizing the average completion time (i.e. latency), or maximizing the number of served requests within a specified time limit. In many variants *preemption* is not allowed, so if the server begins to serve a request, it must do so until completion. On-Line Dial-a-Ride Problems have many practical applications in settings where a vehicle is dispatched to satisfy requests involving pick-up and delivery of people or goods. Important

© Springer Nature Switzerland AG 2020
L. Gąsieniec et al. (Eds.): IWOCA 2020, LNCS 12126, pp. 180–194, 2020.
https://doi.org/10.1007/978-3-030-48966-3_14

examples include ambulance routing, transportation for the elderly and disabled, taxi services including Ride-for-Hire systems (such as Uber and Lyft), and courier services.

We study a variation of OLDARP where in addition to the source, destination and release time, each request also has a priority and there is a time limit within which requests must be served. The server has unit capacity and the goal for the server is to serve requests within the time limit so as to maximize the total priority. A request's priority may simply represent the importance of serving the request in settings such as courier services. In more time-sensitive settings such as ambulance routing, the priority may represent the urgency of a request. In profit-based settings, such as taxi and ride-sharing services, a request's priority may represent the revenue earned from serving the request. For the remainder of this paper, we will refer to the priority as "revenue," and to this variant of the problem as ROLDARP. Note that if revenues are uniform the problem is equivalent to maximizing the number of served requests.

1.1 Related Work

The Online Dial-a-Ride problem was introduced by Feuerstein and Stougie [10] and several variations of the problem have been studied since. For a comprehensive survey on these and many other problems in the general area of *vehicle routing* see [12] and [16]. Feuerstein and Stougie studied the problem for two different objectives: minimizing completion time and minimizing latency. For minimizing completion time, they showed that any deterministic algorithm must have competitive ratio of at least 2 regardless of the server capacity. They presented algorithms for the cases of finite and infinite capacity with competitive ratios of 2.5 and 2, respectively. For minimizing latency, they proved that any algorithm must have a competitive ratio of at least 3 and presented a 15-competitive algorithm on the real line when the server has infinite capacity. Ascheuer et al. [2] studied OLDARP with multiple servers with the goal of minimizing completion time and presented a 2-competitive algorithm. More recently, Birx et al. [5] studied OLDARP on the real line and presented a new upper bound of 2.67 for the SMARTSTART algorithm [2], which improves the previous bounds of 3.41 [14] and 2.94 [4]. For OLDARP on the real line, Bjelde et al. [6] present a preemptive algorithm with competitive ratio 2.41. The Online Traveling Salesperson Problem (OLTSP), introduced by Ausiello et al. [3] and also studied by Krumke [15], is a special case of OLDARP where for each request the source and destination are the same location. There are many studies of variants of OLDARP and OLTSP [3, 11, 13, 15] that differ from the variant that we study which we omit here due to space limitations.

In this paper, we study OLDARP where each request has a revenue that is earned if the request is served and the goal is to maximize the total revenue earned within a specified time limit; the offline version of the problem was shown to be NP-hard in [8]. More recently, it was shown that even the special case of the offline version with uniform revenues and uniform weights is NP-hard [1]. Christman and Forcier [9] presented a 2-competitive algorithm for OLDARP on graphs with uniform edge weights. Christman et al. [8] showed that if edge weights may be arbitrarily large, then regardless of revenue values, no deterministic algorithm can be competitive. They therefore considered graphs where edge weights are bounded by a fixed fraction of the time limit, and gave a 6-competitive algorithm for this problem. Note that this is a natural subclass of inputs since in real-world dial-a-ride systems, drivers would be unlikely to spend a large fraction of their day moving to or serving a single request.

1.2 Our Results

In this work we begin with improved lower and upper bounds for the competitive ratio of the SEGMENTED BEST PATH (SBP) algorithm that was presented in [8]. We study SBP because it has the best known competitive ratio for ROLDARP and is a relatively straightforward algorithm. In [8], it was shown that SBP's competitive ratio has lower bound 4 and upper bound 6, provided that the edge weights are bounded by a fixed fraction of the time limit, i.e. T/f where T is the time limit and $1 < f < T$, and that the revenue earned by the optimal offline solution (OPT) in the last $2T/f$ time units is bounded by a constant. This assumption is imposed because, as we show in Lemma 1, no non-preememptive deterministic online algorithm can be guaranteed to earn this revenue. We note that as T grows, the significance of the revenue earned by OPT in the last two time segments diminishes.

We then close the gap between the upper and lower bounds of SBP by providing an instance where the lower bound is 5 (Sect. 3.1) and a proof for an upper bound of 5 (Sect. 3.2). We note that another interpretation of our result is that under a weakened-adversary model where OPT has two fewer time segments available, while SBP has the full time limit T, SBP is 5-competitive. We then investigate the problem for uniform revenues (so the objective is to maximize the total number of requests served) and prove that SBP earns at least $1/4$ the revenue of OPT, minus an additive term linear in f, the number of time segments (Sect. 4). This variant is useful for settings where all requests have equal priorities such as not-for-profit services that provide transportation to elderly and disabled passengers and courier services where deliveries are not prioritized.

We then consider the problem for complete bipartite graphs; for these graphs every source is from the left-hand side and every destination is from the right-hand side (Sect. 5). These graphs model the scenario where only a subset of locations may be source nodes and a disjoint subset may be destinations, e.g. in the delivery of goods from commercial warehouses only the warehouses may be sources and only customer locations may be destinations. We refer to this problem as ROLDARP-B. We first show that if edge weights are not bounded by a minimum value, then ROLDARP on general graphs reduces to ROLDARP-B. We therefore impose a minimum edge weight of kT/f for some constant k such that $0 < k \leq 1$. We show that if revenues are uniform, SBP has competitive ratio $\lceil 1/k \rceil$. Finally, we show that if revenues are nonuniform SBP has competitive ratio $\lceil 1/k \rceil$, provided that the revenue earned by OPT in the last $2T/f$ time units is bounded by a constant. (This assumption is justified by Lemma 1 which says no non-preemptive deterministic algorithm can be guaranteed to earn any fraction

Table 1. Bounds on the algorithm SBP for ROLDARP variants. † This upper bound assumes the optimal revenue of the last two time segments is bounded by a constant. ‡ This upper bound assumes the number of time segments is constant. § k is a constant where $0 < k \leq 1$ such that the minimum edge weight is kT/f where T is the time limit and $1 < f < T$.

| | Competitive ratio ρ of SBP for ROLDARP | |
	Uniform revenue	Nonuniform revenue
Weighted graphs	$\rho = 4^{\dagger\ddagger}$ ([8], [this work])	$\rho = 5^{\dagger}$ [this work]
Weighted bipartite graphs	$\rho \leq \lceil 1/k \rceil^{\S}$ [this work]	$\rho \leq \lceil 1/k \rceil^{\dagger\S}$ [this work]

of what is earned by OPT in the last $2T/f$ time units.) Table 1 summarizes our results.

2 Preliminaries

The Revenue-Online-Dial-a-Ride Problem (ROLDARP) is formally defined as follows. The input is an undirected complete graph $G = (V, E)$ where V is the set of vertices (or nodes) and $E = \{(u, v) : u, v \in V, u \neq v\}$ is the set of edges. For every edge $(u, v) \in E$, there is a weight $w_{u,v} > 0$, which represents the amount of time it takes to traverse (u, v).[1] One node in the graph, o, is designated as the origin and is where the server is initially located (i.e. at time 0). The input also includes a time limit T and a sequence of requests, σ, that are dynamically issued to the server.

Each request is of the form (s, d, t, p) where s is the source node, d is the destination, t is the time the request is released, and p is the revenue (or priority) earned by the server for serving the request. The server does not know about a request until its release time t. To serve a request, the server must move from its current location x to s, then from s to d. The total time for serving the request is equal to the length (i.e. travel time) of the path from x to s to d, and the earliest time a request may be released is at $t = 0$. For each request, the server must decide whether to serve the request and if so, at what time. A request may not be served earlier than its release time and at most one request may be served at any given time. Once the server decides to serve a request, it must do so until completion. The goal for the server is to serve requests within the time limit so as to maximize the total earned revenue. (The server need not return to the origin and may move freely through the graph at any time, even if it is not traveling to serve a request.)

The algorithm SEGMENTED BEST PATH (SBP) [8] starts by splitting the total time T into f segments each of length T/f (recall that f is fixed and $1 < f < T$).

---

[1] We note that any simple, undirected, connected, weighted graph is allowed as input, with the simple pre-processing step of adding an edge wherever one is not present whose weight is the length of the shortest path between its two endpoints. We further note that the input can be regarded as a metric space if the weights on the edges are expected to satisfy the triangle-inequality.

Algorithm 1: Algorithm SEGMENTED BEST PATH (SBP). Input is complete graph G with time limit T and maximum edge weight T/f.

1: Let $t_1, t_2, \ldots t_f$ be the time segments ending at times $T/f, 2T/f, \ldots, T$, resp.
2: Let $i = 1$.
3: **if** f is odd **then**
4:    At t_1, do nothing. Increment $i = 2$.
5: **end if**
6: **while** $i < f$ **do**
7:    At the start of t_i, find the *max-revenue-request-set*, R.
8:    **if** R is non-empty **then**
9:       Move to the source location of the first request in R.
10:      At the start of t_{i+1}, serve request-set R.
11:   **else**
12:      Remain idle for t_i and t_{i+1}
13:   **end if**
14:   Let $i = i + 2$.
15: **end while**

At the start of a time segment, the server determines the *max-revenue-request-set*, i.e. the maximum revenue set of unserved requests that can be served within one time segment, and moves to the source of the first request in this set. During the next time segment, it serves the requests in this set. It continues this way, alternating between moving to the source of first request in the max-revenue-request-set during one time segment, and serving this request-set in the next time segment. To find the max-revenue-request-set, the algorithm maintains a directed auxiliary graph, G' to keep track of unserved requests (an edge between two vertices u,v represents a request with source u and destination v). It finds all paths of length at most T/f between every pair of nodes in G' and returns the path that yields the maximum total revenue (please refer to [8] for full details).

It was observed in [8] that no deterministic online algorithm can be guaranteed to serve the requests served by OPT during the last time segment and the authors proved that SBP is 6-competitive barring an additive factor equal to the revenue earned by OPT during the last two time segments. More formally, let $\mathrm{rev}(\mathrm{SBP}(t_j))$ and $\mathrm{rev}(\mathrm{OPT}(t_j))$ denote the revenue earned by SBP and OPT respectively during the j-th time segment. Then if $\mathrm{rev}(\mathrm{OPT}(t_f)) + \mathrm{rev}(\mathrm{OPT}(t_{f-1})) \leq c$ for some constant c, then $\sum_{j=1}^{f} \mathrm{rev}(\mathrm{OPT}(t_j)) \leq 6 \sum_{j=1}^{f} \mathrm{rev}(\mathrm{SBP}(t_j)) + c$. It was also shown in [8] that as T grows, the competitive ratio of SBP is at best 4 (again with the additive term equal to $\mathrm{rev}(\mathrm{OPT}(t_f)) + \mathrm{rev}(\mathrm{OPT}(t_{f-1})))$, resulting in a gap between the upper and lower bounds.

2.1   General Lower Bound

We first present a general lower bound for this problem and show that *no* non-preemptive deterministic online algorithm (e.g. SBP) can be better than

2-competitive with respect to the revenue earned by the offline optimal schedule (ignoring the last two time segments; see Lemma 1, below).

Theorem 1. *No non-preemptive deterministic online algorithm for OLDARP can be guaranteed to earn more than half the revenue earned by* OPT *in the first* $T - 2T/f$ *time units. This is the case whether revenues are uniform or nonuniform.*

Proof (Sketch). The adversary repeatedly releases requests such that depending on which request(s) the algorithm serves, other request(s) are released that the algorithm cannot serve in time. This scheme requires carefully constructed edge weights, release times, and revenues so that the optimal offline revenue is always twice that of any online algorithm. Please see the full version of the paper for details [7].

We now show that *no* non-preemptive deterministic online algorithm (e.g. SBP) can be competitive with the revenue earned by OPT in the last two segments of time. We note that this claim applies to the version of non-preemption where, as in real-world systems like Uber/Lyft, once the server decides to serve a request, it must move there and serve it to completion.

Lemma 1. *No non-preemptive deterministic online algorithm can be guaranteed to earn any fraction of the revenue earned by* OPT *in the last* $2T/f$ *time units. This is the case whether revenues are uniform or nonuniform.*

Proof (). The adversary releases a request in the last two time segments and if the online algorithm chooses not to serve it, no other requests will be released. If the algorithm chooses to serve it, another batch of requests will be released elsewhere that the algorithm cannot serve in time. Please see the full version of the paper for details [7].

3   Nonuniform Revenues

In this section we improve the lower and upper bounds for the competitive ratio of the SEGMENTED BEST PATH algorithm [8]. In particular, we eliminate the gap between the lower and upper bounds of 4 and 6, respectively, from [8], by providing an instance where the lower bound is 5 and a proof for an upper bound of 5. Note that throughout this section we assume the revenue earned by OPT in the last two time segments is bounded by some constant. We must impose this restriction on the OPT revenue of the last two time segments because, as we showed in Lemma 1, *no* non-preemptive deterministic online algorithm can be guaranteed to earn any constant fraction of this revenue.

3.1   Lower Bound on SBP

Theorem 2. *If the revenue earned by* OPT *in the last two time segments is bounded by some constant, and* SBP *is* γ-*competitive, then* $\gamma \geq 5$.

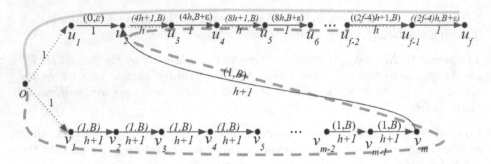

Fig. 1. An instance where OPT (whose path is shown in dashed green below) earns $5 - 4/(f - 2)$ times the revenue of SBP (shown in solid yellow above). In this instance, $T = 2hf$, and edges that represent requests are shown as solid edges. For each such edge the release time followed by revenue of the corresponding request is shown in parenthesis above the edge. The weight of an edge is shown below the edge. Dashed edges represent empty moves. (Color figure online)

Proof (Sketch). For the formal details, please refer to the proof of Theorem 2 in the full version [7]. Consider the instance depicted in Fig. 1. Since $T = 2hf$ in this instance, h represents "half" the length of one time segment, so only one request of length $h + 1$ fits within a single time segment for SBP. The general idea of the instance is that while SBP is serving every other request across the top row of requests (since the other half across the top are not released until after SBP has already passed them by), OPT is serving the entire bottom row in one long chain, then also has time to serve the top row as one long chain.

3.2   Upper Bound on SBP

We now show that SBP is 5-competitive by creating a modified, hypothetical SBP schedule that has additional copies of requests. First, we note that SBP loses a factor of 2 due to the fact that it serves requests during only every other time segment. Then, we lose another factor of two to cover requests in OPT that overlap between time segments. Finally, by adding at most one more copy of the requests served by SBP to make up for requests that SBP "incorrectly" serves prior to when they are served by OPT, we end up with 5 copies of SBP being sufficient for bounding the total revenue of OPT. Note that while this proof uses some of the techniques of the proof of the 6-competitive upper bound in [8], it reduces the competitive ratio from 6 to 5 by cleverly extracting the set of requests that SBP serves prior to OPT before making the additional copies.

Let rev(OPT) and rev(SBP) denote the total revenue earned by OPT and SBP over all time segments t_j from $j = 1 \ldots f$.

Theorem 3. *If the revenue earned by OPT in the last two time segments is bounded by some constant c, then SBP is 5-competitive, i.e., if $\mathrm{rev}(\mathrm{OPT}(t_f)) + \mathrm{rev}(\mathrm{OPT}(t_{f-1})) \le c$, then $\sum_{j=1}^{f} \mathrm{rev}(\mathrm{OPT}(t_j)) \le 5\sum_{j=1}^{f} \mathrm{rev}(\mathrm{SBP}(t_j)) + c$. Note* *that another interpretation of this result is that under a resource augmentation model where* SBP *has two more time segments available than* OPT, SBP *is 5-competitive.*

Proof. We analyze the revenue earned by SBP by considering the time segments in pairs (recall that the length of a time segment is T/f for some $1 < f < T$). We refer to each pair of consecutive time segments as a time window, so if there are f time segments, there are $\lceil f/2 \rceil$ time windows. Note that the last time window may have only one time segment.

For notational convenience we consider a modified version of the SBP schedule, that we refer to as SBP′, which serves exactly the same set of requests as SBP, but does so one time window earlier. Specifically, if SBP serves a set of requests during time window $i \geq 2$, SBP′ serves this set during time window $i-1$ (so SBP′ ignores the set served by SBP in window 1). We note that the schedule of requests served by SBP′ may be infeasible, and that it will earn at most the amount of revenue earned by SBP.

Let B_i denote the set of requests served by OPT in window i that SBP′ already served *before* in some window $j < i$. And let B be the set of all requests that have already been served by SBP′ in a previous window by the time they are served in the OPT schedule. Formally, $B = \bigcup_{i=2}^{\lceil f/2 \rceil} B_i$. Consider a schedule $\overline{\text{OPT}}$ that contains all of the requests in the OPT schedule minus the requests in B. So $\overline{\text{OPT}}$ earns total revenue rev(OPT) − rev(B), where rev(B) denotes the total revenue of the set B.

Let $\overline{\text{OPT}}(t_j)$ denote the set of requests served by $\overline{\text{OPT}}$ in time segment t_j. Let $\overline{\text{OPT}}_i$ denote the set of requests served by $\overline{\text{OPT}}$ in the time segment of window i with greater revenue, i.e. $\overline{\text{OPT}}_i = \arg\max\{\text{rev}(\overline{\text{OPT}}(t_{2i-1})), \text{rev}(\overline{\text{OPT}}(t_{2i}))\}$. Note this set may include a request that was started in the prior time segment, as long as it was completed in the time segment of $\overline{\text{OPT}}_i$. Let rev($\overline{\text{OPT}}_i$) denote the revenue earned in $\overline{\text{OPT}}_i$.

Let SBP′$_i$ denote the set of requests served by SBP′ in window i and let rev(SBP′$_i$) denote the revenue earned by SBP′$_i$. Let H denote the chronologically ordered set of time windows w where rev($\overline{\text{OPT}}_w$) > rev(SBP′$_w$), and let h_j denote the jth time window in H. We refer to each window of H as a window with a "hole," in reference to the fact that SBP′ does not earn as much revenue as $\overline{\text{OPT}}$ in these windows. In each window h_j there is some amount of revenue that $\overline{\text{OPT}}$ earns that SBP′ does not. In particular, there must be a set of requests that $\overline{\text{OPT}}$ serves in window h_j that SBP′ does not serve in h_j. Note that this set must be available for SBP′ in h_j since $\overline{\text{OPT}}$ does not include the set B.

Let $\overline{\text{OPT}}_{h_j} = A_j \cup C_j^*$, where A_j is the subset of requests served by both $\overline{\text{OPT}}$ and SBP′ in h_j and C_j^* is the subset of $\overline{\text{OPT}}$ requests available for SBP′ to serve in h_j but SBP′ chooses not to serve. Let us refer to the set of requests served by SBP′ in h_j as SBP′$_{h_j} = A_j \cup C_j$ for some set of requests C_j. Note that if $\overline{\text{OPT}}_{h_j} = A_j \cup C_j^*$ can be executed within a single time segment, then rev(C_j) $\geq$ rev(C_j^*) by the greediness of SBP′. However, since h_j is a hole we know that the set $\overline{\text{OPT}}_{h_j}$ cannot be served within one time segment.

Our plan is to build an infeasible schedule $\overline{\text{SBP}}$ that will be similar to SBP$'$ but contain additional "copies" of some requests such that no windows of $\overline{\text{SBP}}$ contain holes. We first initialize $\overline{\text{SBP}}$ to have the same schedule of requests as SBP$'$. We then add additional requests to h_j for each $j = 1 \ldots |H|$, based on $\overline{\text{OPT}}_{h_j}$.

Consider one such window with a hole h_j, and let k be the index of the time segment corresponding to $\overline{\text{OPT}}_{h_j}$. We know $\overline{\text{OPT}}$ must have begun serving a request of $\overline{\text{OPT}}_{h_j}$ in time segment t_{k-1} and completed this request in time segment t_k. Let us use r^* to denote this request that "straddles" the two time segments.

After the initialization of $\overline{\text{SBP}} = $ SBP$'$, recall that the set of requests served by $\overline{\text{SBP}}$ in h_j is $\overline{\text{SBP}}_{h_j} = A_j \cup C_j$ for some set of requests C_j. We add to $\overline{\text{SBP}}$ a copy of a set of requests. There are two sub-cases depending on whether $r^* \in C_j^*$ or not.

Case $r^* \in C_j^*$. In this case, by the greediness of SBP, and the fact that both r^* alone and $C_j^* \setminus \{r^*\}$ can separately be completed within a single time segment, we have: $\text{rev}(C_j) \geq \max\{\text{rev}(r^*), \text{rev}(C_j^* \setminus \{r^*\})\} \geq \frac{1}{2}\text{rev}(C_j^*)$. We then add a copy of the set C_j to the $\overline{\text{SBP}}$ schedule, so there are two copies of C_j in h_j. Note that for $\overline{\text{SBP}}$, h_j will no longer be a hole since: $\text{rev}(\overline{\text{OPT}}_{h_j}) = \text{rev}(A_j) + \text{rev}(C_j^*) \leq \text{rev}(A_j) + 2 \cdot \text{rev}(C_j) = \text{rev}(\overline{\text{SBP}}_{h_j})$.

Case $r^* \notin C_j^*$. In this case C_j^* can be served within one time segment but SBP$'$ chooses to serve $A_j \cup C_j$ instead. So we have $\text{rev}(A_j) + \text{rev}(C_j) \geq \text{rev}(C_j^*)$, therefore we know either $\text{rev}(A_j) \geq \frac{1}{2}\text{rev}(C_j^*)$ or $\text{rev}(C_j) \geq \frac{1}{2}\text{rev}(C_j^*)$. In the latter case, we can do as we did in the first case above and add a copy of the set C_j to the $\overline{\text{SBP}}$ schedule in window h_j, to get $\text{rev}(\overline{\text{OPT}}_{h_j}) \leq \text{rev}(\overline{\text{SBP}}_{h_j})$, as above. In the former case, we instead add a copy of A_j to the $\overline{\text{SBP}}$ schedule in window h_j. Then again, for $\overline{\text{SBP}}$, h_j will no longer be a hole, since this time: $\text{rev}(\overline{\text{OPT}}_{h_j}) = \text{rev}(A_j) + \text{rev}(C_j^*) \leq 2 \cdot \text{rev}(A_j) + \text{rev}(C_j) = \text{rev}(\overline{\text{SBP}}_{h_j})$.

Note that for all windows $w \notin H$ that are not holes, we already have $\text{rev}(\overline{\text{SBP}}_w) \geq \text{rev}(\overline{\text{OPT}}_w)$. So we have

$$\sum_{i=1}^{\lceil f/2 \rceil - 1} \text{rev}(\overline{\text{OPT}}_i) \leq \sum_{i=1}^{\lceil f/2 \rceil - 1} \text{rev}(\overline{\text{SBP}}_i) \leq 2 \sum_{i=1}^{\lceil f/2 \rceil - 1} \text{rev}(\text{SBP}_i'). \tag{1}$$

where the second inequality is because $\overline{\text{SBP}}$ contains no more than two instances of every request in SBP$'$. Combining (1) with the fact that SBP$'$ earns at most what SBP does yields

$$\sum_{i=1}^{\lceil f/2 \rceil} \text{rev}(\overline{\text{OPT}}_i) \leq 2 \sum_{i=1}^{\lceil f/2 \rceil} \text{rev}(\text{SBP}_i) + \text{rev}(\overline{\text{OPT}}(t_{f-1})) + \text{rev}(\overline{\text{OPT}}(t_f)). \tag{2}$$

Since SBP serves in only one of two time segments per window, we have $\sum_{i=1}^{\lceil f/2 \rceil} \text{rev}(\text{SBP}_i) = \sum_{j=1}^{f} \text{rev}(\text{SBP}(t_j))$. Hence, by the definition of $\overline{\text{OPT}}$, and by (2) we can say

$$\sum_{j=1}^{f} \operatorname{rev}(\overline{\mathrm{OPT}}(t_j)) \leq 2 \sum_{i=1}^{\lceil f/2 \rceil} \operatorname{rev}(\overline{\mathrm{OPT}}_i)$$

$$\leq 4 \sum_{j=1}^{f} \operatorname{rev}(\mathrm{SBP}(t_j)) + \operatorname{rev}(\overline{\mathrm{OPT}}(t_{f-1})) + \operatorname{rev}(\overline{\mathrm{OPT}}(t_f)). \quad (3)$$

Now we must add in any request in B, such that OPT serves the request in a time window after SBP$'$ serves that request. By definition of B (as the set of all requests that have been served by SBP$'$ in a previous window) B may contain at most the same set of requests served by SBP$'$. Therefore $\operatorname{rev}(B) \leq \operatorname{rev}(\mathrm{SBP}')$, so $\operatorname{rev}(B) \leq \operatorname{rev}(\mathrm{SBP})$. By the definition of OPT, OPT $= \overline{\mathrm{OPT}} + B$, so

$$\sum_{j=1}^{f} \operatorname{rev}(\mathrm{OPT}(t_j)) = \operatorname{rev}(B) + \sum_{j=1}^{f} \operatorname{rev}(\overline{\mathrm{OPT}}(t_j)) \quad (4)$$

And by combining (3)–(4) with the fact that $\operatorname{rev}(B) \leq \operatorname{rev}(\mathrm{SBP})$, we have $\sum_{j=1}^{f} \operatorname{rev}(\mathrm{OPT}(t_j)) \leq \sum_{j=1}^{f} \operatorname{rev}(\mathrm{SBP}(t_j)) + 4\sum_{j=1}^{f} \operatorname{rev}(\mathrm{SBP}(t_j)) + \operatorname{rev}(\overline{\mathrm{OPT}}(t_{f-1})) + \operatorname{rev}(\overline{\mathrm{OPT}}(t_f)) \leq 5\sum_{j=1}^{f} \operatorname{rev}(\mathrm{SBP}(t_j)) + \operatorname{rev}(\overline{\mathrm{OPT}}(t_{f-1})) + \operatorname{rev}(\overline{\mathrm{OPT}}(t_f))$.

4   Uniform Revenues

We now consider the setting where revenues are uniform among all requests, so the goal is to maximize the total number of requests served. This variant is useful for settings where all requests have equal priorities, for example for not-for-profit services that provide transportation to elderly and disabled passengers. The proof strategy is to carefully consider the requests served by SBP in each window and track how they differ from that of OPT. The final result is achieved through a clever accounting of the differences between the two schedules, and bounding the revenue of the requests that are "missing" from SBP.

We note that the lower bound instance of Theorem 2 can be modified to become a uniform-revenue instance that has ratio $5 - 14/f$. We further note that the lower bound instance provided in [8] immediately establishes a lower bound instance for SBP that has a ratio of 4. We now show that OPT earns at most 4 times the revenue of SBP in this setting if we assume the revenue earned by OPT in the last two time segments is bounded by a constant, and allow SBP an additive bonus of f. Note that even when revenues are uniform, *no* non-preemptive deterministic online algorithm can earn the revenue earned by OPT in the last two time segments (see Lemma 1). We begin with several definitions and lemmas.

As in the proof of Theorem 3, we consider a modified version of the SBP schedule, that we refer to as SBP$'$, which serves exactly the same set of requests as SBP, but does so one time window earlier. For all windows $i = 1, 2, ..., m$, where $m = \lceil f/2 \rceil - 1$, we let S_i' denote the set of requests served by SBP$'$ in window

i and S_i^* denote the set of requests served by OPT during the time segment of window i with greater revenue, i.e. $S_i^* = \arg\max\{\mathrm{rev}(\mathrm{OPT}(t_{2i-1})), \mathrm{rev}(\mathrm{OPT}(t_{2i}))\}$ where $\mathrm{rev}(\mathrm{OPT}(t_j))$ denotes the revenue earned by OPT in time segment t_j. We define a new set J_i^* as the set of requests served by OPT during the time segment of window i with less revenue, i.e. $J_i^* = \arg\min\{\mathrm{rev}(\mathrm{OPT}(t_{2i-1})), \mathrm{rev}(\mathrm{OPT}(t_{2i}))\}$.

Let $S_i^* = A_i \cup X_i^* \cup Y_i^*$, and $S_i' = A_i \cup X_i \cup Y_i$, where: (1) A_i is the set of requests that appear in both S_i^* and S_i'; (2) X_i^* is the set of requests that appear in S_w' for some $w = 1, 2, ..., i-1$. Note there is only one possible w for each individual request $r \in X_i^*$, because each request can be served only once; (3) Y_i^* is the set of requests such that no request from Y_i^* appears in S_w' for any $w = 1, 2, ..., i-1, i$; (4) X_i is the set of requests that appear in S_w^* for some $w = 1, 2, ..., i-1$. Note there is only one possible w for each individual request $r \in X_i$, because each request can be served only once; (5) Y_i is the set of requests such that no request from Y_i appears in S_w^* for any $w = 1, 2, ..., i-1, i$.

Note that elements in Y_i can appear in a previous J_w^* for any $w = 1, 2, ..., i-1, i$ or in a future S_v^* or J_v^* for any $v = i+1, i+2, ..., m$, or may not appear in any other sets. Also note that since each request can be served at most once, we have: $A_1 \cap X_1^* \cap Y_1^* \cap A_2 \cap X_2^* \cap Y_2^* \cap ... \cap A_m \cap X_m^* \cap Y_m^* = \emptyset$ and $A_1 \cap X_1 \cap Y_1 \cap A_2 \cap X_2 \cap Y_2 \cap ... \cap A_m \cap X_m \cap Y_m = \emptyset$.

Given the above definitions, we have the following lemma whose proof has been deferred to the full version of the paper [7]. It states that at any given time window, the cumulative requests of OPT that were earlier served by SBP are no more than the number that have been served by SBP but not yet by OPT.

Lemma 2. *For all $i = 1, 2, ..., m$ we have $\sum_{j=1}^{i} |X_j^*| \le \sum_{j=1}^{i} |Y_i|$.*

We are now ready to prove our main theorem of this section.

Theorem 4. *If the revenue earned by OPT in the last two time segments is bounded by some constant c, i.e., if $\mathrm{rev}(\mathrm{OPT}(t_f)) + \mathrm{rev}(\mathrm{OPT}(t_{f-1})) \le c$, then SBP earns at least $1/4$ the revenue of OPT, minus an additive term linear in f, where T/f is the length of one time segment. (So if f is also bounded by some constant, then SBP is 4-competitive). I.e., $\sum_{j=1}^{f} \mathrm{rev}(\mathrm{OPT}(t_j)) \le 4\sum_{j=1}^{f} \mathrm{rev}(\mathrm{SBP}(t_j)) + 2\lceil f/2 \rceil + c$.*

Proof. Note that since revenues are uniform, the revenue of a request-set U is equal to the size of the set U, i.e., $\mathrm{rev}(U) = |U|$. Consider each window i where $\mathrm{rev}(S_i^*) > \mathrm{rev}(S_i')$. Note that the set S_i^* may not fit within a single time segment. We consider two cases based on S_i^*.

1. The set S_i^* can be served within one time segment. Note that within $S_i^* = A_i \cup X_i^* \cup Y_i^*$, X_i^* is not available for SBP' to serve because SBP' has served the requests in X_i^* prior to window i. Among requests that are available to SBP', SBP' greedily chooses to serve the maximum revenue set that can be served within one time segment. Therefore, we have $\mathrm{rev}(X_i) + \mathrm{rev}(Y_i) \ge \mathrm{rev}(Y_i^*)$. Since revenues are uniform, we also have $|X_i| + |Y_i| \ge |Y_i^*|$.
   If this is not the case, then SBP' would have chosen to serve Y_i^* instead of $X_i \cup Y_i$ since it is feasible for SBP' to do so because the entire S_i^* can be served within one time segment.

2. The set S_i^* cannot be served within one time segment. This means there must be one request in S_i^* that OPT started serving in the previous time segment. We refer to this straddling request as r^*. There are three sub-cases based on where r^* appears.

(a) If $r^* \in Y_i^*$, then due to the greediness of SBP$'$, we know that

$$\text{rev}(X_i) + \text{rev}(Y_i) \geq \text{rev}(r^*) \tag{5}$$

since otherwise SBP$'$ would have chosen to serve r^*. We also know

$$\text{rev}(X_i) + \text{rev}(Y_i) \geq \text{rev}(Y_i^* \backslash \{r^*\}) \tag{6}$$

since otherwise SBP$'$ would have chosen to serve $Y_i^* \backslash \{r^*\}$.
From (5), we have $|X_i| + |Y_i| \geq 1$ and from (6), we have $|X_i| + |Y_i| \geq |Y_i^*| - 1$.

(b) If $r^* \in X_i^*$, then r^* is not available to SBP$'$ and only A_i, X_i, Y_i, and Y_i^* are available to SBP$'$. Therefore we know that $\text{rev}(X_i) + \text{rev}(Y_i) \geq \text{rev}(Y_i^*)$ since otherwise, by its greediness, SBP$'$ would have chosen to serve A_i and Y_i^* instead of A_i, X_i and Y_i, because A_i and Y_i^* can be served within one time segment. Therefore, we have $|X_i| + |Y_i| \geq |Y_i^*|$.

(c) $r^* \in A_i$. Then r^* is served by both OPT and SBP$'$. We know that $A_i \cup Y_i^* \backslash \{r^*\}$ can be served within one time segment since r^* is the only request that causes S_i^* to straddle between two time segments. Again by the greediness of SBP$'$, we have $\text{rev}(A_i) + \text{rev}(X_i) + \text{rev}(Y_i) \geq \text{rev}(A_i) + \text{rev}(Y_i^*) - \text{rev}(r^*)$ which means $\text{rev}(X_i) + \text{rev}(Y_i) > \text{rev}(Y_i^*) - \text{rev}(r^*)$ and $|X_i| + |Y_i| \geq |Y_i^*| - 1$.

Therefore, for all cases, for window i, we have $|X_i| + |Y_i| \geq |Y_i^*| - 1$, which means $|Y_i^*| - |X_i| \leq 1 + |Y_i|$, and with $m = \lceil f/2 \rceil - 1$,

$$\sum_{i=1}^{m}(|Y_i^*| - |X_i|) \leq m + \sum_{i=1}^{m} |Y_i|. \tag{7}$$

Now we will build an infeasible schedule $\overline{\text{SBP}}$ that will be similar to SBP$'$ but contain additional "copies" of some requests such that no windows of $\overline{\text{SBP}}$ contain holes, i.e. such that $\text{rev}(\overline{\text{SBP}}) \geq \sum_{i=1}^{m} \text{rev}(S_i^*)$.

We define a modified OPT schedule which we refer to as OPT$'$ such that OPT$' = \cup_{i=1}^{m} S_i^*$ and observe that $\text{rev}(\text{OPT}') = \sum_{i=1}^{m} |A_i| + \sum_{i=1}^{m} |X_i^*| + \sum_{i=1}^{m} |Y_i^*|$, while $\text{rev}(\text{SBP}') = \sum_{i=1}^{m} |A_i| + \sum_{i=1}^{m} |X_i| + \sum_{i=1}^{m} |Y_i|$.

By Lemma 2 and Eq. (7), we can say $\text{rev}(\text{OPT}') - \text{rev}(\text{SBP}') = \sum_{i=1}^{m} |Y_i^*| - \sum_{i=1}^{m} |X_i| + \sum_{i=1}^{m} |X_i^*| - \sum_{i=1}^{m} |Y_i| < \sum_{i=1}^{m} |Y_i^*| - \sum_{i=1}^{m} |X_i| \leq m + \sum_{i=1}^{m} |Y_i|$. This tells us that to form an $\overline{\text{SBP}}$ whose revenue is at least that of OPT$'$, we must "compensate" SBP$'$ by adding to it at most copies of all requests in the set Y_i for all $i = 1, 2, ..., m$, plus m "dummy requests." In other words,

$$\text{rev}(\overline{\text{SBP}}) = \text{rev}(\text{SBP}') + m + \sum_{i=1}^{m} |Y_i| \geq \text{rev}(\text{OPT}'). \tag{8}$$

We know the total revenue of all Y_i can not exceed the total revenue of SBP', hence we have

$$\text{rev}(\overline{\text{SBP}}) = \text{rev}(\text{SBP}') + m + \sum_{i=1}^{m} |Y_i| \le 2\,\text{rev}(\text{SBP}') + m. \tag{9}$$

Combining (8) and (9), we get $\text{rev}(\text{OPT}') \le 2\,\text{rev}(\text{SBP}') + m$, which means

$$\sum_{i=1}^{m} \text{rev}(S_i^*) \le 2 \sum_{i=1}^{m} \text{rev}(S_i') + m. \tag{10}$$

Recall that S_i^* is the set of requests served by OPT during the time segment of window i with greater revenue. In other words, $\sum_{j=1}^{2m} \text{rev}(S^*(t_j)) \le 2\sum_{i=1}^{m} \text{rev}(S_i^*)$, which, combined with (10), gives us

$$\sum_{j=1}^{2m} \text{rev}(S^*(t_j)) \le 4 \sum_{i=1}^{m} \text{rev}(S_i') + 2m. \tag{11}$$

We assumed that the total revenue of requests served in the last two time segments by OPT is bounded by c. From (11), we get

$$\sum_{j=1}^{f} \text{rev}(S^*(t_j)) \le \sum_{j=1}^{2m} \text{rev}(S^*(t_j)) + \text{rev}(S^*(t_{f-1})) + \text{rev}(S^*(t_f)) \le 4 \sum_{i=1}^{m} \text{rev}(S_i') + 2m + c. \tag{12}$$

We also know that the total revenue of requests served by SBP' during the first m windows is less than or equal to the total revenue of SBP. Therefore, from (12), we have $\sum_{j=1}^{f} \text{rev}(S^*(t_j)) \le 4\sum_{j=1}^{f} \text{rev}(S(t_j)) + 2m + c$.

5   Bipartite Graphs

In this section, we consider ROLDARP for complete bipartite graphs $G = (V = V_1 \cup V_2, E)$, where only nodes in V_1 maybe be source nodes and only nodes in V_2 may be destination nodes. One node is designated as the origin and there is an edge from this node to every node in V_1 (so the origin is a node in V_2). Due strictly to space limitations, most proofs of theorems in this section are deferred to the full version of the paper [7].

We refer to this problem as ROLDARP-B and the offline version as RDARP-B. We first show that if edge weights of the bipartite graph are not bounded by a minimum value, then the offline version of ROLDARP on general graphs, which we refer to as RDARP, reduces to RDARP-B. Since RDARP has been show in [1,8] to be NP-hard (even if revenues are uniform), this means RDARP-B is NP-hard as well.

Theorem 5. *The problem RDARP is poly-time reducible to RDARP-B. Also, RDARP with uniform revenues is poly-time reducible to RDARP-B with uniform revenues.*

Proof (Sketch). The idea of the reduction is to split each node into two nodes connected by an edge in the bipartite graph with a distance of ϵ. Then we turn each edge in the original graph into two edges in the bipartite graph. Please see the full version for details [7].

5.1   Uniform Revenue Bipartite

We show that for bipartite graph instances, if revenues are uniform, we can guarantee that SBP earns a fraction of OPT equal to the ratio between the minimum and maximum edge-length.

Theorem 6. *For any instance of ROLDARP-B where the revenues are uniform for all requests, if edge weights are upper and lower bounded by T/f and kT/f, respectively, for some constant $0 < k \leq 1$, then* $\text{rev}(\text{OPT}) \leq \lceil 1/k \rceil \cdot \text{rev}(\text{SBP}) + \lceil 1/k \rceil$.

Proof (Sketch). The proof idea is akin to that of Theorem 7 below. Please see the full version of the paper for details [7].

5.2   Nonuniform Revenue Bipartite

In this section we show that even if revenues are nonuniform, we can still guarantee that SBP earns a fraction of OPT equal to the ratio between the minimum and maximum edge-length, minus the revenue earned by OPT in the last window. Recall that we refer to each pair of consecutive time segments as a time window. Note that *no* non-preemptive deterministic online algorithm can be competitive with any fraction of the revenue earned by OPT in the last $2T/f$ time units (i.e. Lemma 1 also holds for ROLDARP-B with nonuniform revenues). Due space limitations, please refer to the full version of this work [7] for the proof of the following theorem.

Theorem 7. *For any instance of ROLDARP-B where the revenues of requests are nonuniform, if edge weights are upper and lower bounded by T/f and kT/f, respectively, for some constant $0 < k \leq 1$, and if the revenue earned by OPT in the last time window is bounded by some constant c, then* $\text{rev}(\text{OPT}) \leq \lceil 1/k \rceil \cdot \text{rev}(\text{SBP}) + c$.

References

1. Anthony, B., et al.: Maximizing the number of rides served for dial-a-ride. In: 19th Workshop on Algorithmic Approaches for Transportation Modelling, Optimization, and Systems (ATMOS 2019). Schloss Dagstuhl-Leibniz-Zentrum fuer Informatik (2019)
2. Ascheuer, N., Krumke, S.O., Rambau, J.: Online dial-a-ride problems: minimizing the completion time. In: Reichel, H., Tison, S. (eds.) STACS 2000. LNCS, vol. 1770, pp. 639–650. Springer, Heidelberg (2000). https://doi.org/10.1007/3-540-46541-3_53

3. Ausiello, G., Feuerstein, E., Leonardi, S., Stougie, L., Talamo, M.: Algorithms for the on-line travelling salesman 1. Algorithmica **29**(4), 560–581 (2001)
4. Birx, A., Disser, Y.: Tight analysis of the smartstart algorithm for online dial-a-ride on the line. In: 36th International Symposium on Theoretical Aspects of Computer Science (2019)
5. Birx, A., Disser, Y., Schewior, K.: Improved bounds for open online dial-a-ride on the line. In: Approximation, Randomization, and Combinatorial Optimization. Algorithms and Techniques (APPROX/RANDOM 2019). Schloss Dagstuhl-Leibniz-Zentrum fuer Informatik (2019)
6. Bjelde, A., et al.: Tight bounds for online TSP on the line. In: Proceedings of the Twenty-Eighth Annual ACM-SIAM Symposium on Discrete Algorithms, pp. 994–1005. Society for Industrial and Applied Mathematics (2017)
7. Christman, A., et al.: New bounds for maximizing revenue in online dial-a-ride. arXiv preprint arXiv:1912.06300 (2020)
8. Christman, A., Chung, C., Jaczko, N., Milan, M., Vasilchenko, A., Westvold, S.: Revenue maximization in online dial-a-ride. In: 17th Workshop on Algorithmic Approaches for Transportation Modelling, Optimization, and Systems (ATMOS 2017). Schloss Dagstuhl-Leibniz-Zentrum fuer Informatik (2017)
9. Christman, A., Forcier, W.: Maximizing revenues for on-line dial-a-ride. In: Zhang, Z., Wu, L., Xu, W., Du, D.-Z. (eds.) COCOA 2014. LNCS, vol. 8881, pp. 522–534. Springer, Cham (2014). https://doi.org/10.1007/978-3-319-12691-3_38
10. Feuerstein, E., Stougie, L.: On-line single-server dial-a-ride problems. Theoret. Comput. Sci. **268**(1), 91–105 (2001)
11. Jaillet, P., Wagner, M.R.: Generalized online routing: new competitive ratios, resource augmentation, and asymptotic analyses. Oper. Res. **56**(3), 745–757 (2008)
12. Jaillet, P., Wagner, M.R.: Online vehicle routing problems: a survey. In: Golden, B., Raghavan, S., Wasil, E. (eds.) The Vehicle Routing Problem: Latest Advances and New Challenges. ORCS, vol. 43, pp. 221–237. Springer, Boston (2008). https://doi.org/10.1007/978-0-387-77778-8_10
13. Jawgal, V.A., Muralidhara, V.N., Srinivasan, P.S.: Online travelling salesman problem on a circle. In: Gopal, T.V., Watada, J. (eds.) TAMC 2019. LNCS, vol. 11436, pp. 325–336. Springer, Cham (2019). https://doi.org/10.1007/978-3-030-14812-6_20
14. Krumke, S.O.: Online optimization: competitive analysis and beyond (2002)
15. Krumke, S.O., de Paepe, W.E., Poensgen, D., Lipmann, M., Marchetti-Spaccamela, A., Stougie, L.: On minimizing the maximum flow time in the online dial-a-ride problem. In: Erlebach, T., Persinao, G. (eds.) WAOA 2005. LNCS, vol. 3879, pp. 258–269. Springer, Heidelberg (2006). https://doi.org/10.1007/11671411_20
16. Molenbruch, Y., Braekers, K., Caris, A.: Typology and literature review for dial-a-ride problems. Ann. Oper. Res. 295–325 (2017). https://doi.org/10.1007/s10479-017-2525-0

Iterated Type Partitions

Gennaro Cordasco[1(✉)], Luisa Gargano[2], and Adele A. Rescigno[2]

[1] University of Campania "L.Vanvitelli", Caserta, Italy
gennaro.cordasco@unicampania.it
[2] University of Salerno, Fisciano, Italy

Abstract. This paper introduces a novel parameter, called iterated type partition, that can be computed in polynomial time and nicely places between modular-width and neighborhood diversity. We prove that the Equitable Coloring problem is W[1]-hard when parametrized by the iterated type partition. This result extends to modular-width, answering an open question on the complexity of Equitable Coloring when parametrized by modular-width. On the contrary, we show that the Equitable Coloring problem is FPT when parameterized by neighborhood diversity. Furthermore, we present a scheme for devising FPT algorithms parameterized by iterated type partition, which enables us to find optimal solutions for several graph problems. While the considered problems are already known to be FPT with respect to modular-width, the novel algorithms are both simpler and more efficient. As an example, in this paper, we give an algorithm for the Dominating Set problem that outputs an optimal set in time $O(2^t + poly(n))$, where n and t are the size and the iterated type partition of the input graph, respectively.

Keywords: Parameterized complexity · Fixed-parameter tractable algorithms · W[1]-hardness · Neighborhood diversity · Modular-width

1 Introduction

Some NP-hard problems can be solved by algorithms that are exponential only in the size of a parameter while they are polynomial in the size of the input. Such problems are called fixed-parameter tractable, because the problem can be solved efficiently for small values of the parameter [10,33]. Formally, a parameterized problem with input size n and parameter t is called *fixed parameter tractable (FPT)* if it can be solved in time $f(t) \cdot n^c$, where f is a computable function only depending on t and c is a constant.

An important quality of a parameter is that it is easy to compute. Unfortunately there are several parameters whose computation is an NP-hard problem. As an example computing treewidth, rankwidth, and vertex cover are all NP-hard problems but they are computable in FPT time when their respective parameters are bounded; moreover, the parameterized complexity of computing the clique-width of a graph exactly is still an open problem [11].

© Springer Nature Switzerland AG 2020
L. Gąsieniec et al. (Eds.): IWOCA 2020, LNCS 12126, pp. 195–210, 2020.
https://doi.org/10.1007/978-3-030-48966-3_15

We start from two recently introduced parameters: modular-width [22] and neighborhood diversity [31]. Both parameters received much attention [1,2,5,7,12,17,18,21,24,25,29] also due to their property of being computable in polynomial time [22,31].

As the main contribution of this paper we introduce a novel parameter called Iterated Type Partition, which nicely places between the two above parameters and allows to obtain new algorithms and hardness results.

1.1   Modular-Width

The notion of modular decomposition of graphs was introduced by Gallai in [23], as a tool to define hierarchical decompositions of graphs. It has been recently considered in [22] to define the modular-width parameter in the area of parameterized computation.

Consider graphs obtainable by an algebraic expression that uses the operations:

1) Creation of an isolated vertex.
2) Disjoint union of 2 graphs, i.e., the graph with vertex set $V(G_1) \cup V(G_2)$ and edge set $E(G_1) \cup E(G_2)$.
3) Complete join of 2 graphs, i.e., the graph with vertex set $V(G_1) \cup V(G_2)$ and edge set $E(G_1) \cup E(G_2) \cup \{(v, w) : v \in V(G_1), w \in V(G_2)\}$.
4) Substitution operation $G(G_1, \ldots, G_m)$ of the vertices $v_1, \ldots, v_m$ of G by the modules $G_1, \ldots, G_m$, i.e., the graph with vertex set $\bigcup_{1 \leq \ell \leq m} V(G_\ell)$ and edge set

$$\bigcup_{1 \leq \ell \leq m} E(G_\ell) \cup \{(u, v) : u \in V(G_i), v \in V(G_j), (v_i, v_j) \in E(G)\}.$$

As defined in [22], the *modular-width* of a graph G, denoted $mw(G)$, is the least integer m such that G can be obtained by using only the operations 1)–4) (in any number and order) and where each operation 4) has at most m modules.

1.2   Neighborhood Diversity

Given a graph $G = (V, E)$, two nodes $u, v \in V$ have the same *type* iff $N(v) \setminus \{u\} = N(u) \setminus \{v\}$. The *neighborhood diversity* of a graph G, introduced by Lampis in [31] and denoted by $nd(G)$, is the minimum number t of sets in a partition $V_1, V_2, \ldots, V_t$, of the node set V, such that all the nodes in V_i have the same type, for $i \in [t]$[1].

The family $\mathcal{V} = \{V_1, V_2, \ldots, V_t\}$ is called the *type partition* of G.

Let $G = (V, E)$ be a graph with type partition $\mathcal{V} = \{V_1, V_2, \ldots, V_t\}$. By definition, each V_i induces either a *clique* or an *independent set* in G. We treat singleton sets in the type partition as cliques. For each $V_i, V_j \in \mathcal{V}$, we get that

---

[1] For a positive integer n, we use $[n]$ to denote the set of the first n integers, that is $[n] = \{1, 2, \ldots, n\}$.

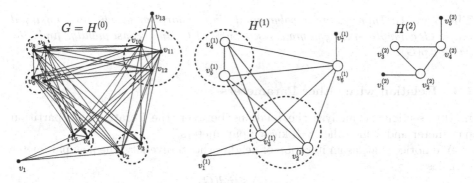

Fig. 1. A graph G with iterated type partition 5 and its corresponding type graph sequence: $G = H^{(0)}, H^{(1)}, H^{(2)}$. Dashed circles group nodes having the same type.

either each node in V_i is a neighbor of each node in V_j or no node in V_i has a neighbor in V_j. Hence, between each pair $V_i, V_j \in \mathcal{V}$, there is either a complete bipartite graph or no edges at all.

Starting from a graph G and its type partition $\mathcal{V} = \{V_1, \ldots, V_t\}$, we can see each element of $\mathcal{V}$ as a *metavertex* of a new graph H, called the *type graph* of G, with

- $V(H) = \{1, 2, \cdots, t\}$
- $E(H) = \{(x, y) \mid x \neq y$ and for each $u \in V_x, v \in V_y$ it holds that $(u, v) \in E(G)\}$.

We say that G is a *base graph* if it matches its type graph, that is, the type partition of G consists of singletons, each representing a node in $V(G)$, and $nd(G) = |V(G)|$.

We introduce a new graph parameter, which generalizes neighborhood diversity. Given a graph G, the *Iterated Type Partition* of G is defined by iteratively constructing type graphs until a base graph is obtained. It is worth mentioning that our base graphs correspond to prime graphs for modular decomposition [1].

Definition 1. *Given a graph* $G = (V, E)$, *let* $H^{(0)} = G$ *and* $H^{(i)}$ *denote the type graph of* $H^{(i-1)}$, *for* $i \geq 1$. *Let* d *be the smallest integer such that* $H^{(d)}$ *is a base graph. The* iterated type partition *of* G, *denoted by* $itp(G)$, *is the number of nodes of* $H^{(d)}$. *The sequence of graphs* $H^{(0)} = G, H^{(1)}, \cdots, H^{(d)}$ *is called the* type graph sequence *of* G *and* $H^{(d)}$ *is denoted as the* base graph *of* G.

An example of a graph and its type graph sequence is given in Fig. 1. For each type graph $H^{(i)}$ each vertex (henceforth metavertex) describes an element of the type partition of $H^{(i-1)}$.

It is well-known that determining $nd(G)$ and the corresponding type partition, can be done in polynomial time [31]. As an immediate consequence, we have that

Theorem 1. *There exists a polynomial time algorithm which, for any input graph G computes the type graphs sequence of G and, consequently, finds the value $itp(G)$.*

1.3   Relation with Other Parameters

In this section we analyze the relations between the iterated type partition parameter and some other well known parameters.

We notice that, as an iteration of neighborhood diversity, the new parameter satisfies

$$itp(G) \leq nd(G). \tag{1}$$

Actually $itp(G)$ can be much smaller than $nd(G)$. Indeed consider the following:

– Choose a positive integer d and a connected base graph $H^{(d)}$ having k nodes;
– For $i = d, d-1, \ldots, 1$, a new graph $H^{(i-1)}$ is obtained as follows:

   • replace each node of $H^{(i)}$, with an independent set of at least two nodes (if $d - i$ is even) or a clique of size at least two (if $d - i$ is odd).
   • for each edge of $H^{(i)}$, put a complete bipartite graph between the nodes of the graphs that replace the endpoints of the edge.

The value $nd(H^{(0)})$ is the number of nodes in $H^{(1)}$, that is at least $k2^{d-1}$, while $itp(H^{(0)})$ is the size k of $H^{(d)}$.

We stress that iterated type partition is a "special case" of modular-width in which the modules in operation 4) can only be independent sets or cliques. Hence, it is not difficult to see that for every graph G

$$mw(G) \leq itp(G). \tag{2}$$

We know from [31] that $nd(G) \leq 2^{vc(G)} + vc(G)$. Hence, by (1), we have $itp(G) \leq 2^{vc(G)} + vc(G)$. Moreover, using the same arguments as in [31] is it possible to show that $cw(G) \leq itp(G) + 1$. Finally, as for the neighborhood diversity we can easily show that the iterated type partition is incomparable to the treewidth by comparing the values of such parameters on a complete graph K_n and a path on n nodes. A summary of the relations holding between some popular parameters is given in Fig. 2. We refer to [19] for the formal definitions of treewidth and clique-width parameters.

1.4   Our Results and Related Work

We give both tractability and hardness results for the new parameter.

The *Equitable Coloring* (EQC) Problem. If the nodes of a graph G are colored with k colors such that no adjacent nodes receive the same color (i.e., properly colored) and the sizes of any two color classes differ by at most one, then G is called to be *equitably k-colorable* and the coloring is called an *equitable k-coloring*. The goal is to minimize the number of used colors. The EQC problem

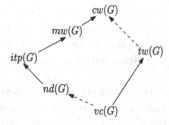

Fig. 2. A summary of the relations holding among some popular parameters. In addition to the previously defined parameters, we use $tw(G)$, $cw(G)$ and $vc(G)$ to denote treewidth, clique-width and minimum vertex cover of a graph G, respectively. Solid arrows denote generalization, e.g., modular-width generalizes iterated type partition. Dashed arrows denote that the generalization may exponentially increase the parameter.

is a well-studied problem, which has been analyzed in terms of parameterized positive or negative results with respect to many different parameters [26].

In particular, Fellows et al. [14] have shown that EQC problem parameterized by treewidth and number of colors is $W[1]$-hard. A series of reductions proving that Equitable Coloring is $W[1]$-hard for different subclasses of chordal graphs are given in [26]: The problem is shown to be W[1]-hard if parameterized by the number of colors for block graphs and for the disjoint union of split graphs; moreover, it remains W[1]-hard for $K_{1,4}$-free interval graphs even when parameterized by treewidth, number of colors and maximum degree. In [3] an XP algorithm parameterized by treewidth is given. We notice that an XP algorithm for Equitable Coloring parametrized by iterated type partition can be obtained by using Theorem 17 in [28]. On the other side, Fiala *et al.* show that the Equitable Coloring problem is FPT when parameterized by the vertex cover number [16]. However, it was an open problem to establish the parameterized complexity of the Equitable Coloring problem parameterized by neighborhood diversity or modular-width. In Sect. 2 we answer to these questions by proving the following results.

Theorem 2. *The Equitable Coloring problem is $W[1]$-hard parametrized by* itp.

Recalling (2), Theorem 2 immediately gives that the Equitable Coloring Problem is $W[1]$-hard w.r.t. modular-width.

Corollary 1. *The EQC problem is $W[1]$-hard parametrized by modular-width.*

We also show that the Equitable Coloring $W[1]$-hardness drops when parameterized by the neighborhood diversity.

Theorem 3. *The EQC problem is FPT when parameterized by neighborhood diversity.*

FPT Algorithms w.r.t. itp. In the last section we deal with FPT algorithms with respect to iterated type partition. Some of the considered problems are already known to be FPT w.r.t modular-width. Nonetheless, we think that the new algorithms, parameterized by iterated type partition, are worthy to be considered, since they are much simpler, faster, and allow to easily determine not only the value, but also the optimal solution. As an example, we consider here the Dominating Set (DS).

Table 1 summarizes our contribution, in relation to known results. Due to space constraints, some proofs are omitted or sketched; full proofs as well as the algorithms for the Vertex Coloring (Coloring) and the Vertex Cover (VC) problems appear in the extended version of this work [6].

Table 1. The table summarizes the results known in literature for several problems parametrized by iterated type partition and related parameters. t denotes the value of the considered parameter and [*] denotes the result obtained in this paper.

	DS, VC	Coloring	EQC
cw	FPT [9]	W[1]-hard [19]	W[1]-hard [20]
mw	FPT [34]	FPT [22]	W[1]-hard [*]
itp	FPT($O(2^t + poly(n))$)[*]	FPT($O(t^{2.5t+o(t)}\log n + poly(n))$)[*]	W[1]-hard [*]
nd	FPT [31]	FPT [31]	FPT[*]
vc	FPT [31]	FPT [31]	FPT [16]

2   Equitable Coloring (EQC)

In this section we prove Theorems 2 and 3.

Equitable Coloring
Instance: A graph $G = (V, E)$ and an integer k.
Question: Is it possible to color the nodes of G with exactly k colors in such a way that nodes connected by an edge receive different colors and each color class has either size $\lfloor |V|/k \rfloor$ or $\lceil |V|/k \rceil$?

2.1   Hardness

In order to prove that Equitable Coloring problem is $W[1]$-hard if parameterized by iterated type partition, we present a reduction from the following Bin packing problem, which has been shown to be W[1]-hard when parameterized by the number of bins [27].

Bin-Packing
Instance: A collection of ℓ items having sizes $a_1, a_2, \cdots, a_\ell$, a number k of bins, and a bin capacity B.
Question: $\exists$ a k-partition $P_1, \cdots, P_k$ of $A = \{a_1, a_2, \cdots, a_\ell\}$ such that $\sum_{a_j \in P_i} a_j = B$, $\forall i \in [k]$?

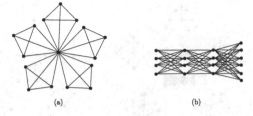

Fig. 3. (a) $(4,3)$–flower; (b) $(3,5,4)$–chain.

In general the Bin-Packing problem asks for the sum of the items of each bin to be *at most B*; however, the above version is equivalent to the general one (even from the parameterized point of view) as it is sufficient to add $kB - \sum_{j=1}^{\ell} a_j$ unitary items [26]. In order to describe our reduction, we introduce two useful gadgets. The first one is the flower gadget also used in [26]. Let a and k be positive integers. An (a,k)–*flower* $F_{a,k}$ is a graph obtained by joining $a+1$ cliques of size k to a central node y. Figure 3(a) shows the $(4,3)$–flower. Formally, let K_k^i be a copy of a cliques of size k, for each $i \in [a+1]$,

- $V(F_{a,k}) = \{y\} \cup \bigcup_{i \in [a+1]} V(K_k^i)$, and
- $E(F_{a,k}) = \{(y,x) \mid x \in \bigcup_{i \in [a+1]} V(K_k^i)\} \cup \bigcup_{i \in [a+1]} E(K_k^i)$.

The second gadget is defined starting from three positive integers: k, ℓ and B. It is a sequence of independent sets $S_1, \cdots, S_k, S_{k+1}$ with $|S_i| = B$, for $i \in [k]$, and $|S_{k+1}| = \ell + 1$ where between each pair of consecutive sets in the sequence S_i, S_{i+1} there is a complete bipartite graph. We call such a gadget a (k, ℓ, B)–*chain* Q. Figure 3(b) shows the $(3,5,4)$–chain. Formally,

- $V(Q) = \bigcup_{i \in [k+1]} S_i$, and
- $E(Q) = \bigcup_{i \in [k]} \{(u,v) \mid u \in S_i, v \in S_{i+1}\}$.

We can now describe our reduction. Let $\langle A = \{a_1, \cdots, a_\ell\}, k, B \rangle$ be an instance of Bin-Packing. Define a graph G as follows: The set of nodes is composed by the disjoint union of two (k, ℓ, B)-chains, Q' and Q'' plus the flowers $F_{a_1,k}, \cdots, F_{a_\ell,k}, F_{B,k}$. Then join each node in the flowers to each node in the chains. In the following, whenever the number of bins k is clear by the context, we use F_a instead of $F_{a,k}$. Formally,

- $V(G) = V(Q') \cup V(Q'') \cup V(F_B) \cup \left(\bigcup_{j \in [\ell]} V(F_{a_j}) \right)$, and

$E(G) = E(Q') \cup E(Q'') \cup E(F_B) \cup \left(\bigcup_{j \in [\ell]} E(F_{a_j}) \right) \cup$
$$\left\{ (x,u) \mid x \in V(F_B) \cup \left(\bigcup_{j \in [\ell]} V(F_{a_j}) \right), \ u \in V(Q') \cup V(Q'') \right\}.$$

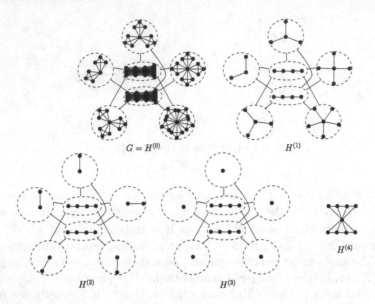

Fig. 4. The type graph sequence of G when $A = \{2, 1, 2, 3\}$, $B = 4$, and $k = 3$. The line connecting dashed circles indicates a complete bipartite graph between the nodes in the circles.

Figure 4 shows the graph G when $A = \{2, 1, 2, 3\}$, $B = 4$ and $k = 3$. The number of nodes in the resulting graph G is

$$|V(G)| = |V(Q')| + |V(Q'')| + |V(F_B)| + \sum_{j \in [\ell]} |V(F_{a_j})| = (k+3)(Bk+\ell+1). \quad (3)$$

Lemma 1. $\langle A = \{a_1, \cdots, a_\ell\}, k, B \rangle$ *is a YES instance of Bin-Packing if and only if G is equitably $(k+3)$–colorable.*

Proof. (*Sketch.*) We first show that, given a k-partition $P_1, \cdots, P_k$ of A that solves the given instance of Bin-Packing, i.e., $\sum_{a_j \in P_i} a_j = B$ for each $i \in [k]$, we can construct an equitable (k+3)-coloring c of the nodes of G.

- Coloring of the nodes in Q': For each $i \in [k+1]$ and $u \in S'_i$ (where S'_i is the i-th set of independent nodes in the (k, ℓ, B)-chain Q') assign

$$c(u) = \begin{cases} k+3 & \text{if } i \text{ is odd,} \\ k+2 & \text{if } i \text{ is even.} \end{cases} \quad (4)$$

- Coloring of the nodes in Q'': For each $i \in [k+1]$ and $u \in S''_i$, (where S''_i is the i-th set of independent nodes in the (k, ℓ, B)-chain Q'') assign

$$c(u) = \begin{cases} k+3 & \text{if } i \text{ is even,} \\ k+2 & \text{if } i \text{ is odd.} \end{cases} \quad (5)$$

- Coloring of the nodes in F_B: Let z be the central node in F_B. Assign $c(z) = k + 1$. Then, assign to each of the k nodes of the $B + 1$ cliques joined to z the remaining k colors (e.g. $1, 2, \cdots k$), so that the nodes of the clique have different colors.
- Coloring of the nodes in F_{a_j}, for $j \in [\ell]$: Let y_j be the central node in F_{a_j}. Assign $c(y_j) = i$ if $a_j \in P_i$. Then, as before assign to each of the k nodes of the $a_j + 1$ cliques joined to y_j the remaining k colors, i.e., those in $\{1, 2, \cdots k, k + 1\} - \{i\}$, so that the nodes of the clique have different colors.

The above coloring c can be proved to be proper and such that each class of colors contains exactly $Bk + \ell + 1$ nodes. By (3) this proves that c is an equitable $(k + 3)$-coloring of G.

Now, let c be an equitable $(k + 3)$-coloring of G. We can prove that exactly two colors among the $k + 3$ are used by c to color only the nodes in the chains Q' and Q''. Furthermore, the color used by c to color the central node of the flower F_B is not used to color the central nodes of any other flowers $F_{a_1}, \cdots, F_{a_\ell}$. By using this result, we can prove that the k classes of colors involving the central nodes of the $F_{a_1}, \cdots, F_{a_\ell}$ induce a k-partition of A that solves our instance of Bin-Packing. □

Lemma 2. *The iterated type partition $itp(G)$ of G is $2k + 3$.*

Proof. (Sketch.) The graph G has type graph sequence $H^{(0)} = G, H^{(1)}, H^{(2)}, H^{(3)}, H^{(4)}$. We derive the above graphs and show that the number of nodes of the base $H^{(4)}$ is $2k + 3$. Figure 4 shows the type graph sequence when $A = \{2, 1, 2, 3\}$, $B = 4$ and $k = 3$. □

Proof of Theorem 2. Given an instance $\langle A = \{a_1, \cdots, a_\ell\}, k, B \rangle$ of Bin-Packing, we use the above construction to create an instance $\langle G = (V, E), itp(G) \rangle$ of Equitable Coloring parameterized by iterated type partition. Lemma 1 show the correctness of our reduction and Lemma 2 provides the iterated type partition of the constructed graph, showing that our new parameter $itp(G)$ is linear in the original parameter k. □

2.2 Neighborhood Diversity: An FPT Algorithm

We prove here that the Equitable Coloring problem admits an FPT algorithm with respect to neighborhood diversity. W.l.o.g. we assume that the number of nodes in the input graph $G = (V, E)$ is a multiple of the number of colors k (this can be attained by adding an independent clique of $k\lceil |V|/k \rceil - |V|$ nodes to G in such a way the answer to the equitable k-coloring question remains unchanged).

Let then $r = |V|/k$. Any equitable k-coloring of G partitions V into k classes of colors, say $C_1, \ldots, C_k$, s.t. C_ℓ is an independent set of G of size $|C_\ell| = r$, for $\ell = 1, \ldots, k$.

If we consider now the type partition $\{V_1, \ldots, V_t\}$ of G and the corresponding type graph $H = (V(H) = \{1, \ldots, t\}, E(H))$, we trivially have that: *Two nodes $u, v \in V$ are independent in G iff $v \in V_i$ and $u \in V_j$, with $i, j \in V(H)$, such that* *either i and j are independent nodes of H or $i = j$ and V_i induces an independent set in G.* This immediately implies that for each color class C_ℓ of the equitable coloring of G there exists an independent set $I_\ell = \{\ell_1, \ldots, \ell_\rho\}$ of H such that

$\sum_{s=1}^{\rho} |C_\ell \cap V_{\ell_s}| = r$ and
$|C_\ell \cap V_{\ell_s}| = 1$ for each $s = 1, \ldots, \rho$ such that V_{ℓ_s} induces a clique.

Let now $\mathcal{I}$ denote the family of all independent sets in H. From the above reasoning, we have that, given any equitable k-coloring of G, we can associate to each $I \in \mathcal{I}$ an independent set of $z_I \geq 0$ colors. We can then define, for each $I \in \mathcal{I}$ and $i \in I$, an integer $z_{I,i}$ representing the number of nodes in V_i that (in the coloring of G) are colored with one of the z_I colors associated to I. Clearly, the value of $z_{I,i}$ is at most z_I if V_i induces a clique in G, but can be larger if V_i induces an independent set. An equitable k-coloring of G satisfies the following conditions:

1. $\sum_{I \in \mathcal{I}} z_I = k$.
2. For each $i \in V(H)$ it holds that the sum of the values $z_{I,i}$, over all $I \in \mathcal{I}$ such that $i \in I$, is exactly $|V_i|$.
3. For each $I \in \mathcal{I}$ it holds that the sum over all $i \in V(H)$ of the number of nodes of V_i that are colored in G with one of the z_I colors associated to I is $r \cdot z_I$.

The above conditions can be expressed by the following linear program on the variables z_I for each $I \in \mathcal{I}$ and $z_{I,i}$ for each $I \in \mathcal{I}$ and for each $i \in I$.

1. $\sum_{I \in \mathcal{I}} z_I = k$;
2. $\sum_{I : i \in I} z_{I,i} = |V_i|$, for each $i \in V(H)$;
3. $\sum_{i \in I} z_{I,i} - r \cdot z_I = 0$, for each $I \in \mathcal{I}$;
4. $z_I - z_{I,i} \geq 0$ for each $I \in \mathcal{I}$ and $i \in I$ such that V_i is a clique;
5. $z_{I,i} \geq 0$ for each $I \in \mathcal{I}$ and $i \in V(H)$.

From the above reasoning, it is clear that if the graph G admits an equitable k-coloring, then there exists an assignation of values to the variables z_I and $z_{I,i}$, for each $I \in \mathcal{I}$ and $i \in I$, that satisfies the above system.

We show now that from any assignation of values to the variables z_I and $z_{I,i}$ that satisfies the above system, we can obtain an equitable k-coloring of G.

- For each independent set $I \in \mathcal{I}$, such that $z_I > 0$, repeat the following procedure:
    - Select a set of z_I new colors, say $c_1^I, \ldots, c_{z_I}^I$ (to be used only for nodes in I);
    We notice that (by 3.) the total number of nodes to be colored is $r \cdot z_I$;
    - Consider the list of colors $c_1^I, c_2^I, \ldots, c_{z_I}^I, c_1^I, c_2^I, \ldots, c_{z_I}^I, \ldots, c_1^I, c_2^I, \ldots, c_{z_I}^I$ (obtained by repeating the sequence $c_1^I, \ldots, c_{z_I}^I$ r times). Assign the colors starting from the beginning of the list as follows: For each $i \in V(H)$, select $z_{I,i}$ uncolored nodes in V_i (it can be done by 2.) and assign to them the next unassigned $z_{I,i}$ colors in the list.

In this way each color is used exactly r times. Moreover, since each independent set uses a separate set of colors, the total number of colors is $\sum_{I \in \mathcal{I}} z_I = k$ (crf. 1.). Furthermore, in each V_i that induces a clique in G, we color $z_{I,i} \leq z_I$ nodes (this holds by 4.). Such nodes get colors which are consecutive in the list, hence they are different. Summarizing, the desired equitable k-coloring of G has been obtained.

Finally, we evaluate the time to solve the above system. We use the well-known result that Integer Linear Programming is FPT parameterized by the number of variables.

ℓ-Variable Integer Linear Programming Feasibility
Instance: A matrix $A \in Z^{m \times \ell}$ and a vector $b \in Z^m$.
Question: Is there a vector $x \in Z^\ell$ such that $Ax \geq b$?

Proposition 1. *[32] ℓ-Variable Integer Linear Programming Feasibility can be solved in time $O(\ell^{2.5t+o(\ell)} \cdot L)$ where L is the number of bits in the input.*

Since $|V(H)| = nd(G)$, our system uses at most $O(nd(G)2^{nd(G)})$ variables: z_I for $I \in \mathcal{I}$ and $z_{I,i}$ for $I \in \mathcal{I}$ and $i \in I$. We have $O(nd(G)2^{nd(G)})$ constraints and the coefficients are upper bounded by $r = |V|/k$. Therefore, Theorem 3 holds.

3   Algorithms

In this section, we deal with FPT algorithms with respect to iterated type partition. In order to solve a problem P on an input graph G, the general algorithm scheme is:

1) Iterate by generating the whole type graph sequence of G.
2) On each graph G' in the type graph sequence, a generalized version P' of the original problem is defined (with P' in G' being equivalent to P in G).
3) Optimally solve P' on the base graph and reconstruct the solution on the reverse type graph sequence (hence solving P in G).

If the construction of the solution for P' (at step 2), can be done in polynomial time and the time to solve P' on the base graph is f, then the whole algorithm needs $O(f + poly(n))$ time.

Using the scheme above we are able to prove that the Dominating Set and Vertex Cover problems can be solved in time $O(2^t + poly(n))$, while the Vertex Coloring problem is solvable in time $O(t^{2.5t+o(t)} \log n + poly(n))$, where n and t are the size and the iterated type partition of the input graph, respectively. In the following, we present the algorithm for the Dominating Set problem. Due to space constraints the proofs for the remaining problems are given in the extended version of this paper [6].

In the following, we assume that the input graph is connected and it is not a clique. Indeed, the domination problem in disconnected graphs can be separately solved on each connected component. Moreover, in the case of a complete graph, the solution trivially consists of one vertex. Notice that the assumption of G

being a non complete connected graph, implies that the base graph of G is connected and $itp(G) \geq 2$.

In order to present our FPT algorithm, we consider the following generalized dominating set problem.

Definition 2. *Given a graph $G = (V, E)$ (connected and not complete) and a set of nodes $Q \subseteq V$, a semi-total Dominating Set of G with respect to Q, called Q-stds of G, is a set $D \subseteq V$ such that every node in Q is adjacent to a node in D, and every other node is either a node in D or is adjacent to a node in D. The set D is called an optimal Q-stds of G, if its size is minimum among all the Q-stds of G.*

Clearly, when $Q = V$, the semi-total Dominating Set problem is the Total Domination problem [4]; if $Q = \emptyset$ it becomes the Dominating Set problem.

Lemma 3. *Let $G = (V, E)$ be a graph and let $\mathcal{V} = \{V_1, \cdots, V_t\}$ be the type partition of G. Let $Q \subseteq V$. There exists an optimal Q-stds D of G such that*

$$|V_x \cap D| \leq 1 \qquad \text{for each } x \in [t]. \tag{6}$$

Proof. Let D be an optimal Q-stds of G. Assume there exists $x \in [t]$ such that $|V_x \cap D| \geq 2$. We distinguish two cases according to V_x being a clique or an independent set.

Let V_x be a clique. Let u and v be two nodes in $V_x \cap D$. Let $u \notin Q$. Since u is a neighbor of v and since u and v share the same neighborhood, we have that the set $D' = D - \{v\}$ is a Q-stds of G. Furthermore, $|D'| < |D|$ and this is not possible since D is optimal. Assume now that $u \in Q$. If there exists a neighbor w of u with $w \in V_y \cap D$, for some $y \neq x$, then as above $D' = D - \{v\}$ is a Q-stds of G and $|D'| < |D|$. If, otherwise, node u has no neighbor in D except for those in V_x, then we can choose any neighbor w of u with $w \in V_y \cap D$, for some $y \neq x$, and $D' = D - \{v\} \cup \{w\}$ is a Q-stds of G and $|D'| = |D|$.

Let V_x be an independent set. Let u be any node in $V_x \cap D$. If there exists a neighbor w of u with $w \in V_y \cap D$, for some $y \neq x$, then the set D' obtained from D removing all the nodes in V_x except for u is again a Q-stds since the neighbors of nodes in V_x are dominated by u and all the nodes in V_x are dominated by $w \in V_y$. Furthermore, $|D'| < |D|$. Otherwise, we have that $V_x \subset D$ and for each neighbor w of u it holds $w \in V_y$, for $y \neq x$, and $w \notin D$. Hence, the set D' obtained from D removing all the nodes in V_x except for u and adding to it a node $w \in V_y$, where y is such that $V_y \cap D = \emptyset$, is a Q-stds of G. Furthermore, $|D'| \leq |D|$.

Repeating the above argument for each $x \in [t]$ such that $|V_x \cap D| \geq 2$, we obtain an optimal solution satisfying (6). $\qquad \square$

The FPT algorithm Dom recursively constructs the graphs in the type graph sequence of G, until the base graph is obtained. It is initially called with $\text{Dom}(G, \emptyset)$. At each recursive step, the algorithm $\text{Dom}(H, Q)$, on a graph H and a set $Q \subseteq V(H)$ of nodes that need to have a neighbor in the solution set, checks if H is a base graph or not. In case H is a base graph, then the algorithm searches by brute force the Q-stds of H and returns it. If H is not a base graph

---

Algorithm 1.   Algorithm $\mathsf{Dom}(H, Q)$

Input: A graph $H = (V(H), E(H))$, a set $Q \subseteq V(H)$.
1 **if** H *is a base graph* **then**
2 $\quad$ $D = V(H)$
3 $\quad$ **for each** $S \subseteq V(H)$ **do if** $((S$ is Q-stds of $H)$ **and** $(|S| < |D|))$ **then** $D = S$
4 **else**
5 $\quad$ Let $V_1, \cdots, V_t$ be the type partition of H and let H' be the type graph of H.
6 $\quad$ $Q' = \{x \in V(H') \mid (V_x \cap Q \neq \emptyset$ or V_x is an independent set)\}$
7 $\quad$ $D' = \mathsf{Dom}(H', Q')$
8 $\quad$ $D = \bigcup_{x \in D'} \{u_x\}$, where u_x is an arbitrarily chosen node in V_x
9 **return** D

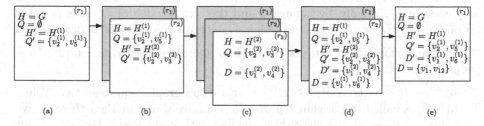

Fig. 5. The recursive execution of the Algorithm 1 on the graph G depicted in Fig. 1: ((a) and (b)), since the input graph is not a base graph, their type partition as well as the set Q' are computed and passed to the next recursive level; (c), H is a base graph and then an optimal solution is computed exploiting a brute force approach; ((d) and (e)), an optimal solution $D = \{v_1, v_{12}\}$ is reconstructed using the solution D' obtained on the reverse type graph sequence.

then the algorithm first constructs the type graph H' and selects nodes in $V(H')$ to assemble a set Q' of nodes that need to have a neighbor in the solution set, then it uses the set D' of nodes in $V(H')$ returned by $\mathsf{Dom}(H', Q')$ to construct the output set $D \subseteq V(H)$. The nodes of the returned set D are chosen by selecting exactly one node from each metavertex V_x having $x \in D'$. Figure 5 gives an example of the execution of Algorithm 1 on the graph G in Fig. 1.

Lemma 4. *Let H be not a base graph and let $Q \subseteq V(H)$. Let $V_1, \cdots, V_t$ be the type partition of H and let H' be its type graph. If $Q' = \{x \in V(H') \mid V_x \cap Q \neq \emptyset$ or V_x is an independent set\} and D' is an optimal Q'-stds of H' then the set D returned by $\mathsf{Dom}(H, Q)$ is an optimal Q-stds of H.*

Proof. We first prove that the set D returned by $\mathsf{Dom}(H, Q)$ is a Q-stds of H, then we prove its optimality. We distinguish two cases according to that a node $v \in V(H)$ is a node in Q or not. W.l.o.g. assume that $v \in V_x$, for some $x \in [t]$.

– If $v \in Q$ then $V_x \cap Q \neq \emptyset$ and by the definition of Q' we have $x \in Q'$. Hence, since D' is a Q'-stds of H', there exists $y \in D'$ that is a neighbor of x in H'. By Algorithm 1 (see line 8) there exists a node $u_y \in V_y \cap D$. Considering that each node in V_y is a neighbor of each node in V_x (since $(x, y) \in E(H')$), we have that v is dominated by $u \in D$.

- Let $v \in V - Q$. We know that D' is a Q'-stds of H'. Hence, if either $x \in Q'$ or $x \notin Q' \cup D'$ we can prove, as in the previous case, that there exists $u \in D$ that dominates v. Assume now that $x \notin Q'$ and $x \in D'$ (i.e., x can be not dominated in H'). By the definition of Q' we have that $V_x \cap Q = \emptyset$ and V_x is a clique. Hence, since by Algorithm 1 (see line 8) there exists a node $u_x \in V_x \cap D$, we have that v is a neighbor of $u_x \in D$ in the clique V_x.

Now, we prove that D is an optimal Q-stds of H whenever D' is an optimal Q'-stds of H'. By contradiction, assume that D is not optimal and let $\tilde{D}$ be an optimal Q-stds of H. By Lemma 3 we can assume that, for each $x \in [t]$, at most one node in V_x is a node in $\tilde{D}$. Let $\tilde{D}' = \{x \mid V_x \cap \tilde{D} \neq \emptyset\}$. We claim that $\tilde{D}'$ is a Q'-stds of H'. Finally, by Lemma 3 and the construction of $\tilde{D}'$, it is possible to prove that $|\tilde{D}'| < |D'|$ thus obtaining a contradiction since D' is optimal. $\square$

Theorem 4. $\mathrm{Dom}(G, \emptyset)$ *returns a minimum dominating set in time* $O(2^{itp(G)} + poly(n))$.

Proof. Let $H^{(0)} = G, H^{(1)}, \cdots, H^{(d)}$ be the type graph sequence of G. When $\mathrm{Dom}(G, \emptyset)$ is called, Algorithm 1 proceeds recursively, and at the i-th recursive step, for $i = 0, \cdots, d$, the algorithm is called with input graph $H^{(i)}$ and input node set $Q_i \subseteq V(H^{(i)})$, where Q_i is constructed at line 3 of the previous step $i - 1$, for $i = 1, \cdots, d$, and it is the empty set when $i = 0$, i.e., $Q_0 = \emptyset$. At step d, the optimal Q_d-stds of the base graph $H^{(d)}$ is established by brute force.

By Lemma 4, the set returned at the end of each recursive step i, for $i = d - 1, \cdots, 0$, is the optimal Q_i-stds of $H^{(i)}$. Hence, at the end (when $i = 0$) the returned set is the optimal $\emptyset$-stds of $H^{(0)}$, that by the definition is the minimum dominating set of G.

Considering that $|V(H^{(d)})| = itp(G)$, the brute search of the solution set at step d requires time $O(2^{itp(G)})$. Furthermore, since the construction of the type partition of $H^{(i)}$ and of its type graph can be done in polynomial time, and that both the construction of Q_i and the selection of the nodes in the solution set are easily obtained in linear time, we have $O(2^{itp(G)} + poly(n))$ time. $\square$

4   Conclusion

We introduced a novel parameter, named iterated type partition, and studied some of its properties. We show that the Equitable Coloring problem is W[1]-hard when parametrized by the iterated type partition. This result extends also to the modular-width parameter. We also prove that the hardness drops for the neighborhood diversity parameter, when the problem becomes FPT. Moreover, we presented a general strategy that enables to find FPT algorithms for several problems when parameterized by iterated type partition. The Algorithm for the Dominating Set problems has been presented, while algorithms for Vertex coloring and Vertex Cover problems appear in the extended version of the work.

It would be interesting to investigate whether the proposed strategy can be applied on other problems and if some meta-algorithm an be devised. Moreover, it would be interesting to analyze the Edge Dominating Set problem, which has been shown to be FPT with the neighborhood diversity parameter [31].

References

1. Abu-Khzam, F.N., Li, S., Markarian, C., Meyer auf der Heide, F., Podlipyan, P.: Modular-width: an auxiliary parameter for parameterized parallel complexity. In: Xiao, M., Rosamond, F. (eds.) FAW 2017. LNCS, vol. 10336, pp. 139–150. Springer, Cham (2017). https://doi.org/10.1007/978-3-319-59605-1_13

2. Belmonte, R., Fomin, F.V., Golovach, P.A., Ramanujan, M.S.: Metric dimension of bounded width graphs. In: Italiano, G.F., Pighizzini, G., Sannella, D.T. (eds.) MFCS 2015. LNCS, vol. 9235, pp. 115–126. Springer, Heidelberg (2015). https://doi.org/10.1007/978-3-662-48054-0_10

3. Bodlaender, H.L., Fomin, F.V.: Equitable colorings of bounded treewidth graphs. Theoret. Comput. Sci. **349**, 22–30 (2005). https://doi.org/10.1016/j.tcs.2005.09.027

4. Cockayne, E.J., Dawes, R.M., Hedetniemi, S.T.: Total domination in graphs. Networks **10**(3), 211–219 (1980)

5. Cordasco, G., Gargano, L., Rescigno, A.A., Vaccaro, U.: Evangelism in social networks: algorithms and complexity. Networks **71**(4), 346–357 (2018)

6. Cordasco, G., Gargano, L., Rescigno, A.A.: Iterated Type Partitions. arXiv 2001.08122, https://arxiv.org/abs/2001.08122 (2020)

7. Coudert, D., Ducoffe, G., Popa, A.: Fully polynomial FPT algorithms for some classes of bounded clique-width graphs. In: Proceedings of the Twenty-Ninth Annual ACM-SIAM Symposium on Discrete Algorithms (SODA 2018), pp. 2765–2784 (2018)

8. Courcelle, B.: The monadic second-order logic of graphs. I. Recognizable sets of finite graphs. Inf. Comput. **85**(1), 12–75 (1990)

9. Courcelle, B., Makowsky, J.A., Rotics, U.: Linear time solvable optimization problems on graphs of bounded clique-width. Theory Comput. Syst. **33**(2), 125–150 (2000)

10. Downey, R.G., Fellows, M.R.: Parameterized Complexity. Springer, Heidelberg (2012)

11. Doucha, M., Kratochvíl, J.: Cluster vertex deletion: a parameterization between vertex cover and clique-width. In: Rovan, B., Sassone, V., Widmayer, P. (eds.) MFCS 2012. LNCS, vol. 7464, pp. 348–359. Springer, Heidelberg (2012). https://doi.org/10.1007/978-3-642-32589-2_32

12. Dvořák, P., Knop, D., Toufar, T.: Target set selection in dense graph classes. In: Proceedings of 29th International Symposium on Algorithms and Computation (ISAAC 2018) (2018). https://doi.org/10.4230/LIPIcs.ISAAC.2018.18

13. Fellows, M.R., Lokshtanov, D., Misra, N., Rosamond, F.A., Saurabh, S.: Graph layout problems parameterized by vertex cover. In: Hong, S.-H., Nagamochi, H., Fukunaga, T. (eds.) ISAAC 2008. LNCS, vol. 5369, pp. 294–305. Springer, Heidelberg (2008). https://doi.org/10.1007/978-3-540-92182-0_28

14. Fellows, M.R., et al.: On the complexity of some colorful problems parameterized by treewidth. Inf. Comput. **209**(2), 143–153 (2011)

15. Fellows, M.R., Rosamond, F.A., Rotics, U., Szeider, S.: Clique-width is NP-complete. SIAM J. Discr. Math. **23**(2), 909–939 (2009)

16. Fiala, J., Golovach, P.A., Kratochvil, J.: Parameterized complexity of coloring problems: treewidth versus vertex cover. Theor. Comput. Sci. **412**, 2513–2523 (2011)
17. Fiala, J., Gavenciak, T., Knop, D., Koutecky, M., Kratochvíl, J.: Fixed parameter complexity of distance constrained labeling and uniform channel assignment problems. http://arxiv.org/abs/1507.00640arXiv:1507.00640 (2015)
18. Gavenciak, T., Knop, D., Koutecký, M.: Integer programming in parameterized complexity: three miniatures. In: Proceedings of 13th International Symposium on Parameterized and Exact Computation, IPEC 2018 (2018). https://doi.org/10.4230/LIPIcs.IPEC.2018.21
19. Fomin, F.V., Golovach, P.A., Lokshtanov, D., Saurabh, S.: Clique-width: on the price of generality. In: Proceedings of SODA (2009)
20. Fomin, F.V., Golovach, P., Lokshtanov, D., Saurabh, S.: Intractability of clique-width parameterizations. SIAM J. Comput. **39**(5), 1941–1956 (2010)
21. Fomin, F.V., Liedloff, M., Montealegre, P., Todinca, I.: Algorithms parameterized by vertex cover and modular width, through potential maximal cliques. In: Ravi, R., Gørtz, I.L. (eds.) SWAT 2014. LNCS, vol. 8503, pp. 182–193. Springer, Cham (2014). https://doi.org/10.1007/978-3-319-08404-6_16
22. Gajarský, J., Lampis, M., Ordyniak, S.: Parameterized algorithms for modular-width. In: Gutin, G., Szeider, S. (eds.) IPEC 2013. LNCS, vol. 8246, pp. 163–176. Springer, Cham (2013). https://doi.org/10.1007/978-3-319-03898-8_15
23. Gallai, T.: Transitiv orientierbare Graphen. Acta Math. Acad. Sci. Hung. **18**, 26–66 (1967)
24. Ganian, R.: Using neighborhood diversity to solve hard problems. arXiv:1201.3091 (2012)
25. Gargano, L., Rescigno, A.A.: Complexity of conflict-free colorings of graphs. Theoret. Comput. Sci. **566**, 39–49 (2015)
26. de C. M. Gomes, G., Lima, C.V.G.C., dos Santos, V.F.: Parameterized complexity of equitable coloring. Discrete Math. Theoret. Comput. Sci. **21**(1) (2019)
27. Jansen, K., Kratsch, S., Marx, D., Schlotter, I.: Bin packing with fixed number of bins revisited. J. Comput. Syst. Sci. **79**(1), 39–49 (2013)
28. Knop, D.: Partitioning graphs into induced subgraphs. Discrete Appl. Math. **272**, 31–42 (2019)
29. Knop, D., Koutecký, M., Masarík, T., Toufar, T.: Simplified algorithmic metatheorems beyond MSO: treewidth and neighborhood diversity. Logical Methods Comput. Sci. **15**(4) (2019)
30. Koutecký, M.: Solving hard problems on neighborhood diversity. Master thesis, Charles University in Prague (2013)
31. Lampis, M.: Algorithmic meta-theorems for restrictions of treewidth. Algorithmica **64**, 19–37 (2012). In: Proc. Eur. Sym. on Alg. (ESA), 549–560 (2010)
32. Lenstra, H.W.: Integer programming with a fixed number of variables. Math. Oper. Res. **8**(4), 538–548 (1983)
33. Niedermeier, R.: Invitation to Fixed-Parameter Algorithms. Oxford University Press, Oxford (2006)
34. Románek, M.: Parameterized algorithms for modular-width. Bachelor thesis, Masaryk University, Brno (2016). https://is.muni.cz/th/tobmd/Thesis.pdf
35. Tedder, M., Corneil, D., Habib, M., Paul, C.: Simpler linear-time modular decomposition via recursive factorizing permutations. In: Aceto, L., Damgård, I., Goldberg, L.A., Halldórsson, M.M., Ingólfsdóttir, A., Walukiewicz, I. (eds.) ICALP 2008. LNCS, vol. 5125, pp. 634–645. Springer, Heidelberg (2008). https://doi.org/10.1007/978-3-540-70575-8_52

Two Robots Patrolling on a Line: Integer Version and Approximability

Peter Damaschke$^{(\boxtimes)}$

Department of Computer Science and Engineering,
Chalmers University, 41296 Göteborg, Sweden
ptr@chalmers.se

Abstract. Suppose that two robots can move at unit speed on a line and must visit certain points called stations infinitely often. Every station allows some maximal waiting time between two visits. The problem is to construct an optimal schedule for the robots. While the one-robot problem is easy to solve in linear time, already for two robots the complexity is open. Chuangpishit, Czyzowicz, Gasieniec, Georgiou, Jurdzinski, and Kranakis (SOFSEM 2018) found a $\sqrt{3}$-approximation algorithm. Here we provide a PTAS, accomplished by rounding and (perhaps more surprisingly) by using the well-quasi ordering of vectors of positive integers. The result is not very practical in the present form, but further investigation of the integer version may make it more usable.

Keywords: Patrolling · Approximation scheme · Well-quasi ordering

1 Introduction

Patrolling problems where mobile robots must visit certain points at least with prescribed frequencies are interesting for monitoring and maintenance. Various cases and aspects have been studied: environments with different topologies, unreliable robots, with equal and different speeds, etc. [3–6]. See also a recent survey in [2]. Pinwheel scheduling [8,10,12] is also a special case of patrolling where all points have equal pairwise distances. More recently, patrolling problems received new attention by observing that different individual frequencies make them difficult, even on the simplest topologies [1,9].

The problem called PUF (patrolling with unbalanced frequencies) in [1] is the following (with somewhat changed notation): n stations are deployed at fixed points s_i $(i = 1, \ldots, n)$ on the real line L. For every station i, a duration $t_i > 0$ is also specified. Two identical robots move on L, at some given maximum speed. We say that station i is *visited* at some moment if at least one robot is at s_i. The problem is to construct a *schedule*, i.e., a pair of trajectories of two robots such that, during an unlimited period of time, every station i gets repeatedly visited, and the time between two consecutive visits never exceeds t_i. (But it does not matter which robots visit the station.) Of course, the same problem

© Springer Nature Switzerland AG 2020
L. Gąsieniec et al. (Eds.): IWOCA 2020, LNCS 12126, pp. 211–223, 2020.
https://doi.org/10.1007/978-3-030-48966-3_16

may be defined for any number of robots and for other topologies. Unless said otherwise, in the present paper PUF refers to the case of two robots on a line.

For any number $c \geq 1$, an instance of PUF is called *c-feasible* if it has a c-feasible schedule, i.e., for every i, the time between two consecutive visits of station i never exceeds ct_i. A 1-feasible instance or solution is simply called *feasible*. One may also view PUF as an optimization problem with the goal to find a schedule with minimum c. A c-approximation algorithm is one that outputs a c-feasible schedule when a feasible schedule exists.

While PUF for one robot on a line is easy to solve in $O(n)$ time, PUF becomes surprisingly difficult already for two robots: Only a $\sqrt{3}$-approximation is achieved in [1]. To our best knowledge, this is the state-of-the-art polynomial-time approximation, and the complexity status of PUF is open (both for deciding existence of feasible solutions and minimizing the times). It is far from obvious how one should divide the visits of certain stations among the two robots.

Overview. In Sect. 2 we introduce the integer version called IntPUF, with $m+1$ stations at points $0, \ldots, m$, and the notion of instance vectors that encode the instances. Since fixed-length vectors of positive integers are well-quasi ordered (WQO), only finitely many minimal feasible vectors exist for any fixed m. Using this fact plus some elementary graph theory, we can solve IntPUF in a time depending on m only, but not on the t_i. (However, the time as a function of m remains open.) In Sect. 3 we switch from m to the parameter $k := \min_i t_i$. The reason is that IntPUF can be approximated with a ratio arbitrarily close to 1, in a time depending on k only: If k/m is small, this is simple, and otherwise the result of Sect. 2 is applied. This insight is used to construct a PTAS in Sect. 4, where k is a discretization parameter. As far as we know, this might be the first example of using WQO in a PTAS. Next, rounding of the s_i and t_i to integers causes yet another approximation error. (The issue is that a station shifted to the next integer point can escape a turning point of a robot's trajectory.) Larger k yield better approximations but also higher time complexity. We remark that several ideas on the way, especially the WQO argument, can be generalized to more robots and to other topologies, but we keep the focus on two robots on a line.

2   The Integer Version of PUF

We introduce a variant of PUF that we name IntPUF. We define it precisely as PUF (see Sect. 1) but with the following additional demands:

- All s_i and t_i are integer (and $t_i > 0$).
- Every robot is, at any time, in one of two possible modes: either it stays at some point with integer coordinate or it moves at unit speed.
- Every robot can change its speed or its moving direction only at integer times and at integer points.

We remark that PUF allows real values, and this might lead to subtle effects for irrational numbers. Thus, algorithms for IntPUF may not completely solve

PUF. However, we have not further considered such questions. Note that, in practice, input numbers are usually rational.

Lemma 1. *Let r and s be real numbers and t an integer, with $0 \leq s - r \leq t$. Then $0 \leq \lfloor s \rfloor - \lfloor r \rfloor \leq t$.*

Proof. Non-negativity is obvious. To see $\lfloor s \rfloor - \lfloor r \rfloor \leq t$, we write the numbers as $r = \lfloor r \rfloor + r'$ and $s = \lfloor s \rfloor + s'$. Observe that $\lfloor s \rfloor - \lfloor r \rfloor = (s - s') - (r - r') = (s - r) + (r' - s')$. If $r' \leq s'$ then $s - r \leq t$ yields the assertion. If $r' > s'$ then $(\lfloor s \rfloor + s') - (\lfloor r \rfloor + r') \leq t$ implies that still $(\lfloor s \rfloor + s') - (\lfloor r \rfloor + s') \leq t$, since t is integer and $r' - s' < 1$. Again the assertion follows. $\square$

Theorem 1. *Every feasible instance of PUF, where all s_i and t_i are integers, is also a feasible instance of IntPUF, and vice versa.*

Proof. As the converse is trivial, we only need to consider a feasible solution to an instance of PUF, and transform it into a feasible solution, for the same numbers s_i and t_i, that enjoys the additional properties of an IntPUF solution.

Whenever a robot leaves a station i and moves back to i without visiting another station, it can just stay at station i. Whenever a robot moves from a station i to a neighboring station j, it can just move first at unit speed and then stay at station j for the remaining time.

Now we can partition the trajectory of each robot into 0-epochs and 1-epochs where the robot has speed 0 and 1, respectively. During a 1-epoch, a robot may change its moving direction at stations.

Let t be the start time of any 1-epoch. If t is not integer, we let the epoch start already at time $\lfloor t \rfloor$. That is, we move the entire 1-epoch back in time by $t - \lfloor t \rfloor$ time units. Because of the unit speed, every arrival and departure time of the robot at any station during the whole 1-epoch is rounded to the next smaller integer, too. This modification is done for all 1-epochs independently.

Applying Lemma 1 to the arrival and departure times at any station i we see that the solution remains valid (i.e., no robot departs before it arrives), and some waiting times between consecutive visits may increase, but they does not exceed the given integer bound t_i. $\square$

For formal reasons we assume from now on that we have $m + 1$ stations, at the integer points $0, 1, \ldots, m$. That is, s_i simply becomes i. If there is no station at point i, we formally set $t_i := \infty$, however, t_0 and t_m are finite. Thus an instance of IntPUF is characterized by an *instance vector* $\mathbf{t} = (t_0, \ldots, t_m)$ whose $m + 1$ entries are positive integers or ∞ symbols. We may use the terms instance vector and instance interchangeably.

Two instance vectors $\mathbf{x} = (x_0, \ldots, x_m)$ and $\mathbf{y} = (y_0, \ldots, y_m)$ are in relation $\mathbf{x} \leq \mathbf{y}$ if $x_i \leq y_i$ for all i. We then say that $\mathbf{x}$ is smaller than $\mathbf{y}$, and $\mathbf{y}$ is larger than $\mathbf{x}$, in the non-strict sense. We call $\mathbf{x}$ strictly smaller than $\mathbf{y}$, and $\mathbf{y}$ strictly larger than $\mathbf{x}$, if $\mathbf{x} \leq \mathbf{y}$ but $\mathbf{x} \neq \mathbf{y}$. Trivially, if $\mathbf{x} \leq \mathbf{y}$ and $\mathbf{x}$ is feasible, then $\mathbf{y}$ is feasible, too. We call a feasible instance vector with only finite entries a *minimal feasible vector* if no strictly smaller vector is feasible. These concepts can be defined in literally the same way for IntPUF with one robot. The following theorem is merely the known solution to the one-robot case [1] adapted to IntPUF. A *zigzag route* between two points i and j is a trajectory that perpetually goes from i to j and back, at unit speed.

Theorem 2. *[1] The only minimal feasible instance vector of IntPUF with one robot is given by $t_i = \max\{2i, 2(m - i)\}$ for all i. Moreover, if an instance is feasible, then a zigzag route between 0 and m is a feasible (and optimal) solution.*

Proof. The robot must sometimes visit point 0, say at time t. Then the last visit of point i was at time $t - i$ or earlier, and the next visit of point i will be at time $t + i$ time or later. Hence $t_i \geq 2i$ is necessary for feasibility. By symmetry we also get $t_i \geq 2(m - i)$. Conversely, in the mentioned zigzag route, the maximum waiting time between two consecutive visits of any point i equals $\max\{2i, 2(m - i)\}$.   □

We denote the vector in Theorem 2 by $F(m)$. This means: $F(1) = (2, 2)$, $F(2) = (4, 2, 4)$, $F(3) = (6, 4, 4, 6)$, $F(4) = (8, 6, 4, 6, 8)$, $F(5) = (10, 8, 6, 6, 8, 10)$, and so on. Characterizing the minimal feasible vectors for IntPUF with two robots appears to be far more complicated. However, suppose for the moment that, for some fixed size m, we know the list of all these minimal feasible vectors, along with a feasible solution for each of them. In fact, this list is finite, as a consequence of the famous Dickson's lemma, as explained below.

Vectors $\mathbf{x}$ and $\mathbf{y}$ are *incomparable* if neither $\mathbf{x} \leq \mathbf{y}$ nor $\mathbf{y} \leq \mathbf{x}$. Dickson's lemma (attributed to Dickson due to some result in [7]) states that every set of pairwise incomparable vectors of some fixed length (an antichain), with positive integers as entries, is finite. In other words, these vectors form a well-quasi ordering (WQO). Dickson's lemma has later been generalized, leading to a rich theory of WQO; see [11] for a historical note.

We could now solve any instance of IntPUF of a fixed size m as follows. First check whether there exists a solution where the two robots move in disjoint intervals $[0, v]$ and $[u, m]$, where $v < u$. For every fixed u and v, these are just two independent instances of the one-robot problem solved in Theorem 2. Even a naive implementation takes only $O(m^3)$ time. In all other cases, the intervals visited by the two robots intersect in some nonempty *shared interval* $[u, v]$, $u \leq v$. Note that every solution with a shared interval visits all $i \in [0, m]$. Hence, every instance vector $\mathbf{t}$ that admits a solution with a shared interval is larger than some feasible instance vector $\mathbf{t}'$ where all entries are finite, and trivially, $\mathbf{t}'$ is larger than some minimal feasible vector $\mathbf{t}''$. Thus it remains to check for the given instance vector $\mathbf{t}$ and every minimal feasible vector $\mathbf{t}''$ whether $\mathbf{t} \geq \mathbf{t}''$, and in the positive case, take a solution for $\mathbf{t}''$.

Not only the list of minimal partial solutions $\mathbf{t} = (t_0, \ldots, t_m)$ is finite, but each of them also has a solution with a finite description. The argument is as follows. Let us describe the situation at any integer time by the *state vector* $\mathbf{p} = (p_0, \ldots, p_m)$, where $p_i < t_i$ is the (integer) time that has passed since the last visit of i. Note that there is some robot at i if and only if $p_i = 0$. Since all t_i are finite, the number of state vectors is finite, too. Define the *state graph* of

t as the directed graph where the vertices are the state vectors, and a directed edge from **p** to **q** indicates that **q** is reachable from **p** in one time unit. The possible solutions to **t** are exactly the infinite directed paths in the state graph. But since the graph is finite, every solution contains some simple directed cycle (i.e., without repeated vertices). Conversely, every simple directed cycle is a solution. Thus we can choose any simple directed cycle, and such a solution is periodic. (A remark is that these matters are not so clear for PUF, as irrational numbers might have bizarre effects.) We have arrived at the following result, where we presume that arithmetic operations with integers, e.g., comparisons, cost constant time:

Lemma 2. *For any fixed m, once we know all minimal feasible instance vectors, we can solve every instance of IntPUF on $[0, m]$ in constant time. Moreover, every feasible instance admits a periodic solution, with a period bounded by some constant. (That is, time and period length are bounded by functions of m only.)*

We may effectively solve any instance $\mathbf{t} = (t_0, \ldots, t_m)$ also in the following way: Construct the state graph from **t** and find a simple directed cycle, or recognize that **t** is not feasible otherwise. However, the size of the state graph depends on **t**. Therefore, Lemma 2 that moves some work to a preprocessing phase is a step of progress. (As a side remark, also in a logarithmic cost model, comparing **t** against a fixed finite list is much faster than using the state graph.) It remains the question whether we can construct the list of all minimal feasible instances effectively. The WQO argument yields only its finiteness, but in fact, we can provide an effective algorithm. However, we will not care about its running time as a function of m, which is a separate (and apparently difficult) matter.

Lemma 3. *There exists an algorithm that effectively constructs all minimal feasible instance vectors for any given m, along with some periodic solution for each of them, in a time that depends on m only.*

Proof. Trivially, some minimal feasible vector exists, and since there are only finitely many of them, there is some finite upper bound on all entries in them. Thus, if we try all vectors $(t, \ldots, t)$ for $t = 0, 1, 2, \ldots$, we will eventually find some feasible vector **t**. Recall that every fixed vector can be tested for feasibility via its state graph. We test all vectors being smaller than our **t**, thus identifying at least one minimal feasible vector. But we cannot stop here, as there may exist further minimal feasible vectors being incomparable to **t**.

Next, assume that we have some nonempty set M of minimal feasible vectors, and we want to decide whether we have already found them all. Assume that $\mathbf{u} = (u_0, \ldots, u_m) \notin M$ is some further, yet unknown minimal feasible vector. Then, for every $\mathbf{t} = (t_0, \ldots, t_m) \in M$ there must exist some i with $u_i < t_i$. This observation suggests the following procedure: For every $\mathbf{t} \in M$ we select some i and set $v_i := t_i - 1$. If the same i is selected several times, we take the minimum of these v_i. If i is never selected, we set $v_i := \infty$. There exist only finitely many such selections, generating finitely many different vectors $\mathbf{v} = (v_0, \ldots, v_m)$. Now, every minimal feasible vector $\mathbf{u} \in M$ has the following property: There exists some $\mathbf{v}$ such that $u_i \leq v_i$ holds for all i.

It remains to test for every vector $\mathbf{v}$ whether it has a feasible solution such that the waiting times of all stations i with finite v_i are actually bounded by these v_i, and the waiting times of all stations i with infinite v_i are bounded by some finite (but unspecified) integers. (Note carefully that the latter condition is also needed to obtain some minimal feasible vector smaller than $\mathbf{v}$; it is not enough to demand and test feasibility of $\mathbf{v}$, because this entails no condition on the points with $v_i = \infty$.) More compactly, the property to be tested is that $\mathbf{v}$ is larger than some feasible vector with only finite entries.

Therefore, we proceed similarly as earlier in the feasibility test for vectors with only finite entries, but we must generalize our notion of state graph; see the details below. If we find such a solution to some $\mathbf{v}$, we also examine all smaller vectors and obtain a new minimal feasible vector that we add to M. If no vector $\mathbf{v}$ has such a solution, we know that M was already complete.

Now we give the details of testing a vector $\mathbf{v}$. We define a state by the passed times $p_i < v_i$ for all i with finite v_i, and by the positions of the two robots. (The passed times for points i with $v_i = \infty$ are not recorded, and the robots' positions are now given explicitly.) As earlier, a directed edge from one state to another one indicates reachability in one time unit. As seen above, every instance that admits a solution, with finite bounds on the waiting times of all stations i, also has a solution with a finite period. In our generalized state graph, such periodic solutions correspond exactly to directed cycles C (not necessarily simple!) such that every i appears as a robot position on some vertex of C.

Let V_i denote the set of states where some robot is visiting i. Then, deciding the existence of a periodic solution boils down to the following graph problem: Given a directed graph and a family of subsets V_i of vertices, find some directed cycle that intersects every V_i. However, this is an easy problem: Such a cycle exists if and only if some strongly connected component of the graph intersects all V_i. The "only if" direction holds because every directed cycle is entirely in some strongly connected component, The "if" direction holds because we can freely navigate in a strongly connected component, and thus connect some vertices from every V_i to some directed cycle. In conclusion, we only need to compute the strongly connected components of the generalized state graph and check their intersections with all V_i.   □

From Lemma 2 and 3 it follows:

Theorem 3. *There exists an algorithm solving IntPUF in a time that depends on m only. Moreover, every feasible instance admits a periodic solution, with a period bounded by some function of m.*

3   Short Waiting Times

The presence of some station j with a small t_j drastically restricts the robots' possible movements, as we will discuss now. Fix some j and define $h := \lfloor t_j/2 \rfloor$, that is, $t_j = 2h$ or $t_j = 2h + 1$. Also define the interval $J := [j - h, j + h]$. At any integer time, some robot must be present in J. (Otherwise, the last visit of j was more than h time units ago, and the next visit will be more than h time units from now, such that j would have to wait at least $2h + 2 > t_j$ time units between two visits.) In other words, at any integer time, at most one robot can be outside J. We call it the *outer robot*, whereas the robot in J is called the *inner robot*. When both robots are in J, the assignment of these two roles is arbitrary, and we can swap the roles of the two robots if we want or need.

Suppose furthermore that $0 \notin J$ and $m \notin J$. Since the outer robot must repeatedly visit stations on both sides of J, it must repeatedly cross J in both directions. In the best case, the outer robot needs one time unit to skip J. In detail: At time t, the outer robot is at point $j - h - 1$, and at time $t + 1$, the outer robot (actually now the other one) is at point $j + h + 1$, or vice versa. It is equivalent to say that, even in this best case, the outer robot must solve the one-robot instance $(t_0, \ldots, t_{j-h-1}, t_{j+h+1}, \ldots, t_m)$ obtained by cutting out J. Now Theorem 2 implies that this vector must be larger than $F(m - 2h - 1)$. If $m \in J$ (or similarly, if $0 \in J$), we have $(t_0, \ldots, t_{j-h-1}) \geq F(j - h - 1)$ by a simpler argument: only the outer robot can visit the stations to the left of J.

We are ready to solve another special case of IntPUF to optimality:

Theorem 4. *For every m and j there exists exactly one minimal feasible vector with $t_j = 1$, which is $F(m - 1)$ with a 1 inserted after the first j entries.*

Proof. If $t_0 = 1$ then the inner robot must stay at point 0 all the time, while the outer robot must solve the one-robot instance $(t_1, \ldots, t_m)$, and the assertion follows from Theorem 2. The argument for $t_m = 1$ is similar. If $t_j = 1$ for some j with $0 < j < m$, we have the lower bound $(t_0, \ldots, t_{j-1}, t_{j+1}, \ldots, t_m) \geq F(m-1)$ for every feasible vector, as shown above. To show that the claimed vector is feasible, note that the outer robot can zigzag between 0 and m and always skip $J = \{j\}$ in only one time unit. Now the maximal waiting time of every station outside J is equal to the corresponding entry of $F(m - 1)$. Hence, together with $t_j = 1$, these integers form a feasible vector matching the lower bound. $\square$

Similarly it should be possible to characterize the minimal feasible vectors with some $t_j = k$ also for any fixed $k > 1$, but it turns out that we run into many case distinctions regarding the times needed to skip J and the lengths of the outer robot's trajectories. However, the above observations still lead to the approximation result below. (Note that we assume that the instance is already given in "compact" form, as the list of all n finite values t_i.)

Theorem 5. *There exists an algorithm for IntPUF which, for every feasible instance with $n \leq m + 1$ stations (with finite t_i) and with $k = \min_i t_i \leq m/4$, outputs some $(1 + O(k/m))$-feasible solution in $O(n)$ time.*

Proof. Fix some j with $t_j = k$, and define $h := \lfloor k/2 \rfloor$ and $J := [j - h, j + h]$ as before (even in the case when $0 \in J$ or $m \in J$). As we have seen, the instance vector after cutting out J must be feasible for one robot. With Theorem 2 we get $t_i \geq m - k$ for all $i \notin J$.

Let m' denote the distance between J and the farthest station (either 0 or m), that means, $m' := \max\{j - h, m - (j + h)\}$. Note that $m' \geq (m - k)/2$. Since the outer robot must visit this farthest station sometimes, the outer robot cannot be in J during some time interval I of duration at least $2m'$.

Next, let J' be the set of all $i \in J$ with $t_i \leq 2m' - 2k$. Since $4k \leq m$, we have $3k \leq m - k \leq 2m'$, hence $t_j = k \leq 2m' - 2k$, thus $j \in J'$, that is, $J' \neq \emptyset$. Let u and v be the leftmost and rightmost point, respectively, in J'. Finally, we define I' to be the time interval I truncated by k time units at both ends.

Since the length of I' is at least $2m' - 2k$, the inner robot must visit every station in J' at least once during I'. In particular, it must visit u during I'. Now assume for some $i \in J'$ that $t_i < 2(i - u)$. For the inner robot it is then impossible to visit i before u and again after u, within t_i time units. Since $i - u \leq 2h \leq k$, the two mentioned visits of i must still happen during I. But since the outer robot is not in J during I, it cannot do any of these visits either. This contradiction to feasibility shows $t_i \geq 2(i - u)$ for all $i \in J'$. Similarly we can prove $t_i \geq 2(v - i)$ for all $i \in J'$. Hence, if we simply let the inner robot zigzag in $[u, v]$, it visits all stations in J' frequently enough.

Our solution is now: Let the inner and outer robot zigzag in $[u, v]$ and $[0, m]$, respectively. It remains to analyze the waiting times of stations outside J'.

Consider any station $i \notin J$. Since the instance vector after cutting out J is feasible for one robot, the outer robot would always return to i within t_i time units if it could skip J. But in reality it may need $2k$ additional time units to traverse J twice. Since $t_i \geq m - k$, the waiting time is at most $t_i + 2k = (1 + 2k/t_i)t_i \leq (1 + 2k/(m - k))t_i$.

Consider any station $i \in J \setminus J'$. By definition we have $t_i > 2m' - 2k$. Also remember that $2m' \geq m - k$. The outer robot returns to i in a time at most $2m' + 2k = (2m' - 2k) + 4k < t_i + 4k = (1 + 4k/t_i)t_i < (1 + 4k/(2m' - 2k))t_i < (1 + 4k/(m - 3k))t_i$.

Altogether, the solution is $(1 + O(k/m))$-feasible. The time $O(n)$ is obvious: We must only find the smallest t_i and construct J' for determining u and v. □

For notational convenience we have formulated Theorem 5 for IntPUF, but the proof does not really use integrality, hence we also have immediately:

Theorem 6. *There exists an algorithm for PUF which, for every feasible instance with n stations, distance m between the outermost stations, and $k = \min_i t_i \leq m/4$, outputs some $(1 + O(k/m))$-feasible solution in $O(n)$ time.*

4   Rounding the Coordinates

A natural idea for solving PUF approximately is now to round all s_i and t_i to integers and apply the results for IntPUF. Given an instance P of PUF and an integer parameter k, we scale the time axis such that $k = \min_i t_i$. In other words, $\min_i t_i/k$ becomes the unit of time. The length unit on the line L is chosen such that the given maximum speed of robots is the unit speed. This setting is assumed throughout this section.

Before we discuss rounding, we study more generally what happens if the stations are slightly shifted. Let Q be an instance of PUF, obtained from P by moving every station by less than half the length unit. That is, n and the t_i are preserved, but the stations in Q are at points s_i' where $|s_i' - s_i| \leq 1/2$ for all i.

Let S be any feasible solution to the instance P. In general, S is not feasible for Q: Besides small delays we may even completely miss some visits, since a robot's trajectory may change its direction at some station, but the station may have been moved away from the turning point. Therefore we would like to modify S so as to construct a solution that is c-feasible for Q, with some "small" $c > 1$. We will modify the two robot trajectories independently, that is, in the following we only consider the trajectory of any single robot.

Lemma 4. *Let I be any time interval of duration r, let $J \subset L$ be any interval of length at most $r/2$, and let $a, b \in J$ be any points therein. Then a robot can move during I such that its trajectory starts in a, ends in b, and visits all of J.*

Proof. Assume that $a \leq b$. (The other case is symmetric.) We simply travel from a to the left end of J, then to the right end of J, and finally to b. Obviously, the robot can manage this in at most $2(r/2) = r$ time units. □

Some more special definitions will be needed; note that they refer to real (not integer) intervals: For a given time interval I, let $J(I) \subset L$ denote the interval of points visited by the considered robot during I. For any interval $J = [u, v] \subset L$ we define $J^+ := [u - 1/2, v + 1/2]$. In the following we temporarily allow robots to be faster than the unit speed.

Lemma 5. *Let I be a time interval of duration r, $J \subset L$ an interval of length at least $(r - 2)/2$, and $a, b \in J$. Assume that a robot can move such that its trajectory during I starts in a, ends in b, and visits all of J. Then there also exists a trajectory during I that starts in a, ends in b, visits all of J^+, and has a speed at most $1 + 4/(r - 4)$.*

Proof. Let u and v be the ends of J, that is, $J = [u, v]$, and let T be the assumed trajectory. Since T visits all of J, it must contain a sub-trajectory T_2 going from u to v (or vice versa, but this case is symmetric). Hence we can partition T into three sub-trajectories: T_1 going from a to u, T_2 going from u to v, and T_3 going from v to b. Note that T_1 and T_3 may be empty, if $a = u$ and $v = b$, respectively.

We modify T as follows. Immediately after T_1 we insert a piece going from u to $u - 1/2$ in $1/2$ time units, and immediately before T_3 we insert a piece going from $v + 1/2$ to v in $1/2$ time units. Finally we adjust T_2 such that (i) it goes from $u - 1/2$ to $v + 1/2$ (to connect to the extended T_1 and T_3) and (ii) it needs one time unit less (to be used for the additional $1/2 + 1/2$ time units). We achieve (i) by stretching T_2 parallel to L, and we achieve (ii) by shrinking T_2 parallel to the time axis. Since J has a length at least $(r - 2)/2$, the robot following the original T_2 has to travel a distance at least $(r - 2)/2$, and it also needs at least $(r - 2)/2$ time units. Travelling one length unit more in one time unit less increases the speed by a factor at most $((r - 2)/2 + 1)/((r - 2)/2 - 1) = 1 + 4/(r - 4)$. □

Lemma 6. *Let P be an instance of PUF specified by s_i and t_i $(i = 1, \ldots, n)$, and let Q be an instance of PUF with the same size n and the same durations t_i but with station positions s_i' such that $|s_i' - s_i| \leq 1/2$ for all i. If P is c-feasible, then Q is $(1 + 4\sqrt{2}/\sqrt{k} + O(1/k))c$-feasible.*

Proof. We partition the time axis into intervals of some length r that we fix later. Consider any time interval I in this partitioning, and the trajectory T of either one of the robots, in some feasible solution to P.

If the length of $J(I)$ is at most $(r-2)/2$, then the length of $J(I)^+$ is at most $r/2$. In this case we apply Lemma 4 with $J := J(I)^+$. If $J(I)$ is longer than $(r-2)/2$, then we apply Lemma 5 with $J := J(I)$. Due to the Lemmas, in both cases we can replace the sub-path of the given trajectory T during I with a path that begins and ends in the same points as in T, but visits all of $J(I)^+$. However, in the second case we increase the speed by a factor up to $1 + 4/(r-4)$.

We do the described change independently in all intervals I of our partitioning, and we observe: (i) Since the robot's positions at the start and end moments of these time intervals have not changed, the modified trajectories form together a new trajectory T' overall. (ii) If T visits a station i during some time interval I of the partitioning, then $s_i \in J(I)$, hence $s_i' \in J(I)^+$, hence also T' visits the station i during I. Moreover, since I has duration r, this visit is by at most r time units earlier or later than in T.

The described changes are done independently for both robots. From the above property (ii) it follows that the waiting time between two consecutive visits of any station i increases by $2r$ time units in the worst case. Remembering that $k = \min_i t_i$ and $c \geq 1$, this implies that the waiting time is always at most $(1 + 2r/k)ct_i$. In other words, the modified solution would have been $(1 + 2r/k)c$-feasible if it had respected the speed limit.

In order to get unit speed again, we finally stretch the trajectories along the time axis by a factor $1 + 4/(r-4)$. This yields a valid $(1 + 2r/k)(1 + 4/(r-4))c$-feasible solution. Choosing $r := \sqrt{2k} + 4$ gives the assertion.   □

Now we can describe an algorithm to solve any feasible instance P of PUF approximately. Note that it is not known in advance whether P is feasible; we must discuss this point later, as well as the choice of parameter k:

We decide on an integer k and choose time and length unit such that $k = \min_i t_i$ and the robots have unit speed, as already explained. Recall that m denotes the distance of the outermost stations.

In the following we distinguish two cases regarding k and m. The exact cut-off point is not that important, but we must decide on some suitable one.

If $m > 4k\sqrt{k}$, then we run the algorithm from Theorem 6 to solve P approximately. (Note that its prerequisites are satisfied here.) It yields some $(1 + O(k/m))$-feasible solution, hence some $(1 + O(1/\sqrt{k}))$-feasible solution to P, in $O(n)$ time.

If $m \leq 4k\sqrt{k}$, then we proceed as follows. We round every s_i to the closest integer. Ties are broken arbitrarily if s_i is an integer plus $1/2$. If several stations i end up on the same point, we only keep one of these stations with the smallest t_i and "mask" the others.

Due to Lemma 6, the obtained instance Q is c-feasible, for some $c = 1 + O(1/\sqrt{k})$. Next we replace every t_i with $t'_i := \lceil ct_i \rceil$. The obtained instance R is feasible, and by Theorem 1, R is also a feasible instance of IntPUF. Defining $k' := \min_i t'_i$ we also note that $k' = \Theta(k)$.

We run the algorithm from Theorem 3 to solve R exactly, in a time that depends on m only, and thus on k only. The computed feasible solution to R, which we denote S, is $(1 + O(1/\sqrt{k}))$-feasible for Q.

Finally we move all stations i (also the masked ones) back to their original positions s_i and apply Lemma 6 again in the opposite direction, to translate S into a solution to P. Since $(1 + O(1/\sqrt{k}))^2 = 1 + O(1/\sqrt{k})$, this solution to P is $(1 + O(1/\sqrt{k}))$-feasible, too.

It is crucial that this last step can be done effectively. By Theorem 3, we can always take a periodic solution S, with a period bounded by some function of m. Furthermore, the proof of Lemma 6 does not only show the existence of a $(1 + O(1/\sqrt{k}))$-feasible solution but also describes a construction of this solution from the given one (here: from S). The approximation ratio $1 + O(k/m)$ remains true if we choose $r := \lfloor \sqrt{2k} + 4 \rfloor$ (to have an integer value r). Then it suffices to modify the trajectories on some time interval of finite duration (the least common multiple of r and the period of S) and then to repeat this solution infinitely on the time axis. That is, our approximate solution to P is periodic, too. For any desired $\varepsilon > 0$ we may choose $k = \Theta(1/\varepsilon^2)$ with some suitable constant factor. Altogether this shows:

Theorem 7. *There exists an algorithm that outputs, for any feasible instance of PUF with n stations and for any prescribed $\varepsilon > 0$, some $(1 + \varepsilon)$-feasible solution, in time $O(\max\{n, g(\varepsilon)\})$, where g is some function that depends on ε only.*

One can trivially check afterwards whether a solution is $(1 + \varepsilon)$-feasible. If the algorithm failed to find such a solution, we know that the given instance P was not feasible. In that case we consider instances P_c obtained from P by replacing all t_i with ct_i. We may choose any factor $c > 1$ and apply the same algorithm to P_c. Either we get a $(1 + \varepsilon)$-feasible solution to P_c, or c was too small. Once our c is within a factor $1 + \varepsilon$ of the minimal c^* that makes P_{c^*} feasible, we get a $(1 + O(\varepsilon))$-approximate solution to P.

It remains to find such a near-optimal c efficiently. Trivially, P_c is feasible when $c = 2m/k$. Hence, if $k/m = \Omega(\varepsilon)$, then $O(\log(1/\varepsilon))$ steps of binary search are enough. The case of smaller k/m is more peculiar, but the concepts of Sect. 3 enable us to first find some c within a constant factor of c^* without binary search, in $O(n)$ time: Let J_c be the interval of length ck, having some station with minimum t_i in the center. As we have seen in Sect. 3, P_c is feasible only if the instance P_c after cutting out J_c is feasible for one robot. A necessary condition is that $ct_i \geq m - ck$ for all $i \notin J_c$. Hence we can pick any $i \notin J_c$ with $ct_i < m - ck$ and raise c until either $ct_i \geq m - ck$ or $i \in J_c$. (Calculation details are straightforward.) As c only grows in this process, we successively get rid of all stations $i \notin J_c$ with a too small ct_i. For the final value c' we have that no instance $P_{c'-\delta}$, $\delta > 0$, is feasible, hence $c^* \geq c'$. Assume that still $c'k/m = O(\varepsilon)$;

otherwise we have already reached the former case. Furthermore, $t_i \geq k$ holds for all i by definition. In particular, $c't_i \geq c'k$ holds for all $i \in J_{c'}$. Hence, if we generously set $c := 3c'$ and let the robots zigzag in $[0, m]$ and $J_{c'}$, respectively, we obtain a feasible solution to our current P_c. It follows $1 \leq c/c^* \leq 3$. Now we have also overcome the restriction that P must be feasible, and we arrive at:

Theorem 8. *PUF admits a polynomial-time approximation scheme.*

Concluding Remarks. Our PTAS is not yet practical. We have not bounded the time as a function of $1/\varepsilon$, and large k may be needed to beat the known $\sqrt{3}$-approximation [1]. However, we believe that our approach paves the way. To achieve practicality, we must *efficiently* solve IntPUF instances up to certain values of k and m, using the structure of cycles in the state graph. That is, we need an efficient version of the algorithm from Lemma 3.

References

1. Chuangpishit, H., Czyzowicz, J., Gąsieniec, L., Georgiou, K., Jurdziński, T., Kranakis, E.: Patrolling a path connecting a set of points with unbalanced frequencies of visits. In: Tjoa, A.M., Bellatreche, L., Biffl, S., van Leeuwen, J., Wiedermann, J. (eds.) SOFSEM 2018. LNCS, vol. 10706, pp. 367–380. Springer, Cham (2018). https://doi.org/10.1007/978-3-319-73117-9_26
2. Czyzowicz, J., Georgiou, K., Kranakis, E.: Patrolling. In: Flocchini, P., Prencipe, G., Santoro, N. (eds.) Distributed Computing by Mobile Entities, Current Research in Moving and Computing. LNCS, vol. 11340, pp. 371–400. Springer, Cham (2019). https://doi.org/10.1007/978-3-030-11072-7_15
3. Czyzowicz, J., Gąsieniec, L., Kosowski, A., Kranakis, E.: Boundary patrolling by mobile agents with distinct maximal speeds. In: Demetrescu, C., Halldórsson, M.M. (eds.) ESA 2011. LNCS, vol. 6942, pp. 701–712. Springer, Heidelberg (2011). https://doi.org/10.1007/978-3-642-23719-5_59
4. Czyzowicz, J., Godon, M., Kranakis, E., Labourel, A., Markou, E.: Exploring graphs with time constraints by unreliable collections of mobile robots. In: Min Tjoa, A., Bellatreche, L., Biffl, S., van Leeuwen, J., Wiedermann, J. (eds.) SOFSEM 2018. LNCS, vol. 10706, pp. 381–395. Springer, Cham (2018)
5. Czyzowicz, J., Kosowski, A., Kranakis, E., Taleb, N.: Patrolling trees with mobile robots. In: Cuppens, F., Wang, L., Cuppens-Boulahia, N., Tawbi, N., Garcia-Alfaro, J. (eds.) FPS 2016. LNCS, vol. 10128, pp. 331–344. Springer, Cham (2017). https://doi.org/10.1007/978-3-319-51966-1_22
6. Das, S., Di Luna, G.A., Gasieniec, L.A.: Patrolling on dynamic ring networks. In: Catania, B., Královič, R., Nawrocki, J., Pighizzini, G. (eds.) SOFSEM 2019. LNCS, vol. 11376, pp. 150–163. Springer, Cham (2019). https://doi.org/10.1007/978-3-030-10801-4_13
7. Dickson, L.E.: Finiteness of the odd perfect and primitive abundant numbers with n distinct prime factors. Am. J. Math. **35**, 413–422 (1913)
8. Fishburn, P.C., Lagarias, J.C.: Pinwheel scheduling: achievable densities. Algorithmica **34**, 14–38 (2002)

9. Gąsieniec, L., Klasing, R., Levcopoulos, C., Lingas, A., Min, J., Radzik, T.: Bamboo garden trimming problem (perpetual maintenance of machines with different attendance urgency factors). In: Steffen, B., Baier, C., van den Brand, M., Eder, J., Hinchey, M., Margaria, T. (eds.) SOFSEM 2017. LNCS, vol. 10139, pp. 229–240. Springer, Cham (2017). https://doi.org/10.1007/978-3-319-51963-0_18
10. Holte, R., Rosier, L.E., Tulchinsky, I., Varvel, D.A.: Pinwheel scheduling with two distinct numbers. Theor. Comput. Sci. **100**, 105–135 (1992)
11. Kruskal, J.B.: The theory of well-quasi-ordering: a frequently discovered concept. J. Comb. Theory A **13**, 297–305 (1972)
12. Lin, S.S., Lin, K.J.: A pinwheel scheduler for three distinct numbers with a tight schedulability bound. Algorithmica **19**, 411–426 (1997)

Ordering a Sparse Graph to Minimize the Sum of Right Ends of Edges

Peter Damaschke[1,2]([⊠])

[1] Department of Computer Science and Engineering, Chalmers University,
41296 Göteborg, Sweden
ptr@chalmers.se
[2] Fraunhofer-Chalmers Research Centre for Industrial Mathematics,
41288 Göteborg, Sweden

Abstract. Motivated by a warehouse logistics problem we study mappings of the vertices of a graph onto prescribed points on the real line that minimize the sum (or equivalently, the average) of the coordinates of the right ends of all edges. We focus on graphs whose edge numbers do not exceed the vertex numbers too much, that is, graphs with few cycles. Intuitively, dense subgraphs should be placed early in the ordering, in order to finish many edges soon. However, our main "calculation trick" is to compare the objective function with the case when (almost) every vertex is the right end of exactly one edge. The deviations from this case are described by "charges" that can form "dipoles". This reformulation enables us to derive polynomial algorithms and NP-completeness results for relevant special cases, and FPT results.

Keywords: Minimum linear arrangement · Pick-by-order · Cycle · Tree · Dynamic programming on subsets · Elimination ordering · 2-core · 3-core

1 Introduction

We study the following problem on undirected graphs $G = (V, E)$. Our graphs may contain parallel edges and loops (and any number of loops may be attached to a vertex), but no isolated vertices. A loop at a vertex v may be formally seen as an edge vv.

MinSumEnds

Given: (1) an undirected graph $G = (V, E)$ with n vertices, and (2) n numbers $s_1 < \ldots < s_n$.

Find: A labeling, that is, a bijective mapping λ of V onto $\{s_1, \ldots, s_n\}$ that minimizes $\sum_{e \in E} \mu(e)$, where $\mu(uv) := \max\{\lambda(u), \lambda(v)\}$ for every edge $e = uv$.

We call such a labeling *optimal*, with respect to this objective function. Our objective function can be rephrased as follows. Let $L(k)$ be the number of edges

© Springer Nature Switzerland AG 2020
L. Gąsieniec et al. (Eds.): IWOCA 2020, LNCS 12126, pp. 224–236, 2020.
https://doi.org/10.1007/978-3-030-48966-3_17

uv such that v is the vertex with label s_k, and the label of u is smaller than or equal to k. (This includes possible loops vv.) Then the sum of edge labels is obviously $\sum_{k=1}^{n} s_k \cdot L(k)$. Informally, $L(k)$ is the "left degree" of the vertex at position k, if the vertices are placed on the number line according to their labels. This way, a labeling can be viewed as a linear ordering of the vertices, placed on points with the coordinates $s_1 < \ldots < s_n$. We may use the words labeling and ordering interchangeably.

If the labels are *equidistant*, we can without loss of generality assume that $s_k = k$ for all k.

MINSUMENDS is similar to the well-known linear arrangement problem where we want to minimize the sum of edge lengths (i.e., differences of labels). Like the linear arrangement problem it can be solved straightforwardly in $O^*(2^n)$ time[1] by dynamic programming on subsets [4].

MINSUMENDS can be generalized to hypergraphs. This work was directly inspired by a real-world problem: Items are stored in a shelf in a warehouse, and certain subsets of items are frequently requested. They must be fetched from the shelf, thereby walking from the left end to the place of the rightmost requested item and back. Given a set of data on the frequently requested subsets, the problem is to store the items so as to minimize the average walking distance.

The minimum linear arrangement problem is a classic NP-complete problem [12] and has been intensively studied. Approximation algorithms and inapproximability results are known [1,2,8,14], as well as exact exponential and parameterized algorithms [3,9–11], and efficient algorithms for special graph classes [5–7,13]. MINSUMENDS is much less explored. In [4], the problem is called the product location problem with a single rack and a front end depot. The problem is proved to be strongly NP-complete for equidistant labels, by a reduction from the linear arrangement problem. In fact, the reduction produces graphs with possible loops, but no hypergraphs.

In the present paper we focus on graphs with barely more edges than vertices. In the warehouse application this corresponds to the rather practical case that, typically, only single items or pairs of items are requested, and the requests are not very diverse, that is, only a small number of different pairs occurs. This easily leads to graphs whose connected components are trees or have only a few cycles. Still these graphs are rather special, but this study may serve as a first step in understanding which graph properties make the problem easy or hard. Also, the related linear arrangement problem is nontrivial even for trees [7], and now we continue this line of research for MINSUMENDS.

In Sect. 2 we solve MINSUMENDS for some simple graphs that contain a chain of densest subgraphs. These are induced subgraphs having the maximum number of edges, given a number of vertices. In Sect. 3 we rephrase MINSUMENDS in terms of so called charges and dipoles which measure the difference to an optimal labeling of a tree. Using these concepts, we eliminate vertices of degree 1, provided that the labels are equidistant; see Sects. 3 and 4. Similarly, in Sect. 5

---

[1] We adopt the O^* notation that focuses on the exponential terms and suppresses polynomial factors.

we eliminate connected components that are merely cycles, and we show NP-completeness of MINSUMENDS for general labels, but for a graph class as "trivial" as disjoint unions of cycles. In Sect. 6 we derive an FPT algorithm in the parameter $m - n$, the number of edges minus the number of vertices (after the previously described eliminations). Here the 2-cores and 3-cores of graphs play a prominent role. We think that the structural properties shown can be useful in their own right, not only as a preparation of the FPT result that is applicable only to graphs "slightly exceeding" forests.

2   Nested Densest Subgraphs

Consider an optimal labeling λ of G and a positive integer $k \leq n$. Let G_k be the subgraph of G induced by the vertices with the k smallest labels. Then the labeling induced by λ on G_k is also an optimal labeling of G_k. This is evident by an exchange argument; note that a permutation of the first k vertices does not affect the values $L(j)$ for $j > k$. In other words: Once we have decided on the vertices that receive the labels larger than s_k, it remains to solve the MINSUMENDS problem on G_k. In such a situation we say that we have *eliminated* the other vertices.

An induced subgraph H of G with k vertices is called a *densest subgraph* if H has a maximum number of edges among all induced subgraphs of G with k vertices. We say that a labeling produces *nested densest subgraphs* if, for every k, G_k is a densest subgraph.

Not every graph allows nested densest subgraphs. The smallest counterexample has three vertices u, v, w, where u and v are joined by two parallel edges, and w has one loop. Then the only densest subgraphs with $k = 1$ and $k = 2$ are induced by $\{w\}$ and by $\{u, v\}$, respectively. However, for graphs that do have nested densest subgraphs, we can characterize optimal solutions of MINSUMENDS:

Lemma 1. *Any labeling that produces nested densest subgraphs is an optimal labeling.*

Proof. Consider a labeling λ produced by nested densest subgraphs, with objective value $L = \sum_{k=1}^{n} s_k \cdot L(k)$, and assume that there is a better labeling with similarly defined values $L' = \sum_{k=1}^{n} s_k \cdot L'(k)$, where $L' < L$. Since $\sum_{j=1}^{n} L'(j) = \sum_{j=1}^{n} L(j)$, this would be possible only if there were some k with $L'(k) > L(k)$. Specifically, let k be the smallest such index. Then we have $\sum_{j=1}^{k} L'(j) > \sum_{j=1}^{k} L(j)$. But this contradicts the assumption that G_k (in λ) was already a densest subgraph. □

The converse (every optimal labeling of such graphs produces nested densest subgraphs) also holds true, but we will only use the direction given in Lemma 1. Perhaps the simplest application is the case of trees.

For clarity we remark that loops as well as pairs of parallel edges count as cycles. A *forest* is a graph without cycles. Hence, in particular, a forest must not contain loops and parallel edges. A *tree* is a connected forest. Every subgraph of a tree is, of course, a forest.

Theorem 1. MINSUMENDS *is solvable in linear time on forests.*

Proof. Every tree possesses nested densest subgraphs: We can start with an arbitrary vertex and successively add a vertex that has a neighbor among the previously selected vertices. In this way, for every k, G_k has exactly $k - 1$ edges, which is indeed maximal for subgraphs of a tree.

More generally, every forest possesses nested densest subgraphs: We sort the connected components (which are trees) by decreasing sizes, order the vertices in every tree as described above, and concatenate these orderings of the trees. Then every G_k has exactly $k - c(k)$ edges, where $c(k)$ is the number of connected components of G_k. It is easy to see that sorting the trees by decreasing sizes minimizes all $c(k)$, and thus all G_k in this labeling are indeed densest subgraphs. Clearly, the procedure can be implemented to run in linear time, where the sorting is done by bucketsort.   □

A slightly larger graph class can still be managed in this way:

Theorem 2. MINSUMENDS *is solvable in linear time on graphs with at most one cycle.*

Proof. Forests are settled by Theorem 1, hence we can suppose that the input graph G has exactly one cycle. Let k denote its length (where $k = 1$ if the cycle is a loop, and $k = 2$ if the cycle consists of two parallel edges).

First consider the case when G is connected and has exactly one cycle. Then the only densest subgraphs of G are the following: all subgraphs of $j < k$ vertices being trees, and all connected subgraphs of $j \geq k$ vertices including the cycle. Hence G has nested densest subgraphs: Starting at any vertex of the cycle, assign the k lowest labels to the vertices of the cycle in their natural ordering, and then successively assign the next label to any vertex that has a neighbor among the already labeled vertices.

The case when G is not connected is solved by combining the previous observations (also from Theorem 1). We skip the straightforward verification of the following claims.

The only densest subgraphs of G are now the following: all subgraphs of $j < k$ vertices being trees, and all subgraphs of $j \geq k$ vertices that include the cycle and intersect the smallest possible number of other connected components, where every such intersection is a subtree of the respective component.

This yields some optimal labeling in linear time: Starting at any vertex of the cycle, assign the k lowest labels to the vertices of the cycle in their natural ordering, then successively assign the next label to any vertex that has a neighbor among the already labeled vertices, until the entire connected component containing the cycle is labeled, and finally append optimal orderings of the other connected components (which are trees), sorted by decreasing sizes.   □

As the above example suggests, graphs with several cycles, in general, do not have nested densest subgraphs, and we must combine the idea with other methods, in order to solve instances of MinSumEnds.

3   Charges and Dipoles

For reasons that will become apparent soon, we work from now on with the numbers $M(k) := L(k) - 1$. Obviously, minimizing $\sum_{k=1}^{n} s_k \cdot L(k)$ is equivalent to minimizing $M := \sum_{k=1}^{n} s_k \cdot M(k)$. When v is the vertex with label s_k, we may also write $M(v)$ instead of $M(k)$. Note that the prefix sum $\sum_{j=1}^{k} M(j)$ equals the number of edges in G_k minus k, and that $M(k) \geq -1$ for every k, by definition, and vertices with $M(k) = 0$ can be ignored.

If $M(k) = -1$, then we imagine a *negative charge* at point s_k on the number line. Similarly, if $M(k) > 0$, then we imagine $M(k)$ *positive charges* at point s_k. Equivalently we may imagine that the vertices (rather than the points s_k) are charged.

Next we may pair up some of these charges to *dipoles* according to the following rules. Every dipole consists of a negative charge and a positive charge of a vertex with a higher label, and every charge belongs to at most one dipole. Of course, this pairing is by no means uniquely determined. The *length* of a dipole is defined to be the absolute value of the difference of the labels at the two involved vertices. Hence every dipole contributes exactly its length to the sum M. (It may be fun to notice that the paired-up positive and negative charges "attract each other", in the sense that we want to minimize their distances.)

A labeling such that $\sum_{j=k}^{n} M(j) \geq 0$ holds for all k is said to have the *dipole property*. This is equivalent to the property that we can form dipoles that contain all negative charges. Some surplus positive charges remain outside these dipoles.

For brevity, a *tree component* is a connected component being a tree.

Lemma 2. *Every labeling of any graph without a tree component has the dipole property.*

Proof. We show the contraposition: If some labeling of a graph G fails to have the dipole property, then G has a tree component.

Hence, assume that $\sum_{j=k}^{n} M(j) < 0$ for some k, and specifically, let k be the largest such index. Then we have $M(k) = -1$, and all charges above k can be paired up to dipoles. Let H denote the subgraph of G induced by the vertices with labels $s_k, \ldots, s_n$. Since the sum of all $M(j)$ in H is negative, there also exists some connected component T of H with more negative than positive charges. But this is possible only if T is a tree, and furthermore, no edges exist between vertices of T and vertices outside H. Hence the tree T is a connected component of G as well.   □

We say that two connected components C and D are *separated* in a labeling if all labels in C are smaller than all labels in D, or vice versa.

For any optimal labeling of G, trivially, the labeling restricted to any connected component T of G must be optimal, too. In particular, if T is a tree component, we can without loss of generality assume that T is labeled as in Theorem 1. Hence T contains only one charge which is negative and sits at the vertex with the lowest label in T.

Proposition 1. *Let G be a graph with a total number $t > 0$ of vertices in its tree components. For equidistant labels, there exists an optimal labeling of G where the vertices of the tree components have the t highest labels, the tree components are separated, and they are sorted by decreasing sizes.*

Proof. Given a labeling, we divide the vertex set of G in two sets X and Y consisting of the $|X|$ vertices with the lowest labels and the $|Y|$ vertices with the highest labels, respectively, where Y is a union of tree components. (Y may be empty.) Assume that not yet all tree components are in Y. Then, let $r \in X$ be the unique vertex that has a negative charge, belongs to some tree component T, and has the highest label among all such vertices.

By Lemma 2 and the assumed labeling of tree components, it follows that all charged vertices in X with higher labels than r are positively charged or belong to dipoles. Now we relabel X such that the orderings in both T and $X - T$ are preserved, but the vertices of T receive the highest labels in X. This has the following effects. The tree T is removed from X and included in Y, the negatively charged vertex r gets a higher label, and the labels of positively charged vertices as well as the lengths of the dipoles in $X - T$ can only decrease. Altogether, the objective M cannot get worse. By an inductive argument we achieve a labeling where all tree components are at the end of the ordering and are separated.

Finally, in an optimal labeling, the tree components must also be sorted by decreasing sizes as in Theorem 1. $\square$

For the proof it is crucial that the labels are equidistant. In the case of general labels, a dipole moving to points with other coordinates can get longer, although the number of vertices between the two charges does not increase. (It is easy to produce such counterexamples.) Of course, this cannot happen if the labels are equidistant. Moreover, since the dipoles in X can only move to smaller labels, it would be sufficient to suppose labels with monotone non-decreasing distances $s_{j+1} - s_j$. However, we stick to equidistant labels, which is a more natural assumption in the warehouse application.

We have shown that, in the case of equidistant labels, by Theorems 1 and 2 we can *eliminate* all tree components (see the beginning of Sect. 2). Therefore, from now on we can focus on graphs where every connected component has at least one cycle.

4   Eliminating the Leaves

A *leaf* is a vertex of degree 1. In the case of equidistant labels we can eliminate leaves also from connected components with cycles:

Proposition 2. *Let v be a leaf in a graph without tree components. For equidistant labels, there exists an optimal labeling where v has the highest label.*

Proof. Let u denote the unique neighbor of v, and let $G - v$ denote the graph G without v and the edge uv. Since G has no tree component, neither has $G - v$. We consider any labeling where v has not the highest label.

If the label of v is larger than the label of u, then the charges of vertices in $G - v$ are identical to their charges in G. Hence, due to Lemma 2, the labeling induced on $G - v$ has the dipole property. The leaf v is not charged. Now we simply assign the highest label to v and relabel the vertices of $G - v$ preserving their ordering. This can only decrease the lengths of dipoles (since the labels are equidistant) and the labels of the positively charged vertices outside the dipoles, thus M can only decrease.

The case when the label of v is smaller than the label of u is only slightly more complicated. If $M(u) > 0$, then we put one positive charge at u aside and form a dipole on the edge uv, together with the existing negative charge at v. If $M(v) = 0$, then we create a pair of a negative and a positive charge at v, and again, we form a dipole on the edge uv, whereas the new negative charge is assigned to u. In all cases, the charges not involved in the dipole on uv are identical to those in $G - v$, and these manipulations do not alter M. Precisely as above, we assign the highest label to v and relabel the vertices of $G - v$ preserving their ordering. The dipole at uv disappears, and for the same reasons as above, M can only decrease. $\square$

Using Proposition 2 we can eliminate any one leaf v, and the problem of optimally labeling $G - v$ remains. Of course, we can apply this step successively, until the residual graph has no leaves anymore. Therefore, from now on we can focus on graphs with minimum degree 2.

5   Eliminating and Separating the Cycle Components

A *cycle component* is a connected component which is merely a cycle. Our next observation is quite similar to Proposition 1. First we can optimally label every cycle component independently: An optimal labeling of a cycle was already observed in the proof of Theorem 2. It has one negative and one positive charge, at the vertex with the lowest and highest label, respectively. We declare them a dipole.

Proposition 3. *Let G be a graph with minimum vertex degree 2, and with a total number c > 0 of vertices in its cycle components. For equidistant labels, there exists an optimal labeling of G where the vertices of the cycle components have the c highest labels, and the cycle components are separated.*

Proof. Given a labeling, we divide the vertex set of G in two sets X and Y consisting of the $|X|$ vertices with the lowest labels and the $|Y|$ vertices with the highest labels, respectively, where Y is a union of cycle components. (Y may be empty.) Assume that not yet all cycle components are in Y. Then, let C be any cycle component that is not yet in Y.

Due to the minimum degree 2, the graph G has no tree components. By Lemma 2 it follows that all charged vertices in $X - C$ are positively charged or belong to dipoles.

Now we relabel X such that the orderings in both C and $X - C$ are preserved, but the vertices of C receive the highest labels in X. This has the following effects. The cycle C is removed from X and included in Y, and the labels of positively charged vertices in $X - C$ as well as the lengths of the dipoles in both $X - C$ and in C can only decrease. Altogether, the objective M cannot get worse. By an inductive argument we achieve a labeling as described in the statement.   □

Due to Proposition 3, we can also eliminate cycle components, in the case of equidistant labels. Moreover, the ordering of cycles is irrelevant, since every cycle contributes exactly its length minus 1 to M, regardless of its position in the ordering. Without further ado this settles MinSumEnds for a larger graph class than in Theorem 2, however for equidistant labels only.

Theorem 3. MinSumEnds *with equidistant labels is solvable in linear time on graphs where every connected component has at most one cycle.*

Proof. First eliminate the tree components due to Proposition 1 and the leaves due to Proposition 2, then concatenate optimal labelings of the cycles, where the permutation of the cycles is arbitrary.   □

As we already observed, this approach fails for general labels; we cannot even eliminate the leaves. But let us still consider disjoint unions of cycles for a moment. This is a too special case for applications, but the interesting point is that, with the help of dipoles, we get a rather straightforward NP-completeness proof for MinSumEnds in this very special case, by a reduction from the strongly NP-complete 3-Partition problem. We stress that this reduction does not work for equidistant labels, and the result complements NP-completeness for equidistant labels but general graphs [4].

Theorem 4. MinSumEnds *is NP-complete even for disjoint unions of cycles.*

Proof. We first observe again that every cycle must be optimally labeled, and its lowest and highest labeled vertex form a dipole. Furthermore, we can separate any two cycles that are not yet separated, because this decreases the total length of the dipoles. (We stress that this holds for arbitrary labels.) Hence, an optimal labeling is given by some permutation of the cycles also here.

Let $\{x_1, \ldots, x_{3t}\}$ be an instance of 3-Partition, that is, a multiset of $3t$ positive integers. The problem asks to partition this multiset into t triples, each with the same sum that we denote q. We create $3t$ disjoint cycles of lengths $q + x_i$ ($i = 1, \ldots, 3t$). On the number line we place t disjoint segments, each of length $4q$. In every segment we mark the $4q$ integer points. Let the gap between any two segments be larger than 1 (but otherwise arbitrary). The coordinates of the marked integer points are our $4qt$ labels. The constructed cycles have together $3tq + tq = 4tq$ vertices.

As stated above, there exists an optimal labeling where all cycles are separated and, moreover, every cycle has a dipole with a positive and negative charge at the vertex with lowest and highest label, respectively. The total length of the dipoles is $(4q - 1)t$ if and only if we can embed every cycle entirely in some segment. Since every cycle has a length larger than q, only 3 cycles fit in every segment. Finally, in order to embed all $3t$ cycles in the 3 segments, we must divide them into t triples, each with a total of $4q = 3q + q$ vertices. This establishes the equivalence of the problem instances.   □

6 Paths of Degree-2 Vertices and Cores

In the following, let G be a graph with minimum vertex degree 2 (with the understanding that every loop contributes 1 to the degree of its vertex) and without cycle components. (Recall that a cycle component is a cycle without further edges, both inside and to the rest of G.)

We call every vertex of degree larger than 2 a *principal vertex*. We call a path a *principal path* if it ends in two principal vertices (which may be identical), it has at least one inner vertex, and all its inner vertices are of degree 2. Hence the edge set of graph G can be uniquely partitioned into the edge sets of its principal paths and single edges that do not belong to principal paths as they end in two principal vertices.

Lemma 3. *Let G be a graph of minimum degree 2 and without cycle components. For equidistant labels, there exists an optimal labeling of G where either (1) some principal vertex gets the highest label, or (2) some inner vertex v of some principal path P gets the highest label, followed by all other inner vertices of P getting the next smaller labels. Furthermore, in case (2) and for any fixed P, the choice of v from P is arbitrary.*

Proof. The distinction of cases (1) and (2) is trivial, since other types of vertices do not exist in G. In case (2), where we first eliminate a vertex v from a principal path P, we can apply Proposition 2 repeatedly until the rest of P is eliminated, too. Not only the leaves may be eliminated in any order, it is also immaterial which inner vertex v from P we choose first: In any case, v gets one positive charge, and the other inner vertices of P get no charge, hence the choice of v on P does not affect the objective value M.   □

Lemma 3 enables dynamic programming on subsets of principal vertices and paths (rather than just vertices):

Theorem 5. MINSUMENDS *with equidistant labels can be solved in $O^*(2^p)$ time, where p is the total number of principal vertices and paths after the elimination of tree components, leaves, and cycle components.*

Proof. We eliminate principal vertices and paths as in Lemma 3, in all possible ways, but: Among all partial solutions that assign the labels larger than s_k to the same $n - k$ vertices (that is, retain the same graph G_k), it suffices to keep some solution with minimum $\sum_{i=k+1}^{n} s_k \cdot M(k)$. Furthermore, whenever we eliminate some principal vertex being incident to some principal paths, we next eliminate these paths, leaf by leaf, as in Proposition 2.

Let us call two vertices equivalent if they are inner vertices of the same principal path. That is, every principal path becomes an equivalence class. Every principal vertex is an equivalence class of its own. With this definition we observe that, during the elimination process, equivalence classes are either removed completely or they get merged, but they are never torn apart. This implies that the parameter value p never increases, and the time bound follows.   □

In the following we strengthen Theorem 5 by making the parameter smaller. The next lemmas presume the same type of graphs as before.

Lemma 4. *Let P be some principal path that ends in some principal vertex v (and in some other principal vertex different from v). Then, instead of eliminating v, one can always eliminate the inner vertices of P first, without making the labeling worse.*

Proof. Let u be an arbitrary inner vertex of P, and let $d \geq 3$ denote the degree of v. If we first eliminate v, followed by P, then v receives $d - 1$ positive charges. In fact, we can assume that P is completely eliminated next, as the ordering of eliminating leaves is arbitrary.

If we instead eliminate u first, followed by the rest of P and by v, then we eliminate the same set of vertices and edges as before, until that moment, but u receives only one positive charge, whereas v receives only $d - 2$ positive charges which are located at smaller labels. This makes M strictly smaller, hence it is never advantageous to assign the highest label to v.   □

Lemma 5. *Let P be some principal path with principal vertex v at both ends. Then, instead of eliminating v, one can always eliminate P first, without making the labeling worse.*

Proof. The argument is similar. Let u be an arbitrary inner vertex of P, and let $d \geq 3$ denote the degree of v. If we first eliminate v, followed by P, then v receives $d - 1$ positive charges. Now P becomes a tree component and receives one negative charge at its lowest label.

If we instead eliminate u first, followed by the rest of P and by v, then u receives one positive charge, and v receives $d - 3$ positive charges, making M strictly smaller. Hence it is not advantageous to give the highest label to v.   □

Lemmas 4 and 5 together state that a principal vertex needs to be considered for elimination only if all its neighbors are principal vertices, too. Some of our results can now be nicely expressed using the notion of a core.

For any positive integer d, the *d-core* of the graph G is the graph obtained from G by removing vertices of degree smaller than d, and their incident edges, as long as possible. The result does not depend on the order of removals. Equivalently, the d-core is the uniquely determined largest induced subgraph of G with minimum vertex degree d. Remember that we adopt the convention that a loop at a vertex v contributes only 1 to the degree of v.

Propositions 1 and 2 immediately imply:

Proposition 4. *For every graph G and for equidistant labels, there exists an optimal labeling of G where all vertices in the 2-core of G have smaller labels than all other vertices.* $\square$

A similar statement is not true for the 3-core. A small counterexample is the graph consisting of one vertex with two loops and a clique of four vertices. The clique is the 3-core, but the only optimal labeling gives the lowest label to the two-loop vertex. However, we can somewhat strengthen Theorem 5 using the 3-core. The following parameter q is smaller than p from Theorem 5, because it includes only principal vertices in the 3-core.

Theorem 6. MinSumEnds *with equidistant labels can be solved in $O^*(2^q)$ time, where q is the number of vertices in the 3-core plus the number of principal paths, after the elimination of tree components, leaves, and cycle components.*

Proof. We proceed as in Theorem 5, but according to Lemmas 4 and 5 we never have to eliminate principal vertices outside the 3-core. $\square$

This also implies a bound in a more natural parameter:

Theorem 7. MinSumEnds *with equidistant labels can be solved in $O^*(6^{m-n})$ time, where m and n denotes the number of edges and vertices, respectively.*

Proof. Due to Theorem 6 it suffices to show $q \leq 3(m - n)$.

We can replace every principal path of arbitrary length with a principal path with only one inner vertex, as this changes neither q nor $m - n$. Now every vertex in the 3-core and every inner vertex of a principal path contributes a summand exactly 1 to q, by the definition of q. We also divide edges with two different ends between these two vertices and thus assign fractions of edges to vertices, such that no fraction is erroneously counted twice.

Every principal vertex outside the 3-core contributes zero to q, by the definition of q. We assign $1/3$ of every incident edge to it, hence it contributes a summand at least $3 \cdot (1/3) - 1 \geq 0$ to $m - n$, that is, it does not contribute negatively. Every vertex on a principal path contributes a summand at least $2 \cdot (2/3) - 1 = 1/3$ to $m - n$ via its 2 incident edges. (In the worst case, both ends may be principal vertices that do not belong to the 3-core.) Every vertex in the 3-core contributes a summand at least $1/2 = 3/2 - 1$ to $m - n$, via halves of 3 of its incident edges within the 3-core.

In conclusion, the ratio $(m - n)/q$ is at least $1/3$. $\square$

7   Conclusions

We considered a product location problem in warehouses, with a collection point at the end of a shelf, and with a small number of different requests of at most two items, leading to a labeling problem on sparse graphs. We believe that the FPT results can be further improved: The worst case in Theorem 7 is 3-regular graphs with subdivided edges. Then, eliminations of the principal paths cause mergings of many other principal paths, hence by far not all subsets of principal paths can appear. Also, more can be done for non-equidistant labels, weighted (instead of multiple) edges, and hypergraphs.

Acknowledgments. This work has been done during the author's engagement as scientific advisor at the Fraunhofer-Chalmers Research Centre for Industrial Mathematics, Göteborg (FCC). The author appreciates support from FCC and many discussions with Fredrik Ekstedt and Raad Salman who brought up this type of problems. He also thanks the referees for very careful reading.

References

1. Ambühl, C., Mastrolilli, M., Svensson, O.: Inapproximability results for maximum edge biclique, minimum linear arrangement, and sparsest cut. SIAM J. Comput. **40**, 567–596 (2011)
2. Arora, S., Frieze, A., Kaplan, H.: A new rounding procedure for the assignment problem with applications to dense graphs arrangements. Math. Program. **92**, 1–36 (2002)
3. Bhasker, J., Sahni, S.: Optimal linear arrangement of circuit components. In: HICSS 1987, vol. 2, pp. 99–111 (1987)
4. Boysen, N., Stephan, K.: The deterministic product location problem under a pick-by-order policy. Discrete Appl. Math. **161**, 2862–2875 (2013)
5. Cohen, J., Fomin, F., Heggernes, P., Kratsch, D., Kucherov, G.: Optimal linear arrangement of interval graphs. In: Královič, R., Urzyczyn, P. (eds.) MFCS 2006. LNCS, vol. 4162, pp. 267–279. Springer, Heidelberg (2006). https://doi.org/10.1007/11821069_24
6. Eikel, M., Scheideler, C., Setzer, A.: Minimum linear arrangement of series-parallel graphs. In: Bampis, E., Svensson, O. (eds.) WAOA 2014. LNCS, vol. 8952, pp. 168–180. Springer, Cham (2015). https://doi.org/10.1007/978-3-319-18263-6_15
7. Esteban, J.L., Ferrer-i-Cancho, R.: A correction on shiloach's algorithm for minimum linear arrangement of trees. SIAM J. Comput. **46**, 1146–1151 (2017)
8. Feige, U., Lee, J.R.: An improved approximation ratio for the minimum linear arrangement problem. Inf. Process. Lett. **101**, 26–29 (2007)
9. Fellows, M.R., Hermelin, D., Rosamond, F.A., Shachnai, H.: Tractable parameterizations for the minimum linear arrangement problem. ACM Trans. Comput. Theory **8**, 6:1–6:12 (2016)
10. Fernau, H.: Parameterized algorithmics for linear arrangement problems. Discrete Appl. Math. **156**, 3166–3177 (2008)
11. Fomin, F.V., Kratsch, D.: Split and list. In: Fomin, F.V., Kratsch, D. (eds.) Exact Exponential Algorithms. TTCSAES, pp. 153–160. Springer, Heidelberg (2010). https://doi.org/10.1007/978-3-642-16533-7_9

12. Garey, M.R., Johnson, D.S.: Computers and Intractability. A Guide to the Theory of NPcompleteness. Freeman, New York (1979)
13. Mirzaei, S., Kfoury, A.J.: Linear arrangement of Halin graphs. CoRR abs/1509. 08145 (2015)
14. Tamaki, S., Yoshida, Y.: Approximation guarantees for the minimum linear arrangement problem by higher eigenvalues. In: Gupta, A., Jansen, K., Rolim, J., Servedio, R. (eds.) APPROX/RANDOM-2012. LNCS, vol. 7408, pp. 313–324. Springer, Heidelberg (2012). https://doi.org/10.1007/978-3-642-32512-0_27

On the Complexity of Singly Connected Vertex Deletion

Avinandan Das[1], Lawqueen Kanesh[1], Jayakrishnan Madathil[1(⊠)],
Komal Muluk[1], Nidhi Purohit[2], and Saket Saurabh[1,2]

[1] The Institute of Mathematical Sciences, HBNI, Chennai, India
adas33745@gmail.com, {lawqueen,jayakrishnanm,saket}@imsc.res.in,
komalmuluk15@gmail.com
[2] Department of Informatics, University of Bergen, Bergen, Norway
nidhipurohit95@gmail.com

Abstract. A digraph D is singly connected if for all ordered pairs of
vertices $u, v \in V(D)$, there is at most one path in D from u to v. In this
paper, we study the SINGLY CONNECTED VERTEX DELETION (SCVD)
problem: Given an n-vertex digraph D and a positive integer k, does
there exist a set $S \subseteq V(D)$ such that $|S| \leq k$ and $D - S$ is singly
connected? This problem may be seen as a directed counterpart of the
(UNDIRECTED) FEEDBACK VERTEX SET problem, as an undirected graph
is singly connected if and only if it is acyclic. SCVD is known to be NP-
hard on general digraphs. We study the complexity of SCVD on various classes of digraphs such as tournaments, and various generalisations
of tournaments such as digraphs of bounded independence number, in-
and out-tournaments and local tournaments. We show that unlike the
FEEDBACK VERTEX SET ON TOURNAMENTS (FVST) problem, SCVD
is polynomial time solvable on tournaments. In addition, we show that
SCVD is polynomial time solvable on digraphs of bounded independence
number, and on the class of acyclic local tournaments. We also study
the parameterized complexity of SCVD, with k as the parameter, on the
class of in-tournaments. And we show that on in-tournaments (and out-
tournaments), SCVD admits a fixed-parameter tractable algorithm and
a quadratic kernel. We also show that on the class of local tournaments,
which is a sub-class of in-tournaments, SCVD admits a linear kernel.

Keywords: Singly connected digraphs · FPT algorithm · Kernel ·
Bounded independence number · Tournaments · Local tournaments

This project has received funding from the European Research Council (ERC) under
the European Union's Horizon 2020 research and innovation programme (grant no.
819416), and the Swarnajayanti Fellowship grant DST/SJF/MSA-01/2017-18.

© Springer Nature Switzerland AG 2020
L. Gąsieniec et al. (Eds.): IWOCA 2020, LNCS 12126, pp. 237–250, 2020.
https://doi.org/10.1007/978-3-030-48966-3_18

1   Introduction

A digraph D is said to be singly connected if for every (ordered) pair of vertices u and v of D, there is at most one (directed) path in D from u to v. In this paper, we study the SINGLY CONNECTED VERTEX DELETION (SCVD for short) problem, where the goal is to test if a given digraph can be made singly connected by deleting a few vertices. This problem may be seen as a directed counterpart of the FEEDBACK VERTEX SET problem. To see this, let us first define undirected singly connected graphs. An undirected graph G is said to be singly connected if for every pair of vertices u and v of G, there is at most one path in G between u and v. But note that an undirected graph is singly connected if and only if it is acyclic. So, the problem of checking whether it is possible to delete at most k vertices from a given graph to make it singly connected is the same as the problem of checking whether it is possible to delete at most k vertices to make a graph acyclic. This precisely is the FEEDBACK VERTEX SET (FVS) problem. (A feedback vertex set of a graph is a set of vertices whose deletion will render the graph acyclic.) The complexity of FVS has been studied extensively [3,10,12–14,17,21,26,27,31–34,37]. FVS, in fact, was one of Karp's 21 NP-hard problems [28]. As for its algorithmic tractability, FVS is fixed-parameter tractable (when parameterized by the solution size) [21] and it admits a quadratic kernel [40]. FVS also admits constant factor approximation algorithms [3,8,16,24].

Coming back to digraphs, the DIRECTED FEEDBACK VERTEX SET (DFVS) problem asks if a given digraph can be made acyclic by deleting at most k vertices. Naturally, this problem has been deemed the appropriate directed counterpart of FEEDBACK VERTEX SET, and has been studied in the frameworks of approximation algorithms [39] and parameterized algorithms [15]. Although the parameterized complexity of DFVS had been raised as an open problem since the emergence of parameterized algorithms in the early 90s [20,22], it was settled only in 2008 by Chen et al. [15]. They showed that the problem admits a $4^k k! n^{O(1)}$ time algorithm, and hence is fixed-parameter tractable when parameterized by k. If fixed-parameter tractability of DFVS remained open for years, the kernelization complexity of the problem proved even more elusive. While the question whether DFVS (parameterized by k) admits a polynomial kernel still remains unresolved, several attempts have been made to study the kernelization complexity of "DFVS-adjacent" problems. These include studying the problem with larger parameters [9,35], restricting the input digraph to smaller classes [1,7] and imposing more conditions on the acyclic digraph that results from the deletion of a feedback vertex set [2,36].

While FVS and DFVS generated a large volume of literature, the SCVD problem, already known to be NP-hard [19], received little attention from the parameterized complexity community. In this paper, as a first step, we start an investigation into the complexity of SCVD on various classes of digraphs such as tournaments, local tournaments, digraphs of bounded independence number etc. We formally define the problem below.

SINGLY CONNECTED VERTEX DELETION (SCVD)   **Parameter:** k
Input: A digraph D and a non-negative integer k.
Question: Does there exist a set $S \subseteq V(D)$ such that $|S| \leq k$ and $D - S$ is singly connected?

(a) Obstruction to acyclic tournament.

(b) Obstruction to singly connected tournament.

Fig. 1. Obstructions to acyclic and singly connected tournaments.

As observed earlier, an undirected graph is singly connected if and only if it is acyclic. But notice that this property does not hold for digraphs. A directed cycle, for instance, is singly connected. And consider a digraph on 3 vertices, say, x, y and z, and with arcs $(x, y), (y, z)$ and (x, z). This digraph, while acyclic, is not singly connected. It is not surprising then that SCVD and DFVS show markedly different behaviour. This is perhaps best illustrated by the fact that while DFVS is NP-hard on tournaments, we show that SCVD is polynomial time solvable on tournaments (Lemma 2). This difference in behaviour appears even starker considering the fact that these two problems require that "obstructions" with a "similar structure" be hit. Notice that obstructions to an acyclic tournament are directed triangles, i.e., all triplets of vertices x, y and z with arcs $(x, y), (y, z)$ and (z, x), whereas obstructions to a singly connected tournament are all triplets of vertices x, y and z with arcs $(x, y), (y, z)$ and (x, z) (see Fig. 1).

A digraph D is not singly connected if and only if there exists a pair of vertices u and v such that D contains two paths from u to v. It is not difficult to see that a digraph D is not singly connected if and only if there exists a pair of vertices u and v such that D contains two *internally vertex disjoint* paths from u to v. (See Lemma 1.) Two internally vertex disjoint paths between a pair of vertices of a digraph constitute a cycle in the underlying undirected graph. That is, the obstructions to a singly connected digraph are cycles in the underlying undirected graph. But notice that not every cycle in the underlying undirected graph is necessarily an obstruction. Thus both DFVS and SCVD require us to examine if a subset of the cycles in the underlying undirected graph can be hit with a few vertices.

Our Contribution. We study the SCVD problem on several well-studied classes of digraphs such as tournaments, α-bounded digraphs, local tournaments, etc.

A digraph D is said to be a *tournament* if for every pair of vertices u and v of D, exactly one of the arcs (u, v) and (v, u) is present in D. The class of α-*bounded* digraphs were introduced by Fradkin and Seymour [23] as a generalisation of tournaments. For a fixed positive integer α, a digraph D is said to be α-bounded if the size of a maximum independent set of the underlying undirected graph of D is at most α. Note that tournaments are 1-bounded digraphs. Local tournaments are yet another generalisation of tournaments. A digraph D is said to be an *in-tournament* (resp. *out-tournament*) if for every vertex v of D, the set of in-neighbours (resp. out-neighbours) of v induces a tournament. A digraph D is said to be a *local tournament* if it is both an in-tournament and an out-tournament. A digraph D is said to be a an *acyclic local tournament* if D is both a directed acyclic graph and a local tournament. (See, for example, the chapter on locally semi-complete digraphs [5] in the monograph edited by Bang-Jensen and Gutin [6] for a survey of literature on these classes of digraphs.)

We show that SINGLY CONNECTED VERTEX DELETION

- is polynomial time solvable on tournaments and α-bounded digraphs,
- is polynomial time solvable on acyclic local tournaments,
- has a $2^k n^{\mathcal{O}(1)}$ algorithm and $\mathcal{O}(k^2)$ vertex kernel on in- and out-tournaments,
- and has an $\mathcal{O}(k)$ vertex kernel on local tournaments.

The polynomial time solvability of SCVD on tournaments follows from a simple observation that no tournament with at least four vertices can be singly connected. A similar result holds for α-bounded digraphs as well: no α-bounded digraph with at least $2\alpha^2 + 4\alpha$ vertices can be singly connected. In order to prove this observation, we use the Gallai-Milgram theorem [25], which says that the vertices of a digraph D can be covered by a disjoint collection of paths, such that the number of paths does not exceed the size of a maximum independent set of the underlying undirected graph of D. On acyclic local tournaments, we design a polynomial time algorithm that computes a minimum-sized vertex subset whose deletion will make the digraph singly connected. Our algorithm uses the fact that every connected local tournament has a Hamiltonian path [4], which in turn, implies that every connected acyclic local tournament has a unique topological ordering. We show that SCVD on in-tournaments (and out-tournaments) can be reduced to the 3-HITTING SET problem, and thus admits a simple $3^k n^{\mathcal{O}(1)}$ time branching algorithm and an $\mathcal{O}(k^2)$ vertex kernel. But we use the technique of iterative compression to design a $2^k n^{\mathcal{O}(1)}$ algorithm for SCVD on in and out-tournaments. And our $\mathcal{O}(k)$ vertex kernel for SCVD on local tournaments relies on the fact that for a local tournament D and a set of vertices $S \subseteq V(D)$ such that $D - S$ is singly connected, no vertex in S can have more than a constant number of neighbours in $V(D) \setminus S$.

Related Work on Singly-Connected Digraphs. As noted above, the SCVD problem was shown to be NP-hard by Dietzfelbinger and Jaberi [19]. The reduction in [19], in fact, shows that the problem is NP-hard even on directed acyclic graphs. Their work shows that the arc-deletion version of the problem is also NP-hard, i.e., the problem of testing whether a given digraph can be made singly connected by deleting at most a given number of arcs. As for recognising singly connected digraphs, i.e., the problem of testing whether a given digraph is singly connected, Buchsbaum and Carlisle [11] gave an algorithm that runs in $\mathcal{O}(n^2)$ time, where n is the number of vertices in the input digraph. Khuller [29,30] gave another $\mathcal{O}(n^2)$ algorithm for this problem. Dietzfelbinger and Jaberi [19] presented a refined version of the algorithm of Buchsbaum and Carlisle [11] that runs in time $\mathcal{O}(s \cdot t + m)$, where m is the number of arcs, and s and t respectively are the number of sources and sinks in the input digraph.

2 Preliminaries

For a positive integer n, we denote the set $\{1, 2, \ldots, n\}$ by $[n]$. Let S be a finite set, and let σ be an ordering of the elements of S. For $x, y \in S$, we write $x <_\sigma y$ to mean that x appears before y in the ordering σ. And we write $x \leq_\sigma y$ to mean that either $x = y$ or $x <_\sigma y$.

Digraphs. For a digraph D, $V(D)$ denotes the vertex set and $A(D)$ denotes the arc set of D. For a vertex $v \in V(D)$, $N_D^+(v)$ denotes the set of all out-neighbours of v, and $N_D^-(v)$ denotes the set of all in-neighbours of v, that is, $N_D^+(v) = \{u \in V(D) \mid (v, u) \in A(D)\}$ and $N_D^-(v) = \{u \in V(D) \mid (u, v) \in A(D)\}$. And $N_D(v)$ denotes the set of all neighbours of v in the underlying undirected graph of D, that is, $N_D(v) = N_D^+(v) \cup N_D^-(v)$. Also, we define $N_D^+[v] = N_D^+(v) \cup \{v\}$, $N_D^-[v] = N_D^-(v) \cup \{v\}$ and $N_D[v] = N_D(v) \cup \{v\}$. For a set $X \subseteq V(D)$, we define $N_D(X) = \cup_{v \in X} N_D(v)$.

For a set $A' \subseteq A(D)$, $D - A'$ denotes the digraph $(V(D), A(D) \setminus A')$. For a set $V' \subseteq V(D)$, $D[V']$ denotes the subgraph of D induced by V'. Similarly, for $S \subseteq V(D)$, $D - S$ denotes the digraph $D[V(D) \setminus S]$.

A digraph D is said to be *connected* if the underlying undirected graph of D is connected. A digraph D on 3 vertices, say, x, y and z, is said to be an *acyclic triangle* if $A(D) = \{(x, y), (y, z), (x, z)\}$.

A *path cover* $\mathcal{P}$ of a digraph D is a disjoint collection of paths in D such that for every vertex $v \in V(D)$, there is a path $P \in \mathcal{P}$ such that $v \in V(P)$.

For the sake of convenience, we repeat below some of the definitions we introduced in Sect. 1. Recall that a directed graph D is a tournament if for every pair of distinct vertices $u, v \in V(D)$, either $(u, v) \in A(D)$ or $(v, u) \in A(D)$, but not both.

Definition 1 (Out-tournament and In-tournament). *A directed graph D is an out-tournament (resp. in-tournament) if for all $v \in V(D)$, $D[N_D^+(v)]$ (resp. $D[N_D^-(v)]$) is a tournament.*

Definition 2 (Local tournament). *A directed graph D is a local tournament if D is both an out-tournament and an in-tournament.*

Note that, by the definition of singly connected digraphs, a digraph D is *not* singly connected if there exist two paths from u to v for $u, v \in V(D)$. Note that these two paths need not be internally vertex disjoint. But the following lemma says that we may as well assume that the two paths are internally vertex disjoint.

Lemma 1 ($\star$[1]). *A directed graph D is not singly connected if and only if there exist two vertices $u, v \in V(D)$ such that there exist two internally vertex disjoint paths from u to v.*

3   Singly Connected Vertex Deletion on α-bounded Digraphs and Acyclic Local Tournaments

In this section, we study the optimisation version of SCVD restricted to α-bounded digraphs and acyclic local tournaments, and prove that the problem is polynomial time solvable on both these classes of digraphs. That is, we consider the following problem.

MINIMUM SINGLY CONNECTED VERTEX DELETION (MIN-SCVD)
Input: A digraph D.
Output: A minimum-sized set $S \subseteq V(D)$ such that $D - S$ is singly connected.

3.1   MIN-SCVD on α-bounded Digraphs

In this section, we prove that MIN-SCVD is polynomial time solvable on α-bounded digraphs. Specifically, we prove the following theorem.

Theorem 1. MIN-SCVD *can be solved in time $\mathcal{O}(n^{\alpha(2\alpha+3)})$ on α-bounded digraphs, where n is the number of vertices of the input α-bounded digraph.*

We first consider the problem on tournaments. Although Theorem 1 applies to tournaments as well, as tournaments are 1-bounded digraphs, we consider tournaments separately, and prove that the MIN-SCVD problem can be solved in $\mathcal{O}(n^3)$ time on tournaments. This result follows from a simple observation that no tournament with 4 or more vertices can be singly connected.

Lemma 2 ($\star$). *Any tournament on at least 4 vertices is not singly connected.*

Using Lemma 2 and the fact that tournaments are hereditary, we get the following corollary.

Corollary 1 ($\star$). MIN-SCVD *on tournaments is solvable in $\mathcal{O}(n^3)$ time.*

We now move on to α-bounded digraphs, and prove Theorem 1. We prove below that no α-bounded digraph with at least $\alpha(2\alpha + 4)$ vertices is singly connected. Note that this immediately gives an $\mathcal{O}(n^{\alpha(2\alpha+3)})$ time algorithm for MIN-SCVD on α-bounded digraphs, as solving MIN-SCVD reduces to finding a maximum sized induced subgraph that is singly connected, which can be done in the claimed runtime.

We need the following theorem due to Gallai and Milgram [25] to prove our observation that no α-bounded digraph with at least $\alpha(2\alpha + 4)$ vertices can be singly connected.

---

[1] Due to paucity of space, the proofs of statements marked with a $\star$ have been omitted.

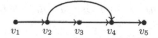   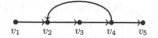

(a) The arc (v_2, v_4) is a forward arc w.r.t. the path $v_1 \cdots v_5$.   (b) The arc (v_4, v_2) is a backward arc w.r.t. the path $v_1 \cdots v_5$.

Fig. 2. Forward and backward arcs w.r.t. a path.

Theorem 2 (Gallai and Milgram [18, 25]). *Every directed graph D has a path cover $\mathcal{P}$ and an independent set $\{v_p \mid P \in \mathcal{P}\}$ of vertices such that $v_p \in P$ for every $P \in \mathcal{P}$.*

We can assume that the set $\{v_p \mid P \in \mathcal{P}\}$ in Theorem 2 is a maximal independent set. If not, we can add more vertices to the set until it becomes maximal, and "break" the paths in $\mathcal{P}$ at those newly added vertices to make new paths. The new collection of paths is a path cover of D such that every path contains a vertex of the maximal independent set. We record this fact below.

Observation 1. *Every directed graph D has a path cover $\mathcal{P}$ and a maximal independent set $\{v_p \mid P \in \mathcal{P}\}$ of vertices such that $v_p \in P$ for every $P \in \mathcal{P}$.*

Let D be a digraph. For a path $P = v_1 \ldots v_\ell$ in D, we define forward arcs and backward arcs with respect to P in D as follows. An $(v_i, v_j) \in A(D)$ is a *forward arc w.r.t.* P if $v_i, v_j \in V(P)$, and $j > i + 1$. And $(v_i, v_j) \in A(D)$ is a *backward arc w.r.t.* P if $v_i, v_j \in V(P)$ and $i > j + 1$ (see Fig. 2).

For a path $P = v_1 \ldots v_\ell$ in a digraph D, if $(v_i, v_j) \in A(D)$ is a forward arc w.r.t. P then note that there are two distinct paths from v_i to v_j in D: $v_i \ldots v_j$ and $v_i v_j$. Therefore, we have the following observation.

Observation 2. *If a digraph D has a path P such that D contains a forward arc w.r.t. P, then D is not singly connected.*

We now prove the following lemma, which, in turn proves Theorem 1.

Lemma 3. *For each fixed $\alpha \in \mathbb{N}$, every α-bounded digraph with at least $\alpha(2\alpha+4)$ vertices is not singly connected.*

Proof. Let D be any α-bounded digraph such that $|V(D)| \geq \alpha(2\alpha + 4)$. Assume that D is singly connected. By Theorem 2 (and Observation 1), there is a maximal independent set I such that D can be decomposed into a collection $\mathcal{P}$ of $|I|$ vertex disjoint paths such that each path contains one vertex from I. Let $|I|(= |\mathcal{P}|) = \alpha'$. Note that $\alpha' \leq \alpha$, as D is an α-bounded digraph. Then, since $|V(D)| \geq \alpha(2\alpha + 4)$, by the pigeonhole principle, there exists a path P in $\mathcal{P}$ with at least $2\alpha + 4$ vertices. Let P be $v_1 \ldots v_\ell$, where $\ell \geq (2\alpha + 4)$, be such a path. Let v_P be a vertex of P such that $v_P \in I$. We now prove the following two claims.

Claim 1 ($\star$). *With respect to the path P, the vertex v_P can have at most two backward arcs and no forward arcs incident on it.*

Claim 2 ($\star$). *For a vertex $v \notin V(P)$, there can be at most two arcs between v and $V(P)$.*

Now, let $I_P = N[v_P] \cap V(P)$, i.e., the set $I_P \subseteq V(P)$ contains v_P and the vertices in $V(P)$ that are adjacent to v_P. Since P is a path and because of Claim 1, $|I_P| \leq 5$. Let $S = V(P) \setminus I_P$. Then, $|S| \geq 2\alpha - 1$, as $|V(P)| \geq 2\alpha + 4$. Also, observe that no vertex in S is adjacent to v_P. Then, every vertex in S is adjacent to some vertex in $I \setminus \{v_P\}$. To see this, consider $x \in S$. Note first that $x \notin I$, as $I \cap V(P) = \{v_P\}$. And now, if x is not adjacent to any vertex in $I \setminus \{v_P\}$, then $I \cup \{x\}$ is an independent set, which contradicts the maximality of I. Therefore, $|N_D(I \setminus \{v_P\}) \cap S| = |S| \geq 2\alpha - 1$. Now, since $|I \setminus \{v_P\}| \leq \alpha - 1$, by the pigeonhole principle, there is a vertex in $I \setminus \{v_P\}$ which is adjacent to at least three vertices in S, which, by Claim 2, is not possible. This completes the proof of Lemma 3. $\square$

3.2   Polynomial Time Algorithm for MIN-SCVD on Acyclic Local Tournaments

In this section, we prove that MIN-SCVD is polynomial time solvable on acyclic local tournaments. Without loss of generality, let us assume that the input acyclic local tournament is connected. Otherwise, we can find an optimal solution in each connected component separately and return the union of the optimal solutions for all the connected components. Specifically, this section is devoted to proving the following theorem.

Theorem 3. MINIMUM SINGLY CONNECTED VERTEX DELETION *can be solved in time $\mathcal{O}(n^{\mathcal{O}(1)})$ on acyclic local tournaments, where n is the total number of vertices in the input acyclic local tournament.*

The proof of Theorem 3 crucially uses the fact that every connected local tournament has a Hamiltonian path [4], which, in turn, implies that every connected acyclic local tournament has a unique topological ordering.

We first state the following lemma. It is so well-known that we omit its proof.

Lemma 4. *Let D be a directed acyclic graph. Then, D has a topological ordering. That is, there exists an ordering $\sigma = (v_1, \ldots, v_n)$ of the vertices of D such that for every arc $(v_i, v_j) \in A(D)$, we have $i < j$, i.e., v_i appears before v_j in the ordering σ. Moreover, there exists a polynomial time algorithm that, given a directed acyclic graph D as input, finds a topological ordering of D.*

It is a folklore result that every tournament contains a Hamiltonian path. Bang-Jensen [4] showed that this applies to connected local tournaments as well. For the sake of completeness, we prove this below.

Lemma 5 ($\star$). *Let D be a connected local tournament. Then D contains a Hamiltonian path.*

The following lemma follows from Lemmas 4 and 5.

Lemma 6 ($\star$). *Let D be a connected acyclic local tournament and $P = v_1 v_2 \ldots v_n$ be a Hamiltonian path of D. Then, $\sigma = (v_1, \ldots, v_n)$ is the unique topological ordering of D.*

Notation. Let D be an acyclic local tournament and $\sigma = (v_1, \ldots, v_n)$ be the unique topological ordering of D. For a vertex $u \in V(D)$, by $\ell(u)$, we denote the last vertex v in the ordering σ such that $(u, v) \in A(D)$. For each $i \in [n]$, we define an ordered set $S_i = \{v_i, v_{i+1}, \ldots, \ell(v_i)\}$.

Lemma 7 ($\star$). *Let D be a connected acyclic local tournament and $\sigma = (v_1, \ldots, v_n)$ be the topological ordering of D. Then, for all $i \in [n]$, the graph $D[S_i]$ is an acyclic tournament. Moreover, $S_i = N_D^+(v_i) \cup \{v_i\}$.*

The following lemma says that any optimal solution to MIN-SCVD on D can exclude at most two vertices from the set S_i for each $i \in [n]$.

Lemma 8 ($\star$). *Let D be an acyclic local tournament and S be an optimal solution to MIN-SCVD on D. Let $\sigma = (v_1, v_2, \ldots, v_n)$ be the topological ordering of D. Then, for every $i \in [n]$, we have $|S_i \setminus S| \leq 2$.*

Lemma 9 ($\star$). *Let D be an acyclic local tournament and $\sigma = (v_1, v_2, \ldots, v_n)$ be the topological ordering of D. Let $v_i, v_j \in V(D)$ such that $i < j$. Let $\ell(v_i) = v_{p_i}$ and $\ell(v_j) = v_{p_j}$. Then, $p_i \leq p_j$.*

The following lemma forms the basis of our algorithm.

Lemma 10 ($\star$). *Let D be an acyclic local tournament and $\sigma = (v_1, \ldots, v_n)$ be the topological ordering of D. Then, there exists an optimal solution to MIN-SCVD on D that does not contain the vertices v_1, v_2.*

Proof (Proof Sketch). Let S be an optimal solution to MIN-SCVD on D. If $v_1, v_2 \notin S$, then the lemma holds. So assume that either $v_1 \in S$ or $v_2 \in S$.

By Lemma 7, the graphs $D[S_1]$ and $D[S_2]$ are acyclic tournaments, and $S_1 = N_D^+(v_1) \cup \{v_1\}$ and $S_2 = N_D^+(v_2) \cup \{v_2\}$. By Lemma 9, we have $\ell(v_1) \leq_\sigma \ell(v_2)$. This implies that $S_1 \setminus \{v_1\} \subseteq S_2$. By Lemma 8, we have $|S_1 \setminus S| \leq 2$ and $|S_2 \setminus S| \leq 2$. We now consider two cases depending on whether $v_1 \in S$ or $v_2 \in S$. We only prove the case when $v_1 \in S$ here.

Case 1: $v_1 \in S$. If $S_1 \setminus S = \emptyset$, then since $N_D^+(v_1) \subseteq S_1$, the digraph $D - (S \setminus \{v_1\})$ is also singly connected, which contradicts the assumption that S is an optimal solution. Therefore, $|S_1 \setminus S| \geq 1$. Let $v_p \in S_1$ be such that $v_p \notin S$. (Note that $p \neq 1$ as we are in the case when $v_1 \in S$.) We shall show that $(S \setminus \{v_1\}) \cup \{v_p\}$ is also an optimal solution to MIN-SCVD on D.

Now, consider the digraph $D - (S \setminus \{v_1\})$. Since S is an optimal solution, $D - (S \setminus \{v_1\})$ is not singly connected. That is, $D - (S \setminus \{v_1\})$ contains a pair of vertices u and v such that there are two internally vertex disjoint paths in $D - (S \setminus \{v_1\})$ from u to v. We refer to such a pair of paths as a forbidden structure. But since $D - S$ is singly connected, any forbidden structure in $D - (S \setminus \{v_1\})$

must contain v_1. Also, note that since v_1 is the first vertex in the topological ordering σ, any forbidden structure in $D - (S \setminus \{v_1\})$ must be a pair of paths that start from v_1.

Now, since, $|S_1 \setminus S| \leq 2$, the vertex v_1 has at most two out-neighbours in the digraph $D - (S \setminus \{v_1\})$, and v_p is one of them. Therefore, if there exists a vertex v_j in $D - (S \setminus \{v_1\})$ such that there are two vertex disjoint paths from v_1 to v_j in $D - (S \setminus \{v_1\})$, then one of those paths must contain the vertex v_p. This implies that $(S \setminus \{v_1\}) \cup \{v_p\}$ is also an optimal solution to MIN-SCVD on D.

$\square$

Algorithm 1: ALGO(D)

1 **Input:** A connected acyclic local tournament D.
2 **Output:** A solution S to MIN-SCVD for D.
3 Let $(v_1, \ldots, v_n)$ be the topological ordering of D.
4 **if** D *is singly connected* **then**
5 $\quad$ | $\quad$ return $S = \emptyset$;
6 **else**
7 $\quad$ | $\quad$ return $S = (S_1 \setminus \{v_1, v_2\}) \cup$ ALGO($D - (S_1 \setminus \{v_2\})$);
8 **end**

We are now ready to describe our algorithm, which works as follows. We greedily construct a solution S as follows. First, we add the set $S_1 \setminus \{v_1, v_2\}$ to S, and by doing this, we cover all the forbidden structures containing v_1. (Note that $D - (S_1 \setminus \{v_1, v_2\})$ could still contain some forbidden structures containing v_2). Next, we recursively find a solution in the digraph $D - (S_1 \setminus \{v_2\})$. A formal description of our algorithm ALGO is in Algorithm 1. It is easy to see that the algorithm runs in polynomial time. The correctness of the algorithm follows from Lemma 10. This completes the proof of Theorem 3.

4 Singly Connected Vertex Deletion on In-Tournaments

In this section, we design an algorithm for SCVD on in-tournaments that runs in time $2^k n^{\mathcal{O}(1)}$. We use the technique of iterative compression, introduced by Reed, Smith and Vetta [38] to design this algorithm. We also show that SCVD on in-tournaments admits a kernel with $\mathcal{O}(k^2)$ vertices.

Remark 1. We note that the classical complexity of SCVD on in-tournaments (and local touranments) is still open. We do not know whether the problem is NP-hard or not on in-tournaments and on local tournaments.

Recall that a directed graph D is said to be an in-tournament if for all vertices $v \in V(D)$, $D[N_D^-(v)]$ is a tournament. We first prove the following preparatory results that will be used to design our algorithm and kernel.

Lemma 11 ($\star$). *Let D be an in-tournament. Then D is singly connected if and only if $|N_D^-(v)| \leq 1$ for all $v \in V(D)$.*

As an immediate consequence of Lemma 11, we get the following result, which says that singly connected in-tournaments are precisely those digraphs that are acyclic triangle-free.

Lemma 12 ($\star$). *Let D be an in-tournament. Then D is singly connected if and only if D does not contain an acyclic triangle as an induced subgraph.*

In light of Lemma 12, it is not difficult to see that the SCVD problem on in-tournaments reduces to the 3-HITTING SET problem. The 3-HITTING SET problem takes as input a set U, a family $\mathcal{F}$ of subsets of U such that $|F| \leq 3$ for every $F \in \mathcal{F}$, and a non-negative integer k. And the question is to determine if there exists $X \subseteq U$ such that $|X| \leq k$ and $X \cap F \neq \emptyset$ for every $F \in \mathcal{F}$. Given an instance (D, k) of SCVD on in-tournaments, where D is an n-vertex in-tournament, we can construct an equivalent instance $(U, \mathcal{F}, k')$ of 3-HITTING SET by taking $U = V(D)$, $\mathcal{F} = \{\{x, y, z\} \mid \{x, y, z\}$ induces an acyclic triangle$\}$, and $k' = k$. The fastest algorithm for 3-HITTING SET, to the best of our knowledge, is due to Wahlström [41, Corollary 69] and runs in time $2.0755^k n^{\mathcal{O}(1)}$. Thus, we can conclude that SCVD problem on in-tournaments can be solved in time $2.0755^k n^{\mathcal{O}(1)}$ as well. In the remaining part of this section, we show that SCVD on in-tournaments can in fact be solved in time $2^k n^{\mathcal{O}(1)}$. Before that we also note that 3-HITTING SET has a $\mathcal{O}(k^2)$-sized kernel [1, Remark 1], which can be adapted to SCVD on in-tournaments as well. We record this fact below.

Observation 3. SCVD *on in-tournaments admits an $\mathcal{O}(k^2)$ kernel.*

We now prove the following theorem.

Theorem 4 ($\star$). SCVD *on in-tournaments admits an algorithm that runs in time $2^k n^{\mathcal{O}(1)}$.*

To prove Theorem 4, we apply the technique of iterative compression, and show that solving SCVD on in-tournaments boils down to solving $2^k n^{\mathcal{O}(1)}$ many instances of the VERTEX COVER (VC) problem on pseudoforests. A pseudo-forest is an undirected graph in which every connected component contains at most one cycle; and VC is polynomial time solvable on pseudoforests. Thus we obtain the runtime claimed in the theorem statement. Theorem 4 implies an analogous result for out-tournaments as well.

Theorem 5 ($\star$). SCVD *on out-tournaments admits an algorithm that runs in time $2^k n^{\mathcal{O}(1)}$.*

5   A Linear Kernel for SCVD on Local Tournaments

In this section, we prove that SCVD admits a linear vertex kernel on local tournaments. Specifically, we prove the following theorem.

Theorem 6. SCVD *on local tournaments admits a kernel with $\mathcal{O}(k)$ vertices.*

Let (D, k) be an instance of SCVD, where D is a local tournament. The basis of our kernelization algorithm is Lemma 12. Recall Lemma 12, which says that an in-tournament (and hence a local tournament) is singly connected if and only if it does not contain an acyclic triangle as a subgraph. We give the following reduction rule in order to simplify the input instance (D, k) of SCVD. We apply this reduction rule exhaustively.

Reduction Rule 1. *If a vertex $v \in V(D)$ is not contained in any acyclic triangle, then delete v from D. Return instance (D', k), where $D' = D - \{v\}$.*

Lemma 13 ($\star$). *Reduction Rule 1 is safe.*

After an exhaustive application of Reduction Rule 1, every vertex in D is contained in some acyclic triangle.

Next, we prove the following lemma that will help us bound the kernel size.

Lemma 14 ($\star$). *Let D be a local tournament and $S \subseteq V(D)$ such that $D - S$ is singly connected. Then, for every vertex $v \in S$, v has at most 3 in-neighbours and at most 3 out-neighbours in $V(D) \setminus S$.*

Next, using Lemma 14, we obtain the following lemma.

Lemma 15 ($\star$). *Let (D, k) be an instance of SCVD on local tournaments and assume that Reduction Rule 1 is no longer applicable. If (D, k) is a yes-instance of SCVD, then $|V(D)| \leq 7k$.*

Reduction Rule 2. *If $|V(D)| \geq 7k+1$, then return that (D, k) is a no-instance of SCVD.*

The safeness of the above reduction rule follows from Lemma 15. When Reduction Rule 2 is no longer applicable, we obtain our required bound in Theorem 6. Observe that both the reduction rules can be applied in polynomial time and are applied only polynomially many times. The correctness of our kernel follows from Lemmas 13 and 15. This completes the proof of Theorem 6.

6   Conclusion

We studied the SCVD problem on various classes of digraphs such as tournaments, α-bounded digraphs, acyclic local tournaments, in-tournaments and local tournaments. Our algorithm for SCVD on in-tournaments runs in time $2^k n^{\mathcal{O}(1)}$. It remains to be seen if this runtime is optimal or can be improved. In particular, as noted in Remark 1, it is open whether SCVD is NP-hard or polynomial time solvable on in-tournaments. Another class of digraphs that one could consider is the class of locally transitive tournaments. A digraph D is said to be a locally transitive tournament if for every vertex $v \in V(D)$, both $N_D^+(v)$ and $N_D^-(v)$ induce transitive tournaments. Note that locally transitive tournaments are a super-class of acyclic local tournaments, and a sub-class of local tournaments. It would be interesting to see if one can extend the polynomial time algorithm for SCVD on acyclic local tournaments to locally transitive tournaments. As for the parameterized complexity of SCVD, the most interesting open problem is to resolve the complexity of SCVD on general digraphs, i.e., whether SCVD, parameterized by the solution size, admits a fixed-parameter tractable algorithm on general digraphs?

References

1. Abu-Khzam, F.N.: A kernelization algorithm for d-hitting set. J. Comput. Syst. Sci. **76**(7), 524–531 (2010)
2. Agrawal, A., Saurabh, S., Sharma, R., Zehavi, M.: Kernels for deletion to classes of acyclic digraphs. J. Comput. Syst. Sci. **92**, 9–21 (2018)
3. Bafna, V., Berman, P., Fujito, T.: A 2-approximation algorithm for the undirected feedback vertex set problem. SIAM J. Discrete Math. **12**(3), 289–297 (1999)
4. Bang-Jensen, J.: Locally semicomplete digraphs: a generalization of tournaments. J. Graph Theory **14**(3), 371–390 (1990)
5. Bang-Jensen, J.: Locally semicomplete digraphs and generalizations. In: Classes of Directed Graphs, pp. 245–296 (2018)
6. Bang-Jensen, J., Gutin, G. (eds.): Classes of Directed Graphs. SMM. Springer, Cham (2018). https://doi.org/10.1007/978-3-319-71840-8
7. Bang-Jensen, J., Maddaloni, A., Saurabh, S.: Algorithms and kernels for feedback set problems in generalizations of tournaments. Algorithmica **76**(2), 320–343 (2016)
8. Bar-Yehuda, R., Geiger, D., Naor, J., Roth, R.M.: Approximation algorithms for the feedback vertex set problem with applications to constraint satisfaction and bayesian inference. SIAM J. Comput. **27**(4), 942–959 (1998)
9. Bergougnoux, B., Eiben, E., Ganian, R., Ordyniak, S., Ramanujan, M.S.: Towards a polynomial kernel for directed feedback vertex set. In: MFCS, pp. 36:1–36:15 (2017)
10. Bodlaender, H.L.: On disjoint cycles. Int. J. Found. Comput. Sci. **5**(1), 59–68 (1994)
11. Buchsbaum, A.L., Carlisle, M.C.: Determining uni-connectivity in directed graphs. Inf. Process. Lett. **48**(1), 9–12 (1993)
12. Cao, Y.: A naive algorithm for feedback vertex set. In: SOSA, pp. 1:1–1:9 (2018)
13. Cao, Y., Chen, J., Liu, Y.: On feedback vertex set: new measure and new structures. Algorithmica **73**(1), 63–86 (2015)
14. Chen, J., Fomin, F.V., Liu, Y., Lu, S., Villanger, Y.: Improved algorithms for feedback vertex set problems. J. Comput. Syst. Sci. **74**(7), 1188–1198 (2008)
15. Chen, J., Liu, Y., Lu, S., O'Sullivan, B., Razgon, I.: A fixed-parameter algorithm for the directed feedback vertex set problem. J. ACM **55**(5), 21:1–21:19 (2008)
16. Chudak, F.A., Goemans, M.X., Hochbaum, D.S., Williamson, D.P.: A primal-dual interpretation of two 2-approximation algorithms for the feedback vertex set problem in undirected graphs. Oper. Res. Lett. **22**(4–5), 111–118 (1998)
17. Dehne, F.K.H.A., Fellows, M.R., Langston, M.A., Rosamond, F.A., Stevens, K.: An $o(2^{o(k)}n^3)$ FPT algorithm for the undirected feedback vertex set problem. Theory Comput. Syst. **41**(3), 479–492 (2007)
18. Diestel, R.: Graph Theory. GTM, vol. 173. Springer, Heidelberg (2017). https://doi.org/10.1007/978-3-662-53622-3
19. Dietzfelbinger, M., Jaberi, R.: On testing single connectedness in directed graphs and some related problems. Inf. Process. Lett. **115**(9), 684–688 (2015)

20. Downey, R.G., Fellows, M.R.: Fixed-parameter intractability. In: Proceedings of the Seventh Annual Structure in Complexity Theory Conference, pp. 36–49 (1992)
21. Downey, R.G., Fellows, M.R.: Fixed parameter tractability and completeness. In: Complexity Theory: Current Research, pp. 191–225 (1992)
22. Downey, R.G., Fellows, M.R.: Fixed-parameter tractability and completeness I: basic results. SIAM J. Comput. **24**(4), 873–921 (1995)
23. Fradkin, A., Seymour, P.: Edge-disjoint paths in digraphs with bounded independence number. J. Comb. Theory Ser. B **110**, 19–46 (2015)
24. Fujito, T.: A note on approximation of the vertex cover and feedback vertex set problems - unified approach. Inf. Process. Lett. **59**(2), 59–63 (1996)
25. Gallai, T., Milgram, A.N.: Verallgemeinerung eines graphentheoretischen satzes von rédei: Ladislaus rédei zum 60. geburtstag. Acta scientiarum mathematicarum **21**(3–4), 181–186 (1960)
26. Guo, J., Gramm, J., Hüffner, F., Niedermeier, R., Wernicke, S.: Compression-based fixed-parameter algorithms for feedback vertex set and edge bipartization. J. Comput. Syst. Sci. **72**(8), 1386–1396 (2006)
27. Kanj, I.A., Pelsmajer, M.J., Schaefer, M.: Parameterized algorithms for feedback vertex set. In: IWPEC, pp. 235–247 (2004)
28. Karp, R.M.: Reducibility among combinatorial problems. In: Miller, R.E., Thatcher, J.W., Bohlinger, J.D. (eds.) Complexity of Computer Computations, pp. 85–103. Springer, Boston (1972). https://doi.org/10.1007/978-1-4684-2001-2_9
29. Khuller, S.: An $o(|v|^2)$ algorithm for single connectedness. Inf. Process. Lett. **72**(3–4), 105–107 (1999)
30. Khuller, S.: Addendum to "an $o(|v|^2)$ algorithm for single connectedness". Inf. Process. Lett. **74**(5–6), 263 (2000)
31. Kociumaka, T., Pilipczuk, M.: Faster deterministic feedback vertex set. Inf. Process. Lett. **114**(10), 556–560 (2014)
32. Li, D., Liu, Y.: A polynomial algorithm for finding the minimum feedback vertex set of a 3-regular simple graph 1. Acta Mathematica Scientia **19**(4), 375–381 (1999)
33. Liang, Y.D.: On the feedback vertex set problem in permutation graphs. Inf. Process. Lett. **52**(3), 123–129 (1994)
34. Liang, Y.D., Chang, M.: Minimum feedback vertex sets in cocomparability graphs and convex bipartite graphs. Acta Informatica **34**(5), 337–346 (1997)
35. Lokshtanov, D., Ramanujan, M.S., Saurabh, S., Sharma, R., Zehavi, M.: Wannabe bounded treewidth graphs admit a polynomial kernel for DFVS. In: WADS, pp. 523–537 (2019)
36. Mnich, M., van Leeuwen, E.J.: Polynomial kernels for deletion to classes of acyclic digraphs. Discrete Optim. **25**, 48–76 (2017)
37. Raman, V., Saurabh, S., Subramanian, C.R.: Faster fixed parameter tractable algorithms for finding feedback vertex sets. ACM Trans. Algorithms **2**(3), 403–415 (2006)
38. Reed, B.A., Smith, K., Vetta, A.: Finding odd cycle transversals. Oper. Res. Lett. **32**(4), 299–301 (2004)
39. Seymour, P.D.: Packing directed circuits fractionally. Combinatorica **15**(2), 281–288 (1995)
40. Thomassé, S.: A $4k^2$ kernel for feedback vertex set. ACM Trans. Algorithms **6**(2), 32:1–32:8 (2010)
41. Wahlström, M.: Algorithms, measures and upper bounds for satisfiability and related problems. Ph.D. thesis, Department of Computer and Information Science, Linköpings universitet (2007)

Equitable *d*-degenerate Choosability
of Graphs

Ewa Drgas-Burchardt[1] ![ID], Hanna Furmańczyk[2]([✉]) ![ID],
and Elżbieta Sidorowicz[1] ![ID]

[1] Faculty of Mathematics, Computer Science and Econometrics,
University of Zielona Góra, Prof. Z. Szafrana 4a, 65-516 Zielona Góra, Poland
{E.Drgas-Burchardt,E.Sidorowicz}@wmie.uz.zgora.pl
[2] Institute of Informatics, Faculty of Mathematics, Physics and Informatics,
University of Gdańsk, 80-309 Gdańsk, Poland
hanna.furmanczyk@ug.edu.pl

Abstract. Let $\mathcal{D}_d$ be the class of d-degenerate graphs and let L be a list assignment for a graph G. A colouring of G such that every vertex receives a colour from its list and the subgraph induced by vertices coloured with one color is a d-degenerate graph is called the $(L, \mathcal{D}_d)$-colouring of G. For a k-uniform list assignment L and $d \in \mathbb{N}_0$, a graph G is equitably $(L, \mathcal{D}_d)$-colorable if there is an $(L, \mathcal{D}_d)$-colouring of G such that the size of any colour class does not exceed $\lceil |V(G)|/k \rceil$. An equitable $(L, \mathcal{D}_d)$-colouring is a generalization of an equitable list coloring, introduced by Kostochka et al., and an equitable list arboricity presented by Zhang. Such a model can be useful in the network decomposition where some structural properties on subnets are imposed. In this paper we give a polynomial-time algorithm that for a given (k, d)-partition of G with a t-uniform list assignment L and $t \geq k$, returns its equitable $(L, \mathcal{D}_{d-1})$-colouring. In addition, we show that 3-dimensional grids are equitably $(L, \mathcal{D}_1)$-colorable for any t-uniform list assignment L where $t \geq 3$.

Keywords: Equitable choosability · *d*-degenerate graph

1 Motivation and Preliminaries

In last decades, a social network graphs, describing relationship in real life, started to be very popular and present everywhere. Understanding key structural properties of large-scale data networks started to be crucial for analyzing and optimizing their performance, as well as improving their security. This topic has been attracting attention of many researches, recently (see [1,6,7,11]). We consider one of problems connected with the decomposition of networks into smaller pieces fulfilling some structural properties. For example, we may desire that, for some security reason, the pieces are acyclic or even independent. This is because of in such a piece we can easily and effectively identify a node failure since the local structure around such a node in this piece is so clear that it

ⓒ Springer Nature Switzerland AG 2020
L. Gąsieniec et al. (Eds.): IWOCA 2020, LNCS 12126, pp. 251–263, 2020.
https://doi.org/10.1007/978-3-030-48966-3_19

can be easily tested using some classic algorithmic tools [11]. Sometimes, it is also desirable that the sizes of pieces are balanced. It helps us to maintain the whole communication network effectively. Such a problem can be modeled by minimization problems in graph theory, called an *equitable vertex arboricity* or an *equitable vertex colourability* of graphs. Sometimes we have some additional requirements on vertices/nodes that can be modeled by a list of available colours. So, we are interested in the list version, introduced by Kostochka, Pelsmajer and West [5] (an independent case), and by Zhang [10] (an acyclic case).

In colourability and arboricity models the properties of a network can be described in the language of the upper bound on the minimum degree, i.e. each colour class induces a graph whose each induced subgraph has the minimum degree bounded from above by zero or one, respectively. In the paper we consider the generalization of these models in which each colour class induces a graph whose each induced subgraph has the minimum degree bounded from above by some natural constant. Let $\mathbb{N}_0 = \mathbb{N} \cup \{0\}$. For $d \in \mathbb{N}_0$, the graph G is *d-degenerate* if $\delta(H) \leq d$ for any subgraph H of G, where $\delta(H)$ denotes the minimum degree of H. The class of all d-degenerate graphs is denoted by $\mathcal{D}_d$. In particular, $\mathcal{D}_0$ is the class of all edgeless graphs and $\mathcal{D}_1$ is the class of all forests. A $\mathcal{D}_d$-*coloring* is a mapping $c : V(G) \to \mathbb{N}$ such that for each $i \in \mathbb{N}$ the set of vertices coloured with i induces a d-degenerate graph. A *list assignment* L, for a graph G, is a mapping that assigns a nonempty subset of $\mathbb{N}$ to each vertex $v \in V(G)$. Given $k \in \mathbb{N}$, a list assignment L is k-*uniform* if $|L(v)| = k$ for every $v \in V(G)$. A colouring $c : V(G) \to \mathbb{N}$ such that $c(v) \in L(v)$ for each $v \in V(G)$ is called an *L-colouring*. Given $d \in \mathbb{N}_0$, a graph G is $(L, \mathcal{D}_d)$-*colourable* if there exists such an L-colouring $c : V(G) \to \mathbb{N}$ that is also $\mathcal{D}_d$-*coloring*. Such a mapping c is called an $(L, \mathcal{D}_d)$-*colouring* of G. If f is any function defined on the set X, then its restriction to Y, $Y \subseteq X$, is denoted by $f|_Y$. For a partially coloured graph G, let $N_G^{col}(d, v) = \{w \in N_G(v) : w \text{ has } d \text{ neighbors coloured with } c(v)\}$, where $N_G(v)$ denotes the set of vertices of G adjacent to v. We refer the reader to [2] for terminology not defined in this paper.

Given $k \in \mathbb{N}$ and $d \in \mathbb{N}_0$, a graph G is *equitably* $(k, \mathcal{D}_d)$-*choosable* if for any k-uniform list assignment L there is an $(L, \mathcal{D}_d)$-colouring of G such that the size of any colour class does not exceed $\lceil |V(G)|/k \rceil$. The notion of equitable $(k, \mathcal{D}_0)$-choosability was introduced by Kostochka et al. [5] whereas the notation of equitable $(k, \mathcal{D}_1)$-choosability was introduced by Zhang [10].

Let $k, d \in \mathbb{N}$. A partition $S_1 \cup \cdots \cup S_{\eta+1}$ of $V(G)$ is called a (k, d)-*partition* of G if $|S_1| \leq k$, and $|S_j| = k$ for $j \in \{2, \ldots, \eta+1\}$, and for each $j \in \{2, \ldots, \eta+1\}$, there is such an ordering $\{x_1^j, \ldots, x_k^j\}$ of vertices of S_j that

$$|N_G(x_i^j) \cap (S_1 \cup \cdots \cup S_{j-1})| \leq di - 1, \tag{1}$$

for every $i \in \{1, \ldots, k\}$. Observe that if $S_1 \cup \cdots \cup S_{\eta+1}$ is a (k, d)-partition of G, then $\eta + 1 = \lceil |V(G)|/k \rceil$. Moreover, immediately by the definition, each (k, d)-partition of G is also its $(k, d+1)$-partition. Surprisingly, the monotonicity of the (k, d)-partition with respect to the parameter k is not so easy to analyze. We illustrate this fact by Example 1. Note that for integers k, d the complexity of deciding whether G has a (k,d)-partition is unknown. The main result of this paper is as follows.

Theorem 1. *Let $k, d, t \in \mathbb{N}$ and $t \geq k$. If a graph G has a (k,d)-partition, then it is equitably $(t, \mathcal{D}_{d-1})$-choosable. Moreover, there is a polynomial-time algorithm that for any graph with a given (k,d)-partition and for any t-uniform list assignment L returns an equitable $(L, \mathcal{D}_{d-1})$-colouring of G.*

The first statement of Theorem 1 generalizes the result obtained in [3] for $d \in \{1, 2\}$. In this paper we present an algorithm that confirms both, the first and second statements of Theorem 1 for all possible d. The algorithm, given in Sect. 2, for a given (k,d)-partition of G with t-uniform list assignment L returns its equitable $(L, \mathcal{D}_{d-1})$-colouring. Moreover, in Sect. 3 we give a polynomial-time algorithm that for a given 3-dimensional grid finds its $(3,2)$-partition, what, in consequence, implies $(t, \mathcal{D}_1)$-choosability of 3-dimensional grids for every $t \geq 3$.

2 The Proof of Theorem 1

2.1 Background

For $S \subseteq V(G)$ by $G - S$ we denote a subgraph of G induced by $V(G) \backslash S$. We start with a generalization of some results given in [5,9,10] for classes $\mathcal{D}_0$ and $\mathcal{D}_1$.

Proposition 1. *Let $k, d \in \mathbb{N}$ and let S be a set of distinct vertices $x_1, \ldots, x_k$ of a graph G. If $G - S$ is equitably $(k, \mathcal{D}_{d-1})$-choosable and*

$$|N_G(x_i) \backslash S| \leq di - 1$$

holds for every $i \in \{1, \ldots, k\}$, then G is equitably $(k, \mathcal{D}_{d-1})$-choosable.

Proof. Let L be a k-uniform list assignment for G and let c be an equitable $(L|_{V(G) \backslash S}, \mathcal{D}_{d-1})$-colouring of $G - S$. Thus each colour class in c has the cardinality at most $\lceil (|V(G)| - k)/k \rceil$ and induces in $G - S$, and consequently in G, a graph from $\mathcal{D}_{d-1}$. We extend c to $(V(G) \backslash S) \cup \{x_k\}$ by assigning to x_k a colour from $L(x_k)$ that is used on vertices in $N_G(x_k) \backslash S$ at most $d - 1$ times. Such a colour always exists because $|N_G(x_k) \backslash S| \leq dk - 1$ and $|L(x_k)| = k$. Next, we colour vertices $x_{k-1}, \ldots, x_1$, sequentially, assigning to x_i a colour from its list that is different from colours of all vertices $x_{i+1}, \ldots, x_k$ and that is used at most $d - 1$ times in $N_G(x_i) \backslash S$. Observe that there are at least i colours in $L(x_i)$ that are different from $c(x_{i+1}), \ldots, c(x_k)$, and, since $|N_G(x_i) \backslash S| \leq di - 1$ for $1 \leq i \leq k - 1$, then such a choice of $c(x_i)$ is always possible. Next, the colouring procedure forces that the cardinality of every colour class in the extended colouring c is at most $\lceil |V(G)|/k \rceil$. Let $G_i = G[(V(G) \backslash S) \cup \{x_i, \ldots, x_k\}]$. It is easy to see that for each i each colour class in $c|_{V(G_i)}$ induces a graph belonging to $\mathcal{D}_{d-1}$, $1 \leq i \leq k$. In particular this condition is satisfied for G_1, i.e. for G. Hence c is an equitable $(L, \mathcal{D}_{d-1})$-colouring of G and G is equitably $(k, \mathcal{D}_{d-1})$-choosable. $\square$

Note that if a graph G has a (k, d)-partition, then one can prove that G is equitably $(k, \mathcal{D}_{d-1})$-choosable by applying Proposition 1 several times. In general, the equitable $(k, \mathcal{D}_{d-1})$-choosability of G does not imply the equitable $(t, \mathcal{D}_{d-1})$-choosability of G for $t \geq k$. Unfortunately, if G has a (k, d)-partition, then G may have neither a $(k + 1, d)$-partition nor a $(k - 1, d)$-partition. The infinite family of graphs defined in Example 1 confirms the last fact.

Example 1. Let $q \in \mathbb{N}$ and let $G_1, \ldots, G_{2q+1}$ be vertex-disjoint copies of K_6 such that $V(G_i) = \{v_1^i, \ldots, v_6^i\}$ for $i \in \{1, \ldots 2q+1\}$. Let $G(q)$ (cf. Fig. 1) be the graph resulted by adding to $G_1, \ldots, G_{2q+1}$ edges that join vertices of G_i with vertices of G_{i-1}, $i \in \{2, \ldots, 2q + 1\}$, in the following way:

for i even:

$$N_{G_{i-1}}(v_1^i) = \emptyset$$
$$N_{G_{i-1}}(v_2^i) = \{v_1^{i-1}\}$$
$$N_{G_{i-1}}(v_3^i) = \{v_2^{i-1}, v_3^{i-1}\}$$
$$N_{G_{i-1}}(v_4^i) = \{v_1^{i-1}, v_2^{i-1}, v_3^{i-1}\}$$
$$N_{G_{i-1}}(v_5^i) = \{v_1^{i-1}, v_4^{i-1}, v_5^{i-1}, v_6^{i-1}\}$$
$$N_{G_{i-1}}(v_6^i) = \{v_2^{i-1}, v_3^{i-1}, v_4^{i-1}, v_5^{i-1}, v_6^{i-1}\}$$

for i odd:

$$N_{G_{i-1}}(v_1^i) = \{v_2^{i-1}, v_3^{i-1}, v_4^{i-1}, v_5^{i-1}, v_6^{i-1}\}$$
$$N_{G_{i-1}}(v_2^i) = \{v_1^{i-1}, v_4^{i-1}, v_5^{i-1}, v_6^{i-1}\}$$
$$N_{G_{i-1}}(v_3^i) = \{v_1^{i-1}, v_2^{i-1}, v_3^{i-1}\}$$
$$N_{G_{i-1}}(v_4^i) = \{v_2^{i-1}, v_3^{i-1}\}$$
$$N_{G_{i-1}}(v_5^i) = \{v_1^{i-1}\}$$
$$N_{G_{i-1}}(v_6^i) = \emptyset$$

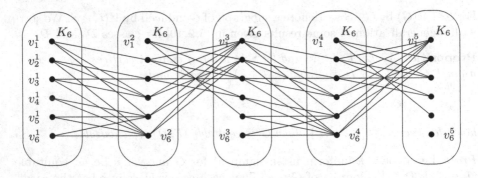

Fig. 1. A draft of the graph $G(2)$ from Example 1.

The construction of $G(q)$ immediately implies that for every $q \in \mathbb{N}$ the graph $G(q)$ has a $(6, 1)$-partition. Also, observe that $deg_{G(q)}(v) \geq 5$ for each vertex v of $G(q)$. Suppose that $G(q)$ has a $(5, 1)$-partition $S_1 \cup \cdots \cup S_{\eta+1}$ with $S_{\eta+1} = \{x_1^{\eta+1}, \ldots, x_5^{\eta+1}\}$ such that $|N_{G(q)}(x_i^{\eta+1}) \cap (S_1 \cup \cdots \cup S_\eta)| \leq i - 1$. Thus $|N_{G(q)}(x_1^{\eta+1}) \cap (S_1 \cup \cdots \cup S_\eta)| = 0$ and consequently $deg_{G(q)}(x_1^{\eta+1}) \leq 4$, contradicting our previous observation. Hence, $G(q)$ has no $(5, 1)$-partition. In [4] we show that $G(q)$ has no $(7, 1)$-partition for $q \geq 2$.

2.2   Algorithm

Now we are ready to present the algorithm that confirms both statements of Theorem 1. Note that Proposition 1 and the induction procedure could be used to prove the first statement of Theorem 1 for $t = k$ but, this approach seems to be useless for $t > k$, as we have observed in Example 1. Mudrock et al. [8] proved the lack of monotonicity for the equitable $(k, \mathcal{D}_0)$-choosability with respect to the parameter k. It motivates our approach for solving the problem. To simplify understanding we give the main idea of the algorithm presented in the further part. It can be expressed in a few steps (see also Fig. 2):

- on the base of the given (k, d)-partition $S_1 \cup \cdots \cup S_{\eta+1}$ of G, we create the list S, consisting of the elements from $V(G)$, whose order corresponds to the order in which the colouring is expanded to successive vertices in each S_j (cf. the proof of Proposition 1);
- let $|V(G)| = \beta t + r_2$, $1 \le r_2 \le t$; we colour r_2 vertices from the beginning of S taking into account the lists of available colours; we delete the colour assigned to v from the lists of available colours for vertices from $N_G^{col}(d, v)$;
- let $|V(G)| = \beta(\gamma k + r) + r_2 = \beta\gamma k + (\rho k + x) + r_2$; observe that $r \equiv t \pmod{k}$ and $\rho k + x \equiv 0 \pmod{r}$; we colour $\rho k + x$ vertices taking into account the lists of available colours in such a way that every sublist of length k is formed by vertices coloured differently (consequently, every sublist of length r is coloured differently); we divide the vertices colored here into β sets each one of cardinality r; we delete $c(v)$ from the lists of vertices from $N_G^{col}(d, v)$;
- we extend the list colouring into the uncoloured $\beta\gamma k$ vertices by colouring β groups of γk vertices; first, we associate each group of γk vertices with a set of r vertices coloured in the previous step (for different groups these sets are disjoint); next, we color the vertices of each of the group using γk different colors that are also different from the colors of r vertices of the set associated with this group;
- our final equitable list colouring is the consequence of a partition of $V(G)$ into $\beta + 1$ coloured sets, each one of size at most t and each one formed by vertices coloured differently.

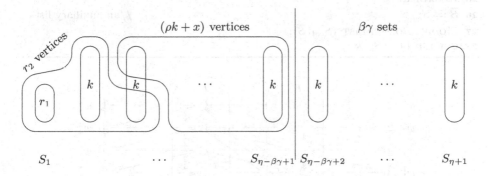

Fig. 2. An exemplary illustration of the input of EQUITABLE $(L, \mathcal{D}_{d-1})$-COLOURING.

Now we illustrate EQUITABLE $(L, \mathcal{D}_{d-1})$-COLOURING using a graph from Example 2.

Algorithm 1: EQUITABLE $(L, \mathcal{D}_{d-1})$-COLOURING(G)

Input : Graph G on n vertices; L - t-uniform list assignment; a
(k, d)-partition $S_1 \cup \cdots \cup S_{\eta+1}$ of G, given by lists
$S_1 = (x_1^1, \ldots, x_{r_1}^1)$ and $S_j = (x_1^j, \ldots, x_k^j)$ for $j \in \{2, \ldots, \eta+1\}$.

Output : Equitable $(L, \mathcal{D}_{d-1})$-colouring of G.

1 initialization;
2 $S := empty$; $L_R := empty$; $L_X := empty$;
3 **for** $j := 1$ **to** $\eta + 1$ **do**
4 $\quad$ add REVERSE(S_j) to S;   //REVERSE is the procedure for reversing lists
5 **end**
6 $\beta := \lceil n/t \rceil$-1;
7 **if** $n \equiv 0 \pmod{t}$ **then**
8 $\quad$ $r_2 := t$
9 **else**
10 $\quad$ $r_2 := n \pmod{t}$;
11 **end**
12 $\gamma := t \div k$; $r := t \pmod{k}$; $\rho := \beta r \div k$; $x := \beta r \pmod{k}$;
13 take and delete r_2 elements from the beginning of S, and add them, vertex
14 $\qquad\qquad$ by vertex, to list L_R;
15 COLOUR_LIST(L_R, r_2);
16 take and delete x elements from the beginning of S, and add them, vertex
17 $\qquad\qquad$ by vertex, to list L_X;
18 COLOUR_LIST(L_X, x);
19 $S_{col} := L_X$;
20 **for** $j = 1$ **to** ρ **do**
21 $\quad$ $S' := empty$; take and delete k elements from the beginning of S, and add
$\qquad$ them, vertex by vertex, to list S';
22 $\quad$ COLOUR_LIST(S', k);
23 $\quad$ $S_{col} := S_{col} + S'$;
24 **end**
25 REORDER(S_{col});
26 $\overline{S} := S$;   //an auxiliary list
27 MODIFY_COLOURLISTS$(S_{col}, \overline{S})$;
28 COLOUR_LIST$(S, \gamma k)$;

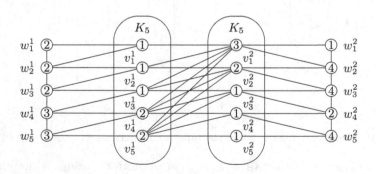

Fig. 3. An exemplary graph G depicted in Example 2 with an exemplary colouring returned by EQUITABLE $(L, \mathcal{D}_{d-1})$-COLOURING(G).

Procedure 2: COLOUR_LIST(S', p)

    Input   : List S' of vertices; integer p.   //the length of S' is multiple of p

    Output : L-colouring of the vertices from S'.

    //The procedure also modifies a global variable of the list assignment L.

1 initialization;

2 **while** $S' \neq empty$ **do**

3 | let S'' be the list of the p first elements of S';

4 | $C := \emptyset$; //set C is reserved for the colours being assigned to the vertices of S''

5 | **while** $S'' \neq empty$ **do**

6 | | let v be the first element of S'';

7 | | $L(v) := L(v) \backslash C$;

8 | | $c(v) :=$COLOUR_VERTEX(v);

9 | | delete the vertex v from S' and S''; $C := C \cup \{c(v)\}$;

10 | **end**

11 **end**

Procedure 3: COLOUR_VERTEX(v)

    Input   : Vertex v of the graph G.

    Output : L-colouring of the vertex v.

    //The procedure modifies also a global variable of the list assignment L.

1 initialization;

2 $c(v) :=$ any colour from $L(v)$;

3 delete $c(v)$ from $L(w)$ for all $w \in N_G^{col}(d, v)$;   //d is a global variable

4 **return** $c(v)$;

Procedure 4: REORDER(S')

    Input   : List S' of coloured vertices of G.

    Output : List S' - reordered in such a way that every its sublist of length k is formed by vertices being coloured with different colours.

1 initialization;

2 $S_{aux} := empty$;   //an auxiliary list

3 take and delete r_1 elements from S', and add them, vertex by vertex, to S_{aux};

4 **for** $j = 1$ **to** $\eta - \beta\gamma$ **do**

5 | $P := \emptyset$;

6 | take and delete first k elements from S', and add them to set P;

7 | **for** $i = 1$ **to** k **do**

8 | | let v be a vertex from P such that $c(v)$ is different from the colours of the last $k - 1$ vertices of S_{aux}; add v to the end of S_{aux};

9 | **end**

10 **end**

11 $S' := S_{aux}$;

Procedure 5: MODIFY_COLOURLISTS(L_1, L_2)

Input : List L_1 of βr coloured vertices and list L_2 of $\beta\gamma k$ uncoloured
vertices.
Output : Modified colour list assignment L for vertices of L_2.
//L is a global variable

1 initialization;
2 $C := \emptyset$;   //C is a set of colours of vertices from the depicted part of L_1
3 **for** $i = 1$ **to** β **do**
4 ⎪   take and delete first r vertices from L_1;
5 ⎪   let C be the set of colours assigned to them;
6 ⎪   **for** $j = 1$ **to** γk **do**
7 ⎪ ⎪   let v be the first vertex from L_2;
8 ⎪ ⎪ $L(v) := L(v)\backslash C$; delete v from L_2;
9 ⎪   **end**
10 **end**

Example 2. Let G_1, G_2 be two vertex-disjoint copies of K_5 and $V(G_i) = \{v_1^i, \ldots, v_5^i\}$ for $i \in \{1, 2\}$. We join every vertex v_j^2 to $v_j^1, v_{j+1}^1, \ldots, v_5^1$ for $j \in \{1, 2, 3, 4, 5\}$. Next, we add a vertex w_j^i and join it with v_j^i for $i \in \{1, 2\}$ $j \in \{1, 2, 3, 4, 5\}$. In addition, we join w_j^i to arbitrary two vertices in $\{v_p^q : (q < i) \vee (q = i \wedge p < j)\} \cup \{w_p^q : (q < i) \vee (q = i \wedge p < j)\}$, $i \in \{2, 3, 4, 5\}$ $j \in \{1, 2\}$. Let G be a resulted graph. Observe that $|V(G)| = 20$ and the partition $S_1 \cup S_2 \cup \ldots \cup S_{10}$ of $V(G)$ such that $S_{p+1} = \{v_{r+1}^{s+1}, w_{r+1}^{s+1}\}$ for $p \in \{0, \ldots, 9\}$, where $s = \lfloor \frac{p}{5} \rfloor$, $r \equiv p$ (mod 5) is a $(2, 3)$-partition of G.

For the purpose of Example 2, we assume the following 3-uniform list assignment for the graph from Fig. 3: $L(v_j^i) = \{1, 2, 3\}$, for $i \in \{1, 2, 3, 4\}$, $L(w_j^1) = \{2, 3, 4\}$, and $L(w_j^2) = \{1, 2, 4\}$, $j \in \{1, 2, 3, 4, 5\}$, while given $(2, 3)$-partition of G is: $S_p = \{w_{r+1}^{s+1}, v_{r+1}^{s+1}\}$, where $s = \lfloor \frac{p}{5} \rfloor$, $r \equiv p$ (mod 5), $p \in [10]$.

Thus EQUITABLE $(L, \mathcal{D}_{d-1})$-COLOURING returns equitable $(L, \mathcal{D}_2)$-coloring of G. Note, that $20 = |V(G)| = \eta \cdot k + r_1 = 9 \cdot 2 + 2$. While on the other hand, we have $20 = |V(G)| = \beta \cdot t + r_2 = 6 \cdot 3 + 2$. Observe that $x = 0$. When we colour a vertex, we always choose the first colour on its list. The list S determined in lines 3–5 of EQUITABLE $(L, \mathcal{D}_{d-1})$-COLOURING and the colours assigned to first part of vertices of S (lines 18–24) are as follows:

$$S = (v_1^1, w_1^1, \big| v_2^1, w_2^1, v_3^1, w_3^1, v_4^1, w_4^1, \big| v_5^1, w_5^1, v_1^2, w_1^2, \ldots, v_5^2, w_5^2)$$

$$\underbrace{}_{r_2} \quad \underbrace{}_{\rho k} \quad \underbrace{}_{\beta\gamma k}$$

$colours$:   1   2 | 1   2   1*   2   2   3 |

*: after colouring v_3^1 with 1, $L(v_1^2) = \{2, 3\}$ - the result of line 3 in COLOUR_VERTEX.

List S_{col} after REORDER(S_{col}): $(v_2^1, w_2^1, v_3^1, w_3^1, w_4^1, v_4^1)$ with corresponding colours: $(1, 2, 1, 2, 3, 2)$.

$\overline{S} =$	$(v_5^1,$	$w_5^1,$	$v_1^2,$	$w_1^2,$	$v_2^2,$	$w_2^2,$	$v_3^2,$	$w_3^2,$	$v_4^2,$	$w_4^2,$	$v_5^2,$	$w_5^2)$
lists after	2	2	3	1	2	2	1	1	1	1	1	1
procedure	3	3		4	3	4	3	4	2	2	3	4
MODIFY_COLOURLIST	4								4			
final $c(v)$	2	3	3	1	2	4	1	4	1	2	1	4

To prove the correctness of the EQUITABLE $(L, \mathcal{D}_{d-1})$-COLOURING algorithm, we give some observations and lemmas.

Observation 2. *The colour function c returned by the* EQUITABLE $(L, \mathcal{D}_{d-1})$-COLOURING *algorithm is constructed step by step. In each step, $c(v)$ is a result of* COLOUR_VERTEX(v) *and this value is not changed further.*

Observation 3. *The list assignment L, as a part of the input of* EQUITABLE $(L, \mathcal{D}_{d-1})$-COLOURING, *is modified for a vertex v by* COLOUR_LIST *or by* MODIFY_COLOURLISTS.

Lemma 1. *Every time when* EQUITABLE $(L, \mathcal{D}_{d-1})$-COLOURING(G) *calls* COLOUR_VERTEX(v), *$L(v) \neq \emptyset$ holds, i.e.* COLOUR_VERTEX(v) *is always executable.*

Proof. Note that COLOUR_VERTEX is called by COLOUR_LIST. Let
$R = \{v \in V(G) :$ COLOUR_VERTEX(v) is called when COLOUR_LIST(L_R, r_2) in line 15 of EQUITABLE $(L, \mathcal{D}_{d-1})$-COLOURING is executed$\}$,
$X = \{v \in V(G) :$ COLOUR_VERTEX(v) is called when COLOUR_LIST(L_X, x) in line 18 of EQUITABLE $(L, \mathcal{D}_{d-1})$-COLOURING is executed$\}$.
Let $V_1 := S_1 \cup \cdots \cup S_{\eta+1-\beta\gamma}$, $V_2 := V(G) \setminus V_1 = S_{\eta+1-(\beta\gamma-1)} \cup \cdots \cup S_{\eta+1}$. Note that
$V_1 \setminus (R \cup X) = \{v \in V(G) :$ COLOUR_VERTEX(v) is called when COLOUR_LIST(S', k) in line 22 of EQUITABLE $(L, \mathcal{D}_{d-1})$-COLOURING is executed$\}$
$V_2 = \{v \in V(G) :$ COLOUR_VERTEX(v) is called when COLOUR_LIST$(S, \gamma k)$ in line 28 of EQUITABLE $(L, \mathcal{D}_{d-1})$-COLOURING is executed$\}$.
Observe that $|R| = r_2, |X| = x, |V_1 \setminus (R \cup X)| = \rho k, |V_2| = \beta\gamma k$.

Case 1. $v \in R$.
In this case, the vertex v is coloured by COLOUR_LIST(L_R, r_2) in line 15 of EQUITABLE $(L, \mathcal{D}_{d-1})$-COLOURING. Since $|R| = r_2$, the **while** loop in line 2 of COLOUR_LIST is executed only once. The **while** loop in line 5 of COLOUR_LIST is executed r_2 times. Suppose, v is a vertex such that COLOUR_VERTEX(v) is called in the i-th execution of the **while** loop in line 5 of COLOUR_LIST. By Observation 3, the fact that EQUITABLE $(L, \mathcal{D}_{d-1})$-COLOURING has not called MODIFY_COLOURLISTS so far, and because it is the first time when COLOUR_LIST works, we have $|C| = i - 1$, and $L(v) \setminus C$ is the current list of v. Since $t \geq r_2$, the list of v is non-empty.

Case 2. $v \in X$.
This time, the vertex v is coloured by COLOUR_LIST(L_X, x) called in line 18 of EQUITABLE $(L, \mathcal{D}_{d-1})$-COLOURING. Similarly as in *Case 1*, the **while** loop in line 2 of COLOUR_LIST is executed only once and the **while** loop in line

5 of COLOUR_LIST is executed x times. Suppose that v is a vertex such that COLOUR_VERTEX(v) is called in the i-th iteration of the **while** loop in line 5 of COLOUR_LIST. Observe that properties of the (k,d)-partition $S_1 \cup \cdots \cup S_{\eta+1}$ of G (given as the input of EQUITABLE $(L, \mathcal{D}_{d-1})$-COLOURING) and the REVERSE procedure from line 4 of EQUITABLE $(L, \mathcal{D}_{d-1})$-COLOURING imply that v has at most $(x - i + 1)d - 1 + (k - x)$ neighbors w for which COLOUR_VERTEX(w) was executed earlier than COLOUR_LIST(L_X, x). More precisely, by the definition of the (k,d)-partition, v has at most $(x - i + 1)d - 1$ neighbours in $R \setminus Y$, where Y consists of the last $k - x$ vertices w for which COLOUR_VERTEX(w, d) is executed, being called by COLOUR_LIST(L_R, r_2). Thus, at most $k - i$ colours were deleted from $L(v)$ before COLOUR_LIST(L_X, x) began. If the **while** loop in line 5 of COLOUR_LIST is called for the i-th time, then $|C| = i - 1$ and so, from the current list $L(v)$ at most $i - 1$ elements were deleted. Furthermore, EQUITABLE $(L, \mathcal{D}_{d-1})$-COLOURING has not called MODIFY_COLOURLISTS so far. Thus the current size of $L(v)$ is at least $t - k + 1$, by Observation 3. Since $t \geq k$, the list of v is non-empty.

In a similar way we prove the remaining two cases, namely when $v \in V_1 \setminus (R \cup X)$ or $v \in V_2$. The full proof is given in [4]. ☐

Lemma 2. *An output of* EQUITABLE $(L, \mathcal{D}_{d-1})$-COLOURING(G) *is an* $(L, \mathcal{D}_{d-1})$-*colouring of* G.

Proof. We will show that if COLOUR_VERTEX(v) is executed, then an output $c(v)$ has always the following property. For each subgraph H of G induced by vertices x for which COLOUR_VERTEX(x) was executed so far with the output $c(x) = c(v)$, the condition $\delta(H) \leq d - 1$ holds. By Observation 2 and Lemma 1, it will imply that an output c of EQUITABLE $(L, \mathcal{D}_{d-1})$-COLOURING is an $(L, \mathcal{D}_{d-1})$-colouring of G. Note that it is enough to show this fact for H satisfying $v \in V(H)$. By a contradiction, let v be a vertex for which the output $c(v)$ does not satisfy the condition, i.e. v has at least d neighbors in the set of vertices for which COLOUR_VERTEX was already executed with the output $c(v)$. But it is not possible because $c(v)$ was removed from $L(v)$ when the last (in the sense of the algorithm steps) of the neighbors of v, say x, obtained the colour $c(v)$ (COLOUR_VERTEX(x) removed $c(x)$ from $L(v)$ since $v \in N_G^{col}(d, x)$). ☐

Lemma 3. *An output colour function c of* EQUITABLE $(L, \mathcal{D}_{d-1})$-COLOURING (G) *satisfies* $|C_i| \leq \lceil |V(G)|/t \rceil$, *where t is the part of the input of* EQUITABLE $(L, \mathcal{D}_{d-1})$-COLOURING(G) *and* $C_i = \{v \in V(G) : c(v) = i\}$.

Proof. Recall that $\lceil |V(G)|/t \rceil = \beta + 1$. We will show that there exists a partition of $V(G)$ into $\beta + 1$ sets, say $W_1 \cup \cdots \cup W_{\beta+1}$, such that for each $i \in \{1, \ldots, \beta + 1\}$ any two vertices x, y in W_i satisfy $c(x) \neq c(y)$. It will imply that the cardinality of every colour class in c is at most $\beta + 1$, giving the assertion.

Note that after the last, ρ-th execution of the **for** lopp in line 20 of EQUITABLE $(L, \mathcal{D}_{d-1})$-COLOURING the list S_{col} consists of the coloured vertices of the set $V_1 \setminus R$ (observe that $|V_1 \setminus R| = \beta r$). The elements of S_{col} are ordered in such a way that the first x ones have different colours and for every $i \in \{1, \ldots, \rho\}$ the i-th next k elements have different colours. Now the REORDER(S_{col}) procedure in line 25 of EQUITABLE $(L, \mathcal{D}_{d-1})$-COLOURING changes the ordering of elements of S_{col} in such a way that every k consecutive elements have different colours. Since $r \in \{0, \ldots, k-1\}$, it follows that also every r consecutive elements of this list have different colours. The execution of REORDER(S_{col}) is always possible because of the previous assumptions on S_{col}.

For $i \in \{1, \ldots, \beta\}$ let $H_i = S_{\eta+1-((\beta-i+1)\gamma-1)} \cup S_{\eta+1-((\beta-i+1)\gamma-2)} \cup \cdots \cup S_{\eta+1-(\beta-i)\gamma}$. Thus $H_1 \cup \cdots \cup H_\beta$ is a partition of V_2 into β sets, each of the cardinality γk. Note that the vertices of H_i are coloured when COLOUR_LIST$(S, \gamma k)$ in line 28 of EQUITABLE $(L, \mathcal{D}_{d-1})$-COLOURING is executed. More precisely, it is during the i-th execution of the **while** loop in line 2 of COLOUR_LIST. It guarantees that the vertices of H_i obtain pairwise different colours. Moreover, in line 27 of EQUITABLE $(L, \mathcal{D}_{d-1})$-COLOURING the lists of vertices of H_i were modified in such a way that the colours of i-th r elements from the current list S_{col} are removed from the list of each element in H_i. Hence, after the execution of COLOUR_LIST$(S, \gamma k)$ in line 28 of EQUITABLE $(L, \mathcal{D}_{d-1})$-COLOURING the elements in H_i obtain colours that are pairwise different and also different from all the colours of i-th r elements from the list S_{col} (recall that S_{col} consists of the ordered vertices of $V_1 \setminus R$). Hence, for every $i \in \{1, \ldots, \beta\}$ the elements of H_i and the i-th r elements of S_{col} have pairwise different colours in c and can constitute W_i. Moreover, the elements of R constitute $W_{\beta+1}$. Thus $|W_{\beta+1}| = r_2$, which finishes the proof. $\qquad\square$

Theorem 4. *For a given graph G on n vertices, a t-uniform list assignment L, a (k, d)-partition of G the EQUITABLE $(L, \mathcal{D}_{d-1})$-COLOURING(G) algorithm returns $(L, \mathcal{D}_{d-1})$-colouring of G in polynomial time.* $\qquad\square$

The full analysis of the computational complexity of the algorithm is given in [4].

3   Grids

Given two graphs G_1 and G_2, the *Cartesian product* of G_1 and G_2, $G_1 \square G_2$, is defined to be a graph whose the vertex set is $V(G_1) \times V(G_2)$ and the edge set consists of all edges joining vertices (x_1, y_1) and (x_2, y_2) when either $x_1 = x_2$ and $y_1 y_2 \in E(G_2)$ or $y_1 = y_2$ and $x_1 x_2 \in E(G_1)$. Note that the Cartesian product is commutative and associative. Hence the graph $G_1 \square \cdots \square G_d$ is unambiguously defined for any $d \in \mathbb{N}$. Let P_n denote a path on n vertices. If each factor G_i is a path on at least two vertices then $G_1 \square \cdots \square G_d$ is a d-*dimensional grid*. Note that the d-dimensional grid $P_{n_1} \square \cdots \square P_{n_d}$, $d \geq 3$, may be considered as n_1 layers and each layer is the $(d-1)$-dimensional grid $P_{n_2} \square \cdots \square P_{n_d}$. We assume $n_1 \geq \cdots \geq n_d$. Let $P_{n_1} \sqsupset \ldots \sqsupset P_{n_d}$ denote an *incomplete* d-dimensional grid, i.e. a connected graph being a subgraph of $P_{n_1} \square \ldots \square P_{n_d}$ such that its some initial layers may be empty, the first non-empty layer may be incomplete, while any next layer is complete. Note that every grid is particular incomplete grid.

In this subsection we construct a polynomial-time algorithm that for each 3-dimensional grid finds its $(3, 2)$-partition (PARTITION3D(G)). Application of Theorem 1 implies the main result of this subsection.

Theorem 5. *Let $t \geq 3$ be an integer. Every 3-dimensional grid is equitably $(t, \mathcal{D}_1)$-choosable. Moreover, there is a polynomial-time algorithm that for every t-uniform list assignment L of the 3-dimensional grid G returns an equitable $(L, \mathcal{D}_1)$-colouring of G.* ☐

Procedure 6: CORNER(G)

> **Input**   : Incomplete non-empty d-dimensional grid $G = P_{n_1} \sqsupset \ldots \sqsupset P_{n_d}$, $d \geq 2$.
> **Output** : Vertex $y = (a_1, \ldots, a_d) \in V(G)$ such that $\deg_G(y) \leq d$.
>
> 1 initialization;
> 2 let a_1 be the number of the incomplete layer of G;
> 3 **for** $i = 2$ **to** $d - 1$ **do**
> 4 | $\quad a_i = \min\{x_i : \exists_{x_{i+1}, \ldots, x_d}(a_1, \ldots, a_{i-1}, x_i, \ldots, x_d) \in V(G)\}$.
> 5 **end**
> 6 $a_d := \min\{x_d : (a_1, \ldots, a_{d-1}, x_d) \in V(G)\}$.
> 7 **return** $(a_1, \ldots, a_d)$;

Algorithm 7: PARTITION3D(G)

> **Input**   : 3-dimensional grid $G = P_{n_1} \square P_{n_2} \square P_{n_3}$.
> **Output** : A $(3, 2)$-partition $S_1 \cup \cdots \cup S_{\alpha+1}$ of G.
>
> 1 initialization;
> 2 $\alpha := \lceil \frac{n_1 n_2 n_3}{3} \rceil - 1$;
> 3 **if** $\alpha \geq 1$ **then**
> 4 | **for** $j := \alpha + 1$ **downto** 2 **do**
> 5 | | $y_1^j = (a_1, a_2, a_3) := $Corner($G$);
> 6 | | **if** $\deg(y_1^j) = 1$ **then**
> 7 | | | $y_2^j := $Corner($G - y_1^j$);
> 8 | | | **if** y_1^j *is the only vertex on a_1 layer* **then**
> 9 | | | | let y_3^j be any vertex on layer $a_1 + 1$ such that $y_3^j \neq y_2^j$
> 10 | | | **else**
> 11 | | | | let y_3^j be any vertex on layer a_1 such that $y_3^j \neq y_1^j$ and $y_3^j \neq y_2^j$, if exists, otherwise y_3^j is any vertex on layer $a_1 + 1$
> 12 | | | **end**
> 13 | | **end**
> 14 | | **if** $\deg(y_1^j) = 2$ **then**
> 15 | | | let y_2^j be the neighbour of y_1^j lying on the same layer as y_1^j;
> 16 | | | let y_3^j be any vertex on layer a_1, if exists, otherwise, y_3^j is any vertex on layer $a_1 + 1$
> 17 | | **end**
> 18 | | **if** $\deg(y_1^j) = 3$ **then**
> 19 | | | $y_2^j := (a_1, a_2, a_3 + 1); y_3^j := (a_1, a_2 + 1, a_3)$;
> 20 | | **end**
> 21 | | $S_j := \{y_1^j, y_2^j, y_3^j\}; G := G - S_j$;
> 22 | **end**
> 23 **end**
> 24 $S_1 := V(G)$;

Theorem 6. *For a given 3-dimensional grid G the* PARTITION3D(G) *algorithm returns a $(3,2)$-partition of G in polynomial-time.*

The proof of Theorem 6 is presented in [4]. As a consequence of the above theorem and Theorem 1 we get the statement of Theorem 5.

4 Concluding Remarks

In Subsect. 2.2 we have proposed the polynomial-time algorithm that finds an equitable $(L, \mathcal{D}_{d-1})$-colouring of a given graph G assuming that we know a (k,d)-partition of G (L is a t-uniform list assignment for G, $t \geq k$). In this context the following open question seems to be interesting: What is the complexity of recognition of graphs having a (k,d)-partition?

Acknowledgment. The authors thank their colleague Janusz Dybizbański for making several useful suggestions improving the presentation.

References

1. Abu-Ata, M., Dragan, F.F.: Metric tree-like structures in real-world networks: an empirical study. Networks **67**(1), 49–68 (2016)
2. Diestel, R.: Graph Theory. Graduate Texts in Mathematics, vol. 173, 2nd edn. Springer, New York (2000)
3. Drgas-Burchardt, E., Dybizbański, J., Furmańczyk, H., Sidorowicz, E.: Equitable list vertex colourability and arboricity of grids. Filomat **32**(18), 6353–6374 (2018)
4. Drgas-Burchardt, E., Furmańczyk, H., Sidorowicz, E.: Equitable d-degenerate choosability of graphs. arxiv:2003.09722 (2020)
5. Kostochka, A.V., Pelsmajer, M.J., West, D.B.: A list analogue of equitable colouring. J. Graph Theory **44**(3), 166–177 (2003)
6. Lei, H., Li, T., Ma, Y., Wang, H.: Analyzing lattice networks through substructures. Appl. Math. Comput. **329**, 297–314 (2018)
7. Miao, T., Chen, A., Xu, Y.: Optimal structure of damaged tree-like branching networks for the equivalent thermal conductivity. Int. J. Therm. Sci. **102**, 89–99 (2016)
8. Mudrock, J.A., Chase, M., Thornburgh, E., Kadera, I., Wagstrom, T.: A note on the equitable choosability of complete bipartite graphs. DMGT (2019). https://doi.org/10.7151/dmgt.2232
9. Pelsmajer, M.J.: Equitable list-colouring for graphs of maximum degree 3. J. Graph Theory **47**(1), 1–8 (2004)
10. Zhang, X.: Equitable list point arboricity of graphs. Filomat **30**(2), 373–378 (2016)
11. Zhang, X., Niu, B., Li, Y., Li, B.: Equitable vertex arboricity of d-degenerate graphs. arxiv: 1908.05066v1 (2019)

On the Complexity of Broadcast Domination and Multipacking in Digraphs

Florent Foucaud[1,2], Benjamin Gras[2,3], Anthony Perez[2(✉)], and Florian Sikora[4]

[1] Univ. Bordeaux, Bordeaux INP, CNRS, LaBRI, UMR5800, 33400 Talence, France
[2] Univ. Orléans, INSA Centre Val de Loire, LIFO EA 4022, 45067 Orléans, France
anthony.perez@univ-orleans.fr
[3] Universität Trier, Fachbereich IV, Informatikwissenschaften, 54296 Trier, Germany
[4] Univ. Paris-Dauphine, PSL University, CNRS, LAMSADE, 75016 Paris, France

Abstract. We study the complexity of the two dual covering and packing distance-based problems Broadcast Domination and Multipacking in digraphs. A *dominating broadcast* of a digraph D is a function $f : V(D) \to \mathbb{N}$ such that for each vertex v of D, there exists a vertex t with $f(t) > 0$ having a directed path to v of length at most $f(t)$. The cost of f is the sum of $f(v)$ over all vertices v. A *multipacking* is a set S of vertices of D such that for each vertex v of D and for every integer d, there are at most d vertices from S within directed distance at most d from v. The maximum size of a multipacking of D is a lower bound to the minimum cost of a dominating broadcast of D. Let Broadcast Domination denote the problem of deciding whether a given digraph D has a dominating broadcast of cost at most k, and Multipacking the problem of deciding whether D has a multipacking of size at least k. It is known that Broadcast Domination is polynomial-time solvable for the class of all undirected graphs (that is, symmetric digraphs), while polynomial-time algorithms for Multipacking are known only for a few classes of undirected graphs. We prove that Broadcast Domination and Multipacking are both NP-complete for digraphs, even for planar layered acyclic digraphs of small maximum degree. Moreover, when parameterized by the solution cost/solution size, we show that the problems are respectively W[2]-hard and W[1]-hard. We also show that Broadcast Domination is FPT on acyclic digraphs, and that it does not admit a polynomial kernel for such inputs, unless the polynomial hierarchy collapses to its third level. In addition, we show that both problems are FPT when parameterized by the solution cost/solution size together with the maximum out-degree. Finally, we give for both problems polynomial-time algorithms for some subclasses of acyclic digraphs.

Keywords: Broadcast domination · Dominating set · Multipacking · Directed graphs · Parameterized complexity

© Springer Nature Switzerland AG 2020
L. Gąsieniec et al. (Eds.): IWOCA 2020, LNCS 12126, pp. 264–276, 2020.
https://doi.org/10.1007/978-3-030-48966-3_20

1   Introduction

We study the complexity of the two dual problems BROADCAST DOMINATION and MULTIPACKING in digraphs. These concepts were previously studied only for undirected graphs (which can be seen as *symmetric* digraphs, where for each arc (u, v), the symmetric arc (v, u) exists). Unlike most standard packing and covering problems, which are of local nature, these two problems have more global features since the covering and packing properties are based on arbitrary distances. This difference makes them algorithmically very interesting.

Broadcast Domination. Broadcast domination is a concept modeling a natural covering problem in telecommunication networks: imagine we want to cover a network with transmitters placed on some nodes, so that each node can be reached by at least one transmitter. Already in his book in 1968 [20], Liu presented this concept, where transmitters could broadcast messages but only to their neighboring nodes. It is however natural that a transmitter could broadcast information at distance greater than one, at the price of some additional power (and cost). In this setting, for a given non-zero integer cost d, a transmitter placed at node v covers all nodes within radius d from its location. If the network is directed, it covers all nodes with a directed path of length at most d from v. For a feasible solution, the function $f : V(G) \to \mathbb{N}$ assigning its cost to each node of the graph G (a cost of zero means the node has no transmitter placed on it) is called a *dominating broadcast* of G, and the total cost c_f of f is the sum of the costs of all vertices of G. The *broadcast domination number* $\gamma_b(G)$ of G is the smallest cost of a dominating broadcast of G. When all costs are in $\{0, 1\}$, this notion coincides with the well-studied DOMINATING SET problem. The concept of broadcast domination was introduced in 2001 (for undirected graphs) by Erwin in his doctoral dissertation [13] (see also [11,12] for some early publications on the topic), in the context of advertisement of shopping malls – which could nowadays be seen as targeted advertising via "influencers" in social networks. Note that in these contexts, directed arcs make sense since the advertisement or the influence is directed towards someone. The associated computational problem is as follows.

BROADCAST DOMINATION
- **Input**: A digraph $D = (V, A)$, an integer $k \in \mathbb{N}$.
- **Question**: Does there exist a dominating broadcast of D of cost at most k?

Multipacking. The dual notion for BROADCAST DOMINATION, studied from the linear programming viewpoint, was introduced in [5,24] and called *multipacking*. A set S of vertices of a (di)graph G is a *multipacking* if for every vertex v of G and for every possible integer d, there are at most d vertices from S at (directed) distance at most d from v. The *multipacking number* $\mathrm{mp}(G)$ of G is the maximum size of a multipacking in G. Intuitively, if a graph G has a multipacking S, any dominating broadcast of G will require to have cost at least $|S|$ to cover the vertices of S. Hence the multipacking number of G is a lower bound to its broadcast domination number [5]. Equality holds for many graphs, such as strongly chordal graphs [4]. Consider the following computational problem.

MULTIPACKING
- **Input**: A digraph $D = (V, A)$, an integer $k \in \mathbb{N}$.
- **Question**: Does there exist a multipacking $S \subseteq V$ of D of size at least k?

Known Results. In contrast with most graph covering problems, which are usually NP-hard, Heggernes and Lokshtanov designed in [17] (see also [21]) a sextic-time algorithm for BROADCAST DOMINATION in undirected graphs. This intriguing fact has motivated research on further algorithmic aspects of the problem. For general undirected graphs, no faster algorithm than the original one is known. A quintic-time algorithm exists for undirected series-parallel graphs [2]. An analysis of the algorithm for general undirected graphs gives quartic time when it is restricted to chordal graphs [17,18], and a cubic-time algorithm exists for undirected strongly chordal graphs [4]. The problem is solvable in linear time on undirected interval graphs [7] and undirected trees [4,9] (the latter was extended to undirected block graphs [18]).

Regarding MULTIPACKING, to the best of our knowledge, its complexity is currently unknown, even for undirected graphs (an open question posed in [24,25]). However, there exists a polynomial-time $(2 + o(1))$-approximation algorithm for all undirected graphs [1]. MULTIPACKING can be solved with the same complexity as BROADCAST DOMINATION for undirected strongly chordal graphs, see [4]. Improving upon previous algorithms from [22,24], the authors of [4] give a simple linear-time algorithm for undirected trees.

Our Results. We study BROADCAST DOMINATION and MULTIPACKING for directed graphs (digraphs), which form a natural setting for not necessarily symmetric telecommunication networks. In contrast with undirected graphs, we show that BROADCAST DOMINATION is NP-complete, even for planar layered acyclic digraphs (defined afterwards) of maximum degree 4. This holds for MULTIPACKING, even for planar layered acyclic digraphs of maximum degree 3, or acyclic digraphs with a single source and maximum degree 5. Moreover, when parameterized by the solution cost/solution size, we prove that BROADCAST DOMINATION is W[2]-hard (even for bipartite digraphs without directed 2-cycles) and MULTIPACKING is W[1]-hard. On the positive side, we show that BROADCAST DOMINATION is FPT on acyclic digraphs (DAGs for short) but does not admit a polynomial kernel for layered DAGs, unless the polynomial hierarchy collapses to its third level. Moreover, we show that both BROADCAST DOMINATION and MULTIPACKING are polynomial-time solvable for layered DAGs with a single source. We also design FPT algorithms for both problems when parameterized by the solution cost/solution size together with the maximum out-degree. The resulting complexity landscape is represented in Fig. 1. We start with some definitions in Sect. 2. We prove our results for BROADCAST DOMINATION in Sect. 3. The results for MULTIPACKING are presented in Sect. 4. We conclude in Sect. 5. Due to lack of space, we omit some proofs that can be found in the full version of the paper [14].

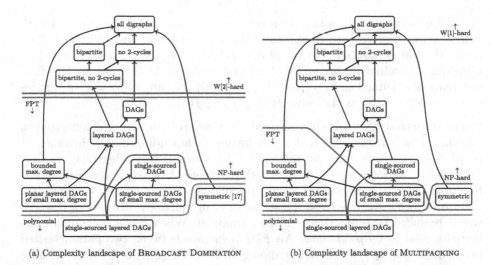

(a) Complexity landscape of BROADCAST DOMINATION   (b) Complexity landscape of MULTIPACKING

Fig. 1. Complexity landscape of BROADCAST DOMINATION and MULTIPACKING for some classes of digraphs. An arc from class A to class B indicates that A is a subset of B. Parameterized complexity results are for parameter solution cost/solution size.

2 Preliminaries

Directed Graphs. We mainly consider digraphs, usually denoted $D = (V, A)^1$, where V is the set of vertices and A the set of arcs. For an arc $uv \in A$, we say that v is an *out-neighbor* of u, and u an *in-neighbor* of v. Given a subset of vertices $V' \subseteq V$, we define the digraph *induced* by V' as $D' = (V', A')$ where $A' = \{uv \in A : u \in V' \text{ and } v \in V'\}$. We denote such an induced subdigraph by $D[V']$. A directed path from a vertex p_1 to p_l is a sequence $\{p_1, \ldots, p_l\}$ such that $p_i \in V$ and $p_i p_{i+1} \in A$ for every $1 \leqslant i < l$. When $p_1 = p_l$, it is a directed cycle. A digraph is *acyclic* whenever it does not contain any directed cycle as an induced subgraph. An acyclic digraph is called a *DAG* for short. The *(open) out-neighborhood* of a vertex $v \in V$ is the set $N^+(v) = \{u \in V : vu \in A\}$, and its *closed out-neighborhood* is $N^+[v] = N^+(v) \cup \{v\}$. We define similarly the open and closed in-neighborhoods of v and denote them by $N^-(v)$ and $N^-[v]$, respectively. A *source* is a vertex v such that $N^-(v) = \emptyset$. For the sake of readability, we always mean out-neighborhood when speaking of the *neighborhood* of a vertex. A DAG $D = (V, A)$ is *layered* when its vertex set can be partitioned into $\{V_0, \ldots, V_t\}$ such that $N^-(V_0) = \emptyset$ and $N^+(V_t) = \emptyset$ (vertices of V_0 and V_t are respectively called *sources* and *sinks*), and $uv \in A$ implies that $u \in V_i$ and $v \in V_{i+1}$, $0 \leqslant i < t$. A *single-sourced layered DAG* is a layered DAG with only one source, that is, satisfying $|V_0| = 1$. A digraph is *bipartite* or *planar* if its underlying undirected graph has the corresponding property. Every layered

---

[1] Our reductions will also use undirected graphs, denoted $G = (V, E)$ with $V = \{v_1, \ldots, v_n\}$ and $E = \{e_1, \ldots, e_m\}$.

digraph is bipartite. Given two vertices u and v, we denote by $d(u, v)$ the length of a shortest directed path from u to v. For a vertex $v \in V$ and an integer d, we define the *ball of radius d centered at v* by $B_d^+(v) = \{u \in V : d(v, u) \leqslant d\} \cup \{v\}$. Consider a dominating broadcast $f : V(D) \to \mathbb{N}$ on D. The set of *broadcast dominators* is defined as $V_f = \{v \in V : f(v) > 0\}$. For any set $S \subseteq V$ of vertices of D, we define $f(S)$ as the value $f(S) = \sum_{u \in S} f(u)$.

Parameterized Complexity. A *parameterized problem* is a decision problem together with a *parameter*, that is, an integer k depending on the instance. A problem is *fixed-parameter tractable* (FPT for short) if it can be solved in time $f(k) \cdot |I|^c$ for an instance I of size $|I|$ with parameter k, where f is a computable function and c is a constant. Given a parameterized problem P, a *kernel* is a function which associates to each instance of P an equivalent instance of P whose size is bounded by a function h of the parameter. When h is a polynomial, the kernel is said to be *polynomial*. An *FPT-reduction* between two parameterized problems P and Q is a function mapping an instance (I, k) of P to an instance $(f(I), g(k))$ of Q, where f and g are computable in FPT time with respect to parameter k, and where I is a YES-instance of P if and only if $f(I)$ is a YES-instance of Q. When moreover f can be computed in polynomial time and g is polynomial in k, we say that the reduction is a *polynomial time and parameter transformation* [3]. Both reductions can be used to derive conditional lower bounds: if a parameterized problem P does not admit an FPT algorithm (resp. a polynomial kernel) and there exists an FPT-reduction (resp. a polynomial time and parameter transformation) from P to a parameterized problem Q, then Q is unlikely to admit an FPT algorithm (resp. a polynomial kernel). Both implications rely on certain standard complexity hypotheses; we refer the reader to the book [8] for details.

3   Complexity of BROADCAST DOMINATION

3.1   Hardness Results

Theorem 1. BROADCAST DOMINATION *is NP-complete, even for planar layered DAGs of maximum degree* 4.

Theorem 2. BROADCAST DOMINATION *parameterized by solution cost k is W[2]-hard, even on bipartite digraphs without directed 2-cycles.*

Proof (sketch). We provide a reduction from the W[2]-hard MULTICOLORED DOMINATING SET problem [6].

MULTICOLORED DOMINATING SET
- **Input**: A graph $G = (V, E)$ with V partitioned into sets $\{V_1, \ldots, V_k\}$, $k \in \mathbb{N}$.
- **Question**: Does there exist a dominating set S of G s.t. $|S \cap V_i| = 1$ for $1 \leqslant i \leqslant k$?

Construction. We build an instance $(D' = (V', A'), k')$ of BROADCAST DOMI-NATION as follows. To obtain the vertex set V', we multiplicate V into four sets V^0, V^1, V^2 and V^3 and we will have a set M of *subdivided* vertices. The set $V^0 \cup V^1$ will induce an oriented complete bipartite graph, while $V^2 \cup V^3$ will induce a matching. For a vertex $v \in V$, $0 \leqslant i \leqslant 3$, its copy in V^i is denoted v^i. We assume that $|V_i| \geqslant 2$, since otherwise one must take the only vertex in V_i. For each $1 \leqslant i \leqslant k$ we then add the following arcs:

- for every pair v, w of distinct vertices of V_i, we add an arc from v^0 to w^1;
- for every $v \in V_i$, we add an arc from v^1 to v^0;
- for every $v \in V_i$, we add an arc from v^2 to v^3.

Moreover, for every edge vw in G, we add an arc from v^1 to w^2, and we subdivide it once. The set of all subdivision vertices is called M. Finally, we set $k' = 3k$. Digraph D' has no directed 2-cycles, and is bipartite with sets $V^0 \cup M \cup V^3$ and $V^1 \cup V^2$. One can check that G has a multicolored dominating set of size k if and only if D' has a dominating broadcast of cost $3k$. □

3.2 Complexity and Algorithms for (Layered) DAGs

We now address the special cases of (layered) DAGs. Note that DOMINATING SET remains W[2]-hard on DAGs by a reduction from [23, Theorem 6.11.2]. In contrast, we now give an FPT algorithm for BROADCAST DOMINATION on DAGs that counterbalances the W[2]-hardness result.

Theorem 3. BROADCAST DOMINATION *parameterized by solution cost k is FPT for DAGs.*

Proof. The proof relies on the following proposition, which is reminiscent of a stronger statement of Dunbar et al. [11] for undirected graphs (stating that there always exists an optimal dominating broadcast where each vertex is covered exactly once, which is false for digraphs).

Proposition 4. *For any digraph $D = (V, A)$, there exists an optimal dominating broadcast such that every broadcast dominator is covered by itself only.*

Proof. Let f be an optimal dominating broadcast of D, and assume there exists two vertices $u, v \in V$ such that $f(v) \geqslant 1$ and $f(u) \geqslant d(u, v)$. In this case, v is covered by both u and itself. Notice that $d(u, v) + f(v) > f(u)$, since otherwise setting $f(v)$ to 0 would result in a better dominating broadcast. We claim that setting $f(u)$ to $d(u, v) + f(v)$ and $f(v)$ to 0 yields an optimal dominating broadcast f_u. Notice that since $d(u, v) + f(v) > f(u)$, any vertex covered by u in f is still covered in f_u. Similarly, any vertex covered by v in f is now covered by u in f_u. Finally, we have $f(u) + f(v) \geqslant f_u(u) + f_u(v)$ since $f_u(u) = d(u, v) + f(v) \leqslant f(u) + f(v)$ and $f_u(v) = 0$, implying that the cost of f_u is at most the cost of f. □

We can now prove Theorem 3. Let $D = (V, A)$ be a DAG. We consider the set V_0 of sources of D. Observe that for every $s \in V_0$, $f(s) \geqslant 1$ must hold. In particular, this means that $|V_0| \leqslant k$ (otherwise we return NO). We provide a branching algorithm based on this simple observation and on Proposition 4. We start with an initial broadcast f consisting of setting $f(s) = 1$ for every vertex s in V_0. At each step of the branching algorithm, we let $N_f = \cup_{v \in V_f} B^+_{f(v)}(v)$ be the set of currently covered vertices, and we consider the digraph $D_f = D[V \setminus N_f]$. Notice that D_f is acyclic and hence contains a source u. Since every vertex of $N_f \setminus V_f$ is covered, we may assume by Proposition 4 that in the sought optimal solution, u is only covered by itself or by a vertex in V_f. This means that one needs to branch on at most $k+1$ distinct cases: either setting $f(u) = 1$, or increasing the cost of one of its at most k broadcasting ancestors in V_f. At every branching, the parameter k decreases by 1, which ultimately gives an $O^*(2^{k \log k})$-time algorithm and completes the proof of Theorem 3. $\qquad\square$

We will now complement the previous result by a negative one, which can be proved using a reduction from HITTING SET, defined as follows.

HITTING SET
- **Input**: A *universe* U of elements, a collection $\mathcal{F}$ of subsets of U, an integer $k \in \mathbb{N}$.
- **Question**: Does there exist a *hitting set* S of size k, that is, a set of k elements from U such that each set of $\mathcal{F}$ contains an element of S?

HITTING SET has no polynomial kernel when parameterized by $k+|U|$, unless the polynomial hierarchy collapses to its third level [10, Theorem 5.1].

Theorem 5. BROADCAST DOMINATION *parameterized by solution cost k does not admit a polynomial kernel even on layered DAGs, unless the polynomial hierarchy collapses to its third level.*

Theorem 6. BROADCAST DOMINATION *is polynomial-time solvable on single-sourced layered DAGs.*

Proof (sketch). Let $D = (V, A)$ be a single-sourced layered DAG with layers $\{V_0, \ldots, V_t\}$. For the sake of readability, sets V_i such that $|V_i| = 1$ are denoted s_i, for $0 \leqslant i \leqslant t$. Due to lack of space, the following claim is given without proof.

Claim 1. There always exists an optimal dominating broadcast f of D such that:
(i) $V_f \subseteq \bigcup_{i=0}^{t} s_i$
(ii) every $s_i \in V_f$, $0 \leqslant i \leqslant t$, covers exactly $B^+_l(s_i)$, where $l = j - i - 1$ and j is the smallest index such that $j \geqslant i + 2$ and $|V_j| = 1$.

We thus deduce a simple top-down procedure to compute an optimal dominating broadcast f. We start with setting $i = 0$. While there remain uncovered vertices, we let $f(s_i) = j - i - 1$ for the smallest value j such that s_j exists and $j \geqslant i + 2$. In other words, s_i will cover all vertices below it, until the closest vertex of the set $\bigcup_{j=0}^{t} s_j$ that is not a neighbour of s_i. We then carry on by setting $i = j$. By Claim 1, this leads to f being an optimal dominating broadcast. $\qquad\square$

We finally give an FPT algorithm for two parameters. By Theorems 1 and 2, such a result probably does not hold for each of them individually.

Theorem 7. BROADCAST DOMINATION *parameterized by solution cost k and maximum out-degree is FPT.*

4   Complexity of MULTIPACKING

We will need the following results to prove our results for MULTIPACKING. The first one was proved for undirected graphs in [16].

Lemma 8. *Let $D = (V, A)$ be a digraph with a shortest directed path of length $3k - 3$ vertices. Then, D has a multipacking of size k.*

Proof. It suffices to select every third vertex on the path.   □

Lemma 9. *Let $D = (V, A)$ be a digraph. There always exists a multipacking of maximum size containing every source of D.*

The following lemma is the central result of both our polynomial-time algorithm (Theorem 14) and NP-completeness reduction (Theorem 12).

Lemma 10. *Let $D = (V, A)$ be a single-sourced layered DAG. There always exists a multipacking $S \subseteq V$ of maximum size such that for every $1 \leqslant i \leqslant t$, $|S \cap V_i| \leqslant 1$.*

Proof. Let $S \subseteq V$ be a multipacking of D of maximum size. By definition of a multipacking, considering each ball centered at the source s, the following holds for every $1 \leqslant i \leqslant t$:

$$|S \cap \cup_{j=0}^{i} V_j| \leqslant i \qquad (1)$$

We will prove the result inductively, by locally modifying S in a top-down manner until it has the desired property. Let $j \geqslant 2$ be the smallest index such that $|S \cap V_j| \geqslant 2$, and $i < j$ be the largest index such that $|S \cap V_i| = 0$. Notice that i is well-defined due to (1). Moreover, let s_j^1 and s_j^2 be two vertices of $S \cap V_j$.

Case 1. We assume first that $i = j - 1$. Let u_i^1 and u_i^2 be vertices of V_i such that $u_i^1 s_j^1$ and $u_i^2 s_j^2$ belong to A (note that in a layered DAG every non-source vertex has a predecessor in the previous layer). Since S is a multipacking, we have $u_i^1 \neq u_i^2$ and neither u_i^1 nor u_i^2 is adjacent to both s_j^1 and s_j^2. Moreover, a vertex s_{i-1} in $S \cap V_{i-1}$ cannot be adjacent to both u_i^1 and u_i^2, since otherwise we would have $|B_2^+(s_{i-1}) \cap S| > 2$. Moreover by minimality of the index j, there is at most one vertex of S in V_{i-1}. Assuming w.l.o.g. that u_i^1 has no predecessor in S, the set $(S \setminus \{s_j^1\}) \cup \{u_i^1\}$ is a multipacking having the same size than S.

Case 2. We now consider the case where $i < j - 1$. First, we will prove that there is a vertex v_i in V_i with no in-neighbor in S. If $S \cap V_{i-1} = \emptyset$, any vertex of V_i can be chosen as vertex v_i. Otherwise, by choice of j we have $|S \cap V_{i-1}| = 1$. Assume $S \cap V_{i-1} = \{s_{i-1}\}$. We claim that s_{i-1} is not adjacent to every vertex of V_i.

Assume for a contradiction that this is the case. This means that s_{i-1} is within distance $j - (i-1)$ of every vertex contained in $\cup_{l=i}^{j} V_l$. By the choice of indices i and j we know that $\cup_{l=i}^{j} V_l$ contains at least $j - (i-1)$ vertices from S, which in turn implies that $\left| B_{j-(i-1)}^{+}(s_{i-1}) \cap S \right| = j - (i-1) + 1$, contradicting (1). Thus, there is a vertex v_i in V_i that has no in-neighbor in S. Now, we know by choices of i and j that $|S \cap V_p| = 1$ for $i < p < j$. Hence the set $(S \setminus \{s_{i+1}\}) \cup \{v_i\}$, where $\{s_{i+1}\} = S \cap V_{i+1}$, is a multipacking of D having the same size than S. By iterating the above argument, we end up with $i = j - 1$, in which case we can apply the argument from Case 1. Overall, after each iteration of Case 1, j strictly increases. The procedure terminates when the value of j reaches t.   □

4.1   Hardness Results

Theorem 11. MULTIPACKING *is NP-complete, even for planar layered DAGs of maximum degree* 3.

Theorem 12. MULTIPACKING *is NP-complete on single-sourced DAGs of maximum degree* 5.

Proof (sketch). We provide a reduction from the NP-complete INDEPENDENT SET problem [15]. We define the function $f : V(G) \to E$ such that for $v \in V(G)$, $f(v) = e_i$ if and only if e_i is the first edge in which v appears (recall that $E = \{e_1, \ldots, e_m\}$). We create the digraph $D = (V', A)$ as follows (see Fig. 2):

$$V' = \{u_i, v_i, w_i, x_i, y_i, z_i : 1 \leqslant i \leqslant m\} \cup V \cup \{s, p\}$$

$$A = \{u_i w_i, u_i x_i : 1 \leqslant i \leqslant m\} \cup \{v_i x_i : 1 \leqslant i \leqslant m\} \cup \{w_i y_i, w_i z_i : 1 \leqslant i \leqslant m\} \bigcup$$

$$\{z_i u_{i+1}, z_i v_{i+1} : 1 \leqslant i \leqslant m-1\} \cup \{x_i u, x_i v : 1 \leqslant i \leqslant m \text{ and } e_i = uv\} \bigcup$$

$$\{u_i u : 1 \leqslant i \leqslant m \text{ and } f(u) = e_i\} \cup \{sp, pu_1, pv_1\}$$

To conclude, one can see that G has an independent set of size k if and only if D has a multipacking of size $k' = k + 2m + 1$.   □

Theorem 13. MULTIPACKING *parameterized by solution size* k *is W[1]-hard.*

Proof (sketch). We provide an FPT-reduction from MULTICOLORED INDEPENDENT SET, which is W[1]-hard when parameterized by k [8].

MULTICOLORED INDEPENDENT SET
• **Input**: A graph $G = (V, E)$ with V partitioned into sets $\{V_1, \ldots, V_k\}$, $k \in \mathbb{N}$.
• **Question**: Does there exist an independent set S of G s.t. $|S \cap V_i| = 1$ for $1 \leqslant i \leqslant k$?

Construction. We construct an instance $(D = (V', A'), k')$ of MULTIPACKING as follows. We consider the bipartite incidence graph of G, that is we add $V \cup E$ to V'. To construct A', we add an arc from a vertex $e \in E$ to a vertex $v \in V$ if and only if e contains v. We next group vertices of E into $\binom{k}{2}$ sets $E_{i,j}$, $1 \leqslant i < j \leqslant k$

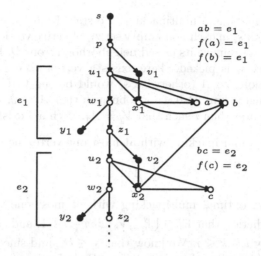

Fig. 2. Sketch of the construction in the proof of Theorem 12 for edges $e_0 = ab$ and $e_1 = bc$ with $f(a) = f(b) = e_0$ and $f(c) = e_1$.

according to the colors of their corresponding endpoints, and add every possible arc within each set $E_{i,j}$. We next duplicate the vertices of each set V_i into a set V_i' such that there is an arc from each vertex $v_i \in V_i$ to its corresponding copy v_i' in V_i'. Finally, we add k vertices $\{s_1, \ldots, s_k\}$ such that there is an arc from s_i to every vertex of V_i. Notice in particular that the maximum finite distance is 3. To conclude, one can see that the graph G has a multicolored independent set of size k if and only if the digraph D has a multipacking of size $k' = 2k + \binom{k}{2}$.

□

4.2 Algorithms

Theorem 14. MULTIPACKING *can be solved in linear time on single-sourced layered DAGs.*

Proof. Let $D = (V, A)$ be a single-sourced layered DAG. By Lemma 10, in every single-sourced layered DAG there is a multipacking of maximum size that is a maximum-size set of vertices with at most one vertex per layer such that two chosen vertices of consecutive layers are not adjacent. We give a polynomial-time bottom-up procedure to find such a set. At each step of the procedure, a layer V_i is partitioned into a set of *active* vertices and a set of *universal* ones, denoted respectively A_i and U_i. Our goal is to select exactly one vertex in each set of active vertices. We initiate the algorithm by setting $A_t = V_t$ and $U_t = \emptyset$. Now, for every i with $0 \leqslant i < t$, we set $U_i = \{u \in V_i : A_{i+1} \subseteq N^+(u)\}$ and $A_i = V_i \setminus U_i$. In other words, U_i contains the vertices of layer V_i that are adjacent to all active vertices of V_{i+1}. During the procedure, if some layer V_i satisfies $A_i = \emptyset$, we let $A_{i-1} = V_{i-1}$ and repeat this process until V_0 is reached.

To construct a maximum multipacking, we start from V_0, and for each $0 \leqslant i \leqslant t$ we pick a vertex s_i in each non-empty set A_i of active vertices. Every time a vertex s_i is picked, we remove its closed neighborhood from D. By construction, every time a vertex s_i is picked, there exists a vertex $s_{i+1} \in A_{i+1}$ such that $s_i s_{i+1}$ does not belong to A (otherwise s_i would belong to U_i). To prove the optimality of our algorithm, let $0 \leqslant i < t$ be such that $A_i = \emptyset$, and $j > i$ be the smallest integer greater than i such that $V_j = A_j$. Such a j exists since $A_t = V_t$.

Claim 2. Let S be a multipacking with at most one vertex per layer. Then:

$$\left| S \cap \cup_{k=i}^{j} V_k \right| \leqslant j - i \tag{2}$$

Proof. Let S be an optimal multipacking with at most one vertex per layer. Assume by contradiction that $\left| S \cap \bigcup_{k=i}^{j} V_k \right| > j - i + 1$, and call s_k the vertex in $V_k \cap S$ for every $i \leqslant k \leqslant j$. We know that $s_i \in U_i$, and since every vertex in A_{i+1} is an out-neighbor of s_i, then $s_{i+1} \in U_{i+1}$. By induction, for every $i \leqslant k \leqslant j$, we have $s_k \in U_k$, but $U_j = \emptyset$ by choice of j, leading to a contradiction.

Note that Claim 2 gives one less vertex than what Lemma 10 implies, and that it is the value reached by our algorithm, since for $i \leqslant k \leqslant j$, the only layer with $U_k = V_k$ is V_i. The sets of active and universal vertices can be constructed by standard graph searching, so the whole algorithm takes $O(|V| + |A|)$ time. $\square$

Theorem 15. MULTIPACKING *parameterized by solution size k and maximum out-degree d is FPT.*

Proof. Let $(D = (V, A), k)$ be an instance of MULTIPACKING such that D has maximum out-degree d. By Lemma 8, if D has a shortest directed path of length $3k - 3$, we can accept the input (this can be checked in polynomial time). Thus, we can assume that the length of any shortest path is at most $3k - 2$. If a vertex u has a directed path to a vertex v, we say that u *absorbs* v, and a set S of vertices is *absorbing* if every vertex in D is absorbed by some vertex of S. If D has a set of k vertices, no two of which are absorbed by some common vertex (*e.g.* a set of k sources), we can accept, since this set forms a valid solution. Note that this property is satisfied by any minimum-size absorbing set S: indeed, if some vertex w absorbs two vertices u, v of S, we may replace them by w and obtain a smaller absorbing set, a contradiction. We claim that we can find a minimum-size absorbing set in FPT time. Indeed, we can reduce this problem to HITTING SET (defined for the proof of Theorem 5) as follows. We let $U = V(D)$, and $\mathcal{F}$ contains a set F_v for every vertex v, where F_v comprises every vertex which absorbs v (including v itself). Because D has out-degree at most d and the length of any shortest path is at most $3k - 2$, every vertex of U is contained in at most $d_U = \sum_{i=0}^{3k-2}(d - 1)^i + 1$ sets of $\mathcal{F}$. Moreover, a set of vertices of $U = V(D)$ is a hitting set of $(U, \mathcal{F})$ if and only if it is an absorbing set of D. We can solve HITTING SET in FPT time when parameterized by d_U and solution size k [19], which proves the above claim. As mentioned before, if the obtained minimum-size absorbing set of D has size at least k, since it forms a valid multipacking, we can accept. Otherwise, D can be covered by $k - 1$ balls of radius at most $3k - 2$. Each ball has at most $\sum_{i=0}^{3k-2}(d-1)^i + 1 = d^{O(k)}$ vertices, so in total D has at most $d^{O(k)}$ vertices and a brute-force algorithm is FPT. $\square$

5 Conclusion

We have studied BROADCAST DOMINATION and MULTIPACKING on various subclasses of digraphs, with a focus on DAGs. It turns out that they behave very differently than for undirected graphs. We feel that MULTIPACKING is slightly more challenging. Indeed, some problems that we solved for BROADCAST DOMINATION are open for MULTIPACKING. For example, it would be interesting to see whether MULTIPACKING is FPT for DAGs, and whether it remains W[1]-hard for digraphs without directed 2-cycles. It is also unknown whether MULTIPACKING is NP-hard on undirected graphs, as asked in [24,25]. On the other hand, we showed that MULTIPACKING is NP-complete for single-sourced DAGs, but we do not know whether the same holds for BROADCAST DOMINATION. It is not difficult to show that both problems are FPT when parameterized by the vertex cover number. What about smaller parameters such as tree-width or DAG-width?

Acknowledgements. FF is partially funded by the IFCAM project "Applications of graph homomorphisms" (MA/IFCAM/18/39) and by the ANR project HOSIGRA (ANR-17-CE40-0022). FS is partially supported by the project ESIGMA (ANR-17-CE23-0010).

References

1. Beaudou, L., Brewster, R.C., Foucaud, F.: Broadcast domination and multipacking: bounds and the integrality gap. Australas. J. Comb. **71**(1), 86–97 (2019)
2. Blair, J.R.S., Heggernes, P., Horton, S., Manne, F.: Broadcast domination algorithms for interval graphs, series-parallel graphs, and trees. In: Proceedings of the 35th South-eastern International Conference on Combinatorics, Graph Theory, and Computing. Congressus Numerantium, vol. 169, pp. 55–77 (2004)
3. Bodlaender, H.L., Thomassé, S., Yeo, A.: Kernel bounds for disjoint cycles and disjoint paths. Theor. Comput. Sci. **412**(35), 4570–4578 (2011)
4. Brewster, R.C., MacGillivray, G., Yang, F.: Broadcast domination and multipacking in strongly chordal graphs. Discrete Appl. Math. **261**, 108–118 (2019)
5. Brewster, R.C., Mynhardt, C.M., Teshima, L.E.: New bounds for the broadcast domination number of a graph. Cent. Eur. J. Math. **11**, 1334–1343 (2013)
6. Casel, K.: Resolving Conflicts for lower-bounded clustering. In: 13th International Symposium on Parameterized and Exact Computation, IPEC 2018, Helsinki, Finland, 20–24 August 2018, pp. 23:1–23:14 (2018)
7. Chang, R.Y., Peng, S.L.: A linear-time algorithm for broadcast domination problem on interval graphs. In: Proceedings of the 27th Workshop on Combinatorial Mathematics and Computation Theory, pp. 184–188. Providence University Taichung, Taiwan (2010)
8. Cygan, M., et al.: Parameterized Algorithms. Springer, Switzerland (2015). https://doi.org/10.1007/978-3-319-21275-3

9. Dabney, J., Dean, B.C., Hedetniemi, S.T.: A linear-time algorithm for broadcast domination in a tree. Networks **53**(2), 160–169 (2009)

10. Dom, M., Lokshtanov, D., Saurabh, S.: Kernelization lower bounds through colors and IDs. ACM Trans. Algorithms **11**(2), 13:1–13:20 (2014)

11. Dunbar, J.E., Erwin, D.J., Haynes, T.W., Hedetniemi, S.M., Hedetniemi, S.T.: Broadcasts in graphs. Discrete Appl. Math. **154**(1), 59–75 (2006)

12. Erwin, D.J.: Dominating broadcasts in graphs. Bull. ICA **42**, 89–105 (2004)

13. Erwin, D.J.: Cost domination in graphs. Ph.D. thesis, Western Michigan University (2001)

14. Foucaud, F., Gras, B., Perez, A., Sikora, F.: On the complexity of broadcast domination and multipacking in digraphs. CoRR abs/2003.10570 (2020). https://arxiv.org/abs/2003.10570

15. Garey, M.R., Johnson, D.S.: Computers and Intractability. Freeman, San Francisco (1979)

16. Hartnell, B.L., Mynhardt, C.M.: On the difference between broadcast and multipacking numbers of graphs. Utilitas Mathematica **94**, 19–29 (2014)

17. Heggernes, P., Lokshtanov, D.: Optimal broadcast domination in polynomial time. Discrete Math. **306**(24), 3267–3280 (2006)

18. Heggernes, P., Sæther, S.H.: Broadcast domination on block graphs in linear time. In: Hirsch, E.A., Karhumäki, J., Lepistö, A., Prilutskii, M. (eds.) CSR 2012. LNCS, vol. 7353, pp. 172–183. Springer, Heidelberg (2012). https://doi.org/10.1007/978-3-642-30642-6_17

19. Hermelin, D., Wu, X.: Weak compositions and their applications to polynomial lower bounds for kernelization. In: SODA 2012, Kyoto, Japan, 17–19 January 2012, pp. 104–113 (2012)

20. Liu, C.L.: Introduction to Combinatorial Mathematics. McGraw-Hill, New York (1968)

21. Lokshtanov, D.: Broadcast domination. Master's thesis, University of Bergen (2007)

22. Mynhardt, C.M., Teshima, L.E.: Broadcasts and multipackings in trees. Utilitas Mathematica **104**, 227–242 (2017)

23. Ordyniak, S., Kreutzer, S.: Width-measures for directed graphs and algorithmic applications. In: Quantitative Graph Theory, pp. 195–245. Chapman and Hall/CRC (2014)

24. Teshima, L.: Broadcasts and multipackings in graphs. Master's thesis, University of Victoria (2012)

25. Yang, F.: New results on broadcast domination and multipacking. Master's thesis, University of Victoria (2015)

A Parameterized Perspective
on Attacking and Defending Elections

Kishen N. Gowda, Neeldhara Misra[⊠], and Vraj Patel

Indian Institute of Technology, Gandhinagar, Gandhinagar, India
{kishen.gowda,neeldhara.m,vraj.patel}@iitgn.ac.in
http://www.iitgn.ac.in

Abstract. We consider the problem of protecting and manipulating elections by recounting and changing ballots, respectively. Our setting involves a plurality-based election held across multiple districts, and the problem formulations are based on the model proposed recently by [Elkind et al., IJCAI 2019]. It turns out that both of the manipulation and protection problems are NP-complete even in fairly simple settings. We study these problems from a parameterized perspective with the goal of establishing a more detailed complexity landscape. The parameters we consider include the number voters, and the budgets of the attacker and the defender. While we observe fixed-parameter tractability when parameterizing by number of voters, our main contribution is a demonstration of parameterized hardness when working with the budgets of the attacker and the defender.

Keywords: Elections · W-hardness · Parameterized complexity

1 Introduction

Electoral fraud and errors in consolidation of large-scale voting data are fundamental issues in democratic societies. To counteract issues with malicious manipulations and accidental errors in the counting of votes, most electoral systems allow for strategic recounting of ballots to verify the election outcome. Recounting is generally an expensive and high-stakes process, and it would be desirable to formalize the problem so as to capture the relevant trade-offs and possibly pursue an algorithmic approach to finding an optimal recounting strategy. Such a framework was recently proposed by Elkind et al. [5], where the authors considered the problems of protecting and manipulating elections by recounting and changing ballots, respectively. These problems are modeled as a Stackelberg game involving an attacker and a defender. Both players work with limited budgets (say $\mathcal{B}_A$ and $\mathcal{B}_D$), and the question is if the players can develop optimal strategies for their desired outcomes.

In this model, the election is spread out across multiple districts, with the voter preferences aggregated according to the plurality voting rule in one of two

Supported by Indian Institute of Technology, Gandhinagar and SERB.

L. Gąsieniec et al. (Eds.): IWOCA 2020, LNCS 12126, pp. 277–288, 2020.
https://doi.org/10.1007/978-3-030-48966-3_21

different ways, which we will explain explicitly in a moment. The manipulation problem is the following. The attacker has to optimize, typically with the goal of turning the election in favor of a particular candidate that he may have in mind, an attack strategy that involves manipulating the votes in at most $\mathcal{B}_A$ districts while ensuring that the impact of the attack persists *even if* the defender restores at most $\mathcal{B}_D$ of these districts to their original state. In the recounting problem, the defender is given complete information about the original and manipulated voting profiles and she can restore the state of at most $\mathcal{B}_D$ districts with the goal of making the "true winner" win the repaired election.

Known Results. The results obtained in [5] already demonstrate the hardness of the attacker's and defender's problems for two natural ways of aggregating the votes: (1) *Plurality over Voters* (PV), where districts are only used for the purpose of collecting the ballots and the winner is selected among the candidates that receive the largest number of votes in total, and (2) *Plurality over Districts* (PD), where each district selects a preferred candidate using the Plurality rule, and the overall winner is chosen among the candidates supported by the largest number of districts, or the set of districts with largest total weight if the districts have weights associated with them. We briefly recall the main highlights from [5], since this provides the context for our contributions. It turns out that the recounting problem is NP-complete for both implementations of the plurality rule, even when there are only three candidates (this result assumes a succinct representation of the votes) or even when the votes are specified in unary. The problem is tractable when PD is employed over unweighted districts. On the other hand, the manipulation problem is NP-hard for PD even with unweighted districts, and in fact, Σ_2^P-complete for PD with succinct input even when there are only three candidates. Further, it is NP-hard for PV, again even when there are only three candidates (in the setting of succinct input) or even when the votes are specified in unary.

Our Contributions. Our main contribution is to establish the parameterized intractability of both the recounting and manipulation problems under both implementations of the plurality voting rule when parameterized by the budget of the players. Our contributions directly address a direction suggested by [5]. In particular, we obtain the following results:

Theorem 1. *The* PV-REC *and* PD-REC *problems are W[2]-hard and W[1]-hard, respectively, when parameterized by the budget of the defender, and are FPT when parameterized by the number of districts.*

Theorem 2. *The* PV-MAN *and* PD-MAN *problems are W[1]-hard when parameterized by the budget of the attacker (even when the defender budget is zero), and are FPT when parameterized by the number of voters.*

Our results rely on reductions from traditional problems such as MULTI-COLORED CLIQUE and DOMINATING SET. Our hardness results work even when the input is specified in unary. It is reasonably natural to imagine that these parameters would be small in practice, since they correspond to real-world budget constraints. To that end, our results here bring mixed news: on the one hand, the hardness of mounting an attack may be viewed as a positive outcome, but on the other hand, it turns out that the problem of optimally reversing damages is hard as well. This triggers the natural question of whether the recounting problem admits good approximations when treated as an optimization problem, either on the criteria of the budget or on the criteria of the quality of the winning candidate that we are able to restore. The FPT algorithms that we present rely mostly on straightforward enumeration, and it would be interesting to improve the running times in question.

Related Work. While we build most closely on the work of Elkind et al. [5], and much of the work on manipulation in the literature of social choice does not consider the possibility of a counter-attack, we note that some recent investigations have been carried out in a spirit that is similar to our present contribution. Dey et al. [3] also consider a parameterized approach to protecting elections, where the voting rule in question is the Condorcet rule. They build on the work of Yin et al. [10], who study a pre-emptive approach to protecting elections. In these models, the defender allocates resources to guard some of the electoral districts, so that the votes there cannot be influenced, and the attacker responds afterward. This is in contrast to our setting, where the defender makes the second move. The social choice literature is rich in studies of manipluation, control, and bribery. For a detailed overview we direct the reader to the surveys [1,6].

2 Preliminaries

We recall the setting from [5]. We consider elections over a *candidate* set C, where $|C| = m$. There are n *voters* who are partitioned into k pairwise disjoint *districts*. The set of all districts is $\mathcal{D}$. For each $i \in \mathcal{D}$, let $n_i = |i|$. We note that in this context, i denotes a subset of voters. For each $i \in \mathcal{D}$, district i has a *weight* w_i, which is a positive integer. We say that an election is *unweighted* if $w_i = 1$ for all $i \in \mathcal{D}$. Each voter votes for a single candidate in C. For each $i \in \mathcal{D}$ and each $c \in C$ let v_{ic} denote the number of votes that candidate c gets from voters in i. We refer to the list $\mathbf{v} = (v_{ic})_{i \in \mathcal{D}, c \in C}$ as the *vote profile*.

Let $\succ$ be a linear order over C; $a \succ b$ indicates that a is favored over b. We consider the following two voting rules, which take the vote profile $\mathbf{v}$ as their input.

- *Plurality over Voters* (PV): We say that a candidate a *beats* a candidate b under PV if $\sum_{i \in \mathcal{D}} v_{ia} > \sum_{i \in \mathcal{D}} v_{ib}$ or $\sum_{i \in \mathcal{D}} v_{ia} = \sum_{i \in \mathcal{D}} v_{ib}$ and $a \succ b$; the winner is the candidate that beats all other candidates. Note that district weights w_i are not relevant for this rule.
- *Plurality over Districts* (PD): For each $i \in \mathcal{D}$ the winner c_i in i is chosen from the set $\arg\max_{c \in C} v_{ic}$, with ties broken according to $\succ$. Then, for each $i \in \mathcal{D}, c \in C$, we set $w_{ic} = w_i$ if $c = c_i$, else $w_{ic} = 0$. We say that a candidate

a *beats* a candidate b under PD if $\sum_{i \in \mathcal{D}} w_{ia} > \sum_{i \in \mathcal{D}} w_{ib}$ or $\sum_{i \in \mathcal{D}} w_{ia} = \sum_{i \in \mathcal{D}} w_{ib}$ and $a \succ b$. The winner is the candidate that beats all other candidates. Given the voting profile $\mathbf{v}$, we take the winner in district i to be $\mathcal{G}_{\mathbf{v}}(i)$. We shall omit the subscript if the voting profile to be used is clear from the context.

For PV and PD, we define the *social welfare* of a candidate $c \in C$ as the total number of votes that c gets and the total weight that c gets, respectively:

$$SW^{PV}(c) = \sum_{i \in \mathcal{D}} \mathbf{v}_{ic}, \quad SW^{PD}(c) = \sum_{i \in \mathcal{D}} w_{ic}$$

Hence, the winner under each voting rule is a candidate with the maximum social welfare. We now define some additional terminology that we use later in this paper:

- *Score:* Given a voting profile u, under the PD rule, the score of a candidate p is defined as the sum of weights of districts in which the candidate p wins. Formally, $sc_u(p) = \sum_{i \in \mathcal{D}} w_i \cdot \delta_{\mathcal{G}(i)p}$ (Here δ is the Kronecker delta function).
- *Rivals:* Given a voting profile u, under the PD rule, we define the rivals of a candidate p to be the set of candidates $\mathcal{C} \subset C$ such that for all candidates $c \in \mathcal{C}$, either $sc_u(c) > sc_u(p)$ or $sc_u(c) = sc_u(p)$ and $c \succ p$.

We now consider the scenario where an election may be manipulated by an attacker, who wants to change the result of the election in favor of his preferred candidate p. The attacker has a budget $B_A \in \mathbb{N}$ which enables him to change the voting profiles in at most B_A districts. For each district $i \in \mathcal{D}$, we define $\gamma_i, 0 \leqslant \gamma_i \leqslant n_i$ to be the number of votes that the attacker can manipulate in i. After the manipulation we have a voting profile $\overline{\mathbf{v}} = (\overline{\mathbf{v}}_{ic})_{i \in \mathcal{D}, c \in C}$. We formalize the notion of a manipulation as a set $M \subseteq \mathcal{D}$ and a voting profile $\overline{\mathbf{v}}$ such that $|M| \leqslant B_A$ where $\overline{\mathbf{v}}_{ic} = \mathbf{v}_{ic}$ for all $i \notin M$ and for all $i \in M$ it holds that $\sum_{c \in C} \overline{\mathbf{v}}_{ic} = n_i$ and $\sum_{c \in C} \frac{|\overline{\mathbf{v}}_{ic} - \mathbf{v}_{ic}|}{2} \leqslant \gamma_i$.

After the attacker, a socially minded defender with budget $B_D \in \{0\} \cup \mathbb{N}$ can demand a recount in at most B_D districts. Formally, a recounting strategy R is a set such that $R \subseteq M$ and $|R| \leqslant B_D$. After the defender recounts, the vote counts of the districts in R are restored to their original values. This results in a new voting profile $u = (u_{ic})_{i \in \mathcal{D}, c \in C}$ where for all $i \in \mathcal{D} \setminus M \cup R, u_{ic} = \mathbf{v}_{ic}$ and for all $i \in M \setminus R, u_{ic} = \overline{\mathbf{v}}_{ic}$. Then the voting rule $\mathcal{V} = \{PV, PD\}$ is applied to the profile u that is obtained to obtain a winner (let it be w') with ties broken according to $\succ$. The defender's objective is to maximize $SW^{\mathcal{V}}(w')$. It is a game of perfect information i.e. both entities know all information about the game.

We say that the attacker wins if he has a manipulation strategy such that after the defender moves optimally, the preferred candidate of attacker i.e. p wins. We define the following two decision problems based on voting rule $\mathcal{V} \in \{PV, PD\}$ and the two entities:

- $\mathcal{V}$-MAN: Given the voting rule $\mathcal{V}$, a voting profile $\mathbf{v}$, the linear order $\succ$, a preferred candidate p, attacker budget B_A, defender budget B_D and weights

w_i and parameter γ_i for each district $i \in \mathcal{D}$, does the attacker have a winning strategy?

- $\mathcal{V}$-REC: Given the voting rule $\mathcal{V}$, a voting profile **v**, a manipulated voting profile $\overline{\mathbf{v}}$, a preferred candidate w, the linear order $\succ$, defender budget B_D and weights w_i for each district $i \in \mathcal{D}$, can the defender make w win by recounting at most B_D districts?

Next we state the definitions and parameterized hardness results for decision problems that are used throughout this paper, and refer the reader to [2,4] for a detailed introduction to parameterized complexity and the framework of parameterized reductions.

- DOMINATING SET: A set of vertices D is a dominating set in graph G if $V(G) = N_G[D]$. DOMINATING SET asks that given a graph G and a non-negative integer k, does there exist a dominating set of size at most k? DOMINATING SET is known to be W[2]-hard parameterized by k [4].
- MULTI-COLORED CLIQUE: Given a graph G and a partition of the vertex set V into k color classes $V_1, V_2, \ldots V_k$, MULTI-COLORED CLIQUE asks whether there exists a clique of size k with one vertex each from $V_1, V_2, \ldots V_k$. MULTI-COLORED CLIQUE is known to be W[1]-hard parameterized by k [7,9].

3   Plurality over Voters (PV)

In this section, we analyze the parameterized complexity of PV-REC and PV-MAN with different parameters. It is easy to see that PV-REC is FPT when parameterized by the number of districts (k), since we may guess the districts to be recounted. Since $k \leqslant n$, the problem is also FPT parameterized by the number of voters.

Proposition 1. PV-REC *is FPT when parameterized by the no. of districts* k *or the number of voters* n.

We now show that PV-REC is W[2]-hard when parameterized by budget of the defender (B_D). Before describing the construction formally, we briefly outline the main idea. We reduce from the DOMINATING SET, which is well-known to be W[2]-hard parameterized by the size of the solution, which we denote by k. Let $(G = (V, E), k)$ be an instance of DOMINATING SET. We create an instance of PV-REC where we have candidates and districts corresponding to vertices of G, along with a special candidate w who is our desired winner. To begin with, we have an "immutable" district—one where the original and manipulated votes are identical—that sets the baseline score of the special candidate at n. The number votes for any other candidate c from this district is fixed to ensure that the total number of votes for c is *also* n. In a district corresponding to a vertex v, every candidate corresponding to a vertex $u \in N[v]$ gets one vote. In the original scenario, all voters in these districts vote only for some dummy candidates. The key is that a "switch" in a district corresponding to a vertex v *reduces* the vote count for all vertices in $N[v]$. Since w receives no votes from any of the other districts, observe that the only way for w to emerge as a unique winner is if all of the other candidates *lose* votes from the switches. It is not hard to infer from here that the defender has a valid switching strategy if and only if G has a dominating set of size at most k.

Lemma 1. *PV-Rec is W[2]-hard parameterized by* B_D, *the defender's budget.*

Proof. We present an FPT reduction from the DOMINATING SET problem. Let $(G = (V, E), k)$ be an instance of DOMINATING SET. Let $N = |V|$ and $M = |E|$. We begin by describing the construction of the reduced instance.

Districts: We introduce a *baseline* district $\mathcal{D}_0$. Further, for each vertex $v \in V$, we introduce a corresponding *primary* district $\mathcal{D}_v$.

Candidates: For each vertex $v \in V$ we introduce a *main* candidate c_v and a *dummy* candidate d_v. We also have a *special* candidate w.

Voting Outcomes: The voting outcomes are as follows. For ease of presentation, let $v \in V$ be arbitrary but fixed.

1. The special candidate does not receive any votes from the primary districts in either the original or the manipulated settings. In particular, $\overline{v}_{\mathcal{D}_v,w} = v_{\mathcal{D}_v,w} = 0$.
2. In the original election, the main candidates have no votes in the primary districts, that is, $v_{\mathcal{D}_v,c_u} = 0$ for all $u \in V$.
3. In the manipulated election, a main candidate c_u has a single vote in its favor in the primary district $\mathcal{D}_v$ if and only if $u \in N_G[v]$. Formally,

$$\overline{v}_{\mathcal{D}_v,c_u} = \begin{cases} 1 & \text{if } u \in N_G[v] \\ 0 & \text{otherwise} \end{cases}$$

4. In the original election, a dummy candidate d_u has a score of $d(u) + 1$ in the primary district corresponding to u, and a score of zero everywhere else. Formally,

$$v_{\mathcal{D}_v,d_u} = \begin{cases} d(v) + 1 & \text{if } u = v \\ 0 & \text{otherwise} \end{cases}$$

5. The dummy candidates receive no votes in the primary districts in the manipulated elections.
6. In the baseline district, the score of the main candidates is defined to ensure that their total score in the manipulated election is N. In particular, $\overline{v}_{\mathcal{D}_0,c_v} = v_{\mathcal{D}_0,c_v} = N - (d(v) + 1)$.
7. The dummy candidates receive no votes in the baseline district in both the original and manipulated elections.
8. The score of w is N in the baseline district in both the original and manipulated elections. In particular, $\overline{v}_{\mathcal{D}_0,w} = v_{\mathcal{D}_0,w} = N$.

To summarize, the primary districts corresponding to a vertex v have $d(v) + 1$ voters, and all main candidates corresponding to vertices in $N_G[v]$ get one vote each in the manipulated world; while the dummy candidate d_v gets all the votes in the original world. Observe that in the manipulated election, all the candidates except the dummy candidates have a total score of N, while the dummy candidates have a score of zero.

We set $B_\mathcal{D} = k$. The preferred candidate is w. We also work with the following tie-breaking order: $\ldots c_v \ldots \succ w \succ \ldots d_{vu} \ldots$, where the main candidates are preferred over the special candidate, but the special candidate dominates the dummy candidates. This completes the description of the constructed instance. Due to lack of space, we defer the proof of equivalence to a full version [8] of this paper.   □

We now turn our attention to PV-MAN. First, we prove that PV-MAN is FPT parameterized by the number of voters. This follows by first observing that $m \leqslant 2n$ without loss of generality, since at most $2n$ candidates can have a non-trivial score across the original and manipulated instances combined, and the remaining candidates are irrelevant to the instance. The algorithm can then proceed by guessing a manipulation strategy—note that the space of all possible strategies is bounded once the candidates are bounded—and then invoking the PD-REC algorithm from the previous section as a subroutine to verify the validity of the guessed strategy. Thus, we have the following.

Proposition 2. PV-MAN *is FPT parameterized by* n, *the number of voters.*

We now show that PV-MAN is W[1]-hard parameterized by budget of the attacker $(B_\mathcal{A})$. Before describing the construction formally, we briefly outline the main idea. We reduce from the MULTI-COLORED CLIQUE problem, which is well-known to be W[1]-hard parameterized by the number of color classes, which we denote by k. Let $(G = (V = V_1 \uplus \cdots \uplus V_k, E), k)$ be an instance of MULTICOLORED-CLIQUE. In the reduced instance, we introduce a special candidate w who is the preferred candidate of the defender. The rival candidates are candidates corresponding to color classes $\mathcal{R}_i$, ordered pairs of color classes $\mathcal{R}_{ij}$ and vertices c_v. We also introduce some *dummy* candidates. We introduce districts corresponding to each $v \in V$ and each $e \in E$. Also, there exists a special district which is "immutable", which sets up the initial scores of all candidates such that they are equal to a large number (say F). Initially, w has 0 votes. The scores are set up in such a way that the attacker has to transfer votes of k^2 districts to w to make her win. The scores are engineered to ensure that the attacker has a successful manipulation strategy if and only if these k^2 districts correspond to k vertices and $\binom{k}{2}$ edges that form a multicolored clique in G.

Lemma 2. PV-MAN *is W[1]-hard parameterized by* $B_\mathcal{A}$, *the attacker's budget.*

Proof. We demonstrate a parameterized reduction from the MULTI-COLORED CLIQUE problem. Let $(G = (V = V_1 \uplus \cdots \uplus V_k, E), k)$ be an instance of MULTI-COLORED CLIQUE. We begin by describing the construction.

Districts: There are two types of districts. We have a *primary district* $\mathcal{D}_v$ for each vertex $v \in V$ and two *secondary districts* $\mathcal{D}_{uv}$ and $\mathcal{D}_{vu}$ for each edge $e = (u, v) \in E$. Apart from these, there is a *baseline district* $\mathcal{D}_0$.

Candidates: For each vertex $v \in V$ we will have a *main* candidate c_v. Also, we have *challenger* candidates corresponding to color classes $\mathcal{R}_i$'s and ordered pairs of color classes $\mathcal{R}_{ij}$'s. We introduce some *dummy* candidates of an unspecified number, whose role is equalize the number of votes across primary and secondary districts. Finally, we have a *special* candidate w.

Voting Profiles: We introduce the following voting outcomes for the candidates.

1. The score of the special candidate w is zero in all districts.
2. A main candidate c_v for $v \in V$ has a score of $k - 1$ in the primary district corresponding to v, and a score of zero in all other primary districts.
3. A main candidate c_v for $v \in V$ has a score of one in every secondary district corresponding to an edge $e = \{u, w\}$ that it is *not* incident to, provided u and v share the same color class. In particular, if $v \in V_i$ and there are t edges incident on $V_i \setminus \{v\}$, then c_v has a score of one in t secondary districts. Formally, assuming $v \in V_i$, we have:

$$
\mathbf{v}_{\mathcal{D}_{uw}, c_v} = \begin{cases} 1 & \text{if } u \neq v \text{ and } u \in V_i \\ 0 & \text{otherwise} \end{cases}
$$

4. A challenger candidate corresponding to color class i has one vote from any primary district corresponding to a vertex $v \in V_i$ and a score of zero from all other primary districts. In other words:

$$
\mathbf{v}_{\mathcal{D}_v, \mathcal{R}_i} = \begin{cases} 1 & \text{if } v \in V_i \\ 0 & \text{otherwise} \end{cases}
$$

5. A challenger candidate corresponding to an ordered pair of color classes (V_i, V_j) has one vote from any secondary district $\mathcal{D}_{uv}$ corresponding to an edge $e \in E$ whose endpoints u and v are in color classes V_i and V_j respectively, and a score of zero from all other secondary districts. Note that the candidates $\mathcal{R}_{ij}$ and $\mathcal{R}_{ji}$ receive scores of one from distinct secondary districts. Specifically, we have: $\mathbf{v}_{\mathcal{D}_{uv}, \mathcal{R}_{ij}} = 1$ if $u \in V_i$ and $v \in V_j$.

Now, let ℓ be the size of the largest—in terms of the number of voters—among the primary and secondary districts constructed so far. For every primary or secondary district D with $v(D)$ voters, we add $\ell - v(D)$ dummy voters and dummy candidates, and we let each dummy voter vote for a distinct dummy candidate. We let $F = \ell k^2$.

We are now ready to specify the voting outcomes from the baseline district. these are simply designed to ensure that all primary candidates c_v get $F + k - 2$ votes and the challenger candidates get F votes, which is easy to verify from the proposed outcomes below:

1. $\mathbf{v}_{\mathcal{D}_0,c_v} = F - 1 - \sum_{e=\{u,v\}\in E} (\mathbf{v}_{\mathcal{D}_{uv},c_v} + \mathbf{v}_{\mathcal{D}_{vu},c_v})$

2. $\mathbf{v}_{\mathcal{D}_0,\mathcal{R}_i} = F - \sum_{u\in V} \mathbf{v}_{\mathcal{D}_u,\mathcal{R}_i} - \sum_{e=\{u,v\}\in E} (\mathbf{v}_{\mathcal{D}_{uv},\mathcal{R}_i} + \mathbf{v}_{\mathcal{D}_{vu},\mathcal{R}_i})$

3. $\mathbf{v}_{\mathcal{D}_0,\mathcal{R}_{ij}} = F - \sum_{u\in V} \mathbf{v}_{\mathcal{D}_u,\mathcal{R}_{ij}} - \sum_{e=\{u,v\}\in E} (\mathbf{v}_{\mathcal{D}_{uv},\mathcal{R}_{ij}} + \mathbf{v}_{\mathcal{D}_{vu},\mathcal{R}_{ij}})$

Note that apart from the above, the dummy candidates get 1 vote each, and the special candidate w has 0 votes. Also, the attacker has no room to manipulate in the baseline district, that is, $\gamma_{\mathcal{D}_0} = 0$. On the other hand, the attacker can modify up to ℓ votes in the primary and secondary districts. Further, we set $B_A = k^2$ and $B_D = 0$. The preferred candidate is w. Finally, we impose the following tie-breaking order:

$$\ldots \mathcal{R}_i \ldots \succ \ldots \mathcal{R}_{ij} \ldots \succ \ldots c_v \ldots \succ w \succ \ldots \text{ dummies} \ldots$$

This completes the description of the constructed instance. We now turn to the proof of equivalence.

Forward Direction. Let a multi-colored clique S of size k be given. The attacker chooses the k primary and $2\binom{k}{2}$ secondary districts corresponding to the vertices and edges of S, and transfers *all* the ℓ votes in these k^2 districts to the desired candidate w. The score of w is now F. Further, note that the scores of all challenger candidates has decreased by one to $F - 1$, and the scores of all main candidates have decreased by $k - 1$ as well, but for different reasons: for main candidates corresponding to vertices of the clique, the drop is directly from the recounting in the primary districts, while for any other main candidate, the drop is cumulative across $(k - 1)$ relevant secondary districts. In particular, suppose $S \cap V_i := \{v_i\}$. Consider any $u \in V_i$ such that $u \neq v_i$, and observe that c_u had a score of one in the following $(k - 1)$ districts:

$$D_{v_i v_1}, \ldots, D_{v_i v_{i-1}}, D_{v_i v_{i+1}}, \ldots, D_{v_i v_k},$$

which have indeed been attacked, and therefore the score of c_u reduces by $k - 1$. This leaves all candidates ranked ahead of the special candidate w in the tie-breaking order with a score less than the final score of w, and the scores of the dummy candidates is either zero or one, thus they pose no threat to w. Therefore, w wins the election under this attack, concluding the argument in the forward direction.

Reverse Direction. In the reverse direction, we essentially argue that any valid solution must resemble the structure of the solution that we obtained while demonstrating the forward direction. This follows from the construction, and due to lack of space, the details are deferred to a full version [8] of this paper. □

4 Plurality over Districts (PD)

In this section, we analyze the parameterized complexity of PD-REC and PD-MAN. As with PV-REC, it is easy to see that PD-REC is also FPT when parameterized by the number of districts (k), since we may guess the districts to be recounted. Since $n \geq k$, it is also FPT parameterized by the number of voters.

Proposition 3. PD-REC *is FPT when parameterized by the number of districts* k *or the number of voters* n.

We now show that PD-REC is W[1]-hard parameterized by budget of the defender $B_{\mathcal{D}}$. Before describing the construction formally, we briefly outline the main idea. We reduce from the MULTICOLORED-CLIQUE PROBLEM, which is well-known to be W[1]-hard parameterized by the no. of color classes, which we denote by k. Let $(G = (V = V_1 \uplus \cdots \uplus V_k, E), k)$ be an instance of MULTICOLORED-CLIQUE. In the reduced instance, we introduce a special candidate w who is the preferred candidate of the defender. The rival candidates are candidates corresponding to color classes $\mathcal{R}_i$ and ordered pairs of color classes $\mathcal{R}_{ij}$. Further, we introduce candidates encoding the vertices c_v and edges h_e & a_e. We also introduce some *dummy* candidates. We introduce *two* districts corresponding to each $v \in V$ and *five* districts corresponding to each $e \in E$. Also, there exists a baseline district for each candidate which is "immutable", which sets up the initial scores of all candidates such that they are equal to a large number (say λ). The scores are set up in such a way that there is no way to increase the score of w, thus we require to reduce the score of all the rivals by at least one while not increasing the scores of other candidates. But the districts and voting profiles are engineered so as to enforce that any recounting solution must have a certain structure, from which we can draw a correspondence to a subset of vertices which must in fact form a multi-colored clique of size k in G.

Lemma 3. PD-REC *is W[1]-hard parameterized by* $B_{\mathcal{D}}$, *the defender's budget.*

Proof. This hardness result follows from the following reduction from the MULTI-COLORED CLIQUE. The given instance is the graph $G = (V, E)$ and the number of unique color classes, k, where V_i denotes the ith color class. We begin by describing the construction of the reduced instance.

Candidates: For every color class $1 \leqslant i \leqslant k$ there is a candidate $\mathcal{R}_i$ corresponding to V_i. Further, for every pair of color classes (i, j) such that $1 \leqslant i < j \leqslant k$, we introduce *two* candidates $\mathcal{R}_{ij}$ and $\mathcal{R}_{ji}$. These will be the rival candidates of the reduced instance. Now we introduce candidates that encode the vertices and edges of the graph G. To begin with, for each vertex $v \in V$ we introduce two candidates c_v and d_v, which we will refer to as the *main* and *dummy* candidates, respectively. Also, for every edge $e \in E$, we introduce two candidates h_e and a_e, which we refer to as the *helper* and *auxiliary* candidates, respectively. Finally, we have a *special* candidate w. To summarize, the overall set of candidates is:

$$C = \{R_i \mid i \in [k]\} \cup \left\{ R_{ij}, R_{ji} \mid (i, j) \in \binom{[k]}{2} \right\} \cup \{c_v, d_v \mid v \in V\} \cup \{a_e, h_e \mid e \in E\}$$

Districts: We introduce the following districts.

1. For each $v \in V$ we introduce a *primary district* D_v with weight one and a *critical district* $D_v^\star$ with weight k.
2. For each $e = (u, v) \in E$, we introduce two *edge districts* D_{uv} and D_{vu}, one *support district* S_e, and two *transfer districts* $T_{e,u}$ and $T_{e,v}$. The support districts have weight two, while the remaining districts have weight one.

3. For each candidate $c \in C \setminus \{d_v \mid v \in V\}$, we introduce a *baseline district* B_c with a weight of λ_c, which will be specified in due course.

Voting Outcomes: The voting outcomes in the original and manipulated districts are depicted in the Table 1 below.

Table 1. For the winners depicted above assume that $v \in V_j$ and that $e = (u, v)$, and further that $u \in V_i$ and $v \in V_j$. The subscript c in the last column corresponding to the baseline districts denotes an arbitrary non-dummy candidate.

District type	D_v	$D_v^\star$	D_{uv}	D_{vu}	S_e	$T_{e,u}$	$T_{e,v}$	B_c
Original winner	c_v	d_v	h_e	h_e	a_e	c_u	c_v	c
Manipulated winner	R_j	c_v	$\mathcal{R}_{ij}$	$\mathcal{R}_{ji}$	h_e	a_e	a_e	c
Weight	1	k	1	1	2	1	1	λ_c

Note that the voting outcome in the baseline districts is the same in the original and manipulated settings. It only remains to specify explicitly the weights of the baseline districts. We let $\lambda_w := (n + 1)k + 6m$. Recall that this is the weight of the baseline district corresponding to the special candidate. For any non-dummy candidate c, let $s(c)$ be its score from the manipulated districts. We then set $\lambda_c := \lambda_w - s(c)$. With these weights, we ensure that all the non-dummy candidates tie for the same score (i.e, λ_w) in the manipulated election, and all dummy candidates have a score of zero. We note that all the weights introduced here are polynomially bounded. We set $B_{\mathcal{D}} = 2k + 5 \times \binom{k}{2}$. The preferred candidate is w.

We enforce the following tie-breaking order: $\ldots \mathcal{R}_i \ldots \succ \ldots \mathcal{R}_{ij} \ldots \succ w \succ \ldots c_v \ldots \succ \ldots a_e \ldots \succ \ldots h_e \ldots \succ \ldots d_v \ldots$.

This completes the description of the constructed instance. We defer the proof of equivalence to a full version [8] of this paper due to lack of space. $\quad\square$

We now turn our attention to PD-MAN. We observe that PD-MAN can be shown to be W[1]-hard by a reduction from PD-REC. We briefly sketch the main idea: to begin with, we switch the roles of the manipulated profiles and the original ones, set the defender budget to zero and the attacker budget to the defender budget. We would also set up the votes in the districts to be such that the only meaningful manipulation by the attacker is to move the manipulated profile to the original one. The equivalence is based on repurposing a recounting strategy to an attacking one and vice-versa. We defer the details of this argument to a full version [8] of the paper.

5 Concluding Remarks

Our main contribution was to settle the parameterized complexity for the problems of recounting and manipulation when parameterized by the player budgets, for both the PD and PV implementations of the plurality voting rules. We also observed that these problems are FPT when parameterized by the number of voters, and that the recounting problem is FPT when parameterized by the number of districts as well.

We make some remarks about directions for future work. In the setting of succinct input, the problems of recounting and manipulation are already para-NP-hard because of the NP-completeness for three candidates. When the votes are specified in unary is an interesting direction for future work. The dynamic programming algorithm proposed by [5] already shows that the problem is in XP, parameterized by the number of candidates, and we leave open the issue of whether the problem is FPT. The problem of manipulation parameterized by the number of districts is another unresolved case. More broadly, it would be interesting to challenge the theoretical hardness results obtained here against heuristics employed on real world data sets. The issue of identifying and working with structural parameters is also an interesting direction for further thought.

References

1. Conitzer, V., Walsh, T.: Barriers to manipulation in voting. In: Brandt, F., Conitzer, V., Endriss, U., Lang, J., Procaccia, A.D. (eds.) Handbook of Computational Social Choice, pp. 127–145. Cambridge University Press, Cambridge (2016)
2. Cygan, M., et al.: Parameterized Algorithms. Springer, Cham (2015). https://doi.org/10.1007/978-3-319-21275-3
3. Dey, P., Misra, N., Nath, S., Shakya, G.: A parameterized perspective on protecting elections. In: Proceedings of the Twenty-Eighth International Joint Conference on Artificial Intelligence, IJCAI 2019, pp. 238–244 (2019)
4. Downey, R.G., Fellows, M.R.: Parameterized Complexity. Springer, New York (1999). https://doi.org/10.1007/978-1-4612-0515-9
5. Elkind, E., Gan, J., Obraztsova, S., Rabinovich, Z., Voudouris, A.A.: Protecting elections by recounting ballots. In: Proceedings of the Twenty-Eighth International Joint Conference on Artificial Intelligence (IJCAI), pp. 259–265 (2019)
6. Faliszewski, P., Rothe, J.: Control and bribery in voting. In: Brandt, F., Conitzer, V., Endriss, U., Lang, J., Procaccia, A.D. (eds.) Handbook of Computational Social Choice, pp. 146–168. Cambridge University Press, Cambridge (2016)
7. Fellows, M.R., Hermelin, D., Rosamond, F.A., Vialette, S.: On the parameterized complexity of multiple-interval graph problems. Theor. Comput. Sci **410**(1), 53–61 (2009)
8. Gowda, K.N., Misra, N., Patel, V.: A parameterized perspective on attacking and defending elections, CoRR abs/2005.03176 (2020). https://arxiv.org/abs/2005.03176
9. Pietrzak, K.: On the parameterized complexity of the fixed alphabet shortest common supersequence and longest common subsequence problems. J. Comput. Syst. Sci **67**(4), 757–771 (2003)
10. Yin, Y., Vorobeychik, Y., An, B., Hazon, N.: Optimally protecting elections. In: Proceedings of the Twenty-Fifth International Joint Conference on Artificial Intelligence, IJCAI, pp. 538–545. IJCAI/AAAI Press (2016)

Skyline Computation with Noisy Comparisons

Benoît Groz[1]([✉]), Frederik Mallmann-Trenn[2], Claire Mathieu[3],
and Victor Verdugo[4,5]

[1] Université Paris-Saclay, CNRS, LRI, Gif-sur-Yvette, France
benoit.groz@lri.fr
[2] King's College London, London, UK
frederik.mallmann-trenn@kcl.ac.uk
[3] CNRS & IRIF, Paris, France
claire.m.mathieu@gmail.com
[4] London School of Economics and Political Science, London, UK
v.verdugo@lse.ac.uk
[5] Universidad de O'Higgins, O'Higgins, Chile
victor.verdugo@uoh.cl

Abstract. Given a set of n points in a d-dimensional space, we seek to compute the *skyline*, i.e., those points that are not strictly dominated by any other point, using few comparisons between elements. We adopt the noisy comparison model [15] where comparisons fail with constant probability and confidence can be increased through independent repetitions of a comparison. In this model motivated by Crowdsourcing applications, Groz and Milo [18] show three bounds on the query complexity for the skyline problem. We improve significantly on that state of the art and provide two output-sensitive algorithms computing the skyline with respective query complexity $O(nd\log(dk/\delta))$ and $O(ndk\log(k/\delta))$, where k is the size of the skyline and δ the expected probability that our algorithm fails to return the correct answer. These results are tight for low dimensions.

Keywords: Skyline · Noisy comparisons · Fault-tolerance

1 Introduction

Skylines have been studied extensively, since the 1960s in statistics [6], then in algorithms and computational geometry [22] and in databases [7,12,16,21]. Depending on the field of research, the *skyline* is also known as the set of *maximum vectors*, the *dominance frontier*, *admissible* points, or *Pareto frontier*. The skyline of a set of points consists of those points which are not strictly dominated by any other point. A point p is *dominated* by another point q if $p_i \leq q_i$ for every coordinate (attribute or dimension) i. It is *strictly dominated* if in addition the inequality is strict for at least one coordinate; see Fig. 1.

© Springer Nature Switzerland AG 2020
L. Gąsieniec et al. (Eds.): IWOCA 2020, LNCS 12126, pp. 289–303, 2020.
https://doi.org/10.1007/978-3-030-48966-3_22

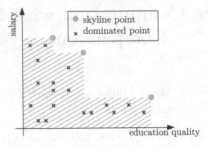

Fig. 1. Given a set of points X, the goal is to find the set of *skyline points*, i.e.,points are not dominated by any other points.

Noisy Comparison Model, and Parameters. In many contexts, comparing attributes is not straightforward. Consider the example of finding *optimal* cities from [18].

> To compute the skyline with the help of the crowd we can ask people questions of the form *"is the education system superior in city x or city y?"* or *"can I expect a better salary in city x or city y"*. Of course, people are likely to make mistakes, and so each question is typically posed to multiple people. Our objective is to minimize the number of questions that need to be issued to the crowd, while returning the correct skyline with high probability.

Thus, much attention has recently been given to computing the skyline when information about the underlying data is uncertain [25], and comparisons may give erroneous answers. In this paper we investigate the complexity of computing skylines in the noisy comparison model, which was considered in [18] as a simplified model for crowd behaviour: we assume queries are of the type *is the i-th coordinate of point p (strictly) smaller than that of point q?*, and the outcome of each such query is independently correct with probability greater than some constant better than $1/2$ (for definiteness we assume probability $2/3$). As a consequence, our confidence on the relative order between p and q can be increased by repeatedly querying the pair on the same coordinate. Our complexity measure is the number of comparison queries performed.

This noisy comparison model was introduced in the seminal paper [15] and has been studied in [8,18]. There are at least 2 straightforward approaches to reduce noisy comparison problems to the noiseless comparison setting. One approach is to take any "noiseless" algorithm and repeat each of its comparisons $\log(f(n))$ times, where n is the input size and $f(n)$ is the complexity of the algorithm. The other approach is to sort the n items in all d dimensions at a cost of $nd\log(nd)$, then run some noiseless algorithm based on the computed orders. The algorithms in [15,18] and this paper thus strive to avoid the logarithmic overhead of these straightforward approaches.

Three algorithms were proposed in [18] to compute skylines with noisy comparisons. Figure 2 summarizes their complexity and the parameters we consider.

The first algorithm is the reduction through sorting discussed above. But sky-lines often contain only a small fraction of the input items (points), especially when there are few attributes to compare (low dimension). This leads to more efficient algorithms because smaller skylines are easier to compute. Therefore, [18] and the present paper expresses the complexity of computing skylines as a function of four parameters that appear on Fig. 2: δ, the probability that the algorithm could fail to return the correct output, and three parameters wholly determined by the input set X: the number of input points $n = |X|$, the dimension d of those points, and the size $k = |\text{skyline}(X)|$ of the skyline (output). There is a substantial gap between the lower bounds and the upper bounds achieved by the skyline algorithms in [18]. In particular, the authors raised the question whether the skyline could be computed in $o(nk)$ for any constant d. In this paper, we tighten the gap between the lower and upper bounds and settle this open question.

Contributions. We propose 2 new algorithms that compute skylines with prob-ability at least $1 - \delta$ and establish a lower bound:

- Algorithm **SkyLowDim**(X, δ) computes the skyline in $O(nd\log(dk/\delta))$ query complexity and $O(nd\log(dk\delta) + ndk)$ overall running time.
- Algorithm **SkyHighDim**(X, δ) computes the skyline in $O(ndk\log(k/\delta))$
- $\Omega(nd\log k)$ queries are necessary to compute the skyline when $d \le k$.
- Additionally, we show that Algorithm **SkyLowDim** can be adapted to com-pute the skyline with $O(nd\log(dk))$ comparisons in the noiseless setting.

Our first algorithm answers positively the above question from [18]. Together with the lower bound, we thus settle the case of low dimensions, i.e., when there is a constant c such that $d \le k^c$. Our 2 skyline algorithms both shave off a factor k from the corresponding bounds in the state of the art [18], as illustrated in Fig. 2 with respect to query complexity. **SkyLowDim** is a randomized algorithm that samples the input, which means it may fail to compute the skyline within the bounds even when comparisons are guaranteed correct. However, we show that our algorithm can be adapted to achieve deterministic $O(nd\log(dk))$ for this specific noiseless case.

As a subroutine for our algorithms, we developed a new algorithm to evaluate disjunctions of boolean variables with noise. Algorithm **NoisyFirstTrue** is, we believe, interesting in its own right: it returns the index of the first positive variable in input order, with a running time that scales linearly with the index.

Technical Core of Our Algorithms. The algorithm underlying the two bounds for $k \ll n$ in [18] recovers the skyline points one by one. It iteratively adds to the skyline the maximum point, in lexicographic order, among those not dominated by the skyline points already found.[1] However, the algorithm in [18] essentially considers the whole input for each iteration. Our two algorithms, on the opposite, can identify and discard some dominated points early. The idea behind our

---

[1] The difference between those two bounds is due to different subroutines to check dominance.

[18]	$O(nd\log(nd/\delta))^\dagger$	$O(ndk\log(dk/\delta))$	$O(ndk^2\log(k/\delta))$	d : dimension
this paper	—	$O(nd\log(dk/\delta))^\dagger$	$O(ndk\log(k/\delta))$	n : # input points
				k : # skyline points
		SkyLowDim	**SkyHighDim**	δ : error rate tolerated

best when: $k \in \Omega(\log(dn))$ $d < k^c < n$ $k \ll d$

Fig. 2. Query complexity of skyline algorithms depending on the values of k. For $\dagger$-labeled bounds, the running time is larger than the number of queries.

algorithm **SkyHighD** – **param** is that it is more efficient to separate the two tasks: (i) finding a point p not dominated by the skyline points already found, on the one hand, and (ii) computing a maximum point (in lexicographic order) *among those dominating p*, on the other hand. Whenever a point is considered for step (i) but fails to satisfy that requirement, the point can be discarded definitively. The $O(ndk)$ skyline algorithm from [13] for the noiseless setting also decomposes the two tasks, although the point they choose to add to the skyline in each of the k iteration is not the same as ours.

Our algorithm **SkyLowD** – **param** can be viewed as a 2-steps algorithm where the first step prunes a huge fraction of dominated points from the input through discretization, and the second step applies a cruder algorithm on the surviving points. We partition the input into buckets for discretization, identify "skyline buckets" and discard all points in dominated buckets. The bucket boundaries are defined by sampling the input points and sorting all sample points in each dimension. In the noisy comparison model, the approach of sampling the input for some kind of discretization was pioneered in [8] for selection problems, but with rather different techniques and objectives. One interesting aspect of our discretization is that a fraction of the input will be, due to the low query complexity, incorrectly discretized yet we are able to recover the correct skyline.

Our lower bound constructs a technical reduction from the problem of identifying null vectors among a collection of vectors, each having at most one non-zero coordinate. That problem can be studied using a two-phase process inspired from [15].

Related Work. The noisy comparison model was considered for sorting and searching objects [15]. While any algorithm for that model can be reduced to the noiseless comparison model at the cost of a logarithmic factor (boosting each comparison so that by union bound all are correct), [15] shows that this additional logarithmic factor can be spared for sorting and for maxima queries, though it cannot be spared for median selection. [17,26] and [8] investigate the trade-off between the total number of queries and the number of rounds for (variants of) top-k queries in the noisy comparison model and some other models. The noisy comparison model has been refined in [14] for top-k queries, where the probability of incorrect answers to a comparison increase with the distance between the two items.

Other models for uncertain data have also been considered in the literature, where the location of points is determined by a probability distribution, or when data is incomplete. Some previous work [3, 27] model uncertainty about the output by computing a ρ-skyline: points having probability at least ρ to be in the skyline. We refer to [5] for skyline computation using the crowd and [23] for a survey in crowdsourced data management.

Our paper aims to establish the worst-case number of comparisons required to compute skylines with output-sensitive algorithms, i.e., when the cost is parametrized by the size of the result set. While one of our algorithm is randomized, we do not make any further assumption on the input (we do not assume input points are uniformly distributed, for instance). In the classic *noiseless* comparison model, the problem of computing skylines has received a large amount of attention [7, 20, 22]. For any constant d, [20] show that skylines can be computed in $O(n\log^{d-2} k)$. In the RAM model, the fastest algorithms we are aware of run in $O(n\log^{d-3} n)$ expected time [10], and $O(n\log\log_{n/k} n(\log_{n/k} n)^{d-3}$ deterministic time [2]. When $d \in \{2, 3\}$, the problem even admits "instance-optimal" algorithms [4]. [11] investigates the constant factor for the number of comparisons required to compute skyline, when $d \in \{2, 3\}$. The technique does not seem to generalize to arbitrary dimensions, and the authors ask among open problems whether arbitrary skylines can be computed with fewer than $dn\log n$ comparisons. To the best of our knowledge, our $O(nd\log(dk))$ is the first non-trivial output-sensitive upper bound that improves on the folklore $O(dnk)$ for computing skylines in arbitrary dimensions. Many other algorithms have been proposed that fit particular settings (big data environment, particular distributions, etc), as evidenced in the survey [19], but those works are further from ours as they generally do not investigate the asymptotic number of comparisons. Other skyline algorithms in the literature for the noiseless setting have used bucketing. In particular, [1] computes the skyline in a massively parallel setting by partitioning the input based on quantiles along each dimension. This means they define similar buckets to ours, and they already observed that the buckets that contain skyline points are located in hyperplanes around the "bucket skyline", and therefore those buckets only contain a small fraction of the whole input.

Organization. In Sect. 2, we recall standard results about the noisy comparison model and introduce some procedure at the core of our algorithms. Section 3 introduces our algorithm for high dimensions (Theorem 4) and Sect. 4 introduces the counterpart for low dimensions (Theorem 6). Section 5 establishes our lower bound (Theorem 7).

2 Preliminaries

The complexity measured is the number of comparisons in the worst case. Whenever the running time and the number of comparisons differ, we will say so. With respect to the probability of error, our algorithms are supposed to fail with probability at most δ. Following standard practice we only care to prove that our algorithms have error in $O(\delta)$: 5δ, for instance, because the asymptotic complexity of our algorithms would remain the same with an adjusted value for the parameter: $\delta' = \delta/5$.

Given two points, $p = (p_1, p_2 \ldots, p_d)$ and $q = (q_1, q_2 \ldots, q_d)$ point p is *lexico-graphically* smaller than q, denoted by $p \leq_{\text{lex}} q$, if $p_i < q_i$ for the first i where p_i and q_i differ. If there is no such i, meaning that the points are identical, we use the id of the points in the input as a tie-breaker, ensuring that we obtain a total order. We next describe and name algorithms that we use as subroutines to compute skylines.

Algorithm **NoisySearch** takes as input an element y, an ordered list $(y_1, y_2, \ldots, y_m)$, accessible by comparisons that each have error probability at most p, and a parameter δ. The goal is to output the interval $I = (y_{i-1}, y_i]$ such that $y \in I$.

Algorithm **NoisySort** relies on **NoisySearch** to solve the **noisy sort problem**. It takes as input an unordered set $Y = \{y_1, y_2, \ldots, y_m\}$, and a parameter δ. The goal is to output an ordering of Y that is the correct non-decreasing sorted order. In the definition above, the order is kept implicit. In our algorithms, the input items are d-dimensional points, so **NoisySort** will take an additional argument i indicating on which coordinate we are sorting those points.

Algorithm **NoisyMax** returns the maximum item in the unordered set Y whose elements can be compared, but we will rather use another variant: algorithm **MaxLex** takes as input an unordered set $Y = \{y_1, y_2, \ldots, y_m\}$, a point x and a parameter δ. The goal is to output the maximum point in lexicographic order among those that dominate x. Algorithm **SetDominates** is the boolean version whose goal is to output whether there exists a point in Y that dominates x.

Algorithm **NoisyFirstTrue** takes as input a list $(y_1, y_2, \ldots, y_m)$ of boolean elements that can be compared to **true** with error probability at most p (typically the result of some comparison or subroutines such as **SetDominates**). The goal is to output the index of the first element with value **true** (and $m + 1$, which we assimilate to **false**, if there are none).[2]

Theorem 1 ([15,18]). *When the input comparisons have error probability at most $p = 1/3$, the table below lists the number of comparisons performed by the algorithms to return the correct answer with success probability $1 - \delta$:*

Algorithm	NoisyMax	NoisySort	NoisySearch	SetDominates	MaxLex
Comparisons	$O(m\log\frac{1}{\delta})$	$O(m\log\frac{m}{\delta})$	$O(\log\frac{m}{\delta})$	$O(md\log\frac{1}{\delta})$	$O(md\log\frac{1}{\delta})$

We denote by CheckVar(x, δ) the procedure that checks if $x = $ **true** with error probability δ by majority vote, and returns the corresponding boolean.

Theorem 2. *Algorithm **NoisyFirstTrue** solves the first positive variable problem with success probability $1 - \delta$ in $O(j \cdot \log(1/\delta))$ where j is the index returned.*

---

[2] As in [18] (but with stronger bounds), this improves upon an $O(m\log(1/\delta))$ algorithm from [15] that only answers whether at least one of the elements is true.

Proof. The proof, left for the long version [24], shows that the error (resp. the cost) of the whole algorithm is dominated by the error (resp. the cost) of the last iteration.

Algorithm NoisyFirstTrue$(x_1, \ldots, x_n, \delta)$ (see Theorem 2)
input: $\{x_1, \ldots, x_n\}$ set of boolean random variables, δ error probability
output: the index j of the first positive variable, or $n + 1$ (=false).

1: $i \leftarrow 1$
2: $\delta' \leftarrow \delta/2$
3: **while** $i \leq n$ **do**
4:     $j \leftarrow$ **NoisyOr**$(x_1, \ldots, x_i, \delta')$
5:     **if** CheckVar$(x_j, \delta'/2^i)$ **then**
6:         **return** j
7:     **else**
8:         $i \leftarrow 2 \cdot i$
9: **return false**

---

3   Skyline Computation in High Dimension

We first introduce Algorithm **SkyHighD** − **param** which assumes that an estimate $\hat{k}$ of k is known in advance. We will show afterwards how we can lift that assumption.

Theorem 3. *Given* $\delta \in (0, 1/2)$ *and a set* X *of data items,* **SkyHighD**− **param**(X, δ) *outputs* $\min(|X|, \hat{k})$ *skyline points, with probability at least* $1 - \delta$. *The running time and number of queries is* $O(nd\hat{k}\log(\hat{k}/\delta))$.

Proof. Each iteration through the loop adds a point to the skyline S with probability of error at most $\delta/\hat{k}$. The final result is therefore correct with success probability $1 - \delta$. The complexity is $O((i' - i) * d\hat{k}\log(\hat{k}/\delta))$ to find a non-dominated point $p_{i'}$ at line 3, and $O(nd\log(\hat{k}/\delta))$ to compute the maximal point above $p_{i'}$ at line 4. Summing over all iterations, the running time and number of queries is $O(nd\hat{k}\log(\hat{k}/\delta))$.

Algorithm **SkyHighD** − **param**(X, δ) needs a good estimate $\hat{k} \in O(k)$ of the skyline cardinality $\hat{k} \in O(k)$ to return the skyline in $O(ndk\log(k/\delta))$. To guarantee that complexity, algorithm **SkyHighDim** exploits the classical trick from Chan [9] of trying a sequence of successive values for $\hat{k}$ − a trick that we also exploit in algorithms **NoisyFirstTrue** and **SkyLowDim**. The sequence grows exponentially to prevent failed attempts from penalizing the complexity.

Theorem 4. *Given* $\delta \in (0, 1/2)$ *and a set* X *of data items,* **SkyHighDim**(X, δ) *outputs a subset of* X *which, with probability at least* $1 - \delta$, *is the skyline. The running time and number of queries is* $O(ndk\log(k/\delta))$.

Proof. The proof is relatively straightforward and left for the long version.

Algorithm SkyHighD – param(k, X, δ) (see Theorem 3)
input: $X = \{p_1, \ldots, p_n\}$ set of points, $\hat{k}$ upper bound on skyline size, δ error probability
output: $\min\{\hat{k}, \text{skyline}(X)\}$ skyline points w.p. $1 - \delta$

1: Initialize $S \leftarrow \emptyset$, $i \leftarrow 1$
2: **while** $i \neq -1$ and $|S| < \hat{k}$ **do**
3:     $i' \leftarrow$ index of the first point $p_{i'}$ not dominated by current skyline points.[a]
        {Find a skyline point dominating p_i}
4:     Compute $p^* \leftarrow \textbf{MaxLex}(p_{i'}, \{p_i, \ldots, p_n\}, \delta/(2\hat{k}))$
5:     $S \leftarrow S \cup \{p^*\}$
6:     $i \leftarrow i'$
7: Output S

[a]This point can be computed using algorithm **NoisyFirstTrue** on the boolean variables: $\neg\textbf{SetDominates}(S, p_i, \delta/(2\hat{k})), \ldots, \neg\textbf{SetDominates}(S, p_n, \delta/(2\hat{k}))$, where we denote by $\neg$ the negation. This means that $\neg\textbf{SetDominates}(S, p_n, \delta/(2\hat{k}))$ returns true when the procedure $\textbf{SetDominates}(S, p_n, \delta/(2\hat{k}))$ indicates that p_n is not dominated.

Algorithm SkyHighDim(X, δ) (see Theorem 4)
input: X set of points, δ error probability
output: $\text{skyline}(X)$ w.p. $1 - \delta$

1: Initialize $j \leftarrow 0$, $\hat{k} \leftarrow 1$
2: **repeat**
3:     $j \leftarrow j + 1$; $\hat{k} \leftarrow 2\hat{k}$; $S \leftarrow \textbf{SkyHighD} - \textbf{param}(\hat{k}, X, \delta/2^j)$
4: **until** $|S| < \hat{k}$
5: Output S

---

4   Skyline Computation in Low Dimension

Let us first sketch our algorithm **SkyLowD** – **param**(k, X, δ). The algorithm works in 3 phases. The first phase partitions input points in buckets. We sort the i-th coordinate of a random sample to define $s + 1$ intervals in each dimension $i \in [d]$, hence $(s + 1)^d$ *buckets*, where each bucket is a product of intervals of the form $\prod_i I_i$; then we assign each point p of X to a bucket by searching in each dimension for the interval I_i containing p_i. Of course we do not materialize buckets that are not assigned any points.

The second phase eliminates irrelevant buckets: those that are dominated by some non-empty bucket and therefore have no chance of containing a skyline point. In short, the idea is to identify the "skyline of the buckets", and use it to discard the dominated buckets, as defined in Sect. 4.2. With high probability the bucketization obtained from the first phase will be "accurate enough" for our purpose: it will allow to identify efficiently the irrelevant buckets, and will also guarantee that the points in the remaining buckets form a small fraction of the input (provided k and d are small).

In phase 3, we thus solve the skyline problem on a much smaller dataset, calling Algorithm **SkyHighD** − **param** to find the skyline of the remaining points.[3] The whole purpose of the bucketization is to discard most input points while preserving the actual skyline points, so that we can then run a more expensive algorithm on the reduced dataset.

Figure 3 illustrates our algorithm: the grey buckets (c, f, g) are dominated by some (non-empty) orange buckets (a, b, d, e) so they cannot contain skyline points: in phase 3, the algorithm solves the reduced problem on the points contained in the orange buckets.

4.1   Identifying "Truly Non-empty" Buckets

Our bucketization does not guarantee that all points are assigned to the proper bucket, because it would be too costly with noisy comparisons. In particular, empty buckets may erroneously be assumed to contain some points (e.g., the buckets above a, b on Fig. 3). Those empty buckets also are irrelevant, even if they are not dominated by the "skyline" buckets. To drop the irrelevant buckets, we thus design a subroutine **First-Nonempty-Bucket** that processes a list of buckets, and returns the first bucket that really contains at least one point. Incidentally, we will not double-check the emptiness of every bucket using this procedure, but will only check those that may possibly belong to the skyline: those that we will define more formally as buckets of type (i), (ii) and (iv) in the proof of Theorem 5. We could not afford to "fix" the whole assignment as it may contain too many buckets.

In the **First-Nonempty-Bucket** problem, the input is a sequence of pairs $[(B_1, X_1), \ldots, (B_n, X_n)]$ where B_i is a bucket and X_i is a set of points. The goal is to return the first i such that $B_i \cap X_i \neq \emptyset$ with success probability $1 - \delta$. The test $B_i \cap X_i \neq \emptyset$ can be formulated as a DNF with $|X_i|$ conjunctions of $O(d)$ boolean variables each. To solve **First-Nonempty-Bucket**, we can flatten the formulas of all buckets into a large DNF with conjunctions of $O(d)$ boolean variables (one conjunction per bucket point). We call **FirstBucket**$([(B_1, X_1), \ldots, (B_n, X_n)], \delta)$ the algorithm that executes **NoisyFirstTrue** to compute the first true conjunction, while keeping tracks of which point belongs to which bucket with pointers:

Lemma 1. *Algorithm* **FirstBucket**$([(B_1, X_1), \ldots, (B_n, X_n)], \delta)$ *solves problem* **First-Nonempty-Bucket** *in* $O(\sum_{i \leq j} d \cdot |X_i| \log(1/\delta))$ *with success probability* $1 - \delta$, *where j is the index returned by the algorithm.*

4.2   Domination Relationships Between Buckets

In the second phase, Algorithm **SkyLowD** − **param**(k, X, δ) eliminates irrelevant buckets. To manage ties, we need to distinguish two kinds of intervals: the

---

[3] Alternatively, one could use an algorithm provided by Groz and Milo [18], it is only important that the size of the input set is reduced to n/k to cope with the larger runtime of the mentioned algorithms.

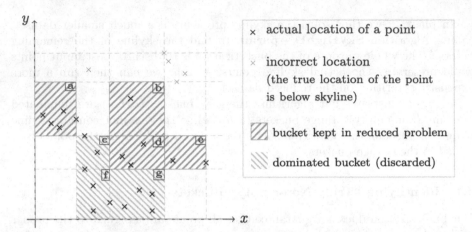

Fig. 3. An illustration of the bucket dominance and its role in **SkyLowD – param**. Here bucket b dominates c and f but not a, d, e or g. Buckets c, f, g are dominated by some non-empty bucket and therefore cannot contain a skyline point. Bucket a does not contain a skyline point, but this cannot be deduced from the bucket assignments, therefore points in bucket a are passed on to the reduced problem. In this figure we may assume to simplify that a bucket contains its upper boundary. But in our algorithm bucket a would actually contain only the 4 leftmost points, and the fifth point would belong to a distinct bucket with a trivial interval on x. (Color figure online)

trivial intervals that match a sample coordinate: $I = [x, x]$, and the non-trivial intervals $I =]a, b[$ $(a < b)$ contained between samples (or above the largest sample, or below the smallest sample). To compare easily those intervals, we adopt the convention that for a non-trivial interval $I =]a, b[$, $\min I = a + \epsilon$ and $\max I = b - \epsilon$ for some infinitesimal $\epsilon > 0$: $\epsilon = (b - a)/3$ would do. We say that a bucket $B = \prod_i I_i$ is *dominated* by a different bucket $B' = \prod_i I'_i$ if in every dimension $\max I_i \leq \min I'_i$. Equivalently: we say that B' dominates B if every point (whether in the dataset or hypothetical) in B' dominates every point in B. The idea is that no skyline point belongs to a bucket dominated by a non-empty bucket. We observe that the relative position of buckets is known by construction, so deciding whether a bucket dominates another one may require time $O(d)$ but does not require any comparison query.

Figure 3 illustrates the relevant and discarded buckets. On that figure, we depicted a few empty buckets above the skyline that are erroneously assumed to contain some points as a result of noise during the assignment. Of course, there are also incorrect assignments of points into empty or non-empty buckets below the skyline, as well as incorrect assignments into the "skyline buckets". These incorrect assignments are not an issue as long as there are not too many of them: dominated buckets will be discarded as such, whether empty or not, and the few irrelevant points maintained into the reduced dataset will be discarded in phase 3, when the skyline of this dataset is computed.

4.3 Algorithm and Bounds for Skyline Computation in Low Dimension

Theorem 5. *Given $\delta \in (0, 1/2)$, $\hat{k} > 0$, and a set X of data items, algorithm* **SkyLowD** $-$ **param**$(\hat{k}, X, \delta)$ *outputs* $\min(|X|, \hat{k})$ *skyline points, with probability at least* $1 - \delta$. *The number of queries is* $O(nd\log(d\hat{k}/\delta))$. *The running time is* $O(nd\log(d\hat{k}/\delta) + nd \cdot \min(\hat{k}, |skyline(X)|))$

Proof. The proof, left for the long version, first shows by Chernoff bounds that the assignment satisfies with high probability some key properties: (1) few points are erroneously assigned to incorrect buckets (2) the skyline points are assigned to the correct bucket, and (3) there are at most $O(n/(d\hat{k}^2))$ points on any hyperplane (i.e., in buckets that are ties on some dimension). The proof then shows that:

- there are at most $O(n/\hat{k})$ points in the reduced problem. This is because those points belong to skyline buckets or buckets that are tied with a skyline bucket on at least one dimension (every other non-empty bucket is dominated), and property (3) of the assignment guarantees that the union of all such buckets has at most $O(n/\hat{k})$ points.
- the buckets above the skyline buckets which are erroneously assumed to contain points can quickly be identified and eliminated since they contain few points.

Algorithm **SkyLowD** $-$ **param**$(\hat{k}, X, \delta)$ needs a good estimate of the skyline cardinality to return the skyline in $O(nd\log(dk/\delta))$: we must have $\hat{k} \geq k$ and $\log(\hat{k}) \in O(\log(k))$. Algorithm **SkyLowDim**(X, δ) (left for the long version) guarantees the complexity by trying a sequence of successive values for $\hat{k}$. The successive values in the sequence grow super exponentially (similarly to [9,18]) to prevent failed attempts from penalizing the complexity.

Theorem 6. *Given $\delta \in (0, 1/2)$ and a set X of data items,* **SkyLowDim**$(\hat{k}, X, \delta)$ *outputs a subset of X which, with probability at least $1 - \delta$, is the skyline. The number of queries is $O(nd\log(dk/\delta))$. The running time is $O(nd\log(dk/\delta) + ndk)$.*

Proof. For iteration j, the algorithm bounds the probability of error by $\delta/2^j$, and the corresponding cost is given by Theorem 5, hence the complexity we claim by summing those terms over all iterations.

Remark 1. In the noiseless setting, we could adopt the same sampling approach to assign points to buckets and reduce the input size. On line 18 we could use any noiseless skyline algorithm such as the $O(ndk)$ algorithm from [13], or our own similar **SkyHighDim** which can clearly run in $O(dnk)$ in the noiseless case. The cost of the bucketing phase remains $O(nd\log(d\hat{k}/\delta))$. The elimination phase becomes rather trivial since all points get assigned to their proper bucket, and therefore there is no need to check buckets for emptiness as in Line 13. By

---

Algorithm SkyLowD – param($\hat{k}, X, \delta$)   (see Theorem 5)
input: $\hat{k}$ integer, X set of points, δ error probability
output: $\min\{\hat{k}, |\text{skyline}(X)|\}$ points of skyline(X)
error probability: δ

1: **if** $\hat{k}^5 \geq n$ or $d^5 \geq n$ or $(\log(1/\delta))^5 \geq n$ **then**
2:     Compute the skyline by sorting every dimension, as in [18]. Return that skyline.
3: $\delta' \leftarrow \delta/(2d\hat{k})^5$ and $s \leftarrow d\hat{k}^2 \log(d^2\hat{k}^2/\delta')$

   {Phase (i): bucketing}
4: **for** each dimension $i \in \{1, 2, \ldots, d\}$ **do**
5:     $S_i \leftarrow$ **NoisySort**(sample of X of size $s, i, \delta'/d$)
6:     Remove duplicates so that, with prob. $1 - \delta'/d$, the values in S_i are all distinct.[a]
7: **for** each point $p \in X$ **do**
8:     Place p in set X_B associated to $B = \prod_{i=1}^{d} I_i$, with $I_i = $ **NoisySearch**($p_i, S_i,$
        $\delta'/(d\hat{k})$).
9: Drop all empty buckets (those that were assigned no point).
10: Sort buckets into a sequence $B_1, \ldots, B_h$ so that each bucket comes before buckets
    it dominates.

    {Phase (ii): eliminating irrelevant buckets}
11: Initialize $X' \leftarrow \emptyset$, $i \leftarrow 1$
12: **while** $i \neq -1$ **do**
13:     $i \leftarrow$ **FirstBucket**($[(B_1, X_{B_1}), \ldots, (B_h, X_{B_h})], \delta'/\hat{k}$)
14:     $X' \leftarrow X' \cup X_{B_i}$
15:     **if** $|X'| > 8n/\hat{k}$ **then**
16:         Raise an error.
17:     Drop from $B_1, \ldots, B_h$ all buckets dominated by B_i, and also buckets B_1 to B_i.

    {Phase (iii): solve reduced problem}
18: Output **SkyHighDim**(X', δ').

[a]Note that X can contain points sharing the same coordinate meaning that the S_i
are not necessarily distinct.

---

setting $\delta = 1/k$ failures are scarce enough so that the higher cost of $O(ndk)$ in case of failure is covered by the cost of an execution corresponding to a satisfying sample. Consequently, the expected query complexity is $O(nd\log(dk))$, and the running time $O(nd\log(dk) + ndk)$.

Better yet: we can replace random sampling with quantile selection to obtain a deterministic algorithm with the same bounds. Algorithms for the *multiple selection* problem are surveyed in [11]. Actually, our algorithm can be viewed as some kind of generalization to higher dimensions of an algorithm from [11] which assigns points to buckets before recursing, the buckets being the quantiles along one coordinate.

5   Skyline Lower Bound

To achieve meaningful lower bounds (that do not reduce to the noiseless setting), we assume here that the input comparisons have a probability of error at least 1/3. Of course, we just need the probability to be bounded away from 0.

Theorem 7. *For any $n \geq k \geq d > 0$, any algorithm that recovers with error probability at most $1/10$ the skyline for any input having exactly k skyline points, requires $\Omega(nd\log(k/\delta))$ queries in expectation on a worst-case input.*

Proof. The proof is left for the long version.

6   Conclusion and Related Work

We introduced 2 algorithms to compute skylines with noisy comparisons. The most involved shows that we can compute skylines in $O(nd\log dk/\delta)$ comparisons. We also show that this bound is optimal when the dimensions is low ($d \leq k^c$ for some constant c), since computing noisy skylines requires $\Omega(dn\log k)$ comparisons. All our algorithms but **SkyLowDim** in $O(nd\log dk/\delta)$ are what we call *trust-preserving*([18]), meaning that when the probability of errors in input comparisons is already at most $\delta < 1/3$, we can discard from the complexity the dependency in δ (replacing δ by some constant).

We leave open the question of the optimal number of comparisons required to compute skylines for arbitrarily large dimensions. Even in the noiseless case, it is not lear whether the skyline could be computed in $O(dn\log k)$ comparisons. Our algorithm is output sensitive (the running time is optimized with respect to the output size) but we did not investigate its instance optimality. However, knowing the input set up to a permutation of the points does not seem to help identifying the skyline points in the noisy comparison model, so we believe that for every k and on any input of skyline cardinality k, even with this knowledge any skyline algorithm would still require $\Omega(dn\log k)$ comparisons. We leave open the question of establishing such a stronger lower bound.

References

1. Afrati, F.N., Koutris, P., Suciu, D., Ullman, J.D.: Parallel skyline queries. In: 15th International Conference on Database Theory, ICDT 2012, Berlin, Germany, 26–29 March 2012, pp. 274–284 (2012). https://doi.org/10.1145/2274576.2274605
2. Afshani, P.: Fast computation of output-sensitive maxima in a word RAM. In: Chekuri, C. (ed.) Proceedings of the Twenty-Fifth Annual ACM-SIAM Symposium on Discrete Algorithms, SODA 2014, Portland, Oregon, USA, 5–7 January 2014, pp. 1414–1423. SIAM (2014). https://doi.org/10.1137/1.9781611973402.104
3. Afshani, P., Agarwal, P.K., Arge, L., Larsen, K.G., Phillips, J.M.: (approximate) uncertain skylines. In: Proceedings of the 14th International Conference on Database Theory, ICDT 2011, pp. 186–196. ACM (2011). https://doi.org/10.1145/1938551.1938576
4. Afshani, P., Barbay, J., Chan, T.M.: Instance-optimal geometric algorithms. J. ACM **64**(1), 3:1–3:38 (2017)
5. Asudeh, A., Zhang, G., Hassan, N., Li, C., Zaruba, G.V.: Crowdsourcing pareto-optimal object finding by pairwise comparisons. In: Proceedings of the 24th ACM International on Conference on Information and Knowledge Management, pp. 753–762. ACM (2015)

6. Barndorff-Nielsen, O., Sobel, M.: On the distribution of the number of admissible points in a vector random sample. Theory Prob. Appl. **11**(2), 249–269 (1966). http://search.proquest.com/docview/915869827?accountid=15867

7. Börzsönyi, S., Kossmann, D., Stocker, K.: The skyline operator. In: Proceedings of the 17th International Conference on Data Engineering, pp. 421–430. IEEE Computer Society (2001). http://dl.acm.org/citation.cfm?id=645484.656550

8. Braverman, M., Mao, J., Weinberg, S.M.: Parallel algorithms for select and partition with noisy comparisons. In: Proceedings of the Forty-eighth Annual ACM Symposium on Theory of Computing, STOC 2016, pp. 851–862 (2016)

9. Chan, T.M.: Optimal output-sensitive convex hull algorithms in two and three dimensions. Discrete Comput. Geom. **16**(4), 361–368 (1996). https://doi.org/10.1007/BF02712873

10. Chan, T.M., Larsen, K.G., Patrascu, M.: Orthogonal range searching on the ram, revisited. In: Hurtado, F., van Kreveld, M.J. (eds.) Proceedings of the 27th ACM Symposium on Computational Geometry, Paris, France, 13–15 June 2011, pp. 1–10. ACM (2011). https://doi.org/10.1145/1998196.1998198

11. Chan, T.M., Lee, P.: On constant factors in comparison-based geometric algorithms and data structures. Discrete Comput. Geom. **53**(3), 489–513 (2015). https://doi.org/10.1007/s00454-015-9677-y

12. Chomicki, J., Ciaccia, P., Meneghetti, N.: Skyline queries, front and back. SIGMOD Rec. **42**(3), 6–18 (2013). https://doi.org/10.1145/2536669.2536671

13. Clarkson, K.L.: More output-sensitive geometric algorithms (extended abstract). In: 35th Annual Symposium on Foundations of Computer Science, Santa Fe, New Mexico, USA, 20–22 November 1994, pp. 695–702 (1994). https://doi.org/10.1109/SFCS.1994.365723

14. Davidson, S.B., Khanna, S., Milo, T., Roy, S.: Top-k and clustering with noisy comparisons. ACM Trans. Database Syst. **39**(4), 35:1–35:39 (2014). https://doi.org/10.1145/2684066

15. Feige, U., Raghavan, P., Peleg, D., Upfal, E.: Computing with noisy information. SIAM J. Comput. **23**(5), 1001–1018 (1994). https://doi.org/10.1137/S0097539791195877

16. Godfrey, P., Shipley, R., Gryz, J.: Algorithms and analyses for maximal vector computation. VLDB J. **16**(1), 5–28 (2007). https://doi.org/10.1007/s00778-006-0029-7

17. Goyal, N., Saks, M.: Rounds vs. queries tradeoff in noisy computation. Theory of Computing **6**(1), 113–134 (2010)

18. Groz, B., Milo, T.: Skyline queries with noisy comparisons. In: Proceedings of the 34th ACM SIGMOD-SIGACT-SIGAI Symposium on Principles of Database Systems, PODS 15, pp. 185–198. ACM (2015). https://doi.org/10.1145/2745754.2745775

19. Kalyvas, C., Tzouramanis, T.: A survey of skyline query processing (2017). CoRR abs/1704.01788, http://arxiv.org/abs/1704.01788

20. Kirkpatrick, D.G., Seidel, R.: Output-size sensitive algorithms for finding maximal vectors. In: Proceedings of the First Annual Symposium on Computational Geometry, SCG 1985, pp. 89–96. ACM (1985). https://doi.org/10.1145/323233.323246

21. Kossmann, D., Ramsak, F., Rost, S.: Shooting stars in the sky: an online algorithm for skyline queries. In: Proceedings of the 28th International Conference on Very Large Data Bases, VLDB 2002, VLDB Endowment, pp. 275–286 (2002). http://dl.acm.org/citation.cfm?id=1287369.1287394

22. Kung, H.T., Luccio, F., Preparata, F.P.: On finding the maxima of a set of vectors. J. ACM **22**(4), 469–476 (1975). https://doi.org/10.1145/321906.321910
23. Li, G., Wang, J., Zheng, Y., Franklin, M.J.: Crowdsourced data management: a survey. In: 33rd IEEE International Conference on Data Engineering, ICDE 2017, San Diego, CA, USA, 19–22 April 2017, pp. 39–40. IEEE Computer Society (2017). https://doi.org/10.1109/ICDE.2017.26
24. Mallmann-Trenn, F., Mathieu, C., Verdugo, V.: Skyline computation with noisy comparisons (2017). CoRR abs/1710.02058, http://arxiv.org/abs/1710.02058
25. Marcus, A., Wu, E., Karger, D., Madden, S., Miller, R.: Human-powered sorts and joins. Proc. VLDB Endow. **5**(1), 13–24 (2011). https://doi.org/10.14778/2047485.2047487
26. Newman, I.: Computing in fault tolerant broadcast networks and noisy decision trees. Rand. Struct. Algorithms **34**(4), 478–501 (2009)
27. Pei, J., Jiang, B., Lin, X., Yuan, Y.: Probabilistic skylines on uncertain data. In: Proceedings of the 33rd International Conference on Very Large Data Bases, VLDB 2007, VLDB Endowment, pp. 15–26 (2007). http://dl.acm.org/citation.cfm?id=1325851.1325858

Strongly Stable and Maximum Weakly Stable Noncrossing Matchings

Koki Hamada[1,2], Shuichi Miyazaki[3](✉), and Kazuya Okamoto[4]

[1] NTT Corporation, 3-9-11, Midori-cho, Musashino-shi, Tokyo 180-8585, Japan
koki.hamada.rb@hco.ntt.co.jp
[2] Graduate School of Informatics, Kyoto University, Yoshida-Honmachi,
Sakyo-ku, Kyoto 606-8501, Japan
[3] Academic Center for Computing and Media Studies, Kyoto University,
Yoshida-Honmachi, Sakyo-ku, Kyoto 606-8501, Japan
shuichi@media.kyoto-u.ac.jp
[4] Division of Medical Information Technology and Administration Planning,
Kyoto University Hospital, 54 Kawaharacho, Shogoin, Sakyo-ku,
Kyoto 606-8507, Japan
kazuya@kuhp.kyoto-u.ac.jp

Abstract. In IWOCA 2019, Ruangwises and Itoh introduced *stable noncrossing matchings*, where participants of each side are aligned on each of two parallel lines, and no two matching edges are allowed to cross each other. They defined two stability notions, *strongly stable noncrossing matching* (*SSNM*) and *weakly stable noncrossing matching* (*WSNM*), depending on the strength of blocking pairs. They proved that a WSNM always exists and presented an $O(n^2)$-time algorithm to find one for an instance with n men and n women. They also posed open questions of the complexities of determining existence of an SSNM and finding a largest WSNM. In this paper, we show that both problems are solvable in polynomial time. Our algorithms are applicable to extensions where preference lists may include ties, except for one case which we show to be NP-complete.

Keywords: Stable marriage · Noncrossing matching ·
Polynomial-time algorithms · NP-completeness

1 Introduction

In the classical *stable marriage problem* [4], there are two sets of participants, traditionally illustrated as men and women, where each person has a *preference list* that orders a subset of the members of the opposite gender. A *matching* is a set of (man, woman)-pairs where no person appears more than once. A *blocking pair* for a matching M is (informally) a pair of a man and a woman who are not matched together in M but both of them become better off if they

Supported by JSPS KAKENHI Grant Numbers JP16K00017 and JP19K12820.

are matched. A matching that admits no blocking pair is a *stable matching*. The stable marriage problem is one of the recently best-studied topics, with a lot of applications to matching and assignment systems, such as high-school match [1,2] and medical resident assignment [13]. See some textbooks [6,11,12,15] for more information.

Recently, Ruangwises and Itoh [16] incorporated the notion of noncrossing matchings to the stable marriage problem. In their model, there are two parallel lines where n men are aligned on one line and n women are aligned on the other line. A matching is *noncrossing* if no two edges of it cross each other. A *stable noncrossing matching* is a matching which is simultaneously stable and noncrossing. They defined two notions of stability: In a *strongly stable noncrossing matching (SSNM)*, the definition of a blocking pair is the same as that of the standard stable marriage problem. Thus the set of SSNMs is exactly the intersection of the set of stable matchings and that of noncrossing matchings. In a *weakly stable noncrossing matching (WSNM)*, a blocking pair has an additional condition that it must be noncrossing with matching edges. Ruangwises and Itoh [16] proved that a WSNM exists for any instance, and presented an $O(n^2)$-time algorithm to find one. They also demonstrated that an SSNM does not always exist, and that there can be WSNMs of different sizes. Concerning these observations, they posed open questions on the complexities of the problems of determining the existence of an SSNM and finding a WSNM of maximum cardinality.

Our Contributions. In this paper, we show that both problems are solvable in polynomial time. The former is solved by exploiting the well-known Rural Hospitals theorem (Proposition 1). For the latter, we design an algorithm based on dynamic programming (Theorem 2). We then consider extended problems where preference lists may include ties. We show that our algorithms are applicable to them without any modification (Corollaries 1, 2, and 3), except for one which we show to be NP-complete (Theorem 1).

Table 1 summarizes previous and our results, where our results are described in bold. In the table, SSNM and WSNM stand for the problems of determining the existence of SSNM and WSNM, respectively. MAX-WSNM is the optimization problem of finding a largest WSNM. SMI and SMTI stand for the stable marriage problems without and with ties, respectively. When ties are allowed in preference lists, there are three stability notions, super, strong, and weak stabilities [7]. Formal definitions of these notions are introduced in Sect. 2.

2   Preliminaries

In this section, we give necessary definitions and notations, some of which are taken from Ruangwises and Itoh [16]. An instance consists of n men $m_1, m_2, \ldots, m_n$ and n women $w_1, w_2, \ldots, w_n$. We assume that the men are lying on a vertical line in an increasing order of indices from top to bottom, and similarly the women are lying in the same manner on another vertical line parallel

Table 1. Previous and our results.

		SSNM	WSNM	MAX-WSNM
SMI		$O(n^2)$ [Proposition 1]	$O(n^2)$ [16]	$O(n^4)$ [Theorem 2]
SMTI	Super	$O(n^2)$ [Corollary 1]		$O(n^4)$ [Corollary 3]
	Strong	$O(n^3)$ [Corollary 2]		$O(n^4)$ [Corollary 3]
	Weak	**NPC** [Theorem 1]		$O(n^4)$ [Corollary 3]

to the first one. Each person has a preference list over a subset of the members of the opposite gender. For now, assume that preference lists are strict, i.e., do not contain ties. We call such an instance an *SMI-instance*. If a person q appears in a person p's preference list, we say that q is *acceptable* to p. If p and q are acceptable to each other, we say that (p, q) is an *acceptable pair*. We assume without loss of generality that acceptability is mutual, i.e., p is acceptable to q if and only if q is acceptable to p. If p prefers q_1 to q_2, then we write $q_1 \succ_p q_2$.

A *matching* is a set of acceptable pairs of a man and a woman in which each person appears at most once. If $(m, w) \in M$, we write $M(m) = w$ and $M(w) = m$. If a person p is not included in a matching M, we say that p is *single* in M and write $M(p) = \emptyset$. Every person prefers to be matched with an acceptable person rather than to be single, i.e., $q \succ_p \emptyset$ holds for any p and any q acceptable to p.

A pair in a matching can be seen as an edge on the plane, so we may use "pair" and "edge" interchangeably. Two edges (m_i, w_j) and (m_x, w_y) are said to *cross* each other if they share an interior point, or formally, if $(x - i)(y - j) < 0$ holds. A matching is *noncrossing* if it contains no pair of crossing edges.

For a matching M, an acceptable pair $(m, w) \notin M$ is called a *blocking pair* for M if both $w \succ_m M(m)$ and $m \succ_w M(w)$ hold. A *noncrossing blocking pair* for M is a blocking pair for M that does not cross with any edge of M. A matching M is a *weakly stable noncrossing matching* (*WSNM*) if M is noncrossing and does not admit any noncrossing blocking pair. A matching M is a *strongly stable noncrossing matching* (*SSNM*) if M is noncrossing and does not admit any blocking pair.

We then extend the above definitions to the case where preference lists may contain ties. A *tie* of a person p's preference list is a set of one or more persons who are equally preferred by p, and p's preference list is a strict order of ties. We call such an instance an *SMTI-instance*. In a person p's preference list, suppose that a person q_1 is in tie T_1, q_2 is in tie T_2, and p prefers T_1 to T_2. Then we say that p *strictly prefers* q_1 to q_2 and write $q_1 \succ_p q_2$. If q_1 and q_2 are in the same tie (including the case that q_1 and q_2 are the same person), we write $q_1 =_p q_2$. If $q_1 \succ_p q_2$ or $q_1 =_p q_2$ holds, we write $q_1 \succeq_p q_2$ and say that p *weakly prefers* q_1 to q_2.

When ties are present, there are three possible definitions of blocking pairs, and accordingly, there are three stability notions, *super-stability*, *strong stability*, and *weak stability* [7]:

– In the super-stability, a blocking pair for a matching M is an acceptable pair $(m, w) \notin M$ such that $w \succeq_m M(m)$ and $m \succeq_w M(w)$.
– In the strong stability, a blocking pair for a matching M is an acceptable pair $(p, q) \notin M$ such that $q \succeq_p M(p)$ and $p \succ_q M(q)$. Note that the person q, who strictly prefers the counterpart p of the blocking pair, may be either a man or a woman.
– In the weak stability, a blocking pair for a matching M is an acceptable pair $(m, w) \notin M$ such that $w \succ_m M(m)$ and $m \succ_w M(w)$.

With these definitions of blocking pairs, the terms "noncrossing blocking pair", "WSNM", and "SSNM" for each stability notion can be defined analogously. In the SMTI case, we extend the names of stable noncrossing matchings using the type of stability as a prefix. For example, a WSNM in super-stability is denoted *super-WSNM*.

Note that, in this paper, the terms "weak" and "strong" are used in two different meanings. This might be confusing but we decided not to change these terms, respecting previous literature.

3 Strongly Stable Noncrossing Matchings

3.1 SMI

In SMI, an easy observation shows that existence of an SSNM can be determined in $O(n^2)$ time:

Proposition 1. *There exists an $O(n^2)$-time algorithm to find an SSNM or to report that none exists, given an SMI-instance.*

Proof. Note that an SSNM is a stable matching in the original sense. In SMI, there always exists at least one stable matching [6], and due to the Rural Hospitals theorem [5,13,14], the set of matched agents is the same in any stable matching. These agents can be determined in $O(n^2)$ time by using the Gale-Shapley algorithm [4]. There is only one way of matching them in a noncrossing manner. Hence the matching constructed in this way is stable if and only if a given instance admits an SSNM. This condition can be checked in $O(n^2)$ time. □

3.2 SMTI

In the presence of ties, super-stable and strongly stable matchings do not always exist. However, there is an $O(n^2)$-time ($O(n^3)$-time, respectively) algorithm that finds a super-stable (strongly stable, respectively) matching or reports that none exists [7,10]. Also, the Rural Hospitals theorem takes over to the super-stability [8] and strong stability [9]. Therefore, the same algorithm as in Sect. 3.1 applies for these cases, implying the following corollaries:

Corollary 1. *There exists an $O(n^2)$-time algorithm to find a super-SSNM or to report that none exists, given an SMTI-instance.*

Corollary 2. *There exists an $O(n^3)$-time algorithm to find a strong-SSNM or to report that none exists, given an SMTI-instance.*

In contrast, the problem becomes NP-complete for weak stability:

Theorem 1. *The problem of determining if a weak-SSNM exists, given an SMTI-instance, is NP-complete, even if each tie is of length at most two.*

Proof. Membership in NP is obvious. We show NP-hardness by a reduction from 3SAT, which is well-known to be NP-complete [3]. Its instance consists of a set of variables and a set of clauses. Each variable takes either true (1) or false (0). A *literal* is a variable or its negation. A *clause* is a disjunction of at most three literals. A clause is *satisfied* if at least one of its literals takes the value 1, and is *unsatisfied* otherwise. A 0/1 assignment to variables that satisfies all the clauses is called a *satisfying assignment*. An instance f of 3SAT is *satisfiable* if it has at least one satisfying assignment; otherwise f is *unsatisfiable*. 3SAT asks if there exists a satisfying assignment. We may assume without loss of generality that each clause contains *exactly* three literals, as if not, we may simply duplicate a literal without affecting the satisfiability of the instance.

Now we show the reduction. Let f be an instance of 3SAT having n variables $x_i (1 \leq i \leq n)$ and m clauses $C_j (1 \leq j \leq m)$. For each variable x_i, we create two men $p_{i,1}, p_{i,2}$ and one woman q_i. These three persons are said to constitute an x_i-*gadget* (or generally a *variable gadget*). For each clause C_j, we create two men $y_{j,1}, y_{j,2}$ and three women $z_{j,1}, z_{j,2}, z_{j,3}$, who are said to constitute a C_j-*gadget* (or generally a *clause gadget*). Additionally, we create a man s and a woman t, who constitute a gadget called the *separator*. Thus, there are $2n + 2m + 1$ men and $n + 3m + 1$ women in the created SMTI instance $I(f)$. We finally add dummy persons who have empty preference lists to make the numbers of men and women equal. They do not play any role in the following arguments, so we omit them.

Suppose that x_i appears a_i times positively in f, and for each k ($1 \leq k \leq a_i$), x_i's kth positive occurrence is in the $d_{i,k}$th clause $C_{d_{i,k}}$ as the $e_{i,k}$th literal ($1 \leq e_{i,k} \leq 3$). Similarly, suppose x_i appears b_i times negatively, and for each k ($1 \leq k \leq b_i$), x_i's kth negative occurrence is in the $g_{i,k}$th clause $C_{g_{i,k}}$ as the $h_{i,k}$th literal ($1 \leq h_{i,k} \leq 3$). Then preference lists of three persons in the x_i-gadget are constructed as shown in Fig. 1. Here, each preference list is denoted as a sequence from left to right according to preference, i.e., the leftmost person(s) is the most preferred and the rightmost person(s) is the least preferred. Tied persons (i.e., persons with the equal preference) are included in parentheses.

Preference lists of five persons in the C_j-gadget are given in Fig. 2. For $k = 1, 2, 3$, suppose that the kth literal of C_j is x_{j_k}, and let $\ell_{j,k} = 1$ (respectively, $\ell_{j,k} = 2$) if x_{j_k} appears negatively (respectively, positively) in C_j.

Finally, each of the man and the woman in the separator includes only the other in the list (Fig. 3). They are guaranteed to be matched together in any stable matching.

$$p_{i,1}: \quad q_i \quad z_{g_{i,1},h_{i,1}} \quad z_{g_{i,2},h_{i,2}} \quad \cdots \quad z_{g_{i,b_i},h_{i,b_i}} \qquad\qquad q_i: \quad (p_{i,1} \ p_{i,2})$$

$$p_{i,2}: \quad q_i \quad z_{d_{i,1},e_{i,1}} \quad z_{d_{i,2},e_{i,2}} \quad \cdots \quad z_{d_{i,a_i},e_{i,a_i}}$$

Fig. 1. Preference lists of persons in x_i-gadget.

$$y_{j,1}: \ (z_{j,1} \ z_{j,2}) \qquad\qquad z_{j,1}: \ y_{j,1} \ p_{j_1,\ell_{j,1}}$$

$$y_{j,2}: \ (z_{j,2} \ z_{j,3}) \qquad\qquad z_{j,2}: \ (y_{j,1} \ y_{j,2}) \ p_{j_2,\ell_{j,2}}$$

$$z_{j,3}: \ y_{j,2} \ p_{j_3,\ell_{j,3}}$$

Fig. 2. Preference lists of persons in C_j-gadget.

Alignment of agents is depicted in Fig. 4. Variable gadgets are placed top, then followed by the separator, clause gadgets come bottom. Within each gadget, people are aligned according to indices. The separator plays a role of prohibiting a person of a variable gadget and a person of a clause gadget to match together; if they are matched, then the corresponding edge crosses with the separator.

Now the reduction is completed. It is not hard to see that the reduction can be performed in polynomial time.

We then show the correctness. First, suppose that f is satisfiable and let A be a satisfying assignment. We construct a weak-SSNM M of $I(f)$ from A. For an x_i-gadget, define two matchings $M_{i,0} = \{(p_{i,1}, q_i)\}$ and $M_{i,1} = \{(p_{i,2}, q_i)\}$. If $x_i = 0$ under A, then add $M_{i,0}$ to M; otherwise, add $M_{i,1}$ to M. For a C_j-gadget, define three matchings $N_{j,1} = \{(y_{j,1}, z_{j,2}), (y_{j,2}, z_{j,3})\}$, $N_{j,2} = \{(y_{j,1}, z_{j,1}), (y_{j,2}, z_{j,3})\}$, and $N_{j,3} = \{(y_{j,1}, z_{j,1}), (y_{j,2}, z_{j,2})\}$. If C_j is satisfied by the kth literal ($k = 1, 2, 3$), then add $N_{j,k}$ to M. (If C_j is satisfied by more than one literal, then choose one arbitrarily.) Finally add the pair (s, t) to M.

It is not hard to see that M is noncrossing. We show that it is weakly stable. Note that all the women in the variable gadgets, all the men in the clause gadgets, and s and t in the separator are matched with the first choice person. Therefore, if there exists a blocking pair, it must be the form of $(p_{i,\ell}, z_{j,k})$ for some i, ℓ, j, and k. Furthermore, any person matched in M is matched with the first choice, so both $p_{i,\ell}$ and $z_{j,k}$ are single in M. Suppose that $\ell = 1$. The reason for $(p_{i,1}, z_{j,k})$ being an acceptable pair is that C_j's kth literal is $\neg x_i$, the negative occurrence of x_i. Since $p_{i,1}$ is single, $M_{i,1} \subset M$ and hence $x_i = 1$ under A. Since $z_{j,k}$ is single, $N_{j,k} \subset M$ and hence C_j is satisfied by its kth literal $\neg x_i$, but this is a contradiction. The other case $\ell = 2$ can be argued in the same manner.

Conversely, suppose that $I(f)$ admits a weak-SSNM M. We construct a satisfying assignment A of f. Before giving construction, we observe structural properties of M in two lemmas:

Lemma 1. *For each i ($1 \leq i \leq n$), either $M_{i,0} \subset M$ or $M_{i,1} \subset M$.*

Proof. Note that preference lists of the three persons of the x_i-gadget include persons of the same x_i-gadget or some persons from clause gadgets. Hence, due to the separator, persons of the x_i-gadget can only be matched within this gadget

s: t   t: s

Fig. 3. Preference lists of the man and the woman in the separator.

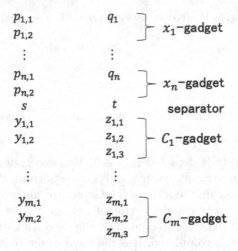

Fig. 4. Alignment of agents.

to avoid edge crossings. Since a stable matching is a maximal matching, either $M_{i,0}$ or $M_{i,1}$ must be a part of M. ☐

Lemma 2. *For each j $(1 \leq j \leq m)$, either $N_{j,1} \subset M$, $N_{j,2} \subset M$, or $N_{j,3} \subset M$.*

Proof. The proof is similar to that of Lemma 1. Note that preference lists of the five persons of the C_j-gadget include persons of the same C_j-gadget or some persons from variable gadgets. The maximal matchings within the C_j-gadget are $N_{j,1}$, $N_{j,2}$, and $N_{j,3}$, so one of them must be in M. ☐

For each i, we know that either $M_{i,0} \subset M$ or $M_{i,1} \subset M$ by Lemma 1. If $M_{i,0} \subset M$ then we set $x_i = 0$ in A, and if $M_{i,1} \subset M$ then we set $x_i = 1$ in A. We show that A satisfies f. Suppose not, and let C_j be an unsatisfied clause. Fix an integer $k \in \{1, 2, 3\}$. Suppose that the kth literal of C_j is a positive occurrence of x_i. Then, by construction of preference lists, $(p_{i,2}, z_{j,k})$ is an acceptable pair. Since C_j is unsatisfied, $x_i = 0$ under A. Then, by construction of A, $M_{i,0} \subset M$ and hence $p_{i,2}$ is single in M. If $N_{j,k} \subset M$, then $z_{j,k}$ is single in M, which contradicts stability of M. For the other case, suppose that the kth literal of C_j is a negative occurrence of x_i. Then, by construction of preference lists, $(p_{i,1}, z_{j,k})$ is an acceptable pair. Since C_j is unsatisfied, $x_i = 1$ under A. Then, by construction of A, $M_{i,1} \subset M$ and hence $p_{i,1}$ is single in M. If $N_{j,k} \subset M$, then $z_{j,k}$ is single in M, which contradicts stability of M.

The above argument holds for any $k \in \{1, 2, 3\}$, so none of $N_{j,1}$, $N_{j,2}$, and $N_{j,3}$ can be a part of M. But this contradicts Lemma 2. Hence A satisfies f. ☐

4   Maximum Cardinality Weakly Stable Noncrossing Matchings

In this section, we present an algorithm to find a maximum cardinality WSNM. For an instance I, let $opt(I)$ denote the size of the maximum cardinality WSNM.

4.1   SMI

Let I' be a given instance with men $m_1, \ldots, m_n$ and women $w_1, \ldots, w_n$. To simplify the description of the algorithm, we translate I' to an instance I by adding a man m_0 and a woman w_0, each of whom includes only the other in the preference list, and similarly a man m_{n+1} and a woman w_{n+1}, each of whom includes only the other in the preference list. It is easy to see that, for a WSNM M' of I', $M = M' \cup \{(m_0, w_0), (m_{n+1}, w_{n+1})\}$ is a WSNM of I. Conversely, any WSNM M of I includes the pairs (m_0, w_0) and (m_{n+1}, w_{n+1}), and $M' = M \setminus \{(m_0, w_0), (m_{n+1}, w_{n+1})\}$ is a WSNM of I'. Thus we have that $opt(I) = opt(I') + 2$. Hence, without loss of generality, we assume that a given instance I has $n + 2$ men and $n + 2$ women, with m_0, w_0, m_{n+1}, and w_{n+1} having the above mentioned preference lists.

Let $M = \{(m_{i_1}, w_{j_1}), (m_{i_2}, w_{j_2}), \ldots, (m_{i_k}, w_{j_k})\}$ be a noncrossing matching of I such that $i_1 < i_2 \cdots < i_k$ and $j_1 < j_2 \cdots < j_k$. We call (m_{i_k}, w_{j_k}) the *maximum pair* of M. Suppose that (m_x, w_y) is the maximum pair of a noncrossing matching M. We call M a *semi-WSNM* if each of its noncrossing blocking pair (m_i, w_j) (if any) satisfies $x \le i \le n + 1$ and $y \le j \le n + 1$. Intuitively, a semi-WSNM is a WSNM up to its maximum pair. Note that any semi-WSNM must contain (m_0, w_0), as otherwise it is a noncrossing blocking pair. For $0 \le i \le n+1$ and $0 \le j \le n + 1$, we define $X(i, j)$ as the maximum size of a semi-WSNM of I whose maximum pair is (m_i, w_j); if I does not admit a semi-WSNM with the maximum pair (m_i, w_j), $X(i, j)$ is defined to be $-\infty$.

Lemma 3. $opt(I) = X(n + 1, n + 1)$.

Proof. Note that any WSNM of I includes (m_{n+1}, w_{n+1}), as otherwise it is a noncrossing blocking pair. Hence it is a semi-WSNM with the maximum pair (m_{n+1}, w_{n+1}). Conversely, any semi-WSNM with the maximum pair (m_{n+1}, w_{n+1}) does not include a noncrossing blocking pair and hence is also a WSNM. Therefore, the set of WSNMs is exactly the set of semi-WSNMs with the maximum pair (m_{n+1}, w_{n+1}). This completes the proof.   □

To compute $X(n + 1, n + 1)$, we shortly define quantity $Y(i, j)$ ($0 \le i \le n + 1, 0 \le j \le n + 1$) using recursive formulas, and show that $Y(i, j) = X(i, j)$ for all i and j. We then show that these recursive formulas allow us to compute $Y(i, j)$ in polynomial time using dynamic programming.

We say that two noncrossing edges (m_i, w_j) and (m_x, w_y) ($i < x, j < y$) are *conflicting* if they contain a noncrossing blocking pair between them; precisely, if the matching $\{(m_i, w_j), (m_x, w_y)\}$ contains a blocking pair (m_s, w_t)

such that $i \leq s \leq x$ and $j \leq t \leq y$. Otherwise, (m_i, w_j) and (m_x, w_y) are *non-conflicting*. Intuitively, two conflicting edges cannot be consecutive elements of a semi-WSNM.

Now we give the definition of $Y(i,j)$. For convenience, we assume that $-\infty + 1 = -\infty$.

$$Y(0,0) = 1 \tag{1}$$

$$Y(0,j) = -\infty \quad (1 \leq j \leq n+1) \tag{2}$$

$$Y(i,0) = -\infty \quad (1 \leq i \leq n+1) \tag{3}$$

$$Y(i,j) = \begin{cases} 1 + \max\limits_{\substack{0 \leq i' \leq i-1 \\ 0 \leq j' \leq j-1}} \{Y(i',j') \mid (m_i, w_j) \text{ and } (m_{i'}, w_{j'}) \text{ are nonconflicting}\} \\ \qquad \text{(if } (m_i, w_j) \text{ is an acceptable pair)} \\ -\infty \quad \text{(otherwise)} \end{cases}$$
$$(1 \leq i \leq n+1, 1 \leq j \leq n+1) \tag{4}$$

Lemma 4. $Y(i,j) = X(i,j)$ *for* $0 \leq i \leq n+1$ *and* $0 \leq j \leq n+1$.

Proof. We prove the claim by induction. We first show that $Y(0,0) = X(0,0)$. The matching $\{(m_0, w_0)\}$ is the unique semi-WSNM with the maximum pair (m_0, w_0), so $X(0,0) = 1$ by definition. Also, $Y(0,0) = 1$ by Eq. (1). Hence we are done. We then show that $Y(0,j) = X(0,j)$ for $1 \leq j \leq n+1$. Since (m_0, w_j) is an unacceptable pair, there is no semi-WSNM with the maximum pair (m_0, w_j), so $X(0,j) = -\infty$ by definition. Also, $Y(0,j) = -\infty$ by Eq. (2). We can show that $Y(i,0) = X(i,0)$ for $1 \leq i \leq n+1$ by a similar argument.

Next we show that $Y(i,j) = X(i,j)$ holds for $1 \leq i \leq n+1$ and $1 \leq j \leq n+1$. As an induction hypothesis, we assume that $Y(a,b) = X(a,b)$ holds for $0 \leq a \leq i-1$ and $0 \leq b \leq j-1$. First, observe that if $X(i,j) \neq -\infty$, then $X(i,j) \geq 2$. This is because two pairs (m_0, w_0) and (m_i, w_j) must present in any semi-WSNM having the maximum pair (m_i, w_j).

We first consider the case that $X(i,j) \geq 2$. Let $X(i,j) = k$. Then, there is a semi-WSNM M with the maximum pair (m_i, w_j) such that $|M| = k$. Let $M' = M \setminus \{(m_i, w_j)\}$ and (m_x, w_y) be the maximum pair of M'. It is not hard to see that M' is a semi-WSNM with the maximum pair (m_x, w_y) and that $|M'| = k-1$. Therefore, $X(x,y) \geq k-1$ by the definition of X, and $Y(x,y) = X(x,y) \geq k-1$ by the induction hypothesis. Since M is a semi-WSNM, (m_i, w_j) and (m_x, w_y) are nonconflicting, so (x,y) satisfies the condition for (i',j') in Eq. (4). Hence $Y(i,j) \geq 1 + Y(x,y) \geq k$. Suppose that $Y(i,j) \geq k+1$. By the definition of Y, this means that there is (i',j') such that $0 \leq i' \leq i-1$, $0 \leq j' \leq j-1$, $(m_{i'}, w_{j'})$ and (m_i, w_j) are nonconflicting, and $Y(i',j') \geq k$. By the induction hypothesis, $X(i',j') = Y(i',j') \geq k$. Then there is a semi-WSNM M' with the maximum pair $(m_{i'}, w_{j'})$ such that $|M'| \geq k$. Since M' is a semi-WSNM, and $(m_{i'}, w_{j'})$ and (m_i, w_j) are nonconflicting, $M = M' \cup \{(m_i, w_j)\}$ is a semi-WSNM with the maximum pair (m_i, w_j) such that $|M| = |M'| + 1 \geq k + 1$. This contradicts the assumption that $X(i, j) = k$. Hence $Y(i, j) \leq k$ and therefore $Y(i, j) = k$ as desired.

Finally, consider the case that $X(i, j) = -\infty$. If (m_i, w_j) is unacceptable, then the latter case of Eq. (4) is applied and $Y(i, j) = -\infty$. So assume that (m_i, w_j) is acceptable. Then the former case of Eq. (4) is applied. It suffices to show that for any (i', j') such that $0 \leq i' \leq i - 1$, $0 \leq j' \leq j - 1$, and $(m_{i'}, w_{j'})$ and (m_i, w_j) are nonconflicting, $Y(i', j') = -\infty$ holds. Assume on the contrary that there is such (i', j') with $Y(i', j') = k$. Then $X(i', j') = k$ by the induction hypothesis, and there is a semi-WSNM M' such that $|M'| = k$, $(m_{i'}, w_{j'})$ is the maximum pair of M', and $(m_{i'}, w_{j'})$ and (m_i, w_j) are nonconflicting. Then $M = M' \cup \{(m_i, w_j)\}$ is a semi-WSNM such that (m_i, w_j) is the maximum pair and $|M| = |M'| + 1 = k + 1$, implying that $X(i, j) = k + 1$. This contradicts the assumption that $X(i, j) = -\infty$ and the proof is completed. $\square$

Now we analyze time-complexity of the algorithm. We assume that, given persons p, q_1, and q_2, whether or not p prefers q_1 to q_2 can be determined in constant time using ranking arrays described in Sec. 1.2.3 of [6]. Computing each $Y(0, 0)$, $Y(0, j)$, and $Y(i, 0)$ can be done in constant time. For computing one $Y(i, j)$ according to Eq. (4), there are $O(n^2)$ candidates for (i', j'). For each (i', j'), checking if $(m_{i'}, w_{j'})$ and (m_i, w_j) are conflicting can be done in constant time with $O(n^4)$-time preprocessing described in the subsequent paragraphs. Therefore one $Y(i, j)$ can be computed in time $O(n^2)$. Since there are $O(n^2)$ $Y(i, j)$s, the time-complexity for computing all $Y(i, j)$s is $O(n^4)$. Adding the $O(n^4)$-time for preprocessing mentioned above, the total time-complexity of the algorithm is $O(n^4)$.

In the preprocessing, we construct three tables S, A, and B.

- S is a $\Theta(n^4)$-sized four-dimensional table that takes logical values 0 and 1. For $0 \leq i' \leq i \leq n+1$ and $0 \leq j' \leq j \leq n+1$, $S(i', i, j', j) = 1$ if and only if there exists at least one acceptable pair (m, w) such that $m \in \{m_{i'}, m_{i'+1}, \ldots, m_i\}$ and $w \in \{w_{j'}, w_{j'+1}, \ldots, w_j\}$. Since $S(i, i, j, j) = 1$ if and only if (m_i, w_j) is an acceptable pair, it can be computed in constant time. In general, $S(i', i, j', j)$ can be computed in constant time as follows.

$$S(i', i, j', j) = \begin{cases} 1 & \text{(if } (m_i, w_j) \text{ is an acceptable pair)} \\ S(i', i-1, j', j) \vee S(i', i, j', j-1) & \text{(otherwise)} \end{cases}$$

  Hence S can be constructed in $O(n^4)$ time by a simple dynamic programming.
- A is a $\Theta(n^3)$-sized table where, for $0 \leq i \leq n+1$ and $0 \leq j' \leq j \leq n+1$, $A(i, j', j)$ stores the woman whom m_i most prefers among $\{w_{j'}, \ldots, w_j\}$. Since $A(i, j, j) = w_j$ and $A(i, j', j)$ is the better of $A(i, j', j-1)$ and w_j in m_i's list, each element can be computed in constant time and hence A can be constructed in $O(n^3)$ time.
- B plays a symmetric role to A; for $0 \leq i' \leq i \leq n+1$ and $0 \leq j \leq n+1$, $B(i', i, j)$ stores the man whom w_j most prefers among $\{m_{i'}, \ldots, m_i\}$. B can also be constructed in $O(n^3)$ time.

It is easy to see that $(m_{i'}, w_{j'})$ and (m_i, w_j) are conflicting if and only if one of the following conditions hold. Condition 1 can be clearly checked in constant time. Thanks to the preprocessing, Conditions 2–4 can also be checked in constant time.

1. $(m_{i'}, w_j)$ or $(m_i, w_{j'})$ is a blocking pair for the matching $\{(m_{i'}, w_{j'}), (m_i, w_j)\}$.
2. $S(i'+1, i-1, j'+1, j-1) = 1$. (If this holds, there is a blocking pair (m, w) such that $m \in \{m_{i'+1}, m_{i'+2}, \ldots, m_{i-1}\}$ and $w \in \{w_{j'+1}, w_{j'+2}, \ldots, w_{j-1}\}$).
3. m_i prefers $A(i, j'+1, j-1)$ to w_j or $m_{i'}$ prefers $A(i', j'+1, j-1)$ to $w_{j'}$. (If this holds, there exists a blocking pair (m, w) such that $m \in \{m_{i'}, m_i\}$ and $w \in \{w_{j'+1}, \ldots, w_{j-1}\}$).
4. w_j prefers $B(i'+1, i-1, j)$ to m_i or $w_{j'}$ prefers $B(i'+1, i-1, j')$ to $m_{i'}$. (If this holds, there exists a blocking pair (m, w) such that $m \in \{m_{i'+1}, \ldots, m_{i-1}\}$ and $w \in \{w_{j'}, w_j\}$).

Theorem 2. *There exists an $O(n^4)$-time algorithm to find a maximum cardinality WSNM, given an SMI-instance.*

4.2 SMTI

The algorithm in Sect. 4.1 can be applied to SMTI straightforwardly. The only difference is the definition of two edges (m_i, w_j) and (m_x, w_y) being nonconflicting, which we need to extend depending on one of the three stability notions. According to the introduction of ties, we also need to extend the definition of the tables A and B. $A(i, j', j)$ holds *one of* the women whom m_i most prefers among $\{w_{j'}, \ldots, w_j\}$, and similarly, $B(i', i, j)$ holds *one of* the men whom w_j most prefers among $\{m_{i'}, \ldots, m_i\}$. With these modifications, checking whether two edges are conflicting or not can be done in constant time in the same manner as mentioned above. Therefore, we have the following corollary:

Corollary 3. *There exists an $O(n^4)$-time algorithm to find a maximum cardinality super-WSNM (strong-WSNM, weak-WSNM), given an SMTI-instance.*

5 Conclusion

In this paper, we have shown algorithms and complexity results for the problems of determining existence of an SSNM and finding a maximum cardinality WSNM, in the settings both with and without ties.

One of interesting future directions is to consider optimization problems. For example, in SMI we have shown that it is easy to determine if there exists an SSNM with zero-crossing. What is the complexity of the problem of finding an SSNM with the minimum number of crossings, and if it is NP-hard, is there a good approximation algorithm for it? Another direction is to modify the alignment of agents to, e.g., on a circle or on general position in 2-dimensional plane.

Acknowledgments. The authors would like to thank the anonymous reviewers for their comments on an earlier version of the paper.

References

1. Abdulkadiroğlu, A., Pathak, P.A., Roth, A.E.: The New York City high school match. Am. Econ. Rev. **95**(2), 364–367 (2005)
2. Abdulkadiroğlu, A., Pathak, P.A., Roth, A.E., Sönmez, T.: The Boston public school match. Am. Econ. Rev. **95**(2), 368–371 (2005)
3. Cook, S.A.: The complexity of theorem-proving procedures. In: Proceedings STOC 1971, pp. 151–158 (1971)
4. Gale, D., Shapley, L.S.: College admissions and the stability of marriage. Am. Math. Mon. **69**(1), 9–15 (1962)
5. Gale, D., Sotomayor, M.: Some remarks on the stable matching problem. Discrete Appl. Math. **11**(3), 223–232 (1985)
6. Gusfield, D., Irving, R.W.: The Stable Marriage Problem: Structure and Algorithms. MIT Press, Boston (1989)
7. Irving, R.W.: Stable marriage and indifference. Discrete Appl. Math. **48**, 261–272 (1994)
8. Irving, R.W., Manlove, D.F., Scott, S.: The hospitals/residents problem with ties. In: Halldórsson, M.M. (ed.) SWAT 2000. LNCS, vol. 1851, pp. 259–271. Springer, Heidelberg (2000). https://doi.org/10.1007/3-540-44985-X_24
9. Irving, R.W., Manlove, D.F., Scott, S.: Strong stability in the hospitals/residents problem. In: Alt, H., Habib, M. (eds.) STACS 2003. LNCS, vol. 2607, pp. 439–450. Springer, Heidelberg (2003). https://doi.org/10.1007/3-540-36494-3_39
10. Kavitha, T., Mehlhorn, K., Michail, D., Paluch, K.: Strongly stable matchings in time $O(nm)$ and extension to the hospitals-residents problem. ACM Trans. Algorithms **3**(2) (2007). Article No. 15
11. Knuth, D.E.: Mariages Stables, Les Presses de l'Université Montréal (1976). (Translated and corrected edition, Stable Marriage and Its Relation to Other Combinatorial Problems, CRM Proceedings and Lecture Notes, Vol. 10, American mathematical Society, 1997)
12. Manlove, D.F.: Algorithmics of Matching Under Preferences. World Scientific, Singapore (2013)
13. Roth, A.E.: The evolution of the labor market for medical interns and residents: a case study in game theory. J. Polit. Econ. **92**(6), 991–1016 (1984)
14. Roth, A.E.: On the allocation of residents to rural hospitals: a general property of two-sided matching markets. Econometrica **54**(2), 425–427 (1986)
15. Roth, A.E., Sotomayor, M.: Two-Sided Matching: A Study in Game-theoretic Modeling and Analysis. Cambridge University Press, Cambridge (1990)
16. Ruangwises, S., Itoh, T.: Stable noncrossing matchings. In: Colbourn, C.J., Grossi, R., Pisanti, N. (eds.) IWOCA 2019. LNCS, vol. 11638, pp. 405–416. Springer, Cham (2019). https://doi.org/10.1007/978-3-030-25005-8_33

Connectivity Keeping Trees
in 2-Connected Graphs
with Girth Conditions

Toru Hasunuma$^{(\boxtimes)}$ (iD)

Department of Mathematical Science, Tokushima University,
2–1 Minamijosanjima, Tokushima 770-8506, Japan
hasunuma@tokushima-u.ac.jp

Abstract. Mader conjectured in 2010 that for any tree T of order m, every k-connected graph G with minimum degree at least $\lfloor \frac{3k}{2} \rfloor + m - 1$ contains a subtree $T' \cong T$ such that $G - V(T')$ is k-connected. This conjecture has been proved for $k = 1$; however, it remains open for general $k \geq 2$; for $k = 2$, partially affirmative answers have been shown, all of which restrict the class of trees to special subclasses such as trees of order at most 8, trees with diameter at most 4, trees with at most 5 internal vertices, and caterpillars. Instead of restricting the class of trees, we consider 2-connected graphs with girth conditions. We then show that Mader's conjecture is true for every 2-connected graph G with $g(G) \geq \delta(G) - 6$, where $g(G)$ and $\delta(G)$ denote the girth of G and the minimum degree of a vertex in G, respectively. Besides, we show that for every 2-connected graph G with $g(G) \geq \delta(G) - 3$, the lower bound of $m + 2$ on $\delta(G)$ in Mader's conjecture can be improved to $m + 1$ if $m \geq 6$. Moreover, the lower bound of $\delta(G) - 6$ (respectively, $\delta(G) - 3$) on $g(G)$ in these results can be improved to $\delta(G) - 7$ (respectively, $\delta(G) - 4$ with $m \geq 7$) if no six (respectively, four) cycles of length $g(G)$ have a common path of length $\left\lceil \frac{g(G)}{2} \right\rceil - 1$ in G. Mader's conjecture is interesting not only from a theoretical point of view but also from a practical point of view, since it may be applied to fault-tolerant problems in communication networks. Our proofs lead to $O(|V(G)|^4)$ time algorithms for finding a desired subtree in a given 2-connected graph G satisfying the assumptions.

Keywords: 2-connected graphs · Connectivity · Girth · Trees

1 Introduction

Throughout this paper, a graph $G = (V, E)$ means a simple undirected graph unless stated otherwise. The minimum degree of a vertex in G is denoted by $\delta(G)$. For a proper subset $S \subsetneq V(G)$, we denote by $G - S$ the graph obtained from G by deleting every vertex in S, where $G - \{v\}$ is abbreviated to $G - v$. For two sets A and B, we denote by $A \setminus B$ the set difference $\{x \mid x \in A, x \notin B\}$.

© Springer Nature Switzerland AG 2020
L. Gąsieniec et al. (Eds.): IWOCA 2020, LNCS 12126, pp. 316–329, 2020.
https://doi.org/10.1007/978-3-030-48966-3_24

For a nonempty subset $S \subseteq V(G)$, the subgraph of G induced by S is denoted by $\langle S \rangle_G$, i.e., $\langle S \rangle_G = G - (V(G) \setminus S)$.

In 1972, Chartrand, Kaigars, and Lick proved the following result on the existence of a vertex whose removal does not influence k-connectedness of a graph.

Theorem 1 (Chartrand, Kaigars, and Lick [1]). *Every k-connected graph G with $\delta(G) \geq \lfloor \frac{3k}{2} \rfloor$ contains a vertex v such that $G - v$ is k-connected.*

After more than 30 years, Fujita and Kawarabayashi considered a similar problem for an edge of a graph and showed the following.

Theorem 2 (Fujita and Kawarabayashi [3]). *Every k-connected graph G with $\delta(G) \geq \lfloor \frac{3k}{2} \rfloor + 2$ contains an edge uv such that $G - \{u, v\}$ is k-connected.*

In the same paper, they conjectured the next statement.

Conjecture 1. There is a function $f(m)$ such that every k-connected graph G with $\delta(G) \geq \lfloor \frac{3k}{2} \rfloor + f(m)$ contains a connected subgraph W of order m such that $G - V(W)$ is k-connected.

Note that the condition that W is connected is essential, since by iteratively applying Theorem 1, we can see that every k-connected graph G with $\delta(G) \geq \lfloor \frac{3k}{2} \rfloor + m - 1$ contains a subgraph X of order m such that $G - V(X)$ is k-connected. In 2010, Mader [8] settled Conjecture 1 by showing the following result. Mader's result in fact improves the lower bound on $\delta(G)$ in Theorem 2 and generalizes Theorem 1.

Theorem 3 (Mader [8]). *Every k-connected graph G with $\delta(G) \geq \lfloor \frac{3k}{2} \rfloor + m - 1$ contains a path P of order m such that $G - V(P)$ is k-connected.*

Based on this result, Mader conjectured the following, i.e., a path in Theorem 3 can be generalized to any tree of the same order.

Conjecture 2 (Mader [8]). *For any tree T of order m, every k-connected graph G with $\delta(G) \geq \lfloor \frac{3k}{2} \rfloor + m - 1$ contains a subtree $T' \cong T$ such that $G - V(T')$ is k-connected.*

Mader's conjecture is a generalization not only from Theorem 1 but also from the next well-known result on the existence of a subtree isomorphic to any given tree.

Proposition 1. *For any tree T of order m, every graph G with $\delta(G) \geq m - 1$ contains a subtree $T' \cong T$.*

Apart from Mader's conjecture, Locke's conjecture concerning nonseparating trees in connected graphs is known. A k-cohesive graph is a non-trivial connected graph in which for any two distinct vertices u and v, the sum of the degrees of u and v and the distance between u and v is at least k.

Conjecture 3 (Locke [5]). For any tree T of order $m \geq 3$, every $2m$-cohesive graph G has a subtree $T' \cong T$ such that $G - V(T')$ is connected.

Motivated by Locke's conjecture, Diwan and Tholiya proved a theorem which is weaker than the conjecture, but it is the same as Mader's conjecture for $k = 1$ (Mader in fact mentioned their result in the paper [8]). Note that if G is connected and $\delta(G) \geq m$, then G is $2m$-cohesive.

Theorem 4 (Diwan and Tholiya [2]). *For any tree T of order m, every connected graph G with $\delta(G) \geq m$ contains a subtree $T' \cong T$ such that $G - V(T')$ is connected.*

For general $k \geq 2$, Mader's conjecture remains open; however for $k = 2$, partially affirmative answers have been shown. Tian et al. [10] first proved that Mader's conjecture for $k = 2$ is true when T is a star or a double-star, and they [11] further extended their results to a path-star or a path-double-star. Hasunuma and Ono [4] showed that for any tree T of order m, every 2-connected graph G with $\delta(G) \geq \max\{m + n(T) - 3, m + 2\}$ contains a subtree $T' \cong T$ such that $G - V(T')$ is 2-connected, where $n(T)$ is the number of internal vertices of T. As a corollary, it follows that Mader's conjecture for $k = 2$ holds for any tree T with $n(T) \leq 5$ and for any tree of order $m \leq 8$. Lu and Zhang [6] also proved that Mader's conjecture for $k = 2$ is true for any tree with diameter at most 4. Very recently, it was reported that Lu and Ye [7] proved that Mader's conjecture for $k = 2$ holds for any caterpillars. Note that every known result which is a partially affirmative answer to Mader's conjecture for $k = 2$ restricts the class of trees to special subclasses. In this paper, we employ another approach to Mader's conjecture for $k = 2$. Namely, we add girth conditions to 2-connected graphs. The *girth* of a 2-connected graph G denoted by $g(G)$ is the length of a smallest cycle in G. We then show that Mader's conjecture is true for every 2-connected graph G with girth at least $\delta(G) - 6$. Note that for any given integers $r \geq 2$ and $g \geq 3$, there exists an r-regular graph with girth g, which has been shown in [12].

Theorem 5. *For any tree T of order m, every 2-connected graph G with $\delta(G) \geq m + 2$ and $g(G) \geq \delta(G) - 6$ contains a subtree $T' \cong T$ such that $G - V(T')$ is 2-connected.*

By increasing the lower bound of $\delta(G) - 6$ on $g(G)$, we can improve the lower bound of $m + 2$ on $\delta(G)$ to $m + 1$ if $m \geq 6$. Namely, a stronger statement holds in such a case.

Theorem 6. *For any tree T of order $m \geq 6$, every 2-connected graph G with $\delta(G) \geq m + 1$ and $g(G) \geq \delta(G) - 3$ contains a subtree $T' \cong T$ such that $G - V(T')$ is 2-connected.*

Moreover, by adding structural conditions, we can improve the girth conditions in Theorems 5 and 6.

Theorem 7. *For any tree T of order m, every 2-connected graph G with $\delta(G) \geq m+2$ and $g(G) \geq \delta(G) - 7$ in which no six cycles of length $g(G)$ have a common path of length $\left\lceil \frac{g(G)}{2} \right\rceil - 1$ contains a subtree $T' \cong T$ such that $G - V(T')$ is 2-connected.*

Theorem 8. *For any tree T of order $m \geq 7$, every 2-connected graph G with $\delta(G) \geq m+1$ and $g(G) \geq \delta(G) - 4$ in which no four cycles of length $g(G)$ have a common path of length $\left\lceil \frac{g(G)}{2} \right\rceil - 1$ contains a subtree $T' \cong T$ such that $G - V(T')$ is 2-connected.*

Mader's conjecture is interesting not only from a theoretical point of view but also from a practical point of view, since it may be applied to fault-tolerant problems in communication networks. That is, it is considered that Mader's conjecture guarantees the reliability of a communication network for a faulty subtree structure rather than a set of faulty vertices. Our proofs are constructive and lead to $O(|V(G)|^4)$ time algorithms for finding a desired subtree in a given 2-connected graph G in Theorems 5 and 6 (respectively, Theorems 7 and 8) if $g(G) \geq \delta(G) - 3$ (respectively, $g(G) \geq \delta(G) - 4$).

This paper is organized as follows. Section 2 presents notations, terminology, and known results used in this paper. Section 3 gives an outline of our proofs. Detailed proofs of Theorems 5 and 6 (respectively, Theorems 7 and 8) are given in Sect. 4 (respectively, Sect. 5). Section 6 concludes the paper with several remarks.

2 Preliminaries

For a nonempty subset $E' \subseteq E(G)$, we denote by $G - E'$ and $\langle E' \rangle$ the graph obtained from G by deleting every edge in E' and the edge-induced subgraph of G by E', respectively. For $v \in V(G)$, we denote by $N_G(v)$ the set of neighbors of v in G, i.e., vertices adjacent to v in G. The cardinality of $N_G(v)$ may be written by $\deg_G(v)$. Let $\Delta(G) = \max_{v \in V(G)} \deg_G(v)$. For $S \subseteq V(G)$, $N_G(S)$ is defined as $(\cup_{v \in S} N_G(v)) \setminus S$. For $G' \subseteq G$, let $N_G(G') = N_G(V(G'))$.

A *component* of G is a maximal connected subgraph of G, while a *block* of G is a maximal connected subgraph of G without a cut vertex. A *cyclic block* is a block with order at least 3. For a tree T, the set of internal vertices, i.e., vertices with degree at least two, is denoted by $V_I(T)$, while the set of leaves, i.e., vertices with degree one, is denoted by $V_L(T)$. For a vertex v of a tree T, if v is adjacent to at least $\deg_T(v) - 1$ leaves, then v is called a *pseudo-leaf* of T. A *caterpillar* is a tree T such that $\langle V_I(T) \rangle_T$ is a path if $V_I(T) \neq \emptyset$.

We denote by $d_G(u, v)$ the distance between two vertices u and v in a connected graph G. The *eccentricity* $\mathrm{ecc}_G(v)$ of v in G is defined as $\max_{w \in V(G)} d_G(v, w)$. A *central vertex* of G is a vertex u with $\mathrm{ecc}_G(u) = \min_{v \in V(G)} \mathrm{ecc}_G(v)$, while a *peripheral vertex* is a vertex u with $\mathrm{ecc}_G(u) = \max_{v \in V(G)} \mathrm{ecc}_G(v)$. The *diameter* of a connected graph G denoted by $\mathrm{diam}(G)$ is the maximum distance for every pair of vertices in G, i.e., $\mathrm{diam}(G) = \max_{u,v \in V(G)} d_G(u, v)$. Let $\mathrm{diam}(G) = 0$ if $|V(G)| = 1$.

Proposition 1 can be stated in a more general form as follows.

Lemma 1 [4]. *Let T be a tree of order m and S a subtree obtained from T by deleting leaves adjacent to a vertex in $V_S \subseteq V_I(T)$. If a graph G contains a subtree $S' \cong S$ such that $\deg_G(u) \geq m - 1$ for any $u \in \{\phi(v) \mid v \in V_S\}$ where ϕ is an isomorphism from $V(S)$ to $V(S')$, then G contains a subtree $T' \cong T$ such that $S' \subseteq T'$.*

Since any tree T of order m with $\mathrm{diam}(T) \geq m - 2$ is a caterpillar and Mader's conjecture holds for any caterpillars [7], the following result is obtained.

Lemma 2. *For any tree T of order m with $\mathrm{diam}(T) \geq m-2$, Mader's conjecture for $k = 2$ is true.*

An *orientation* D of a graph G is a directed graph obtained from G by replacing each edge by an arc (directed edge) with the same end-vertices. The *outdegree* $\deg_D^+(v)$ (respectively, *indegree* $\deg_D^-(v)$) of a vertex v in D is the number of arcs from (respectively, to) v in D. If for any $v \in V(G)$, $\deg_G(v)$ is even, then G is eulerian and has an orientation D in which for any $v \in V(D)$, $\deg_D^+(v) = \deg_D^-(v)$. If G has a vertex with odd degree, we can find a directed walk W connecting two vertices with odd degree, and by inductively applying a similar discussion for $G - E(W)$, we can see the following lemma holds. We here remark that Lemma 3 holds for multigraphs.

Lemma 3. *Every graph G has an orientation D such that $|\deg_D^+(v) - \deg_D^-(v)| \leq 1$ for any $v \in V(D)$.*

3   Outline of Proofs

In this section, we explain the outline of our constructive proofs and the time complexity for the algorithms based on the proofs.

Let T be a tree of order m. Let G be a 2-connected graph with $\delta(G) \geq m+2$. From Proposition 1, G contains a subtree $T' \cong T$. Let B be a maximum block in $G - V(T')$, i.e., a block with the maximum order among all the blocks in $G - V(T')$. Note that B is a cyclic block since $\delta(G - V(T')) \geq 2$. If $B = G - V(T')$, then T' is a desired subtree. Suppose that $B \neq G - V(T')$. Then there is a vertex in $G - V(T') \cup V(B)$. For any vertex $w \in V(G) \setminus (V(T') \cup V(B))$, $|N_G(w) \cap V(B)| \leq 1$. Now let $P = (v_1, v_2, \ldots, v_t)$, where $v_1, v_t \in V(B)$ and $v_i \notin V(B)$ for $1 < i < t$, be a shortest path among all the paths of G connecting two vertices in B such that every internal vertex is not in B. Since G is 2-connected, there are internally disjoint paths from a vertex in $G - V(B)$ to two vertices in B. Thus, P is well-defined. Suppose that $t \geq 4$. Then, we have that $N_G(v_2) \cap V(B) = \{v_1\}$ and $N_G(v_2) \cap V(P) = \{v_1, v_3\}$. Therefore, $|N_G(v_2) \setminus (V(P) \cup V(B))| \geq m + 2 - 2 \geq m$, which implies that $V(G) \setminus (V(P) \cup V(B)) \neq \emptyset$. Let w be any vertex in $G - V(P) \cup V(B)$. By the definition of P, w can be adjacent to at most three vertices in $V(B) \cup V(P)$. Thus, $\delta(G - V(P) \cup V(B)) \geq m + 2 - 3 = m - 1$. Hence, by Proposition 1, $G - V(P) \cup V(B)$ contains a subtree $T'' \cong T$ such that $G - V(T'')$ has a block $B' \supseteq \langle V(B) \cup V(P) \rangle_G$. Thus, we can find a block with order at least $|V(B)| + 2$.

Suppose that $t = 3$. Then $v_2 \in V(T')$. If there exists a subtree T'' in $G -V(B) \cup \{v_2\}$ such that $T'' \cong T$, then $G - V(T'')$ has a block $B' \supseteq \langle V(B) \cup \{v_2\}\rangle_G$, i.e., we can find a block with order at least $|V(B)| + 1$. If we have a manipulation to find such a subtree T'', then by applying the manipulations for $t \geq 4$ or $t = 3$ iteratively, we finally obtain a desired subtree T'', i.e., $T'' \cong T$ such that $G - V(T'')$ is 2-connected. Therefore, if we can show the following statement, then it is concluded that Mader's conjecture for $k = 2$ is true.

Statement 1. Let T be a tree of order m and G a 2-connected graph with $\delta(G) \geq m + 2$. For any subtree $T' \cong T$ in G and a maximum block B in $G - V(T')$, if $B \neq G - V(T')$ and $V_B(T') = \{u \in V(T') \mid |N_G(u) \cap V(B)| \geq 2\} \neq \emptyset$, then there exist a vertex $v \in V_B(T')$ and a subtree $T'' \cong T$ in $G - V(B) \cup \{v\}$.

The above manipulations can be algorithmically described as follows.

1. Compute a subtree $T' \cong T$ in G.
2. Compute a maximum block B in $G - V(T')$.
3. If $B = G - V(T')$ then output T' as a desired subtree of G and stop.
4. If $B \neq G - V(T')$ then compute a shortest path P connecting vertices in B such that every internal vertex is not in B.
5. Compute a subtree T'' in $G - V(B) \cup V(P)$, let $T' = T''$, and return to Step 2.

We here check the complexity of the above algorithm under the assumption that there exists a constructive proof of Statement 1. A subtree $T' \cong T$ in G can be computed in $O(|E(G)|)$ time in Step 1, and a maximum block B can also be found in $O(|E(G)|)$ time in Step 2. In Step 4, a shortest path P can be found by computing all shortest paths for vertices of $V(B)$ in $G - E(B)$. Thus, it takes $O(|V(G)|^3)$ time. Since the number of iterations is $O(|V(G)|)$, if Statement 1 can be shown by a constructive proof from which a procedure within $O(|V(G)|^3)$ time is obtained, we have an $O(|V(G)|^4)$ time algorithm. These observations are summarized as follows.

Lemma 4. *If Statement 1 holds, then G contains a subtree $T' \cong T$ such that $G - V(T')$ is 2-connected. Besides, if there is a procedure for Statement 1 within $O(|V(G)|^3)$ time, we have an $O(|V(G)|^4)$ time algorithm for finding a desired subtree.*

Next, we consider the case that a 2-connected graph G has no triangle, i.e., $g(G) \geq 4$. In such a case, we can show a similar lemma using the following statement. Note that the minimum degree condition $\delta(G) \geq m + 2$ is replaced with $\delta(G) \geq m + 1 \geq 3$.

Statement 2. Let T be a tree of order $m \geq 2$ and G a 2-connected graph with $\delta(G) \geq m+1$ and $g(G) \geq 4$. For any subtree $T' \cong T$ in G and a maximum block B in $G - V(T')$, if $B \neq G - V(T')$ and $V_B(T') = \{u \in V(T') \mid |N_G(u) \cap V(B)| \geq 2\} \neq \emptyset$, then there exist a vertex $v \in V_B(T')$ and a subtree $T'' \cong T$ in $G - V(B) \cup \{v\}$.

Lemma 5. *If Statement 2 holds, then G contains a subtree $T' \cong T$ such that $G - V(T')$ is 2-connected. Besides, if there is a procedure for Statement 2 within $O(|V(G)|^3)$ time, we have an $O(|V(G)|^4)$ time algorithm for finding a desired subtree.*

Proof. We show that the algorithm for Lemma 4 also works well under the assumption that Statement 2 holds.

Let $T' \subset G$ such that $T' \cong T$. Let B be a maximum block in $G - V(T')$. Since $\delta(G - V(T')) \geq 1$, it may happen that B is not a cyclic block, i.e., B is a block with two vertices. Note that if B is not a cyclic block, then B is not 2-connected. Suppose that B is not a cyclic block. Assume that $B = G - V(T')$. Then, $|V(G)| = m + 2$. Since $\delta(G) \geq m + 1$, G must be a complete graph with at least four vertices, which contradicts the girth condition that $g(G) \geq 4$. Therefore, if B is not a cyclic block, then $B \neq G - V(T')$. Hence, in the case that $G - V(T')$ has no cyclic block, the algorithm does not incorrectly output a subtree in Step 3.

Let $P = (v_1, v_2, \ldots, v_t)$ be a shortest path between two vertices in B such that every internal vertex is not in B. Suppose that $t \geq 4$. By the definition of P and the girth condition $g(G) \geq 4$, any vertex w in $G - V(P) \cup V(B)$ can be adjacent to at most two vertices in $V(B) \cup V(P)$. Thus, $\delta(G - V(P) \cup V(B)) \geq m - 1$. Therefore, $G - V(P) \cup V(B)$ contains a subtree $T'' \cong T$. Hence, if $t \geq 4$, then we can find a subtree T'' in $G - V(P) \cup V(B)$ in Step 5. We here remark that the condition $m \geq 2$ is necessary to guarantee that $V(G) \setminus (V(P) \cup V(B)) \neq \emptyset$.

For the time complexity, similarly to Lemma 4, we have an $O(|V(G)|^4)$ time algorithm, if Statement 2 can be shown by a constructive proof which induces a procedure within $O(|V(G)|^3)$ time.   □

Note that in Statement 2, if B is not a cyclic block, then by the girth condition $g(G) \geq 4$, we have that $V_B(T') = \{u \in V(T') \mid |N_G(u) \cap V(B)| \geq 2\} = \emptyset$. Thus, in Statement 2, we may assume that a maximum block B is a cyclic block.

4   Proofs of Theorems 5 and 6

In order to show our main results, we prove the following lemma.

Lemma 6. *Let T be a tree of order m and G a 2-connected graph with $\delta(G) \geq m + 1$ and $g(G) \geq \text{diam}(T) - 1$. For any subtree $T' \cong T$ in G and a maximum block B in $G - V(T')$, if $B \neq G - V(T')$ and $V_B(T') = \{u \in V(T') \mid |N_G(u) \cap V(B)| \geq 2\} \neq \emptyset$, then there exist a vertex $v \in V_B(T')$ and a subtree $T'' \cong T$ in $G - V(B) \cup \{v\}$ such that v and T'' can be found in $O(|E(G)|)$ time.*

Proof. Let $T' \subset G$ such that $T' \cong T$. Let B be a maximum block in $G - V(T')$ such that $B \neq G - V(T')$. Also, let $v \in V_B(T')$ and $H = G - V(T') \cup V(B)$. When $m \leq 2$, the lemma can be easily checked. Let $m \geq 3$. Suppose that v is a leaf of T' and for the neighbor v' of v in T', $v' \notin V_B(T')$, i.e., $|N_G(v') \cap V(B)| \leq 1$. Then, $|N_G(v') \cap V(H)| \geq 1$. For any $v'' \in N_G(v') \cap V(H), T'' = \langle (E(T'-v) \cup \{v'v''\}) \rangle \cong T$

such that $T'' \subset G - V(B) \cup \{v\}$. Thus, w.l.o.g., we may assume that v is not a leaf of T'. Let $S' = \langle V_I(T') \rangle_{T'}$. Then $v \in V(S')$. Since $\operatorname{diam}(S') = \operatorname{diam}(T') - 2$, $g(G) \geq \operatorname{diam}(S') + 1$. We regard S' as a rooted tree at v and denote by $C(u)$ the set of children of a vertex u in S'. Besides, we denote by $h(S')$ the height of S', i.e., $h(S') = \operatorname{ecc}_{S'}(v)$.

Since $\delta(G) \geq m + 1$, it holds that for any vertex $w \in V(H)$, $\deg_{G-V(B)\cup\{v\}}(w) \geq m - 1$. If there exists a subtree $W \subset \langle V(H) \cup V(T' - v) \rangle_G$ such that W is isomorphic to a subtree U obtained from T' by deleting leaves adjacent to a vertex in $V' \subseteq V(S')$ and $\phi(V') = \{\phi(u) \mid u \in V'\} \subseteq V(H)$ where ϕ is an isomorphism from $V(U)$ to $V(W)$, then by Lemma 1, there exists a subtree T'' in $G - V(B) \cup \{v\}$ such that $T'' \cong T$. In particular, if there exists a vertex w in H such that $C(v) \subseteq N_G(w)$, then we can employ the subtree $\langle E(T' - v) \cup \{wu \mid u \in C(v)\} \rangle$ as a desired subtree W where $V' = \{v\}$ and $\phi(V') = \{w\}$. Note that $C(v) = \emptyset$ when $\operatorname{diam}(S') = 0$, i.e., $|V(S')| = 1$. Suppose that v is a leaf of S'. Let $C(v) = \{v'\}$. If there is no vertex in H adjacent to v', i.e., $C(v) \not\subseteq N_G(w)$ for any $w \in V(H)$, then $\delta(H) \geq 1$ and $v' \in V_B(T')$. In such a case, we can employ $\langle \{xy\} \rangle$ as a desired subtree W for $xy \in E(H)$ where $V' = \{v, v'\}$ and $\phi(V') = \{x, y\}$ when $\operatorname{diam}(S') = 1$. From these observations, we may assume that there is no vertex w in H with $N_G(w) \supseteq C(v)$, v is not a leaf of S' (since we can employ v' instead of v if $v' \in V_B(T')$) and $\operatorname{diam}(S') \geq 2$.

Let $x \in V(H)$ and $C(v) \setminus N_G(x) = \{v_1, v_2, \ldots, v_p\}$. Since $|N_G(x) \cap V(B)| \leq 1$ and $|N_G(x) \cap V(T')| \leq m - p$, $|N_H(x)| \geq p$, i.e., there are at least p neighbors of x in H. Let $\{x_1, x_2, \ldots, x_p\} \subseteq N_H(x)$. If $h(S') = 1$, then we can employ $\langle E(T' - v) \cup \{xu \mid u \in C(v) \cap N_G(x)\} \cup \{xx_i \mid 1 \leq i \leq p\} \rangle$ as a desired subtree W where $V' = \{v, v_1, v_2, \ldots, v_p\}$ and $\phi(V') = \{x, x_1, x_2, \ldots, x_p\}$. Suppose that $h(S') \geq 2$. Let $|C(v_i) \setminus N_G(x_i)| = q_i$ for each i. Since $|C(v) \setminus N_G(x_i)| \geq 1$, there are at least $q_i + 1$ neighbors of x_i in H, which means that we can select q_i vertices $y_{i,1}, y_{i,2}, \ldots, y_{i,q_i}$ as children of x_i in the subtree $\langle \{xx_i \mid 1 \leq i \leq p\} \rangle$ rooted at x. By letting these children correspond to the q_i children of v_i in $C(v_i) \setminus N_G(x_i)$ for each i with $q_i > 0$, we can obtain a desired subtree W if $h(S') = 2$. Note that when $\operatorname{diam}(S') = 3$, there is exactly one i such that $C(v_i) \neq \emptyset$, and if $q_i > 0$, then $\{x_1, x_2, \ldots, x_p\} \cap \{y_{i,1}, y_{i,2}, \ldots, y_{i,q_i}\} = \emptyset$, since $g(G) \geq \operatorname{diam}(S') + 1 = 4$. When $\operatorname{diam}(S') = 4$, by the girth condition, we can see that $\{x_1, x_2, \ldots, x_p\} \cap \{y_{i,1}, y_{i,2}, \ldots, y_{i,q_i}\} = \emptyset$ for each i with $q_i > 0$ and $\{y_{i,1}, y_{i,2}, \ldots, y_{i,q_i}\} \cap \{y_{i',1}, y_{i',2}, \ldots, y_{i',q_{i'}}\} = \emptyset$ for any pair of i and i' with $q_i > 0$ and $q_{i'} > 0$. Thus, the subtree defined by $\langle E(T' - \{v, v_1, v_2, \ldots, v_p\}) \cup \{xu \mid u \in C(v) \cap N_G(x)\} \cup \{xx_i \mid 1 \leq i \leq p\} \cup (\cup_{1 \leq i \leq p}(\{x_iu \mid u \in C(v_i) \cap N_G(x_i)\} \cup \{x_iy_{i,j} \mid 1 \leq j \leq q_i\})) \rangle$ can be employed as a desired subtree W. If $h(S') \geq 3$, by inductively applying similar manipulations to descendants of x, we can finally obtain a desired subtree W. Note that in each extension step, disjointness of the sets of new children for descendants of x is guaranteed by the girth condition $g(G) \geq \operatorname{diam}(S') + 1$.

The assumption that v is neither a leaf of T' nor a leaf of S' can be realized by preferentially selecting a vertex in $V_B(T') \setminus (V_L(T') \cup V_L(S'))$ if $V_B(T') \setminus (V_L(T') \cup V_L(S')) \neq \emptyset$. For $v \in V_B(T') \setminus (V_L(T') \cup V_L(S'))$, we apply the manipulations for constructing W in a depth-first search order for S'. The selection process for new children of a descendant of x and the extension process from W to T'' can be done greedily. If $V_B(T') \setminus (V_L(T') \cup V_L(S')) = \emptyset$, then we can directly obtain either W or T''. Therefore, a desired subtree T'' can finally be found in $O(|E(G)|)$ time.   □

Lemma 6 is stronger than Statement 1 under the assumption that $g(G) \geq \mathrm{diam}(T) - 1$. Therefore, by Lemmas 4 and 6, we have the following.

Theorem 9. *For any tree T of order m, every 2-connected graph G with $\delta(G) \geq m + 2$ and $g(G) \geq \mathrm{diam}(T) - 1$ contains a subtree $T' \cong T$ such that $G - V(T')$ is 2-connected.*

For any 2-connected graph G, it holds that $g(G) \geq 3$. Thus, the following result by Lu and Zhang [6] is obtained from Theorem 9.

Corollary 1 [6]**.** *For any tree T of order m with $\mathrm{diam}(T) \leq 4$, every 2-connected graph G with $\delta(G) \geq m + 2$ contains a subtree $T' \cong T$ such that $G - V(T')$ is 2-connected.*

Besides, by combining Lemmas 5 and 6, we have the following.

Theorem 10. *For any tree T of order $m \geq 2$, every 2-connected graph G with $\delta(G) \geq m + 1$ and $g(G) \geq \max\{\mathrm{diam}(T) - 1, 4\}$ contains a subtree $T' \cong T$ such that $G - V(T')$ is 2-connected.*

From Theorem 10, the following result for 2-connected graphs without a triangle is obtained.

Corollary 2. *For any tree T of order $m \geq 2$ with $\mathrm{diam}(T) \leq 5$, every 2-connected graph G with $\delta(G) \geq m + 1$ and $g(G) \geq 4$ contains a subtree $T' \cong T$ such that $G - V(T')$ is 2-connected.*

Now, we are ready to show our main two results stated in the introduction. Let T be a tree of order m. Suppose that G is a 2-connected graph with $\delta(G) \geq m + 2$ and $g(G) \geq \delta(G) - 6$. Then, $g(G) \geq m - 4$. From Lemma 2, it is sufficient to consider a tree T with $\mathrm{diam}(T) \leq m - 3$. That is, we have $g(G) \geq \mathrm{diam}(T) - 1$. Therefore, Theorem 5 follows from Theorem 9. Next, suppose that $m \geq 6$ and G is a 2-connected graph with $\delta(G) \geq m + 1$ and $g(G) \geq \delta(G) - 3$. Then, $g(G) \geq m - 2 \geq 4$, i.e., $g(G) \geq \max\{\mathrm{diam}(T) - 1, 4\}$. Hence, Theorem 6 follows from Theorem 10.

From Lemmas 4, 5 and 6, we can see that a desired subtree T' in Theorem 5 (respectively, Theorem 6) can be found in $O(|V(G)|^4)$ time if $g(G) \geq \delta(G) - 4$ (respectively, $g(G) \geq \delta(G) - 3$). Note that such a restriction on $g(G)$ for Theorem 5 follows from the fact that we use Lemma 2.

5   Proofs of Theorems 7 and 8

In this section, we try to improve the lower bounds on $g(G)$ in Theorems 5 and 6, and show that such improvements are possible if a 2-connected graph G satisfies a structural property on the smallest cycles. Note that for any two cycles C_1 and C_2 of length $g(G)$, it holds that $|E(C_1) \cap E(C_2)| \leq \left\lfloor \frac{g(G)}{2} \right\rfloor$.

Lemma 7. *Let T be a tree of order m. Let G be a 2-connected graph with $\delta(G) \geq m + 2$ and $g(G) \geq \mathrm{diam}(T) - 2$ in which no six cycles of length $g(G)$ have a common path of length $\left\lceil \frac{g(G)}{2} \right\rceil - 1$ in G. For any subtree $T' \cong T$ in G and a maximum block B in $G - V(T')$, if $B \neq G - V(T')$ and $V_B(T') = \{u \in V(T') \mid |N_G(u) \cap V(B)| \geq 2\} \neq \emptyset$, then there exist a vertex $v \in V_B(T')$ and a subtree $T'' \cong T$ in $G - V(B) \cup \{v\}$.*

Proof. We use the notations such as $T', B, v, H, S', C(u), W$, and x with the same meaning in the proof of Lemma 6. If $\mathrm{diam}(S') \leq 2$, then we can easily construct a desired subtree W without an additional structural property. Suppose that $\mathrm{diam}(S') \geq 3$. By the discussion in the proof of Lemma 6, we suppose that v is not a leaf of S' and there is no vertex w in H such that $C(v) \subseteq N_G(w)$. For $u \in C(v)$, we denote by S'_u the subtree rooted at u in S'. Let F be a component of H containing x. Note that $|N_G(F) \cap V(B)| \leq 1$. In the following discussion, w.l.o.g., we may assume that $N_G(F) \cap V(B) \neq \emptyset$. Then, let $N_G(F) \cap V(B) = \{v_B\}$ and $F' = \langle V(F) \cup \{v_B\} \rangle_G$.

Suppose that v is a pseudo-leaf of S' and v' is the non-leaf vertex adjacent to v in S'. If there exists a vertex y in H such that $v' \in N_G(y)$, then by letting the vertex y correspond to v, we can obtain a desired subtree W. If there is no vertex in H which is adjacent to v', then $v' \in V_B(T')$. Thus, we may assume that if $\mathrm{diam}(S') \geq 4$, v is not a pseudo-leaf of S', and if $\mathrm{diam}(S') = 3$, the central vertices v, v' are in $V_B(T')$ such that $\{v, v'\} \cap N_G(w) = \emptyset$ for any $w \in V(H)$. Suppose that $\mathrm{diam}(S') = 3$. Let $xy \in E(F)$. Then, $|N_{F'}(x)| \geq |C(v) \setminus N_G(x)| + 3$ and $|N_{F'}(y) \setminus \{x\}| \geq |C(v') \setminus N_G(y)| + 3$. The assumption on the smallest cycles implies that $|N_{F'}(x) \cap N_{F'}(y)| \leq 5$. Therefore, we can find $y \in C(x) \subset N_F(x)$ and $C(y) \subset N_F(y) \setminus \{x\}$ so that $C(x) \cap C(y) = \emptyset$, $|C(x)| = |C(v) \setminus N_G(x)|$ and $|C(y)| = |C(v') \setminus N_G(y)|$. Thus, a desired subtree W can be constructed. In what follows, we suppose that $\mathrm{diam}(S') \geq 4$.

It is sufficient to consider the case that $g(G) = \mathrm{diam}(T) - 2 = \mathrm{diam}(S')$. Let $P(S')$ and $Q(S')$ be the set of peripheral vertices in S' and the set of parents of a peripheral vertex in S', respectively. Let $S'' = S' - P(S')$. Note that $\mathrm{diam}(S'') = \mathrm{diam}(S') - 2$, and $v \notin P(S') \cup Q(S')$ since any vertex in $P(S')$ is a leaf of S' and any vertex in $Q(S')$ is a pseudo-leaf of S'. For the subtree S'', we apply the manipulations in the proof of Lemma 6. Let W' be the subtree obtained after such manipulations and let $W'_F = \langle V(W') \cap V(F) \rangle_{W'}$. Suppose that $\{z_1, z_2, \ldots, z_q\}$ is the set of vertices in W'_F which are corresponding to vertices in $Q(S')$. Note that q may be less than $|Q(S')|$. Let $\{u_1, u_2, \ldots, u_q\} \subseteq Q(S')$ such that u_i is corresponding to z_i for $1 \leq i \leq q$. For each $1 \leq i \leq q$, let $D(z_i) = N_F(z_i) \setminus \{p(z_i)\}$

where $p(z_i)$ is the parent of z_i in W'_F rooted at x. Also let $r_i = |C(u_i) \setminus N_G(z_i)|$ for each $1 \le i \le q$, where $C(u_i)$ is the set of children of u_i in S'. Since $g(G) = \mathrm{diam}(S')$, it may happen that $D(z_i) \cap D(z_j) \ne \emptyset$ for $1 \le i < j \le q$. It follows from $\delta(G) \ge m + 2$ and $|C(v) \setminus N_G(z_i)| \ge 1$ that $|D(z_i)| \ge r_i + 1$ for each i.

Suppose that $|D(z_k)| = r_k + 1$ for some $k \in \{1, 2, \ldots, q\}$. Then $|C(v) \setminus N_G(z_k)| = 1$ and z_k is adjacent to every vertex in T' except for ones in $(C(v) \cup C(u_k)) \setminus N_G(z_k)$. Thus, we may assume that $V_B(T') \subseteq N_{T'}(C(v) \setminus N_G(z_k)) \cup N_{T'}(C(u_k) \setminus N_G(z_k))$, since otherwise, there exists $v' \in V_B(T')$ such that $N_{T'}(v') \subseteq N_G(z_k)$. Let $C(v) \setminus N_G(z_k) = \{w_k\}$. Instead of x, we let z_k correspond to v and apply the manipulations in the proof of Lemma 6. Let W''_F be the resultant subtree in F. If w_k is a pseudo-leaf of S', then we can immediately obtain a desired subtree W in this setting. Otherwise, there is no pseudo-leaf adjacent to v in S' which corresponds to a vertex in the subtree W''_F. Thus, w.l.o.g., we may assume that $u_k \notin C(v)$. Consider the case that $u_k \in V_B(T')$. Since u_k is a pseudo-leaf of S', by the previous discussion, we may assume that $p(u_k) \in V_B(T')$ where $p(u_k)$ is the parent of u_k in S'. Since $p(u_k) \notin N_{T'}(C(u_k) \setminus N_G(z_k))$, $p(u_k) \in N_{T'}(w_k)$. This means that $w_k = p(p(u_k))$. Next consider the case that $u_k \notin V_B(T')$. In this case, we may assume that no descendant of u_k in T' is in $V_B(T')$. Hence, it is concluded that $V_B(T') \cap (\cup_{u \in C(v) \setminus \{w_k\}} V(S'_u)) = \emptyset$. Note that $w_k \notin V_B(T')$. Let $H' = \langle V(H) \cup (\cup_{u \in C(v) \setminus \{w_k\}} V(S'_u)) \rangle_G$. For every vertex $u' \in \cup_{u \in C(v) \setminus \{w_k\}} V(S'_u)$, $|N_G(u') \cap V(B)| \le 1$. Thus, it holds that $\delta(H') \ge 1 + \sum_{u \in C(v) \setminus \{w_k\}} |V(S'_u)|$. Let $w'_k \in N_G(w_k) \cap V(H)$. Then, there exists a subtree $U'_{H'}$ in H' which is isomorphic to $S' - V(S'_{w_k})$ such that w'_k corresponds to v in an isomorphism from $V(S') \setminus V(S'_{w_k})$ to $V(U'_{H'})$. Then, $\langle E(S'_{w_k}) \cup \{w_k w'_k\} \cup E(U'_{H'}) \rangle$ can be employed as a desired subtree W. Consequently, we may assume that any vertex z_i in $\{z_1, z_2, \ldots, z_q\}$ satisfies that $|D(z_i)| \ge r_i + 2$.

Let $D'(z_i) = N_{F'}(z_i) \setminus \{p(z_i)\}$ for $1 \le i \le q$. Then, $|D'(z_i)| \ge r_i + 3$ for each i. Note that either $D'(z_i) = D(z_i)$ or $D'(z_i) = D(z_i) \cup \{v_B\}$. Define I_G as the (multi)graph with vertex set $\{z_1, z_2, \ldots, z_q\}$ in which z_i and z_j are joined by $|D'(z_i) \cap D'(z_j)|$ edges. Note that I_G may be a multigraph only if $\mathrm{diam}(S') = 4$. The assumption that no six cycles of length $g(G)$ have a common path of length $\left\lceil \frac{g(G)}{2} \right\rceil - 1 = \left\lceil \frac{\mathrm{diam}(S'')}{2} \right\rceil$ in G implies that $\Delta(I_G) \le 5$, i.e., each vertex in I_G is incident to at most five edges. Besides, the intersection of at least seven (respectively, three) sets in $\{D'(z_1), D'(z_2), \ldots, D'(z_q)\}$ is empty if $\mathrm{diam}(S'')$ is even (respectively, odd). Modify the graph I_G as follows, and let J_G be the resultant (multi)graph.

1. Delete every edge generated by a vertex in the intersection of at least three sets $D'(z_{i_1})$, $D'(z_{i_2})$, and $D'(z_{i_3})$.
2. Delete the edge generated by v_B if v_B is contained in exactly two sets $D'(z_{i_1})$ and $D'(z_{i_2})$.

Note that if v_B is contained in exactly one set $D'(z_i)$, then $|D(z_i)| \ge r_i + 2$ and $|D(z_j)| \ge r_j + 3$ for any $j \ne i$. By Lemma 3, J_G has an orientation D_G such that $|\deg^+_{D_G}(z) - \deg^-_{D_G}(z)| \le 1$ for any $z \in V(D_G)$ and if v_B is contained in exactly one set $D'(z_i)$ then $\deg_{D_G}^-(z_i) \leq 2$. Note that if an orientation of J_G satisfying the first condition does not satisfy the second condition, the reverse orientation satisfies both the conditions since $\Delta(I_G) \leq 5$. Based on D_G, we can disjointly select r_i vertices in $D(z_i)$ for $1 \leq i \leq q$ as follows. For each arc from z_{i_1} to z_{i_2} in D_G, we select the vertex in $D(z_{i_1}) \cap D(z_{i_2})$ corresponding to the edge $z_{i_1}z_{i_2}$ as a child of z_{i_1}. Note that we do not select the vertex v_B and any vertex in the intersection of at least three sets $D'(z_{i_1}), D'(z_{i_2})$, and $D'(z_{i_3})$. In this way, we can appropriately extend W_F' for a desired subtree W and finally obtain a subtree $T'' \cong T$ in $G - V(B) \cup \{v\}$. □

Next, we consider the case that $\delta(G) \geq m + 1$. In this case, we need to strengthen the structural condition on the smallest cycles in Lemma 7.

Lemma 8. *Let T be a tree of order m. Let G be a 2-connected graph with $\delta(G) \geq m + 1$ and $g(G) \geq \mathrm{diam}(T) - 2$ in which no four cycles of length $g(G)$ have a common path of length $\left\lceil \frac{g(G)}{2} \right\rceil - 1$ in G. For any subtree $T' \cong T$ in G and a maximum block B in $G - V(T')$, if $B \neq G - V(T')$ and $V_B(T') = \{u \in V(T') \mid |N_G(u) \cap V(B)| \geq 2\} \neq \emptyset$, then there exist a vertex $v \in V_B(T')$ and a subtree $T'' \cong T$ in $G - V(B) \cup \{v\}$.*

Proof. We use the notations in the proof of Lemma 7 with the same meaning. A desired subtree W can be constructed without an additional structural property if $\mathrm{diam}(S') \leq 2$. Suppose that $\mathrm{diam}(S') = 3$. Applying a similar discussion in the proof of Lemma 7, we have that $|N_{F'}(x)| \geq |C(v) \backslash N_G(x)| + 2$ and $|N_{F'}(y) \backslash \{x\}| \geq |C(v') \backslash N_G(y)| + 2$. Since the condition on smallest cycles implies that $|N_{F'}(x) \cap N_{F'}(y)| \leq 3$, we can find $y \in C(x) \subset N_F(x)$ and $C(y) \subset N_F(y) \backslash \{x\}$ so that $C(x) \cap C(y) = \emptyset$, $|C(x)| = |C(v) \backslash N_G(x)|$ and $|C(y)| = |C(v') \backslash N_G(y)|$. Suppose that $\mathrm{diam}(S') \geq 4$. From a similar discussion in the proof of Lemma 7, we may assume that every vertex z_i in $\{z_1, z_2, \ldots, z_q\}$ satisfies that $|D(z_i)| \geq r_i + 1$ and $|D'(z_i)| \geq r_i + 2$. Note that the degree condition $\delta(H') \geq \sum_{u \in C(v) \backslash \{w\}} |V(S_u')|$ is sufficient to construct a subtree $U_{H'}'$ in H'. The assumption that no four cycles of length $g(G)$ have a common path of length $\left\lceil \frac{g(G)}{2} \right\rceil - 1$ in G implies that $\Delta(I_G) \leq 3$. By Lemma 3, J_G has an orientation D_G such that $|\deg_{D_G}^+(z) - \deg_{D_G}^-(z)| \leq 1$ for any $z \in V(D_G)$ and if v_B is contained in exactly one set $D'(z_i)$ then $\deg_{D_G}^-(z_i) \leq 1$. Based on D_G, we can disjointly select r_i vertices in $D(z_i)$ for $1 \leq i \leq q$. Hence, we can appropriately extend W_F' in order to obtain a desired subtree T''. □

From Lemmas 4, 5, 7, and 8, we have the following results.

Theorem 11. *For any tree T of order m, every 2-connected graph G with $\delta(G) \geq m + 2$ and $g(G) \geq \mathrm{diam}(T) - 2$ in which no six cycles of length $g(G)$ have a common path of length $\left\lceil \frac{g(G)}{2} \right\rceil - 1$ contains a subtree $T' \cong T$ such that $G - V(T')$ is 2-connected.*

Theorem 12. *For any tree T of order $m \geq 2$, every 2-connected graph G with $\delta(G) \geq m+1$ and $g(G) \geq \max\{\mathrm{diam}(T)-2, 4\}$ in which no four cycles of length $g(G)$ have a common path of length $\left\lceil \frac{g(G)}{2} \right\rceil - 1$ contains a subtree $T' \cong T$ such that $G - V(T')$ is 2-connected.*

Theorems 7 and 8 follow from Theorem 11 with Lemma 2 and Theorem 12, respectively. Manipulations in the proofs of Lemmas 7 and 8 can be done in $O(|E(G)|)$ time, although they are more complicated than those in the proof of Lemma 6. Therefore, we can find a desired subtree T' in Theorem 7 (respectively, Theorem 8) in $O(|V(G)|^4)$ time if $g(G) \geq \delta(G)-5$ (respectively, $g(G) \geq \delta(G)-4$).

6   Concluding Remarks

In this paper, we have shown that Mader's conjecture for $k = 2$ (with a weak degree condition $\delta(G) \geq m + 1$) holds for graphs with large girth. Mader's conjecture was posed in a purely mathematical interest; however, it has a potential application to fault-tolerant problems in communication networks. We then have shown that our constructive proofs lead to $O(|V(G)|^4)$ time algorithms.

Our lower bounds on the girth in Theorems 5 and 7 can be improved if the upper bound on the diameter of a tree for which Mader's conjecture for $k = 2$ holds is improved. Namely, the following result follows from Theorem 9.

Theorem 13. *If Mader's conjecture for $k = 2$ holds for any tree T with $\mathrm{diam}(T) \geq |V(T)| - \ell$, then Mader's conjecture for $k = 2$ holds for any 2-connected graph G with $g(G) \geq \delta(G) - \ell - 4$.*

In particular, by checking the proof in [7], we can see that Statement 2 holds for any caterpillars; thus, the lower bounds on $g(G)$ in Theorems 6 and 8 can be improved to $\delta(G) - 5$ and $\delta(G) - 6$, respectively. Besides, the restriction that $g(G) \geq \delta(G)-4$ (respectively, $g(G) \geq \delta(G)-5$) for an $O(|V(G)|^4)$ time algorithm can be removed for Theorem 5 (respectively, Theorem 7). On the other hand, in order to improve the lower bounds on the girth in Theorems 9, 10, 11, and 12 directly, we may need some other techniques.

Even though Mader's conjecture for $k = 2$ still remains open, from Lemma 5 and Corollary 2, we may conjecture the following.

Conjecture 4. For any tree T of order $m \geq 2$, every 2-connected graph G with $\delta(G) \geq m + 1$ and $g(G) \geq 4$ contains a subtree $T' \cong T$ such that $G - V(T')$ is 2-connected.

Although we consider Mader's conjecture only for $k = 2$, it would be interesting to approach Mader's conjecture for general $k \geq 2$ by considering girth conditions.

Acknowledgments. The author is grateful to the reviewers for their helpful comments. This work was supported by JSPS KAKENHI Grant Number JP19K11829.

References

1. Chartrand, G., Kaigars, A., Lick, D.R.: Critically n-connected graphs. Proc. Am. Math. Soc. **32**, 63–68 (1972)
2. Diwan, A.A., Tholiya, N.P.: Non-separating trees in connected graphs. Discrete Math. **309**, 5235–5237 (2009)
3. Fujita, S., Kawarabayashi, K.: Connectivity keeping edges in graphs with large minimum degree. J. Comb. Theory Ser. B **98**, 805–811 (2008)
4. Hasunuma, T., Ono, K.: Connectivity keeping trees in 2-connected graphs. J. Graph Theory **94**, 20–29 (2020)
5. Locke, S.C.: Problem 10647. MAA Mon. **105**, 176 (1998)
6. Lu, C., Zhang, P.: Connectivity keeping trees in 2-connected graphs. Discrete Math. **343**, 111677 (2020)
7. Lu, C., Ye, Q.: Connectivity keeping caterpillars in 2-connected graphs. Manuscript (2019)
8. Mader, W.: Connectivity keeping paths in k-connected graphs. J. Graph Theory **65**, 61–69 (2010)
9. Mader, W.: Connectivity keeping trees in k-connected graphs. J. Graph Theory **69**, 324–329 (2012)
10. Tian, Y., Meng, J., Lai, H.-J., Xu, L.: Connectivity keeping stars or double stars in 2-connected graphs. Discrete Math. **341**, 1120–1124 (2018)
11. Tian, Y., Xu, L., Meng, J., Lai, H.-J.: Nonseparating trees in 2-connected graphs and oriented trees in strongly connected digraphs. Discrete Math. **342**, 344–351 (2019)
12. Tutte, W.T.: Connectivity in Graphs. Univ. of Toronto Press, Toronto (1966)

The Steiner Problem for Count Matroids

Tibor Jordán[1]([✉]), Yusuke Kobayashi[2], Ryoga Mahara[2], and Kazuhisa Makino[2]

[1] Department of Operations Research, Eötvös University, and the MTA-ELTE Egerváry Research Group on Combinatorial Optimization, Pázmány Péter sétány 1/C, Budapest 1117, Hungary
jordan@cs.elte.hu
[2] Research Institute for Mathematical Sciences, Kyoto University, Kyoto 606-8502, Japan
{yusuke,ryoga,makino}@kurims.kyoto-u.ac.jp

Abstract. We introduce and study a generalization of the well-known Steiner tree problem to count matroids. In the count matroid $\mathcal{M}_{k,l}(G)$, defined on the edge set of a graph $G = (V, E)$, a set $F \subseteq E$ is independent if every vertex set $X \subseteq V$ spans at most $k|X| - l$ edges of F. The graph is called (k, l)-tight if its edge set is independent in $\mathcal{M}_{k,l}(G)$ and $|E| = k|V| - l$ holds.

Given a graph $G = (V, E)$, a non-negative length function $w : E \to \mathbb{R}$, a set $T \subseteq V$ of terminals and parameters k, l, our goal is to find a shortest (k, l)-tight subgraph of G that contains the terminals. Since $\mathcal{M}_{1,1}(G)$ is isomorphic to the graphic matroid of G, the special case $k = l = 1$ corresponds to the Steiner tree problem. We obtain other interesting problems by choosing different parameters: for example, in the case $k = 2$, $l = 3$ the target is a shortest rigid subgraph containing all terminals.

First we show that this problem is NP-hard even if $k = 2$, $l = 3$, and w is metric, or $w \equiv 1$ and $|T| = 2$. As a by-product of this result we obtain that finding a shortest circuit in $\mathcal{M}_{2,3}(G)$ is NP-hard.

Then we design a $(k + 1)$-approximation algorithm for the metric version of the problem with parameters $(k, k + 1)$, for all $k \geq 2$. In particular, we obtain a 3-approximation algorithm for the Steiner version of the shortest rigid subgraph problem. We also show that the metric version can be solved in polynomial time for $k = 2$, $l = 3$, provided $|T|$ is fixed.

Keywords: Count matroid · Steiner problem · Rigid graph

This work was supported by the Research Institute for Mathematical Sciences, an International Joint Usage/Research Center located in Kyoto University, the JSPS KAKENHI grant no. JP18H05291, and the Hungarian Scientific Research Fund grant no. K 109240. The first author was also supported by Project ED-18-1-2019-030 (Application-specific highly reliable IT solutions), which has been implemented with the support provided from the National Research, Development and Innovation Fund of Hungary, financed under the Thematic Excellence Programme funding scheme.

L. Gąsieniec et al. (Eds.): IWOCA 2020, LNCS 12126, pp. 330–342, 2020.
https://doi.org/10.1007/978-3-030-48966-3_25

1 Introduction

Let k be a positive integer and let l be an integer satisfying $2k - l \geq 1$. We say that a graph $G = (V, E)$ is (k, l)-*sparse* if

$$i_G(X) \leq k|X| - l, \text{ for all } X \subseteq V \text{ with } |X| \geq 2,$$

where $i_G(X)$ denotes the number of edges induced by X in G. The graph is called (k, l)-*tight* if it is (k, l)-sparse and $|E| = k|V| - l$ holds. It is well-known that the edge sets of the (k, l)-sparse subgraphs of a graph G form the independent sets of a matroid, defined on the edge set of G. This matroid, denoted by $\mathcal{M}_{k,l}(G)$, is called the *count matroid* of G, with parameters k, l, see e.g. [5, 17].

For a graph $G = (V, E)$ and set $T \subseteq V$ of terminal vertices, we say that a subgraph $H = (V', E')$ of G is T-(k, l)-*tight* if H is (k, l)-tight and $T \subseteq V'$. Given a graph $G = (V, E)$, a terminal set $T \subseteq V$, a length function $w : E \to \mathbb{R}_+$, and parameters k, l, the *shortest* T-(k, l)-*tight subgraph problem* is to find a T-(k, l)-tight subgraph H of G with minimum total edge-length. If G is a complete graph and w is metric (that is, w satisfies the triangle inequality), this problem is called the *metric shortest* T-(k, l)-*tight subgraph problem*. Note that we use $\mathbb{R}_+$ to denote the set of non-negative real numbers.

Since $\mathcal{M}_{1,1}(G)$ is isomorphic to the graphic matroid of G, the special case $k = l = 1$ corresponds to the Steiner tree problem. Although we may obtain other interesting optimization problems by choosing different parameters (see below), this is the only special case of our general problem - that we call the *Steiner problem for count matroids* - that has been studied before.

1.1 Previous Work

The Steiner tree problem is one of the fundamental problems in combinatorial optimization: given a graph $G = (V, E)$, a terminal set $T \subseteq V$, a length function $w : E \to \mathbb{R}_+$, find a shortest tree in G which contains all terminal vertices. It is NP-hard. It is known that there is an approximation factor preserving reduction to its metric version. The best known approximation factor, due to Byrka et al. [2], is 1.39. It is also well-known that it can be solved in polynomial time if $|T| = 2$ (which is a shortest path problem) and more generally, if $|T|$ is fixed. This problem has numerous other versions and extensions, see e.g. [3, 4].

A related notion, which is also relevant in the context of count matroids, is the Steiner ratio. Consider a metric instance of a Steiner problem, in which we have a complete graph $G = (V, E)$, a terminal set $T \subseteq V$, and a length function $w : E \to \mathbb{R}_+$, and we wish to find a shortest subgraph H of G that contains all terminals and satisfies a given property. For example, we may want to find a connected subgraph, but we can also think of other properties (e.g. k-edge-connected or (k, l)-tight) satisfied by $G[T]$ (i.e. the complete subgraph of G induced by T).

Then the total length of an optimal solution divided by the length of a shortest spanning subgraph of $G[T]$ that satisfies the given property is called the *Steiner ratio* of the instance. The *Steiner ratio* of the (metric) problem is the best possible lower bound on the Steiner ratio that is valid for all instances.

Note that, just like in the Steiner tree problem, the shortest (k, l)-tight spanning subgraph of $G[T]$, if it exists, can be found in polynomial time by a greedy algorithm. It holds for all parameters k, l, due to the matroidal nature of the problem, see e.g. [5].

1.2   Motivation and New Results

Our motivation to introduce and study this problem comes from rigidity theory and its applications. In this area count matroids play an important role. For example, a graph (realized as a generic two-dimensional bar-and-joint structure) is rigid if and only if it contains a $(2, 3)$-tight spanning subgraph (see Sect. 2). Thus, by choosing $k = 2$ and $l = 3$ in our problem, we look for the shortest rigid subgraph of a graph that contains a designated set of vertices. Other well-studied parameters that show up in e.g. parallel drawing and in rigidity problems of body-bar and body-hinge frameworks include the cases when $l = k$ and $l = k+1$, for all $k \geq 2$. See [17] for more on these connections. Approximation algorithms for these counts may also be useful in variants of the sensor network localization problem, where rigidity theory plays a key role, see [7].

Another reason for investigating the complexity of the Steiner problem for count matroids is to have a better understanding of the problem of finding the girth of a (count) matroid, see [14,15]. We shall see that the problem of finding a shortest circuit containing a given element in a matroid $\mathcal{M}_{2,3}(G)$ is equivalent to the corresponding Steiner problem with two terminals.

We first show that the Steiner problem for count matroids is NP-hard, even if $k = 2$, $l = 3$, and w is metric, or $w \equiv 1$ and $|T| = 2$. The latter result settles the complexity status of the girth problem for count matroids with parameters $k = 2$, $l = 3$. It also illustrates that - apart from the graphic matroid (the Steiner tree problem) and the bicircular matroid (see Sect. 6) - the Steiner problem for count matroids is hard even for two terminals.

Then we give a $(k+1)$-approximation algorithm for the metric version for the counts $(k, k + 1)$, for all $k \geq 2$. This specializes to a 3-approximation algorithm for the shortest rigid subgraph problem. As a corollary we obtain that the Steiner ratio of the metric shortest T-$(k, k+1)$-tight subgraph problem is between $\frac{1}{2}$ and $\frac{1}{k+1}$.

We also show that the (metric) shortest T-$(2, 3)$-tight subgraph problem can be solved in polynomial time for fixed $|T|$. The algorithm is based on a structural result: we prove that there always exists an optimal solution H with $|V(H)| \leq 15|T| - 1$. It shows that, unlike in the case of the Steiner tree problem, the behaviour of the metric version is quite different from that of the case of general length functions. It is another new phenomenon for general counts.

We have similar results for the shortest T-(k, k)-tight subgraph problem for all $k \geq 2$. By a result of Nash-Williams (see Theorem 3 below) a graph is (k, k)-tight if and only if its edge set can be decomposed into k disjoint spanning trees. Although these graphs are well-studied and occur in important applications, we omit the results on (k,k)-tight subgraphs from this extended abstract: the $(k,k+1)$-tight case appears to be more involved and the methods used are similar.

2 Preliminary Results

In this section we make some preliminary observations and introduce some notions and earlier results we shall use in this paper.

2.1 The Extension Operation

We shall use the following operation on graphs several times. Let $G = (V,E)$ be a simple graph. The (k,i)-*extension* operation, for some integers $k \geq 1$ and $0 \leq i \leq k$, removes i edges $u_1v_1, u_2v_2, \ldots u_iv_i \in E$ from G, and adds a new vertex r and new edges $ru_1, \ldots, ru_i, rv_1, \ldots, rv_i, rw_1, \ldots, rw_{k-i}$, for some vertices $w_1, \ldots, w_{k-i}$ of G, in such a way that the resulting graph G' remains simple. Notice that the new vertex r has degree $k+i$ in G'.

The following lemma (which is implicit in [6]) is easy to verify. We remark that the lemma – with minor changes – holds for multigraphs, too. In this paper we restrict ourselves to simple graphs.

Lemma 1. *Let $G = (V,E)$ be a $(k,k+1)$-tight simple graph and suppose that G' is obtained from G by a (k,i)-extension operation for some $0 \leq i \leq k$. Then G' is also $(k,k+1)$-tight.*

As the first application of Lemma 1 we show that for every $t \geq 2k+1$ there exist $(k,k+1)$-tight graphs on t vertices.

Lemma 2. *Let k and t be integers with $k \geq 2$ and $t \geq 2k+1$. Define $C_{t,k}$ as the graph whose vertex set and edge set are $\{x_1, \ldots, x_t\}$ and $\{x_ix_{i+1}, x_ix_{i+2}, \ldots, x_ix_{i+k} \mid i \in \{1, \ldots, t\}\}$, respectively, where we denote $x_{t+j} = x_j$ for $j = 1, \ldots, t$. Let $C'_{t,k} := C_{t,k} - \{x_1x_t, x_1x_{t-1}, \ldots, x_1x_{t-k+1}, x_kx_t\}$. Then, $C'_{t,k}$ is a $(k,k+1)$-tight graph.*

Proof. We show that $C'_{t,k}$ is a $(k,k+1)$-tight graph by induction on t. We first consider the case of $t = 2k+1$. Let K_{2k} be the complete graph with $2k$ vertices $x_2, x_3, \ldots, x_{2k}, x_{2k+1}$. Then, $K_{2k} - \{x_kx_{2k+1}\}$ is a $(k,k+1)$-tight graph by a simple counting argument. Since $C'_{2k+1,k}$ is obtained from $K_{2k} - \{x_kx_{2k+1}\}$ by a $(k,0)$-extension operation (which adds a new vertex x_1 and k new edges $x_1x_2, x_1x_3, \ldots, x_1x_k, x_1x_{k+1}$), we have that $C'_{2k+1,k}$ is a $(k,k+1)$-tight graph by Lemma 1. This shows the base case of the induction.

To show the induction step, assume that $C'_{t,k}$ is a $(k,k+1)$-tight graph. We observe that $x_ix_{t-k+i} \in E(C'_{t,k})$ for $i = 2, 3, \ldots, k-1$. Since $C'_{t+1,k}$ is obtained from $C'_{t,k}$ by a $(k,k-2)$-extension operation (which adds a new vertex x_{t+1} together with $2k-2$ new edges $x_{t+1}x_i$ for $i = 2, 3, \ldots, k-1, t-k, t-k+1, \ldots, t-1$ and removes x_ix_{t-k+i} for $i = 2, 3, \ldots, k-1$), we have that $C'_{t+1,k}$ is a $(k,k+1)$-tight graph by Lemma 1. This completes the proof. $\qquad\square$

2.2   Rigid Graphs

We say, somewhat informally, that a graph $G = (V, E)$ is generically rigid in the plane if every bar-and-joint framework in the plane with underlying graph G and with generic vertex coordinates is rigid: that is, every continuous motion of the vertices in the plane that preserves the edge lengths preserves all pairwise distances. Laman [12] proved that G is generally rigid if and only if it has a $(2, 3)$-tight spanning subgraph (or equivalently, its rigidity matroid $\mathcal{M}_{2,3}(G)$ has rank $2|V| - 3$). See [17] for an introduction to rigidity theory and for further count parameters that show up in this field, and [9] for more details on the combinatorial and matroidal aspects of two-dimensional rigidity.

Thus the Steiner problem for count matroids contains the problem of finding a shortest rigid subgraph containing a given set of terminals. Since we shall mostly focus on this special case, for simplicity we shall also use T-*rigid* instead of saying that a subgraph which has a T-$(2, 3)$-tight spanning subgraph. In this context minimally T-rigid corresponds to T-$(2, 3)$-tight.

The extension operations with parameters $(2, 0)$ and $(2, 1)$ introduced above play an important role in rigidity theory. If the parameter $k = 2$ is clear from the context we use 0-extension and 1-extension to mean a $(2, 0)$- or $(2, 1)$-extension operation, respectively.

Lemma 3 [9]. *Let $G = (V, E)$ be a minimally rigid graph and suppose that G' is obtained from G by a 0-extension or a 1-extension operation. Then G' is minimally rigid.*

2.3   Feasibility, Components, and Sparse Input Graphs

In this subsection we consider $(2, 3)$-sparsity (and rigidity), but the results easily extend to all counts studied in this paper.

A basic question concerning an instance of the Steiner problem for count matroids is whether there exists a feasible solution. The answer is based on the concept of *rigid components*: a rigid component of a graph G is a maximal rigid subgraph. It is known that two rigid components have at most one vertex in common and that the family of rigid components can be found in polynomial time [9]. Since $|T| \geq 2$, it follows that all feasible solutions, if they exist, are subgraphs of the same rigid component of G. Furthermore, there is a feasible solution if and only if G has a rigid component which contains all the terminals. In this case we can simply delete the complement of this rigid component and assume that the input graph is rigid.

Next suppose that the input graph $G = (V, E)$ is minimally rigid, that is, rigid and sparse at the same time. A useful observation is that the shortest T-rigid subgraph problem has a simple and efficient solution in this case. It follows from the next lemma, see e.g. [9].

Lemma 4. *Let $G = (V, E)$ be a minimally rigid graph and let G_1, G_2 be minimally rigid subgraphs of G with $|V(G_1) \cap V(G_2)| \geq 2$. Then $G_1 \cap G_2$ is minimally rigid.*

Thus there is a unique smallest rigid subgraph of G that contains T. Since w is non-negative, it is an optimal solution.

The following result shows that we can find this smallest rigid subgraph efficiently. For a given $S \subseteq V$ with $|S| \geq 2$ let $C_S(G)$ be the unique smallest rigid subgraph of G with $S \subseteq V(C_S(G))$. If $S = \{a, b\}$ then we also use the notation $C_{a,b}(G)$.

Lemma 5 [10]. *Let $G = (V, E)$ be a minimally rigid graph and $S \subseteq V$ with $|S| \geq 2$. Then*

$$C_S(G) = \bigcup_{a,b \in S} C_{a,b}(G).$$

Lemma 5 shows that we can compute $C_S(G)$ by computing $C_{a,b}(G)$ for all pairs in S. It is not hard to see that for a given pair $a, b \in S$ the (edge set of) $C_{a,b}(G)$ is either ab (if a and b are adjacent) or it is equal to the fundamental circuit of ab with respect to E (which is a base in the count matroid $\mathcal{M}_{2,3}(G)$). Since we have polynomial time independence oracles (using network flows, bipartite matchings, or graph orientations [1,13]), we can find all $C_{a,b}(G)$'s and $C_S(G)$ in polynomial time.

Finally, consider the case when $p := |E| - (2|V| - 3)$ is a fixed constant for the input graph $G = (V, E)$. Let $m = |E|$. Then G has at most $\binom{m}{p}$ minimally rigid spanning subgraphs.

Since every (minimally rigid) feasible solution can be extended to a minimally rigid spanning subgraph of G, and there is a unique smallest optimal solution whenever the input is minimally rigid, we can find an optimal solution by enumerating all minimally rigid spanning subgraphs of G and computing the unique smallest rigid subgraph containing T in each of them.

Proposition 1. *The shortest T-rigid subgraph problem is polynomial time solvable if $p := |E| - (2|V| - 3)$ is a fixed constant for the input graph $G = (V, E)$.*

3   Hardness Results

The proof of the next lemma is given in the full version of the paper [11].

Lemma 6. *The shortest T-rigid subgraph problem is NP-hard even if $w(e) = 1$ for every $e \in E$.*

We can strengthen Lemma 6 as follows.

Theorem 1. *The shortest T-rigid subgraph problem is NP-hard even if $w(e) = 1$ for every $e \in E$ and $|T| = 2$.*

Proof. Lemma 6 shows that the shortest T-rigid subgraph problem is NP-hard even if $w(e) = 1$ for every $e \in E$. We reduce this problem to the case of $|T| = 2$.

Let $G = (V, E)$ and $T \subseteq V$ be an instance of the shortest T-rigid subgraph problem such that $|T| \geq 3$ and $w(e) = 1$ for every $e \in E$. Pick up two distinct terminals $t_1, t_2 \in T$ arbitrarily. Construct a new graph $G' = (V', E')$ from G by adding a new vertex v together with two edges vt_1 and vt_2. Let $T' = (T \setminus \{t_1, t_2\}) \cup \{v\}$. Then, the obtained instance (G', T') is equivalent to the original instance (G, T) in the following sense. If G contains a T-rigid subgraph $H = (V_H, E_H)$ with k edges, then $H' = (V_H \cup \{v\}, E_H \cup \{vt_1, vt_2\})$ is a T'-rigid subgraph of G' with $k + 2$ edges. Conversely, if G' contains a T'-rigid subgraph $H' = (V_{H'}, E_{H'})$ with $k + 2$ edges, then $H = (V_{H'} \setminus \{v\}, E_{H'} \setminus \{vt_1, vt_2\})$ is a T-rigid subgraph of G with k edges by Lemma 3.

By repeating this procedure $|T| - 2$ times, we obtain a graph $G^* = (V^*, E^*)$ and $T^* \subseteq V^*$ with $|T^*| = 2$ such that G contains a T-rigid subgraph with k edges if and only if G^* contains a T^*-rigid subgraph with $k + 2(|T| - 2)$ edges. This shows that the original shortest T-rigid subgraph problem can be reduced to the case of $|T| = 2$, and hence this problem is NP-hard even when $|T| = 2$. $\square$

A corollary of Theorem 1, which appears to be new (see [15]) that finding a shortest circuit in a matroid $\mathcal{M}_{2,3}(G)$ is NP-hard. To see this consider a graph $G = (V, E)$ and a designated edge $f = uv \in E$. It is known (see e.g. [9]) that if $C \subseteq E$ is a circuit of $\mathcal{M}_{2,3}(G)$ then $(V(C), C)$ is rigid. Furthermore, if H is a rigid subgraph of $E - f$ then $H + f$ contains a circuit. Thus a shortest T-rigid subgraph of $E - f$ with respect to $T = \{u, v\}$ and $w \equiv 1$ corresponds to a shortest circuit containing f in $\mathcal{M}_{2,3}(G)$.

The metric version is also hard - see [11] for the details.

Theorem 2. *The metric shortest T-rigid subgraph problem is NP-hard.*

In the rest of the paper we shall consider the metric version and design approximation algorithms as well as an exact algorithm (for fixed $|T|$).

4   An Approximation Algorithm for the Metric Case

Let $G = (V, E)$, $T \subseteq V$, $w : E \to \mathbb{R}_+$ be an instance of the metric shortest T-$(k, k + 1)$-tight subgraph problem, for some $k \geq 2$. We shall prove that the total length of a shortest T-$(k, k+1)$-tight spanning subgraph of $G[T]$ is at most $(k + 1)OPT$, where OPT denotes the total length of an optimal solution to the shortest T-$(k, k + 1)$-tight subgraph problem. Since a shortest T-$(k, k + 1)$-tight spanning subgraph of $G[T]$ can be found in polynomial time, this leads to a $(k + 1)$-approximation algorithm. In particular, we obtain a 3-approximation algorithm for the shortest T-rigid subgraph problem.

In our analysis we shall use the following theorem of Nash-Williams.

Theorem 3 [16]. *The edge set of a graph $G = (V, E)$ can be partitioned into the edge sets of k forests if and only if $i_G(X) \leq k|X| - k$ for all $\emptyset \neq X \subseteq V$.*

A simple counting argument shows that $G[T]$ does not contain a $(k, k + 1)$-tight spanning subgraph if $|T| \leq 2k - 1$ (except for $k = 2$). Otherwise we do have a feasible solution on vertex set T, c.f. Lemma 2.

Theorem 4. *Let k be an integer with $k \geq 2$. Suppose that we are given a complete graph $G = (V, E)$, a terminal set $T \subseteq V$, and a metric length function $w : E \to \mathbb{R}_+$. If $|T| \geq 2k$, then a shortest T-$(k, k+1)$-tight spanning subgraph of $G[T]$ is a $(k+1)$-approximate solution for the metric shortest T-$(k, k+1)$-tight subgraph problem in G.*

Proof. Let $H = (V_H, E_H)$ be a shortest T-$(k, k+1)$-tight subgraph of G. Our goal is to show that $G[T]$ contains a T-$(k, k+1)$-tight subgraph whose total length is at most $(k+1)w(H)$. For simplicity we shall use $w(J)$ to denote the total length of the edges of some graph J.

Let $e \in \binom{V_H}{2}$ be a shortest edge with both endvertices in V_H. Consider the graph $H + e$ that might have parallel edges. By Theorem 3, the edge set of $H + e$ can be partitioned into k edge-disjoint spanning trees $F_1, \ldots, F_k$ on V_H. By changing the indices if necessary, we may assume that $w(F_1) \leq \frac{w(H)+w(e)}{k}$. Consider the graph obtained from F_1 by duplicating every edge, which is a connected Eulerian graph. Then, it contains an Eulerian walk through all vertices in V_H. Since w is metric and $T \subseteq V_H$, by shortcutting[1] this Eulerian walk, we obtain a cycle C such that $V(C) = T$ and

$$w(C) \leq 2w(F_1) \leq \frac{2}{k}(w(H) + w(e)). \tag{1}$$

Let $x_1, \ldots, x_t$ be the vertices of C that appear in this order along C, where $t = |T|$. For notational convenience, we denote $x_{t+j} = x_j$ for $j = 1, \ldots, t$. We consider the following two cases separately.

Case 1. We first consider the case when $t = 2k$. Let K_{2k} be the complete graph with vertex set $\{x_1, \ldots, x_{2k}\}$ and pick an edge f in K_{2k} arbitrarily. Since the metric property implies that $w(x_i x_{i+h}) \leq w(x_i x_{i+1}) + w(x_{i+1} x_{i+2}) + \cdots + w(x_{i+h-1} x_{i+h})$ for $i \in \{1, \ldots, 2k\}$ and for $h \in \{1, \ldots, k\}$, we have

$$w(K_{2k} - f) = \sum_{i=1}^{2k} (w(x_i x_{i+1}) + \cdots + w(x_i x_{i+k-1})) + \sum_{i=1}^{k} w(x_i x_{i+k}) - w(f)$$

$$\leq \frac{k(k-1)}{2} w(C) + k(w(C) - w(e)) - w(e)$$

$$= \frac{k(k+1)}{2} \left(w(C) - \frac{2}{k} w(e) \right)$$

$$\leq (k+1)w(H),$$

where we use (1) in the last inequality. Furthermore, we see that $K_{2k} - f$ is a $(k, k+1)$-tight spanning subgraph of $G[T]$ by a simple counting argument. Therefore, $G[T]$ contains a T-$(k, k+1)$-tight subgraph whose total length is at most $(k+1)w(H)$.

---

[1] We follow the walk W and we shortcut every maximal subwalk that contains only non-terminal vertices and vertices already visited by W.

Case 2. We next consider the case when $t > 2k$. Let $C'_{t,k}$ be the $(k, k+1)$-tight subgraph of $G[T]$ defined in Lemma 2. Then, by a similar calculation to Case 1, we obtain

$$
\begin{aligned}
w(C'_{t,k}) &= \sum_{i=1}^{2k}(w(x_i x_{i+1}) + \cdots + w(x_i x_{i+k})) - w(\{x_1 x_t, x_1 x_{t-1}, \ldots, x_1 x_{t-k+1}, x_k x_t\}) \\
&\leq \frac{k(k+1)}{2}w(C) - (k+1)w(e) \\
&= \frac{k(k+1)}{2}\left(w(C) - \frac{2}{k}w(e)\right) \\
&\leq (k+1)w(H).
\end{aligned}
$$

Therefore, $G[T]$ contains a T-$(k, k+1)$-tight subgraph whose total length is at most $(k+1)w(H)$.   □

Since a shortest $(k, k+1)$-tight spanning subgraph of $G[T]$ can be computed by a greedy algorithm, this theorem yields a $(k+1)$-approximation algorithm for the metric shortest T-$(k, k+1)$-tight subgraph problem with $|T| \geq 2k$.

By specializing the above result to the case when $k = 2$, we obtain:

Corollary 1. *There is a polynomial time 3-approximation algorithm for the metric shortest T-rigid subgraph problem.*

The following example shows that the approximation factor of the above algorithm is not better than 2. Suppose that every edge in $G[T]$ has length 2, and every other edge has length 1. Then the shortest rigid spanning subgraph of $G[T]$ has total length $4|T| - 6$. On the other hand the optimum is at most $2|T| + 1$: pick two vertices $a, b \in V - T$ and consider the complete bipartite subgraph $K_{X,T}$ with color classes $X = \{a, b\}$ and T. By adding the edge ab to this graph we obtain a feasible solution (a rigid subgraph of G containing T) of total length $2|T| + 1$.

Note that if $w(e) \in \{1, 2\}$ for all $e \in E$ then the approximation ratio of the above algorithm is not worse than 2. Hence, by the same example, it is equal to 2.

Corollary 2. *Let r be the Steiner ratio of the metric shortest T-rigid subgraph problem. Then $\frac{1}{3} \leq r \leq \frac{1}{2}$.*

5   Optimal Solutions for Fixed $|T|$ in the Metric Case

Consider an optimal solution H to some instance of the metric shortest T-rigid subgraph problem. One strategy to show that the number of non-terminal vertices in H is small (compared to $|T|$), or can be made small, is to apply specific shortcutting operations that remove vertices (or sets of vertices) of $V(H) - T$ maintaining rigidity and without increasing the total length.

This strategy works easily in the metric Steiner tree problem since degree-one vertices can be removed, degree-two vertices can be shortcut, and hence an upper bound on $|V(H) - T|$, in terms of $|T|$, follows immediately. This approach, with much more complicated arguments, works in the k-edge-connected Steiner network problem, too, see [8].

In our case H is a minimally rigid graph that contains T. It is easy to eliminate vertices of degree-two and degree-three from H (see Lemma 7 below. The number of vertices of degree at least five can be bounded by using the fact that $|E(H)| = 2|V| - 3$ and hence the average degree of H is (a bit less than) four.

Thus the main question is whether the number of degree-four vertices in H can be bounded by a function of $|T|$. We deal with this question in the next subsection.

5.1 Reductions in Minimally Rigid Graphs

Let $G = (V, E)$ be a minimally rigid graph and let $v \in V$ be a designated vertex with $d(v) = r$, where $d(v)$ denotes the degree of vertex v. The *reduction* operation at v removes v from the graph and adds $r - 2$ disjoint edges connecting vertices in $N_G(v)$ (where $N_G(v)$ denotes the set of neighbours of v in G). We shall be interested in the cases when $2 \leq r \leq 4$. A reduction operation is *admissible* if the resulting graph is also minimally rigid. We call v *admissible* if there exists an admissible reduction at v. Otherwise v is *non-admissible*.

The following lemma is well-known, see e.g. [9]. It shows that vertices of degree two and three are all admissible.

Lemma 7. *Let $G = (V, E)$ be a minimally rigid graph and $v \in V$. Then*

(i) if $d(v) = 2$ then $G - v$ is minimally rigid,
(ii) if $d(v) = 3$ then there is an admissible reduction at v.

Vertices of degree four may be non-admissible. In such a case there is a simple certificate of non-admissibility, as we shall prove below.

We say that $X \subseteq V$ with $|X| \geq 2$ is *critical* if $i_G(X) = 2|X| - 3$ holds. The next lemma is also well-known [9]. Its proof uses the fact that the function $i_G : 2^V \to Z$ is supermodular. For two disjoint sets X, Y we use $d(X, Y)$ to denote the number of edges between X and Y.

Lemma 8. *Let $G = (V, E)$ be a minimally rigid graph and let X, Y, Z be critical sets in G. Then*

(i) if $|X \cap Y| \geq 2$ then $X \cap Y$ and $X \cup Y$ are also critical,
(ii) if $|X \cap Y| = 1$ and $d(X - Y, Y - X) \geq 1$ then $X \cup Y$ is critical,
(iii) if $X \cap Y \cap Z = \emptyset$ and $|X \cap Y| = |X \cap Z| = |Y \cap Z| = 1$ then $X \cup Y \cup Z$ is critical.

Let v be a designated vertex with $d(v) = 4$. We say that three critical sets $X, Y, Z \subseteq V - \{v\}$ and a vertex $p \in V - \{v\}$ form a *flower* $\{X, Y, Z\}$ associated with v, with *core* p, if

(i) $X \cap Y = X \cap Z = Y \cap Z = \{p\}$,

(ii) $vp \in E$, and

(iii) $d(v, X - \{p\}) = d(v, Y - \{p\}) = d(v, Z - \{p\}) = 1$.

Observe that if there is a flower associated with v then v is non-admissible: adding a new edge connecting the core p to any other neighbour of v violates the sparsity condition in $V - \{v\}$.

The proofs of the next two key lemmas can be found in [11]. The first one shows that every non-admissible vertex of degree four has an associated flower.

Lemma 9. *Let $G = (V, E)$ be a minimally rigid graph and let $v \in V$ be a non-admissible vertex with $d(v) = 4$. Then there exists a flower associated with v in G.*

Lemma 10. *Let $G = (V, E)$ be a minimally rigid graph and let v be a non-admissible vertex of degree four. Suppose that $\{X, Y, Z\}$ form a flower associated with v with core p. If $d(p) = 4$ then v and p have three common neighbours.*

Note that if the conditions of Lemma 10 hold then v, p and their (common) neighbours induce a minimally rigid subgraph isomorphic to $K_{2,3}$ (plus the edge pv).

Theorem 5. *Let $G = (V, E)$, $T \subseteq V$, $w : E \to Z_+$ be an instance of the metric shortest T-rigid subgraph problem. Then there exists an optimal solution H with $|V(H)| \leq 15|T| - 1$.*

Proof. Let $H = (V', E')$ be an optimal solution for which $|V'|$ is as small as possible. We may assume that H is minimally rigid. Let $S = V' - T$ and $X = \{v \in S : d_H(v) = 4\}$. Since w is metric, we can use Lemma 7 to deduce that

(i) each vertex in S has degree at least four, and

(ii) each vertex in X is non-admissible.

Claim. Every vertex in X has at least one neighbour in $V' - X$.

Proof. Consider a vertex $v \in X$. Since it is non-admissible, there is a flower $\{X, Y, Z\}$ with core p in H associated with v. We have $pv \in E'$. For a contradiction suppose that $p \in X$. By Lemma 10 this implies that v and p have three common neighbours x, y, z and the set $N_H(v) \cup \{v\}$ induces a minimally rigid subgraph in H isomorphic to $K_{2,3}$ (plus the edge pv). It is not hard to see that $H' := H - \{v, p\} + \{xy, xz, yz\}$ is minimally rigid. Furthermore, $w(H') \leq w(H)$, since $w(xy) \leq w(xv) + w(vy)$, $w(xz) \leq w(xp) + w(pz)$, and $w(yz) \leq w(yp) + w(pv) + w(vz)$. Thus H' is a smaller optimal solution, which contradicts the choice of H. □

By the Claim we have

$$d_H(X) \geq |X|. \tag{2}$$

Let $Y = S - X$, and let T_i be the set, and t_i be the number of vertices of degree i in T, for $2 \leq i \leq 4$. Similarly, let T^+ be the set, and t^+ be the number, of vertices of degree at least five in T. Then we have

$$4|V'| - 6 = 2|E'| = \sum_{v \in V'} d_H(v) \geq 2t_2 + 3t_3 + 4t_4 + 5t^+ + 4|X| + 5|Y|, \quad (3)$$

from which

$$4(t_2 + t_3 + t^+ + |Y|) \geq 2t_2 + 3t_3 + 5t^+ + 5|Y| \quad (4)$$

follows. Thus

$$4t_2 + 4t_3 \geq 2t_2 + 3t_3 + t^+ + |Y|, \quad (5)$$

and hence

$$2|T| \geq 2(t_2 + t_3) \geq 2t_2 + t_3 \geq t^+ + |Y|. \quad (6)$$

So we have $|Y| \leq 2|T|$. Now suppose, for a contradiction, that $|V'| \geq 15|T|$. Since $|T \cup Y| \leq 3|T|$, we have $|X| \geq 12|T|$, and hence (2) gives

$$d_H(X, T \cup Y) = d_H(X) \geq 12|T|. \quad (7)$$

Therefore the average degree of the vertices in $T \cup Y$ is at least four in H. This implies

$$4|V'| - 6 = 2|E'| = \sum_{v \in X} d_H(v) + \sum_{v \in T \cup Y} d_H(v) \geq 4|X| + 4|V' - X| = 4|V'|, \quad (8)$$

a contradiction. Hence $|V'| \leq 15|T| - 1$, completing the proof of the theorem. $\square$

We can use this result to argue that if we compute a shortest rigid subgraph with vertex set V' for every $V' \subseteq V$ with $T \subseteq V'$ and $|V'| \leq 15|T| - 1$, the shortest one will correspond to an optimal solution to the shortest T-rigid subgraph problem. Since we can find a shortest rigid subgraph on V' in polynomial time for each V', we obtain:

Theorem 6. *The metric shortest T-rigid subgraph problem can be solved in polynomial time for fixed $|T|$.*

6 Concluding Remarks

The Steiner problem for count matroids, introduced in this paper, gives rise to numerous open problems. The most obvious ones are about potential improvements of the new results: better approximation factors, better bounds for the Steiner ratio, and extensions to further parameters (k, l).

Two Terminals. For the complexity status of the two-terminal case (with general length functions) there seems to be a clean answer. We conjecture that the proof of Theorem 1 can be extended to all count parameters (k, l) with $k \geq 2$. The remaining cases (assuming $l \geq 0$) are $(1, 1)$ and $(1, 0)$. The former case corresponds to the familiar shortest path problem, which is polynomial time solvable by using Dijkstra's algorithm. The latter case is also tractable. Recall that the count matroid $\mathcal{M}_{1,0}$ is the so-called *bicircular matroid*, in which a graph H is tight if and only if each connected component of H is *unicyclic*, that is, it has exactly one cycle.

Theorem 7. *The shortest* $(1,0)$-*tight subgraph problem with* $|T| = 2$ *is polynomial time solvable.*

The proof of this results as well as further comments and potential research directions are given in [11].

References

1. Berg, A.R., Jordán, T.: Algorithms for graph rigidity and scene analysis. In: Di Battista, G., Zwick, U. (eds.) ESA 2003. LNCS, vol. 2832, pp. 78–89. Springer, Heidelberg (2003). https://doi.org/10.1007/978-3-540-39658-1_10
2. Byrka, J., Grandoni, F., Rothvoss, T., Sanita, L.: An improved LP-based approximation ratio for Steiner tree. In: Proceedings of 42nd STOC, pp. 583–592 (2010)
3. Chung, F.R.K., Graham, R.L.: A new bound for Euclidean Steiner minimal trees. Ann. New York Acad. Sci. **440**, 328–346 (1985)
4. Du, D., Hu, X.: Steiner Tree Problems in Computer Communication Networks. World Scientific Publishing, River Edge (2008)
5. Frank, A.: Connections in Combinatorial Optimization. Oxford University Press, Oxford (2011)
6. Frank, A., Szegő, L.: Constructive characterizations for packing and covering with trees. Discrete Appl. Math. **131**, 347–371 (2003)
7. Jackson, B., Jordán, T.: Graph theoretic techniques in the analysis of uniquely localizable sensor networks. In: Mao, G., Fidan, B. (eds.) Localization Algorithms and Strategies for Wireless Sensor Networks, pp. 146–173. IGI Global, Hershey (2009)
8. Jordán, T.: On minimally k-edge-connected graphs and shortest k-edge-connected Steiner networks. Discrete Appl. Math. **131**, 421–432 (2003)
9. Jordán, T.: Combinatorial rigidity: graphs and matroids in the theory of rigid frameworks. Discrete Geom. Anal. MSJ Memoirs **34**, 33–112 (2016)
10. Jordán, T., Domokos, G., Tóth, K.: Geometric sensitivity of rigid graphs. SIAM J. Discrete Math **27**(4), 1710–1726 (2013)
11. Jordán, T., Kobayashi, Y., Mahara, R., Makino, K.: The Steiner problem for count matroids, TR-2020-03, Egerváry Research Group, Budapest
12. Laman, G.: On graphs and rigidity of plane skeletal structures. J. Eng. Math. **4**, 331–340 (1970)
13. Lee, A., Streinu, I.: Pebble game algorithms and sparse graphs. Disc. Math **308**, 1425–1437 (2008)
14. Lomonosov, A.: Graph and combinatorial algorithms for geometric constraint solving, Ph.D. thesis, U. Florida (2004)
15. Matroids - girth and co-girth. In: Egres Open, an open problem collection of the Egerváry Research Group, Budapest. lemon.cs.elte.hu/egres/open
16. Nash-Williams, C.St.J.A.: Decomposition of finite graphs into forests. J. Lond. Math. Soc. **39**, 12 (1964)
17. Whiteley, W.: Some matroids from discrete applied geometry. Contemp. Math. **197**, 171–311 (1996). In: Matroid theory (Seattle, WA, 1995), Amer. Math. Soc., Providence, RI

Bounded Degree Group Steiner Tree Problems

Guy Kortsarz[1] and Zeev Nutov[2(✉)]

[1] Rutgers University, Camden, Camden, USA
guyk@rutgers.edu
[2] The Open University of Israel, Ra'anana, Israel
nutov@openu.ac.il

Abstract. Motivated by some open problems posed in [13], we study three problems that seek a low degree subtree T of a graph $G = (V, E)$. In the MIN-DEGREE GROUP STEINER TREE problem we are given a collection of node subsets (groups), and T should contain a node from every group. In the MIN-DEGREE STEINER k-TREE problem we are given a set R of terminals and an integer k, and T should contain k terminals. In both problems the goal is to minimize the maximum degree of T.

In the more general DEGREES BOUNDED MIN-COST GROUP STEINER TREE problem, we are also given edge costs and individual degree bounds $\{b_v : v \in V\}$. The output tree T should obey the degree constraints $\deg_T(v) \leq b_v$ for all $v \in V$, and among all such trees we seek one of minimum cost. When the input is a tree, an $O(\log^2 n)$ approximation for the cost is given in [10]. Our first result generalizes [10] – we give a bicriteria $(O(\log^2 n), O(\log^2 n))$-approximation algorithm for DEGREES BOUNDED MIN-COST GROUP STEINER TREE problem on tree inputs. This matches the cost ratio of [10] but also approximates the degrees within $O(\log^2 n)$. Our second result shows that if MIN-DEGREE GROUP STEINER TREE admits ratio ρ then MIN-DEGREE STEINER k-TREE admits ratio $\rho \cdot O(\log k)$. Combined with [12], this implies an $O(\log^3 n)$-approximation for MIN-DEGREE STEINER k-TREE on general graphs, in quasi-polynomial time. Our third result is a polynomial time $O(\log^3 n)$-approximation algorithm for MIN-DEGREE GROUP STEINER TREE on bounded treewidth graphs.

1 Introduction

We study the following three problems:

MIN-DEGREE GROUP STEINER TREE
Input: A graph $G = (V, E)$ and a collection of groups (subsets of V).
Output: A subtree T of G that contains a node from every group and has minimal maximum degree.

L. Gąsieniec et al. (Eds.): IWOCA 2020, LNCS 12126, pp. 343–354, 2020.
https://doi.org/10.1007/978-3-030-48966-3_26

MIN-DEGREE STEINER k-TREE
Input: A graph $G = (V, E)$, a set $R \subseteq V$ of terminals, and an integer $k \leq |R|$.
Output: A subtree T of G that contains at least k terminals and has minimal maximum degree.

BOUNDED DEGREES MIN-COST GROUP STEINER TREE
Input: A graph $G = (V, E)$ with edge costs $\{c_e : e \in E\}$, a collection of groups, and degree bounds $\{b_v : v \in V\}$.
Output: A subtree T that contains a node from every group and obeys the degree constraints $\deg_T(v) \leq b_v$ for all $v \in V$, and has minimum costs among such subtrees.

Note that in the first two problems the edges have no costs, since the objective is to minimize the maximum degree. The third problem is more general and has both costs and degree bounds.

In the *The Eighth Workshop on Flexible Network Design, Amsterdam, 2016*, Hajiaghayi posed the following open problem:
Can we obtain a polylogarithmic approximation ratio (in polynomial time) for the BOUNDED DEGREES MIN-COST GROUP STEINER TREE *problem?*

In [12] is given a *quasi-polynomial time* bi-criteria $(O(\log^4 n), O(\log^2 n))$-approximation algorithm for the BOUNDED DEGREES MIN-COST DIRECTED STEINER TREE problem.[1] Hence the same holds for the group Steiner problems studied here. Our paper is motivated by the need to provide approximation algorithms that run in *polynomial time*, which is a standard definition of approximation. Our results are summarized in the following three theorems.

Theorem 1. BOUNDED DEGREES MIN-COST GROUP STEINER TREE *on tree inputs admits a bicriteria randomized $(O(\log^2 n), O(\log^2 n))$-approximation algorithm. Namely, the algorithm computes a tree T that contains at least one node from every group, has expected cost $O(\log^2 n)$ times the optimum cost, and with probability at least $1 - 1/n$ we have $\deg_T(v) = O(\log^2 n) \cdot b_v$ for all $v \in V$.*

This result generalizes the one of Garg, Konjevod, and Ravi [10] that gave the same expected ratio $O(\log^2 n)$ for the cost, but did not consider degree bounds. We note that a bicriteria $(O(\log^2 n), O(\log^3 n))$ approximation is known to some researchers, but getting ratio $O(\log^2 n)$ for the degrees requires some care.

Theorem 2. *If* MIN-DEGREE GROUP STEINER TREE *admits approximation ratio ρ then* MIN-DEGREE STEINER k-TREE *admits ratio $\rho \cdot O(\log k)$. Thus (by [12])* MIN-DEGREE STEINER k-TREE *admits an $O(\log^3 n)$-approximation algorithm that runs in quasi polynomial time.*

Theorem 3. MIN-DEGREE GROUP STEINER TREE *on bounded treewidth input graphs admits approximation ratio $O(\log^3 n)$.*

---

[1] In private communication, B. Laekhanukit reported that this bi-criteria approximation was recently improved to $(O(\log^2 n), O(\log^2 n))$.

Remark: MIN-DEGREE STEINER k-TREE on bounded treewidth graphs admits an exact polynomial time algorithm using dynamic programming (folklore), but is NP-hard even on planar graphs (by a reduction from HAMILTONIAN PATH). However, MIN-DEGREE GROUP STEINER TREE (without costs) is SET-COVER hard even on stars, and thus is $\Omega(\ln n)$ hard to approximate.

We mention some work on min-costs versions. The best ratio known for MIN-COST GROUP STEINER TREE on tree inputs is $O(\log^2 n)$ [10]; for a combinatorial algorithm with ratio $O(\log^{2+\epsilon} n)$ see [4]. This ratio for tree inputs is essentially tight due to the approximation threshold $\Omega(\log^{2-\epsilon} n)$ of [14]. In the case of general graph inputs, the graph is embedded into a tree distribution with stretch $O(\log n)$ [8,10], which gives ratio $O(\log^3 n)$. The k-MST problem admits ratio 2 [9], and this immediately implies ratio 4 for MIN-COST k-STEINER TREE.

Why the degree bounded versions of these problem are hard to approximate? For many classic problems (without degree bounds), good ratios are achieved using the Iterative Rounding Method, see [16,18]. This often allows to achieve good bicriteria ratios for the degree bounded versions. However, for many other problems, including the problems we consider, the existing approximation algorithms rely on different methods; e.g., in [18] it is mentioned that the Iterative Rounding Method seems to fail for problems when we need to connect only a specific number of terminals, as in the k-STEINER TREE problem. Another example is the MIN-COST DIRECTED STEINER TREE problem – the first step in all known approximation algorithms for this problem [3,11,17] is the height reduction of Zelikovsky [25], see also [15]. This gives unbounded degrees, as it works on the transitive closure of the graph. There is also a difficulty in dealing with the MIN-DEGREE GROUP STEINER TREE problem, because the known algorithms [4,10] for the min-cost case first reduce the graph to a random tree [1,8]. However, this increases the degrees, which means that this technique can not be used.

A logical step is to consider the easiest problems that are open. BOUNDED DEGREES MIN-COST GROUP STEINER TREE on bounded treewidth graphs is one of such problems. However even for this relatively simple problem no polylogarithmic ratio is known, see [12]. The MIN-COST GROUP STEINER TREE problem (without degree bounds) on bounded treewidth graph admits ratio $O(\log^2 n)$ [2]. The min-degree case, which is the MIN-DEGREE GROUP STEINER TREE problem on bounded treewidth graphs (namely, bounding the degrees with no costs), remained open until our paper.

In the rest of the introduction we discuss some additional motivations for studying min-degree problem without edge costs.

VLSI Network Design: The MIN-COST GROUP STEINER TREE problem was motivated by VLSI design. The goal is to connect a set of terminals to a designated root r by a min-cost tree, where each terminal has a set of multiple ports it can be placed at (ports of two different terminals may intersect). The set of different ports in which a terminal may be placed at, defines a group. The different possible location may be due to rotating, or mirroring, or both. While low cost is highly desirable, the cost is payed once, and later the VLSI circuit is applied constantly. In many cases low degrees allow faster computations. In [24], a natural VLSI problem is reduced to iteratively solving instances of the MIN-DEGREE STEINER k-TREE problem. This makes the latency of the VLSI computation low. Low degrees are also important for efficient layout of the VLSI circuit [22]. In the MULTICOMMODITY FACILITY LOCATION UNDER GROUP STEINER ACCESS problem [20], each facility belongs to a group Steiner tree. Short service times requires that such trees have low degrees.

The k-Multicast Problem in the Telephone Model: One of our main motivations for studying the MIN-DEGREE STEINER k-TREE problem is the TELEPHONE k-MULTICAST problem [23]. In this problem we are given an undirected graph, a node r, and a target k of terminals. We want to send a message from the root r, to at least k terminals, under the telephone model. In this model, the nodes that know the message can call at most one neighbor in a round, and send the message to this neighbor. This means that a round is a matching between nodes which know the message to nodes which do not. Note that every broadcasting scheme results in a directed tree in which the parent of a node u, is the node v, which sent u the message. The maximum degree in this multicast tree is a lower bound on the optimum, because at every round we can send the message to at most one child. Hence we need trees with k terminals and low maximum degree. Also note that the MIN-DEGREE STEINER k-TREE problem is the minimum degree (without cost) variant of two important and well studied problems: the k-MST and the k-STEINER TREE problems. Since these two problems are considered important, so are their minimum degree versions.

On-Line Degree Bounded Problems: Recently, MIN-COST/DEGREE GROUP STEINER TREE problems has been studied in the online setting [5–7]. Dehghani et al. [6] showed that it is not possible to approximate both cost and degrees in the on-line model, even when the input graph is a star. Namely, there exists an input demand sequence that forces any algorithm to pay a factor of $\Omega(n)$ for the cost or the degree violation. However the above papers are able to give polylogarithmic competitive ratios if there are only degree bounds but no costs, similarly to the problems we consider.

2   Degrees Bounded Min-Cost Group Steiner Tree Problem on Tree Inputs (Theorem 1)

We will assume that we know a node r that belongs to some optimal solution. We root the input tree T at r. For a group S let $\mathcal{A}_S = \{A \subseteq V : r \notin A, S \subseteq A\}$ be the family of cuts that separate the group S from r. Let $\mathcal{A} = \cup_{S \in \mathcal{S}} \mathcal{A}_S$ be the family of all cuts that separate r from some group. The edges with exactly one endpoint in a set A are denoted by $\delta(A)$. Also recall that c_e is the cost of an edge e, and let x_e be an indicator variable whether e is included in the solution. The algorithm of Garg, Konjevod, and Ravi [10] uses the following natural LP for the MIN-COST GROUP STEINER TREE problem

$$\min c \cdot x$$
$$\text{s.t. } x(\delta(A)) \geq 1 \ \forall A \in \mathcal{A}$$
$$x_e \geq 0 \qquad \forall e \in E$$

The authors of [10] give a special rounding method. For $e \in E$ let $p(e)$ be the parent edge of e, $p^2(e) = p(p(e))$ the parent edge of $p(e)$, and so on; namely, $p^i(e)$ is the ith edge on the path from e to the root. Add a dummy parent edge f of the root r and set $x_f = 1$. The algorithm of [10] connects a fraction of groups to the root by choosing every edge $e \in E$ with probability $x_e/x_{p(e)}$. Then the probability that an edge e of depth i is connected to the root is

$$\frac{x_e}{x_{p(e)}} \cdot \frac{x_{p(e)}}{x_{p(p(e))}} \cdots \frac{x_{p^{i-1}(e)}}{x_{p^i(e)}} \cdot \frac{x_{p^i(e)}}{x_f} = \frac{x_e}{x_f} = x_e.$$

Thus the expected cost of the edges that are connected to r is bounded by the value $c \cdot x$ of the LP solution. The key statement in [10] is:

Theorem 4 ([10])**.** *The probability that a specific group is connected the root by the above random process is $\Omega(1/\log N)$, where N is the maximum group size.*

Thus the expected number of iterations required to connect all groups to the root is $O(\log N \cdot \log k) = O(\log^2 n)$, where k is the number of groups, and therefore, this is the expected approximation ratio.

We use the same rounding as [10]. Since we need to bound the degrees of n nodes, we will require $\Theta(\log^2 n)$ iterations of the basic procedure. Let $\delta(v)$ be all the edges that lead from v to one of its children. Let e_v be the edge entering v from his parent. To deal with the degree bounds, we add the following valid constrains to the [10] LP:

$$x(\delta(v)) \leq x_{e_v} \cdot b_v \qquad \forall v \in V. \tag{1}$$

To see that these are valid inequalities, consider the characteristic vector x of an inclusion minimal feasible solution T. If $x_{e_v} = 0$ then $x(\delta(v)) = 0$, since $v \notin T$. If $x_{e_v} = 1$ then $x(\delta(v)) \leq b_v = x_{e_v} \cdot b_v$.

Corollary 1. *For every node v in the tree, $x(v)/x(e_v) \leq b_v$.*

The rounding process of [10] gives expected degree $x(v)/x(e_v) \leq b_v$ in every iteration. Adding degree constraints do not change the expected cost. We analyze the degrees approximation separately using the Chernoff bound (c.f. [19]). If X is a sum of n independent Bernoulli variables with mean μ, then for any $\rho > 0$

$$\Pr\left[X > (1+\rho)\mu\right] \leq \left(\frac{e^\rho}{(1+\rho)^{1+\rho}}\right)^\mu. \tag{2}$$

The degree of v results by $O(\log^2 n)$ iterations. In each round we have a Bernoulli sum of all the children of v that did not reach the root yet. The difficulty here is that the random Bernoulli variables are dependent. For simplicity of the analysis, we bound the degree by $O(\log^2 n)$ *independent* Bernoulli sums, that contains all neighbors of v in every round. This random variable may be *strictly* larger than the "real" degree. A child u can contribute more than 1 to the degree. However our random process gives a sum of independent Bernoulli variable which makes the analysis simpler. For a node v, we have a sum of $\delta(v) \cdot O(\log^2 n)$ independent Bernoulli variables. The expected degree is $\tau_v = O(\log^2 n) \cdot x(\delta(v))/x_{e_v}$ (see Corollary 1) and note that $x(\delta(v))/x_{e_v} \leq b_v$ is implied by the valid inequalities described above. Thus the expected degree is at most $O(\log^2 n) \cdot b_v$. We now bound the expectation of τ_v by three claims.

Claim. If $\tau_v \geq C \cdot \log n$ for some constant C then with probability $1 - 1/n^2$, $\deg(v) = O(\log^2 n) \cdot b_v$.

Proof. We have

$$\Pr\left[\deg(v) > 2\tau_v\right] \leq \left(\frac{e}{4}\right)^{C \log n} \leq \frac{1}{n^2}.$$

The last inequality holds for large enough C. Note that this implies that with probability $1 - 1/n^2$, $\deg(v) = O(\log^2 n) \cdot b_v$ (see Corollary 1). The ratio $O(\log^2 n)$ follows. $\qquad\square$

We now deal with nodes for which $1 \leq \tau_v \leq C \cdot \log n$ for some constant C.

Claim. If $1 \leq \tau_v \leq C \cdot \log n$, then $\deg(v) = O(\log^2 n)$ with probability $\geq 1 - 1/n^2$.

Proof. We know that $\tau_v \leq C \log n$. Set $(1 + \rho) = \log n$.

First we note that if we prove that $\Pr[\deg(v) \geq (1 + \rho)\tau_v] \leq 1/n^2$, then since $\rho = O(\log n)$ and $\tau_v = O(\log n)$ we get that with probability $1 - 1/n^2$ that $\deg(v) = O(\log^2 n)$. Since $b_v \geq 1$ this gives ratio $O(\log^2 n)$. We now prove the required inequality.

Since $\tau_v \geq 1$ we get from the Chernoff bound that:

$$\Pr[\deg(v) \geq (1 + \rho)\tau_v] \leq \frac{e^{\log n}}{(\log n)^{\log n}}.$$

For large enough n this probability is at most $1/n^2$. $\qquad\square$

The last case is $\tau_v < 1$.

Claim. If $\tau_v < 1$ then with probability $1 - 1/n^2$, $\deg(v) = O(\log n) \cdot b_v$.

Proof. We set $(1+\rho) = \log n/\tau_v$. Note that if $\deg(v) \leq (1+\rho) \cdot \tau_v$ then $\deg(v) = O(\log n)$. As $b_v \geq 1$ the ratio is $O(\log n)$. We now bound

$$\Pr[\deg(v) > (1 + \rho) \cdot \tau_v]$$

Consider the term:

$$\frac{e^\rho}{(1+\rho)^{(1+\rho)}} \leq \frac{e^{\log n/\tau_v}}{(\log n/\tau_v)^{\log n/\tau_v}}.$$

To get the Chernoff bound we should raise to above to the power τ_v. Raising this term to τ_v, the τ_v factor cancels in both exponents. Thus:

$$\Pr\left[\deg(v) \le (1+\rho)\tau_v\right] \le \frac{e^{\log n}}{(\log n/\tau_v)^{\log n}}.$$

Since $\tau_v < 1$ the above is bounded by

$$\frac{e^{\log n}}{(\log n)^{\log n}}.$$

and the above term is bounded by $1/n^2$ for large enough n. □

We got that with probability $1 - 1/n^2$, for a given v, $\deg(v) = O(\log^2 n) \cdot b_v$. By the union bound with probability $1 - 1/n$ for every v, $\deg(v) = O(\log^2 n) \cdot b_v$.

3   A Relation Between Min-Degree Steiner k-Tree and Min-Degree Group Steiner Tree (Theorem 2)

Assume that MIN-DEGREE GROUP STEINER TREE admits ratio ρ. We will show that then MIN-DEGREE STEINER k-TREE admits ratio $\rho \cdot O(\log k)$. We first give a simple randomized algorithm with expected ratio $\rho \cdot O(\log^2 k)$. Given a MIN-DEGREE STEINER k-TREE instance G, R, k, create $k/(5\log k)$ bins; the MIN-DEGREE GROUP STEINER TREE instance groups collection is formed by putting uniformly at random, each terminal to a random bin.

Definition 1. *Fix some optimum solution F for the* MIN-DEGREE STEINER k-TREE *instance with maximum degree d^* and terminal set R^*. Terminals in R^* are called* **true terminals,** *and a bin is* **full** *if it contains a true terminal.*

Lemma 1. *With probability at least $1 - 1/k$ each bin is full.*

Proof. Consider (only) the k true terminals in R^*. For each group S, $|S \cap R^*|$ is a binomial variable with probability $5\log k/k$ and k trials. Thus the expected size of $|S \cap R^*|$ is $\mu = 5\log k$. By the Chernoff bound:

$$\Pr\left[|S \cap R^*| \le (1-\rho)\mu\right] \le \exp(-\rho^2\mu/2).$$

We plug the right ρ so that $(1-\rho)\mu \le 1$. This gives ρ very close to 1. By the Chernoff bound $\Pr[S \cap R^* = \emptyset] \le 1/k^2$. By the union bound we get that with probability at least $1 - 1/k$ each bin is full. □

If we think of a bin as a group, since each group contains a true terminal, the optimum solution F (restricted to the true terminals) is a solution for the Group Steiner instance, with maximum degree d^*. Note that we need to cover only $k/(5\log k)$ groups which is *not* the MIN-DEGREE GROUP STEINER TREE problem. However, here is a trivial reduction to the MIN-DEGREE GROUP STEINER

TREE problem. Attach a complete binary tree to the root, with $k - k/5 \log k$ new leaves (we may need to trim the tree to get exactly $k - k/(5 \log k)$ leaves). Every new leaf belongs to all groups. Thus $k - k/(5 \log k)$ groups are covered for free with maximum degree 3. This still requires covering $k/(5 \log k)$ new terminals completing the reduction. The assumed algorithm will find a tree containing at least $k/(5 \log k)$ terminals, with maximum degree bounded by $\rho \cdot d^*$. Taking $O(\log^2 k)$ iterations gives expected ratio $O(\log^2 k \cdot \rho)$.

We now describe a more complicated deterministic reduction with factor loss $O(\log k)$ in the ratio. Let the terminals be $0, 1, \ldots, q - 1$, $q > k$, and assume that the above order of the terminals is random. We build k bins to serve as groups using two point based sampling (see [19]). Let p be a prime such that $4k \le p \le 8k$.

1. Choose a number a, at random, from $1, 2, \ldots, p - 1$.
2. Choose a number b, at random, from $0, 1, \ldots, p - 1$.
3. Terminal $0 \le i \le q - 1$ is assigned bin $((ai + b) \mod p) \mod k$.

The above construction defines the k groups. Group j contains all terminals that reached bin j.

Any true terminal i is first matched to a *random number* in $0, 1, \ldots, p$. The values that will cause item i to reach bin j are $j, j + k, \ldots, j + \alpha \cdot k$ for the maximum integer α such that $\alpha \cdot k \le p - 1$. In the worst case $j = k - 1$. Thus the question is how large is α in the inequality $(k - 1) + \alpha \cdot k \le p - 1$. Choosing $\alpha = (p - k)/k$ achieves the desired bound. Since α is an integer, clearly, $p/k - 2 \le \alpha < p/k$. Dividing by p, implies that the probability that the true terminal i reaches bin j is at least $1/k - 2/p$ and less than $1/k$.

Let X_{ij} be the event that *a true terminal i reaches bin j*. By the above, $\Pr(X_{ij}) \ge 1/k - 2/p$. The events "$i$ arrived to bin j" and "i' arrived to bin j" for $i \ne i'$ are pairwise independent and so $\Pr(i \text{ and } i' \text{ arrive to bin } j) \le 1/k^2$. We lower bound the probability that j is full, namely contains a true terminal, using the first two terms of the inclusion exclusion formula

$$\Pr\left[\bigcup_{i=0}^{k-1} X_{ij}\right] \ge k \cdot \left(\frac{1}{k} - \frac{2}{p}\right) - \frac{\binom{k}{2}}{k^2} \ge \frac{1}{2} - \frac{2}{p}.$$

Thus for every bin, the probability that it's full is at least $1/3$. The expected number of full bins is at least $k/3$. This gives a solution to the MIN-DEGREE GROUP STEINER TREE problem as follow. Select from every appropriate group (full bin) the true terminal, and connect them using the optimum tree F (restricted to the $k/3$ true terminals). Hence there exists a pair a, b in the sample space for which at least $k/3$ bins are full and this can be found via the assumed ρ ratio approximation for the MIN-DEGREE GROUP STEINER TREE. Our sample space of all a, b pairs has size bounded by $O(p^2) = O(k^2)$. Thus we try all a, b pairs with the goal of covering at least $k/3$ groups. For every pair a, b, we apply the assumed ρ ratio algorithm. For at least one of the a, b we get (with probability 1) a tree with maximum degree at most $\rho \cdot d^*$ that contains at least $k/3$ true terminals. Thus outputting the minimum maximal degree tree over all a, b choices guaranties (with probability 1) that the maximum degree in the tree is at most $\rho \cdot d^*$, and at least $k/3$ groups are covered. The penalty is an additional $O(\log k)$ factor (on top of the ρ factor).

In [12] the BOUNDED DEGREES GROUP STEINER TREE problem is given a polylogarithmic approximation that runs in quasi polynomial time. The best approximation ratio known is $O(\log^2 n)$ (this is slightly better than what appears in [12]. The better bound was reported to us by Bundit Laekhanukit, in a private communication. Thus we get:

Corollary 2. *The* MIN-DEGREE STEINER k-TREE *problem (on general graphs) admits an $O(\log^3 n)$ approximation in quasi-polynomial time.*

4   An $O(\log^3 n)$ Approximation for Min-Degrees Group Steiner Tree on Bounded Treewidth Graphs (Theorem 3)

The high level idea of the algorithm is as follows. We show a new method to reduce the graph into a tree with a loss of an $O(\log n)$ factor in the degrees. This process is similar to the one often applied on min-cost problems, that pay $O(\log n)$ penalty for transforming a general graph into a tree, c.f. [10]. Degree problems are often harder, and in our case we also need to pay an additional additive term of $O(\log n)$ (on the degrees) to get back to a graph solution.

We do not use the formal definition of a treewidth of a graph, but we use the fact that a bounded treewidth graph has a small balanced separator. A subset S of nodes in a graph G with n nodes is an α-**balanced separator** (or just a **balanced separator**, if α is clear from the context) if every connected component in $H \setminus S$, if any, has at most αn nodes. It is known that any graph G has a 2/3-balanced separator S of size $\le k$, where k equals the treewidth of G plus 1. We can use a linear time algorithm of [21] that finds a 4/5-balanced separator.

We may assume that the input graph G is connected and has at least k nodes. We construct an auxiliary rooted tree $\hat{T}$ by repeatedly removing a balanced separator S with $|S| \le k$ from a large enough connected component H.

Algorithm 1: SEPARATOR-TREE$(G = (V, E))$

---
1 $\mathcal{H} \leftarrow \{G\}$, $\mathcal{S} \leftarrow \emptyset$, $\hat{\mathcal{E}} \leftarrow \emptyset$
2 **while** there is unmarked graph $H \in \mathcal{H}$ with at least $k + 1$ nodes **do**
3 $\quad$ find a balanced separator S of H with $|S| \le k$ and add S to $\mathcal{S}$
4 $\quad$ mark H and add to $\hat{\mathcal{E}}$ an edge from H to S
5 $\quad$ **for** every connected component H_i of $H \setminus S$ **do**
6 $\quad\quad$ add H_i to $\mathcal{H}$ and add to $\hat{\mathcal{E}}$ an edge from S to H_i
7 **return** $\hat{T} = (\mathcal{H} \cup \mathcal{S}, \hat{\mathcal{E}})$

Let $\mathcal{L}$ be the set of unmarked components in $\mathcal{H}$ at the end of the algorithm. Every marked component $H \in \mathcal{H} \setminus \mathcal{L}$ has a unique child, and thus can be shortcut (or removed, if $H = G$); we denote the resulting tree by $T = (\mathcal{S} \cup \mathcal{L}, \mathcal{E})$. Note that $\mathcal{S} \cup \mathcal{L}$ is a partition V into sets of size at most k each. Also note that T has height $O(\log n)$ (since we used balanced separators) and that every edge of G either connects nodes in the same part or in parts such that in T one is a descendant of the other. Now we define certain trees and paths in G that are used later.

(a) For $S \in \mathcal{S} \cup \mathcal{L}$ the tree T^S is defined as follows. If $S \in \mathcal{S}$ then T^S is an inclusion minimal subtree of H that contains S, where H and S are in line 3 of the algorithm; and if $S \in \mathcal{L}$ then $H = G[S]$ and then T^S is a spanning tree in H. Note that T^S has max-degree $\leq k$.

(b) For an auxiliary edge $SS' \in \mathcal{E}$ where $S' = S_i$ is a child of S in T, let $P^{SS'}$ be the shortest path from S to S' in the graph induced by $H' \cup S$, where $H' = H_i$ is the connected component of $H \setminus S$ that contains $S' = S_i$ (see lines 5, 6 in the algorithm). Clearly, the path $P^{SS'}$ has max-degree 2 and all its nodes lie in S and descendant of S' in T.

Define a new MIN-DEGREE GROUP STEINER TREE instance with input graph being the tree T, where each node $S \in \mathcal{S} \cup \mathcal{L}$ of T belongs to all groups of the original instance that intersect S. The next two lemmas will enable us to finish the proof of Theorem 3. In what follows, for a node $v \in V$ let S_v denote the node of T that contains v.

Lemma 2. *If the original instance on G has a solution T of max-degree d then the new instance on T has a feasible solution T' of max-degree $d \cdot O(k \log n)$.*

Proof. For an edge $uv \in T$ let T_{uv} denote the unique $S_u S_v$-path in T (possibly $S_u = S_v$). We let $T' = \bigcup_{uv \in T} T_{uv}$. Since T is connected, T' is a tree; otherwise, T' has a partition into two part $\mathcal{C}$ and $\mathcal{C}'$ each containing a node from T', such that no edge of T connects these parts. It is also not hard to verify that T' is a feasible solution for the new instance, since each node $S \in \mathcal{S} \cup \mathcal{L}$ of T belongs to all groups of the original instance that intersect S.

We bound the max-degree of T'. Let S be a node in T'. Note that $\deg_{T'}(S)$ is at most the number β of branches hanging on S in T that have an edge of T going from the branch to an ancestor of S (including S) in T. The number of ancestors of S is $O(\log n)$ and the number of nodes in these ancestors is $O(k \log n)$. The T-degree of each node that lies in an ancestor of S is d, hence $\beta = d \cdot O(k \log n)$, concluding the proof. $\qquad\square$

Lemma 3. *There exists a polynomial time algorithm that given a feasible solution $T' = (\mathcal{S}' \cup \mathcal{L}', \mathcal{E}')$ for the new instance on T of max-degree d' constructs a feasible solution T' for the original instance of max-degree $d' + O(k \log n)$.*

Proof. Let G' be the graph formed by the trees $\{T^S : S \in \mathcal{S}' \cup \mathcal{L}'\}$ and the paths $\{P^{SS'} : SS' \in \mathcal{E}'\}$. Clearly, G' is connected, and any spanning tree T' in G' is a feasible solution for the original instance on G. We bound the max-degree of G'. We view T' as a rooted tree, where the root is the node of T' that is closest to the root of T. Consider a node v of G and the node S_v of T' that contains v. Let $\mathcal{P}_v$ be the path from S_v to the root of T'. The height of T is $O(\log n)$, thus $|\mathcal{P}_v| = O(\log n)$. We count the contribution of the trees T^S and the paths $P^{SS'}$ to the degree $\deg_{G'}(v)$ of v in G'.

- Any tree T^S has max-degree k, and v may appear in T^S only if $S \in \mathcal{P}_v$; thus the contribution of the trees T^S to $\deg_{G'}(v)$ is $O(k|\mathcal{P}_v|) = O(k \log n)$.
- Paths that correspond to edges in $\mathcal{P}_v$ may contain v and each of them may contribute $+2$ to $\deg_{G'}(v)$. An edge of T' that goes from S_v to its child may contribute $+1$ to $\deg_{G'}(v)$. Other paths $P^{SS'}$ have no contribution to $\deg_{G'}(v)$, by the construction. Thus the contribution of the paths to $\deg_{G'}(v)$ is at most $2|\mathcal{P}_v| + \deg_{T'}(S_v) - 1 = \deg_{T'}(S_v) + O(\log n)$.

Overall, we have $\deg_{G'}(v) = \deg_{T'}(S_v) + O(k \log n)$, and the lemma follows. $\square$

The Theorem 3 algorithm will find an $O(\log^2 n)$-approximate solution T' for the new instance T using the algorithm from Theorem 1, and then will convert it into a solution T' for the original instance using the algorithm from Lemma 3. The overall ratio will be the product of $O(k \log n)$ (Lemma 2) and $O(\log^2 n)$ (Theorem 1), plus an additive term $O(k \log n)$ (Lemma 3). Thus the overall ratio is $O(k \log^3 n) = O(\log^3 n)$, as claimed in Theorem 3.

Acknowledgment. We thank an anonymous referee for many useful comments.

References

1. Bartal, Y.: Probabilistic approximations of metric spaces and its algorithmic applications. In: FOCS, pp. 184–193 (1996)
2. Chalermsook, P., Das, S., Laekhanukit, B., Vaz, D.: Beyond metric embedding: approximating group Steiner trees on bounded treewidth graphs. In: SODA, pp. 737–751 (2017)
3. Charikar, M., et al.: Approximation algorithms for directed Steiner problems. J. Algorithms **33**(1), 73–91 (1999)
4. Chekuri, C., Even, G., Kortsarz, G.: A greedy approximation algorithm for the group Steiner problem. Discrete Appl. Math. **154**(1), 15–34 (2006)
5. Dehghani, S., Ehsani, S., Hajiaghayi, M.T., Liaghat, L.: Online degree-bounded Steiner network design. In: SODA, pp. 164–175 (2016)
6. Dehghani, S., Ehsani, S., Hajiaghayi, M.T., Liaghat, V., Räcke, H., Seddighin, S.: Online weighted degree-bounded Steiner networks via novel online mixed packing/covering. In: ICALP, pp. 42:1–42:14 (2016)
7. Dehghani, S., Ehsani, S., Hajiaghayi, M.T., Liaghat, V., Seddighin, S.: Greedy algorithms for online survivable network design. In: ICALP, pp. 152:1–152:14 (2018)
8. Fakcharoenphol, J., Rao, S., Talwar, K.: A tight bound on approximating arbitrary metrics by tree metrics. J. Comput. Syst. Sci. **69**(3), 485–497 (2004)
9. Garg, N.: Saving an epsilon: a 2-approximation for the k-MST problem in graphs. In: STOC, pp. 396–402 (2005)
10. Garg, N., Konjevod, G., Ravi, R.: A polylogarithmic approximation algorithm for the group Steiner tree problem. J. Algorithms **37**(1), 66–84 (2000)

11. Grandoni, F., Laekhanukit, B., Li, S.: $o(\log^2 k/\log \log k)$-approximation algorithm for directed Steiner tree: a tight quasi-polynomial-time algorithm. In: STOC, pp. 253–264 (2019)
12. Guo, X., Laekhanukit, B., Li, S., Xian, J.: Tight approximation for variants of directed Steiner tree via state-tree decomposition and linear programming rounding. CoRR, abs/1907.11404 (2019)
13. Hajiaghayi, M.T.: Open problems on bounded-degree network design. In: The Eighth Workshop on Flexible Network Design, Amsterdam (2016)
14. Halperin, E., Krauthgamer, R.: Polylogarithmic inapproximability. In: STOC, pp. 585–594 (2003)
15. Helvig, C.S., Robins, G., Zelikovsky, A.: An improved approximation scheme for the group Steiner problem. Networks **37**(1), 8–20 (2001)
16. Jain, K.: A factor 2 approximation algorithm for the generalized steiner network problem. Combinatorica **21**(1), 39–60 (2001)
17. Kortsarz, G., Peleg, D.: Approximating the weight of shallow Steiner trees. Discrete Appl. Math. **93**, 265–285 (1999)
18. Lau, L.C., Ravi, R., Singh, M.: Iterative Methods in Combinatorial Optimization. Cambridge University Press, New York (2011)
19. Motwani, R., Raghavan, P.: Randomized Algorithms. Cambridge University Press, New York (1995)
20. Poplawski, L.J., Rajaraman, R.: Multicommodity facility location under group Steiner access cost. In: SODA, pp. 996–1013 (2011)
21. Reed, B.: Finding approximate separators and computing tree width quickly. In: STOC, pp. 221–228 (1992)
22. Sharma, N., Kaur, M.: Survey of VLSI techniques for power optimization and estimation of optimization. Int. J. Emerg. Technol. Adv. Eng. **4**, 351–355 (2014)
23. Slater, P.J., Cockayne, E.J., Hedetniemi, S.T.: Information dissemination in trees. SIAM J. Comput. **10**(4), 692–701 (1981)
24. Wang, Y., Hong, X., Jing, T., Yang, Y., Hu, X., Yan, G.: An efficient low-degree RMST algorithm for VLSI/ULSI physical design. In: PATMOS, pp. 442–452 (2004)
25. Zelikovsky, A.: A series of approximation algorithms for the acyclic directed Steiner tree problem. Algorithmica **18**(1), 99–110 (1997)

Between Proper and Strong
Edge-Colorings of Subcubic Graphs

Hervé Hocquard[1], Dimitri Lajou[1(✉)], and Borut Lužar[2]

[1] Univ. Bordeaux, CNRS, Bordeaux INP, LaBRI, UMR 5800,
33400 Talence, France
{herve.hocquard,dimitri.lajou}@u-bordeaux.fr
[2] Faculty of Information Studies in Novo Mesto, Novo mesto, Slovenia
borut.luzar@gmail.com

Abstract. In a proper edge-coloring the edges of every color form a matching. A matching is *induced* if the end-vertices of its edges induce a matching. A *strong edge-coloring* is an edge-coloring in which the edges of every color form an induced matching. We consider intermediate types of edge-colorings, where some of the colors are allowed to form matchings, and the remaining form induced matchings. Our research is motivated by the conjecture proposed in a recent paper on *S*-packing edge-colorings (N. Gastineau and O. Togni, On S-packing edge-colorings of cubic graphs, Discrete Appl. Math. 259 (2019)). We prove that every graph with maximum degree 3 can be decomposed into one matching and at most 8 induced matchings, and two matchings and at most 5 induced matchings. We also show that if a graph is in class I, the number of induced matchings can be decreased by one, hence confirming the conjecture for this class of graphs.

Keywords: Strong edge-coloring · *S*-packing edge-coloring · Induced matching

1 Introduction

A *proper edge-coloring* of a graph $G = (V, E)$ is an assignment of colors to the edges of G such that adjacent edges are colored with distinct colors. Due to a remarkable result of Vizing [22], we know that the minimum number of colors needed to color the edges of a graph G, the *chromatic index* of G (denoted by $\chi'(G)$), is either $\Delta(G)$ or $\Delta(G) + 1$, $\Delta(G)$ being the maximum degree of G. The graphs with the former value of the chromatic index are commonly said to be in *class I*, and the latter in *class II*.

In this paper, we are interested in graphs with maximum degree 3, to which we will refer as *subcubic graphs*. We need at most 4 colors to color such graphs; the complete graph on four vertices with one edge subdivided being the smallest representative of a class II subcubic graph, and the Petersen graph being the smallest 2-connected class II cubic graph. For subcubic graphs of class II,

© Springer Nature Switzerland AG 2020
L. Gąsieniec et al. (Eds.): IWOCA 2020, LNCS 12126, pp. 355–367, 2020.
https://doi.org/10.1007/978-3-030-48966-3_27

it has been shown that they can be colored in such a way that one of the colors (usually denoted δ) is used relatively rarely (cf. [1,6]). This motivates the question if the edges of color δ can be pairwise distant. Note that we consider the distance between edges as the distance between the corresponding vertices in the line graph, i.e. adjacent edges are said to be at distance 1. Payan [17] and independently Fouquet and Vanherpe [6] proved that every subcubic graph with chromatic index 4 admits a proper edge-coloring such that the edges of one color are at distance at least 3, i.e. the end-vertices of those edges induce a matching in the graph.

Gastineau and Togni [7] investigated a generalization of edge-colorings with the property described above. For a given non-decreasing sequence of integers $S = (s_1, \ldots, s_k)$, an S-packing edge-coloring of a graph is a decomposition of edges into disjoint sets $X_1, \ldots, X_k$, where the edges in the set X_i are pairwise at distance at least $s_i + 1$. A set X_i is called an s_i-packing; a 1-packing is simply a matching, and a 2-packing is an induced matching. To simplify the notation, we denote repetitions of same elements in S using exponents, e.g. $(1, 2, 2, 2)$ can be written as $(1, 2^3)$.

The notion of S-packing edge-colorings is motivated by its vertex counterpart, introduced by Goddard and Xu [9] as a natural generalization of the packing chromatic number [8]. In [7], the authors consider S-packing edge-colorings of subcubic graphs with prescribed number of 1's in the sequence. Vizing's result translated to S-packing edge-coloring gives that every subcubic graph admits a $(1, 1, 1, 1)$-packing edge-coloring, while class I subcubic graphs are $(1, 1, 1)$-packing edge-colorable. Moreover, by Payan's, Fouquet's and Vanherpe's result, we have that there is a $(1, 1, 1, 2)$-packing edge-coloring for any subcubic graph.

Theorem 1 (Payan [17], and Fouquet & Vanherpe [6]). *Every subcubic graph admits a $(1, 1, 1, 2)$-packing edge-coloring.*

Here 2 cannot be changed to 3, due to the Petersen and the Tietze graphs (depicted in Fig. 1): they both have chromatic index 4, and we need at least two edges of each color. Since every pair of edges is at distance at most 3, we have the tightness. However, Gastineau and Togni do believe the following conjecture is true.

Conjecture 1 (Gastineau and Togni [7]). Every cubic graph different from the Petersen and the Tietze graph is $(1, 1, 1, 3)$-packing edge-colorable.

Clearly, reducing the number of 1's in sequences increases the total number of needed colors, i.e. the length of the sequence. In fact, if there is no 1 in a sequence, i.e. the edges of every color class induce a matching, the coloring is called a *strong edge-coloring*. It has been proved by Andersen [3] and independently by Horák, Qing, and Trotter [11] that every subcubic graph admits a strong edge-coloring with at most 10 colors, i.e. a (2^{10})-packing edge-coloring. The number of colors is tight, e.g. the Wagner graph in Fig. 2 needs 10 colors for a strong edge-coloring. Let us remark here that the Wagner graph is in class I, meaning that smallest chromatic index does not necessarily mean less number of colors for a strong edge-coloring of a graph.

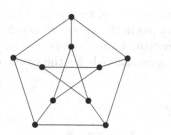 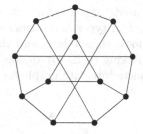

Fig. 1. The Petersen (left) and the Tietze graph (right) admit a $(1,1,1,2)$-packing edge-coloring, and 2 cannot be increased to 3.

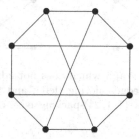

Fig. 2. The Wagner graph is the smallest cubic graph which needs 10 colors for a strong edge-coloring.

Proper and strong edge-coloring of subcubic graphs have been studied extensively already in the previous century. In [7], Gastineau and Togni started filling the gap by considering $(1^k, 2^\ell)$-packing edge-colorings for $k \in \{1, 2\}$. They proved that every cubic graph with a 2-factor admits a $(1, 1, 2^5)$-packing edge-coloring, and the number of required 2-packings reduces by one if the graph is class I. For the case with one 1-packing, they remark that using the bound for the strong edge-coloring one obtains that every subcubic graph admits a $(1, 2^9)$-packing edge-coloring. These bounds are clearly not tight, and they propose a conjecture (the items (a) and (c) in Conjecture 2), which motivated the research presented in this paper. The case (b) has been formulated as a question, and we added the case (d), due to affirmative results of computer tests on subcubic graphs of small orders.

Conjecture 2. Every subcubic graph G admits:

(a) a $(1, 1, 2^4)$-packing edge-coloring [7];
(b) a $(1, 2^7)$-packing edge-coloring [7];
(c) a $(1, 1, 2^3)$-packing edge-coloring if G is in class I [7];
(d) a $(1, 2^6)$-packing edge-coloring if G is in class I.

The conjectured bounds, if true, are tight. For the cases (a) and (b) a subcubic graph that achieves the upper bound is the complete bipartite graph $K_{3,3}$ with one subdivided edge (the left graph in Fig. 3). Recall that this graph is also class II and needs 10 colors for a strong edge-coloring, hence achieving the upper bound for all four types of colorings considered in this paper. For each 1-packing, we have at most three edges, and there remain 4 and 7, respectively, to be in a separate 2-packing each. An analogous argument holds for the cases (c) and (d) on the complete bipartite graph $K_{3,3}$.

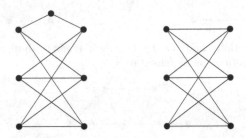

Fig. 3. The smallest subcubic graph which does not admit a $(1,1,2^3)$-packing edge-coloring nor a $(1,2^6)$-packing edge-coloring (left), and the smallest class I subcubic graph which does not admit a $(1,1,2^2)$-packing edge-coloring nor a $(1,2^5)$-packing edge-coloring (right).

Conjecture 2 bridges two of the most important edge-colorings, proper and strong, basicly claiming that each 1-packing can be replaced by three 2-packings. Note that this does not apply to subclasses of graphs, e.g. the Wagner graph needs 10 colors for a strong edge-coloring and it is in class I.

This paper contributes to answering the conjecture by providing upper bounds with one additional color for all four cases of Conjecture 2.

Theorem 2. *Every subcubic graph G admits:*

(a) a $(1,1,2^5)$-packing edge-coloring;
(b) a $(1,2^8)$-packing edge-coloring;
(c) a $(1,1,2^4)$-packing edge-coloring if G is in class I;
(d) a $(1,2^7)$-packing edge-coloring if G is in class I.

The structure of the paper is the following. We begin by presenting notation, definitions and auxiliary results in Sect. 2. In Sect. 3, we give proofs of the cases (a) and (c) of Theorem 2. In Sects. 4 and 5, we proof the cases (b) and (d) of Theorem 2 by proving stronger statements of both. We conclude with an overview of open problems on this topic.

2   Preliminaries

We call a vertex of degree k, at most k, and at least k a k-vertex, a k^--vertex, and a k^+-vertex, respectively. We denote the graph obtained from a graph G by removing a set of vertices X as $G \setminus X$. When $X = \{v\}$ is a singleton, we simply write $G - v$. An analogous notation is used for sets of edges.

As usual, the set of vertices adjacent to a vertex v is denoted $N(v)$, and called the *neighborhood of v*. For a vertex v, we denote the set of edges incident to v by $N'(v)$, and the edges incident to the neighbors of v (including the edges in $N'(v)$) by $N''(v)$. We refer to the former as the *edge-neighborhood of v* and to the latter as the *2-edge-neighborhood of v*. Analogously, we define the edge-neighborhood and the 2-edge-neighborhood of an edge e.

When coloring the edges, we deal with two types of colors. The ones allowing the edges of those colors to be at distance at least 2 we call the *1-colors*, and the one requiring the edges to be at distance at least 3 are called the *2-colors*. An edge colored with a 1-color (resp. a 2-color) is a *1-edge* (resp. a *2-edge*). For an edge uv, we denote by $A_2(uv)$ the number of available 2-colors, i.e., the 2-colors with which the edge can be colored without violating the coloring assumptions.

In our proofs, we will often put lists of colors on some uncoloured edges and try to find a valid assignment that satisfy the color lists. For example, if e has a list L of size k such that all colors of L are available for e and has at most $k-1$ uncolored neighbors then we can ignore e when coloring as there will always be one free color for e in L after coloring all other edges.

Sometimes, we will need a more careful analysis of choosing colors from the lists of available colors. For that purpose, we will use the classical result due to Hall [10].

Theorem 3 (Hall's Theorem). *Let $\mathcal{A} = (A_i, i \in I)$ be a finite family of (not necessarily distinct) subsets of a finite set A. A system of representatives (SDR) for the family $\mathcal{A}$ is a set $\{a_i, i \in I\}$ of distinct elements of A such that $a_i \in A_i$ for all $i \in I$. $\mathcal{A}$ has a system of representatives if and only if $|\bigcup_{i \in J} A_i| \geq |J|$ for all subsets J of I.*

Perhaps the strongest tool for determining if one can always choose colors from the lists of available colors such that given conditions are satisfied is the following result, due to Alon [2].

Theorem 4 (Combinatorial Nullstellensatz). *Let $\mathbb{F}$ be an arbitrary field, and let $P = P(X_1, \ldots, X_n)$ be a polynomial in $\mathbb{F}[X_1, \ldots, X_n]$. Suppose the coefficient of a monomial $X_1^{k_1} \ldots X_n^{k_n}$, where each k_i is a non-negative integer, is non-zero in P and the degree $\deg(P)$ of P equals $\sum_{i=1}^{n} k_i$. If moreover $S_1, \ldots, S_n$ are any subsets of $\mathbb{F}$ with $|S_i| > k_i$ for $i = 1, \ldots, n$, then there are $s_1 \in S_1, \ldots, s_n \in S_n$ so that $P(s_1, \ldots, s_n) \neq 0$.*

In short, P being the chromatic polynomial of a graph G, if there is a monomial (of proper degree) of P with non-zero coefficient, then there exists a coloring of G.

When considering lists of available colors for an edge, we are in fact dealing with the list version of a coloring. We say that L is an *edge-list-assignment* for a graph G if it assigns a list $L(e)$ of possible colors to each edge e of G. If G admits a strong edge-coloring σ such that $\sigma(e) \in L(e)$ for all edges in $E(G)$, then we say that G is *strong L-edge-colorable* or σ is a *strong L-edge-coloring* of G. The graph G is *strong k-edge-choosable* if it is strong L-edge-colorable for every edge-list-assignment L, where $|L(e)| \geq k$ for every $e \in E(G)$. The *list strong chromatic index* $\chi'_{ls}(G)$ of G is the minimum k such that G is strong k-edge-choosable.

We will use the following result, due to Zhang, Liu, and Wang [24] which established a result on an adjacent vertex-distinguishing list edge-coloring of cycles, i.e. proper list edge-coloring where the sets of colors for every pair of adjacent vertices are distinct. It is easy to see that such a coloring is also a strong edge-coloring of a cycle, and we write the statement in this language.

Theorem 5 (Zhang, Liu & Wang, 2002). *Let n be an integer with $n \geq 3$. Then,*

(i) $\chi'_{ls}(C_n) = 5$ *if* $n = 5$;
(ii) $\chi'_{ls}(C_n) = 4$ *if* $n \not\equiv 0 \bmod 3$;
(iii) $\chi'_{ls}(C_n) = 3$ *if* $n \equiv 0 \bmod 3$.

3   Proof of the Cases (a) and (c) of Theorem 2

We begin with the cases of Theorem 2 using two 1-colors. These two cases simply provide a straightforward extensions of the results due to Gastineau and Togni [7], who established them for bridgeless cubic graphs.

The extension comes from the following easy proposition for which we omit the proof (*c.f.* full version of the paper).

Proposition 1. *Let G be a subcubic graph and let X be a set of edges in G such that every two edges in X are at distance exactly 2. Then, X contains at most 5 edges. Moreover, if $|X| = 5$, then G is cubic with 10 vertices.*

Proof (Theorem 2(a) and (c)). We begin with the case (a). Let G be a connected subcubic graph and let π be a $(1,1,1,2)$-packing edge-coloring of G which exists by Theorem 1. To establish the statement, we only need to replace one 1-color in π with four 2-colors. Let X be a set of all the edges in G colored by one 1-color in π, and H be the subgraph of G induced by X. Let G^* be the graph obtained from H by contracting all the edges in X. Clearly, G^* has maximum degree at most 4, and is 4-vertex-colorable by the Brooks' Theorem, unless it is isomorphic to K_5. Observe that vertex coloring of G^* induces a strong edge-coloring of the edges in X. Furthermore, by Proposition 1, the only graphs in which it may happen that five colors are needed to color G^*, are cubic with 10 vertices. For these small graphs we have even determined that they admit a $(1,1,2^4)$-packing edge-coloring computationally, and thus establish the case (a).

The case (b) follows immediately from the argument above, since we do not have an extra 2-color in the coloring π.   □

4   Proof of the Case (b) of Theorem 2

In order to prove Theorem 2(b), we prove a bit stronger result. We say that a $(1,2^8)$-packing edge-coloring of a subcubic graph G with the color set

$\{0, 1, \ldots, 8\}$, where 0 is a 1-color and the others are 2-colors, is a *good* $(1, 2^8)$-*packing edge-coloring* if no 2^--vertex of G is incident with a 0-edge.

Theorem 6. *Every subcubic graph admits a good* $(1, 2^8)$-*packing edge-coloring.*

Proof. We prove Theorem 6 by contradiction. Let G be a minimal counterexample to the theorem in terms of $|V(G)| + |E(G)|$. Clearly, G is connected and has maximum degree 3. In the following lemma, we establish some structural properties of G which will eventually yield a contradiction on the existence of G. In most of the cases, we consider a graph G' smaller than G, which, by the minimality of G, admits a good $(1, 2^8)$-packing edge-coloring π, and we show that π can be extended to G by recoloring some edges of G' and coloring the edges of G not being colored by π.

Lemma 1. *The graph G verifies the following properties:*

1. *G is simple,*
2. *G is cubic,*
3. *G is 2-connected,*
4. *G does not contain 3-cycles,*
5. *G does not contain 4-cycles and*
6. *G contains no cycle of length at least 5.*

Due to size constraint we do not give the proof of Lemma 1 (*c.f.* full version of the paper) except for Lemma 1.6. The main techniques used in the proof of Lemma 1 consist in removing part of the graph and coloring it by minimality. In some cases, we need to use Hall's Theorem.

We nonetheless present the proof of Lemma 1.6 to show why we need the stronger statement of Theorem 6.

Proof (Lemma 1.6). Suppose the contrary, and let $C = u_1 u_2 \ldots u_n$ be a minimal induced n-cycle in G, with $n \geq 5$. For every i, $1 \leq i \leq n$, let u_i' be the neighbor of the vertex u_i not in C, and let $G' = G \setminus V(C)$. Note that the u_i' are pairwise distinct by the minimality of C, 1.4 and 1.5. Then, by the minimality of G, there is a good $(1, 2^8)$-packing edge-coloring π of G'. Since π is good, no u_i' is incident with the color 0. So, in the coloring φ of G induced by π, we can color every edge $u_i u_i'$ with 0. In this way, only the edges of C are left non-colored. Observe that each of those edges has at least 4 available 2-colors. If $n \geq 6$, then we can complete the coloring by Theorem 5, a contradiction.

If $n = 5$ then we can color C, except if all five edges have the same four 2-colors available by Hall's Theorem 3. If we are in this case, then suppose that 1 and 2 are the two colors on the edges incident to u_1', and 3 and 4 are the two colors on the edges incident to u_2'. Then $\{1, 2\}$ must also be on the edges incident to u_3', $\{3, 4\}$ on the edges incident to u_4', and again $\{1, 2\}$ on the edges incident to u_5'. Thus the edge $u_1 u_5$ has five available 2-colors, a contradiction.   ◆

By Lemma 1, we have that G is a cubic bridgeless graph with no cycles. Hence G is a tree, a contradiction with the fact that G is cubic. This concludes the proof of Theorem 6.   □

5   Proof of the Case (d) of Theorem 2

Recall that in the case (d), we assume the graph is in class I. In our proof, this is an important feature which enables us to confirm Conjecture 2(b) for this class of graphs. We again prove a stronger version of the theorem.

Theorem 7. *Let G be a graph of class I. Then for every proper 3-edge-coloring π with colors a, b, and c, and for every color $\alpha \in \{a, b, c\}$ there exists a $(1, 2^7)$-packing edge-coloring σ such that the edges of color α in π are colored with 0 in σ.*

The proof of this theorem is quite involved. Due to size constraints, we only provide the main ideas of the proof of this theorem.

Proof (Ideas only). Let G be a minimal counterexample to the theorem minimizing the sum $|V(G)| + |E(G)|$. Let π be a proper 3-edge-coloring (using colors a, b, and c) and let the color a be the color class for which there is no $(1, 2^7)$-packing edge-coloring σ (using colors $\{0, 1, \dots, 7\}$, 0 being the 1-color) of G such that all edges colored a in π are colored 0 in σ.

We begin by establishing some structural properties of G. First we prove that G is a simple cubic graph. This is done using a case analysis. Recall that G being cubic implies that in π every color appears at every vertex. Then we remove short cycles and prove that G has girth at least 5. The proof here is more complex than for their equivalent in the previous section. We use an additional technique for removing cycles of length 4, that is we apply the Combinatorial Nullstellensatz to color some cases.

Finally, we want to remove long cycles. We do not show this exact fact but a similar one. We call a bc-cycle, a cycle colored only with the two colors b and c in π. These bc-cycles need to be colored with only 2-colors. If u is a vertex of such bc-cycle and u' is one of its neighbour and is not on the cycle then we know that uu' is color with color a. Simply uncoloring the bc-cycle would yield only three available 2-colors for each edge of the cycle which is not enough.

We separate two cases, chordless bc-cycles and bc-cycles with chords. In both cases, we reduce the graph to a smaller one by removing some vertices of the cycle and connecting some neighbours to provide useful properties on the coloring obtained by minimality. These properties will allow us to precolor some edges of the cycle in G. We color the rest of the cycle with the help of the Combinatorial Nullstellensatz. As bc-cycles must be colored with 2-colors it is possible to express the coloring problem as a polynomial to apply the Combinatorial Nullstellensatz. Note that we use the Combinatorial Nullstellensatz in a different way than for small cycles as we have an infinity of cycle lengths. Therefore, we must find a generic non null coefficient in a family of polynomials which depend on the length of the cycle.

Combining the previous facts yields a contradiction.   □

6   Further Work

Conjecture 2 remains open, however, our upper bounds are only by one 2-color off. Unfortunately, we were not able to apply the proving techniques, used to prove tight bounds for proper edge-coloring and strong edge-coloring of subcubic graphs, to the problems considered in this paper. Therefore, since solving Conjecture 2 in the general setting seems to be challenging, we suggest in this section additional problems which arise naturally when dealing with the considered colorings. All of them are supported with computational results on graphs of small orders.

We begin with a general conjecture for strong edge-coloring.

Conjecture 3. Every bridgeless subcubic graph G, not isomorphic to the Wagner graph or the complete bipartite graph $K_{3,3}$ with one edge subdivided, admits a (2^9)-packing edge-coloring.

We proceed with an overview of results in specific graph classes and list open problems for each of them. For that, we follow the conjecture on strong edge-coloring of subcubic graphs proposed by Faudree, Gyárfás, Schelp, and Tuza [5] in 1990.

Conjecture 4 (Faudree, Gyárfás, Schelp & Tuza [5]). For every subcubic graph G it holds:

(1) G admits a (2^{10})-packing edge-coloring;
(2) If G is bipartite, then it admits a (2^9)-packing edge-coloring;
(3) If G is planar, then it admits a (2^9)-packing edge-coloring;
(4) If G is bipartite and each edge is incident with a 2-vertex, then it admits a (2^6)-packing edge-coloring;
(5) If G is bipartite of girth at least 6, then it admits a (2^7)-packing edge-coloring;
(6) If G is bipartite and has girth large enough, then it admits a (2^5)-packing edge-coloring.

All the cases of the conjecture, except (5), are already resolved, and we present the results in what follows.

6.1   Planar Graphs

It was the well-known connection between edge-coloring of bridgeless cubic planar graphs and the Four Color Problem, established by Tait [21], which initiated the research in this area. By the Four Color Theorem, we thus have that every bridgeless cubic planar graph admits a $(1,1,1)$-edge-coloring. The condition of being bridgeless is necessary, since already K_4 with one subdivided edge is in class II. However, not all questions are resolved. The following conjecture of Albertson and Haas [1], which is a special case of Seymour's conjecture [18], is still widely open.

Conjecture 5 (Albertson & Haas [1]*).* Every bridgeless subcubic planar graph with at least two vertices of degree 2 admits a $(1,1,1)$-packing edge-coloring.

The number of required colors for strong edge-coloring of planar graphs is also determined. Just recently, Kostochka et al. [14] proved the following (and resolved the Case (3) of Conjecture 4).

Theorem 8 (Kostochka et al. [14]**).** *Every subcubic planar graph admits a* (2^9)*-packing edge-coloring.*

The upper bound is tight and there are infinitely many bridgeless cubic graphs that need nine 2-colors for strong edge-coloring. An example is e.g. the 3-prism, depicted in Fig. 4.

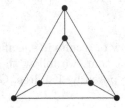

Fig. 4. A bridgeless cubic planar graph which needs nine colors for a strong edge-coloring.

On the other hand, there are no results for planar graphs on the colorings with one or two matchings. We propose the following conjecture.

Conjecture 6. Every subcubic planar graph admits a $(1,2^6)$-packing edge-coloring and a $(1,1,2^3)$-packing edge-coloring.

The conjectured upper bound, if true, is tight and attained by an infinitely many bridgeless subcubic planar graphs for both values. It also appears to be much more demanding as the result of Theorem 8. Thus, also some partial results, with additional restrictions on the structure of planar graphs, might also be interesting, in order to understand the general problem better.

6.2   Bipartite Graphs

In the class of bipartite graphs, the proper and the strong case of the colorings are long solved. In 1916, König [13] proved that every bipartite graph is in class I, and in 1993, Steger and Yu [20] established the following (and resolved the Case (2) of Conjecture 4).

Theorem 9 (Steger & Yu [20]**).** *Every subcubic bipartite graph admits a* (2^9)*-packing edge-coloring.*

Again, these bounds are tight and attained by infinitely many graphs.

Since all bipartite graphs are in class I, the results and conjectures for them apply also in the bipartite case. It is known that as soon as we leave the 'proper' setting, i.e., require some 2-colors instead just 1-colors, the problems become much harder. E.g., a tight upper bound for a strong edge-coloring of bipartite graphs is still not known (c.f. [5,20]). Therefore, the Cases (c) and (d) of Conjecture 2 may be considered just in the bipartite setting. Moreover, we have an infinite number of graphs attaining the conjectured upper bounds also among bipartite graphs.

If we consider subcubic graphs with only edges of weight at most 5, i.e., edges where at least one of the end-vertices is of degree at most 2, the number of required colors decreases substantially. In particular, the Case (4) of Conjecture 4 was resolved by Maydanskiy [16] and independently by Wu and Lin [23].

Theorem 10 (Maydanskiy [16], and Wu & Lin [23]). *Every subcubic bipartite graph, in which each edge has weight at most 5, admits a (2^6)-packing edge-coloring.*

Clearly, an analogous question for coloring such graphs with two 1-colors is if they admit a $(1,1,2^2)$-packing edge-coloring. It is answered in affirmative [19]. The bound is tight already in the class of trees. On the other hand, we do not have the answer for the following.

Question 1. Is it true that every subcubic bipartite graph, in which each edge has weight at most 5, admits a $(1,2^4)$-packing edge-coloring?

This bound is again attained in the class of trees.

6.3   Graphs with Big Girth

Similarly as the bipartiteness, having big girth does not really simplify edge-colorings in which some colors must be 2-colors. Even more, due to Kochol [12] we know, there are graphs with arbitrarily large girth which are in class II! Anyway, if the girth is infinite, i.e., we consider the trees, the following simple observation is immediate.

Observation 1. *Every subcubic tree admits:*

(1) a $(1,1,1)$-packing edge-coloring;
(2) a $(1,1,2,2)$-packing edge-coloring;
(3) a $(1,2^4)$-packing edge-coloring;
(4) a (2^5)-packing edge-coloring.

The bounds are tight already if we just consider a neighborhood of one edge with both end-vertices of degree 3.

In the case of strong edge-coloring, the Case (6) of Conjecture 4 was also rejected just recently by Lužar, Mačajová, Škoviera, and Soták [15], who proved that a cubic graph is a cover of the Petersen graph if and only if it admits a (2^5)-packing edge-coloring.

Before we consider the intermediate colorings, we first recall the result of Gastineau and Togni [7].

Proposition 2 (Gastineau & Togni [7]). *Every cubic graph admitting a* $(1, 1, 2, 2)$-*packing edge-coloring is class I and has order divisible by four.*

Hence, the analogue of the Case (6) of Conjecture 4 when having two 1-colors does not hold. However, the following remains open.

Question 2. Is it true that every subcubic bipartite graph with big enough girth admits a $(1, 2^4)$-packing edge-coloring?

To conclude, we believe that studying properties of the considered edge-colorings will have impact to the initial problem of strong edge-coloring, which is in general case still widely open. Namely, the conjectured upper bound for graphs with maximum degree Δ is $1.25\Delta^2$, while currently the best upper bound is due to Bonamy, Perrett, and Postle [4], set at $1.835\Delta^2$.

Acknowledgement. This research has been done in the scope of the bilateral project between France and Slovenia, BI-FR/19-20-PROTEUS-001. The third author was partly supported by the Slovenian Research Agency Program P1–0383 and the Project N1–0057(B).

References

1. Albertson, M.O., Haas, R.: Parsimonious edge coloring. Discret. Math. **148**, 1–7 (1996)
2. Alon, N.: Combinatorial Nullstellensatz. Comb. Probab. Comput. **8**(1–2), 7–29 (1999)
3. Andersen, L.D.: The strong chromatic index of a cubic graph is at most 10. Discret. Math. **108**, 231–252 (1992)
4. Bonamy, M., Perrett, T., Postle, L.: Colouring graphs with sparse neighbourhoods: bounds and applications. ArXiv Preprint, https://arxiv.org/abs/1810.06704 (2018)
5. Faudree, R.J., Gyárfás, A., Schelp, R.H., Tuza, Z.: The strong chromatic index of graphs. Ars Comb. **29B**, 205–211 (1990)
6. Fouquet, J.-L., Vanherpe, J.-M.: On parsimonious edge-colouring of graphs with maximum degree three. Graphs Comb. **29**(3), 475–487 (2013). https://doi.org/10.1007/s00373-012-1145-3
7. Gastineau, N., Togni, O.: On S-packing edge-colorings of cubic graphs. Discret. Appl. Math. **259**, 63–75 (2019)
8. Goddard, W., Hedetniemi, S.M., Hedetniemi, S.T., Harris, J.M., Rall, D.F.: Broadcast chromatic numbers of graphs. Ars Comb. **86**, 33–49 (2008)
9. Goddard, W., Xu, H.: The S-packing chromatic number of a graph. Discuss. Math. Graph Theory **32**, 795–806 (2012)
10. Hall, P.: On representatives of subsets. J. London Math. Soc. **10**(1), 26–30 (1935)
11. Horák, P., Qing, H., Trotter, W.T.: Induced matchings in cubic graphs. J. Graph Theory **17**(2), 151–160 (1993)
12. Kochol, M.: Snarks without small cycles. J. Comb. Theory Ser. B **67**(1), 34–47 (1996)
13. König, D.: Über graphen und ihre anwendung auf determinantentheorie und mengenlehre. Math. Ann. **77**, 453–465 (1916). https://doi.org/10.1007/BF01456961
14. Kostochka, A.V., Li, X., Ruksasakchai, W., Santana, M., Wang, T., Yu, G.: Strong chromatic index of subcubic planar multigraphs. Eur. J. Comb. **51**, 380–397 (2016)

15. Lužar, B., Mačajová, E., Škoviera, M., Soták, R.: On the conjecture about strong edge-coloring of subcubic graphs (2020, Manuscript)
16. Maydanskiy, M.: The incidence coloring conjecture for graphs of maximum degree 3. Discret. Math. **292**(1–3), 131–141 (2005)
17. Payan, C.: Sur quelques problèmes de couverture et de couplage en combinatoire. PhD thesis, Institut National Polytechnique de Grenoble - INPG, Université Joseph-Fourier - Grenoble I (1977). (in French)
18. Seymour, P.D.: On Tutte's extension of the four-color problem. J. Comb. Theory Ser. B **31**, 82–94 (1981)
19. Soták, R.: Private communication
20. Steger, A., Yu, M.-L.: On induced matchings. Discret. Math. **120**, 291–295 (1993)
21. Tait, P.G.: On the colouring of maps. In: Proceedings of the Royal Society of Edinburgh Section A, vol. 10, no. 729, pp. 501–503 (1880)
22. Vizing, V.G.: On an estimate of the chromatic class of a p-graph. Metody Diskret. Analiz **3**, 25–30 (1964)
23. Wu, J., Lin, W.: The strong chromatic index of a class of graphs. Discret. Math. **308**, 6254–6261 (2008)
24. Zhang, Z., Liu, L., Wang, J.: Adjacent strong edge coloring of graphs. App. Math. Lett. **15**, 623–626 (2002)

Improved Budgeted Connected Domination and Budgeted Edge-Vertex Domination

Ioannis Lamprou, Ioannis Sigalas, and Vassilis Zissimopoulos(✉)

Department of Informatics and Telecommunications, National and Kapodistrian
University of Athens, Zografou, Greece
{ilamprou,sigalasi,vassilis}@di.uoa.gr

Abstract. We consider the *Budgeted* version of the classical *Connected Dominating Set* problem (BCDS). Given a graph G and a budget k, we seek a connected subset of at most k vertices maximizing the number of dominated vertices in G. We improve over the previous $(1 - 1/e)/13$ approximation in [Khuller, Purohit, and Sarpatwar, *SODA 2014*] by introducing a new method for performing tree decompositions in the analysis of the last part of the algorithm. This new approach provides a $(1 - 1/e)/12$ approximation guarantee. By generalizing the analysis of the first part of the algorithm, we are able to modify it appropriately and obtain a further improvement to $(1 - e^{-7/8})/11$. On the other hand, we prove a $(1 - 1/e + \epsilon)$ inapproximability bound, for any $\epsilon > 0$.

We also examine the *edge-vertex domination* variant, where an edge dominates its endpoints and all vertices neighboring them. In *Budgeted Edge-Vertex Domination* (BEVD), we are given a graph G, and a budget k, and we seek a, not necessarily connected, subset of k edges such that the number of dominated vertices in G is maximized. We prove there exists a $(1 - 1/e)$-approximation algorithm. Also, for any $\epsilon > 0$, we present a $(1 - 1/e + \epsilon)$-inapproximability result by a gap-preserving reduction from the *maximum coverage* problem. Finally, we examine the "dual" *Partial Edge-Vertex Domination* (PEVD) problem, where a graph G and a quota n' are given. The goal is to select a minimum-size set of edges to dominate at least n' vertices in G. In this case, we present a $H(n')$-approximation algorithm by a reduction to the *partial cover* problem.

Keywords: Approximation · Budget · Partial · Connected domination · Edge-vertex domination

1 Introduction

The problem of vertices dominating vertices in a graph is very common and has been extensively studied in graph theory and combinatorial optimization

This work was partially supported by the Special Account for Research Grants (ELKE) of the National and Kapodistrian University of Athens (NKUA). A full version of the paper can be found at https://arxiv.org/abs/1907.06576.

© Springer Nature Switzerland AG 2020
L. Gąsieniec et al. (Eds.): IWOCA 2020, LNCS 12126, pp. 368–381, 2020.
https://doi.org/10.1007/978-3-030-48966-3_28

literature. In the classical definition, a dominating set is a subset of vertices such that each vertex is either a member of the subset or adjacent to a member of the subset. Intuitively, a dominating set provides a skeleton for the placement of resources, such that any network node is within immediate reach to them.

However, as it is often the case, there are constraints on the amount of resources available for placement, e.g., due to financial or other management reasons. That is, we are limited to a budget of k resources to be placed on network nodes. The optimization goal is to place the available resources suitably, such that the number of network nodes they dominate is maximized. This problem is known in literature as the *Budgeted Dominating Set* problem.

Budgeted domination has applications especially in ad-hoc wireless (sensor) networks. In this setting, a set of network nodes needs to be identified as the virtual backbone of the network, that is, the structure responsible for routing and packet forwarding. To achieve these tasks, nodes in the backbone must be able to communicate with each other, i.e., form a *connected* set of vertices in the graph capturing the topology of their communication ranges. The resulting optimization problem is the *Budgeted Connected Dominating Set* (BCDS) problem. In this paper, we study BCDS and present an improved guarantee over the previous state of the art [12].

Besides BCDS, we examine other problems where graph edges are selected as dominators. The concept of edges dominating adjacent edges has been well-considered in literature; e.g., see [8,27] for some preliminary results. An example application is in network tomography where probes need to be placed to monitor the health of network links [14].

In this paper, we consider cases where resources must be positioned on the links of a network to dominate network nodes. For instance, consider a power system where a limited number of static var compensators need to be placed on transmission lines' midpoints to locate faults affecting a big proportion of buses [10]. Another example is to identify a limited-size set of friendships, modeled as graph edges, having a big impact in terms of neighborhood in a social network.

More formally, the notion in consideration is *edge-vertex domination*, where an edge dominates its endpoints and any vertices adjacent to its endpoints. We examine the (in)approximability of *Budgeted Edge-Vertex Domination* (BEVD), where we seek a, not necessarily connected, set of k (budget) edges dominating as many vertices as possible. If the edge set is required to be connected, we show that the problem essentially matches BCDS. Finally, we consider the related *Partial Edge-Vertex Domination* (PEVD) problem: a quota of vertices needs to be dominated by utilizing the minimum number of edges possible.

1.1 Related Work

Finding a minimum-size connected set of vertices dominating the whole graph is a classical NP-hard problem. In [7], Guha and Khuller proposed a $\ln \Delta + 3$ approximation algorithm, which is (up to constant factors) the best possible, since the problem is hard to approximate within a factor of $(1 - \epsilon) \log n$ [5]. For a bigger picture of the research landscape, in [4], many connected domination results for special graph classes and other applications are surveyed.

In [21], vertex-vertex and edge-edge budgeted domination are considered. For vertex-vertex, matching upper and lower bounds of $(1-1/e)$ are given, whereas, for edge-edge, a $(1-1/e)$ approximation and a $1303/1304+\epsilon$ hardness are proved.

In the connected case, budgeted and partial versions of domination have their origins in wireless sensor networking [19,26], where a network backbone with good qualities needs to be determined, which must either be limited in resources and/or cover a big-enough proportion of the network. The first, and thus far state of the art, results for the budgeted and partial cases in general graphs appear in [12], where a $(1 - 1/e)/13$-approximation, respectively an $O(\ln \Delta)$-approximation, is proved for the budgeted, respectively partial, case. Other works have followed in particular settings. For example, in [20], a constant factor approximation algorithm for partial connected domination on a superset of unit disk graphs, namely growth-bounded graphs, is proposed. Their result translates to a 27-approximation guarantee on unit disk graphs.

Regarding edge-vertex domination, the graph-theoretic notion was introduced in [22], together with the complementary case of vertex-edge domination, where a vertex dominates all edges incident to it or to a neighbor of it. Some complexity and algorithmic results about the minimal size of an edge-vertex, respectively vertex-edge, dominating set appear in [18]. More recently, some vertex-edge domination open questions posed in [18] were answered in [2]. In [25], an improved bound on the edge-vertex domination number of trees was proved. Except for the vertex-edge and edge-vertex variants, a *mixed* domination variant has been introduced [23], where a minimal subset of both vertices and edges need to be selected so that each vertex/edge of the graph is incident/adjacent to a vertex/edge in the subset. Recent example works in this topic study the problem in special graph classes like trees, cacti, and split graphs [17,28].

1.2   Our Results

In Sect. 2, we present preliminary notions and formally define the problems.

In Sect. 3, we examine the Budgeted Connected Dominating Set (BCDS) problem, see Definition 1, where a connected subset of budget vertices needs to dominate as many vertices as possible. By introducing a new tree decomposition technique in Subsect. 3.2, we prove a $(1 - 1/e)/12 \simeq 0.05267$ approximation, in Theorem 2, which improves over the previous best known $(1-1/e)/13$ guarantee [12]. (We note the same guarantee has recently been achieved independently in [13].) We further improve the ratio to $(1 - e^{-7/8})/11 \simeq 0.05301$ (Theorem 3) by generalizing the first part of the analysis in [12] and then modifying the proposed algorithm accordingly in Subsect. 3.3. On the negative side, for any $\epsilon > 0$, we show a first $(1 - 1/e + \epsilon)$ inapproximability bound; see Theorem 5.

In Sect. 4, we consider edge-vertex domination, where a, not necessarily connected, subset of edges dominates adjacent vertices. If the set of edges is also required to be connected, then the problems essentially reduce to the standard vertex-vertex budgeted/partial dominating set problems; see Proposition 2. In

Subsect. 4.1, we prove there is a $(1-1/e)$-approximation algorithm (Theorem 7). This is the best possible since we prove an $(1 - 1/e + \epsilon)$ inapproximability lower bound, for any $\epsilon > 0$, see Theorem 8. In Subsect. 4.2, we consider the problem of Partial Edge-Vertex Domination. In Theorem 10, we prove that, in the general case, there exists an $H(n')$-approximation, where $H(\cdot)$ is the Harmonic number and n' is the number of vertices requested to be dominated. To do so, we employ a reduction to a partial version of the classical *Set Cover* problem.

Finally, in Sect. 5, we give some concluding remarks.

2   Preliminaries

A graph G is denoted as a pair $(V(G), E(G))$ (or simply (V, E)) of the vertices and edges of G. The graphs considered are simple (neither loops nor multi-edges are allowed), connected and undirected. Besides the aforementioned, no assumptions are made on the topology of the input graphs.

Two vertices $u, v \in V$ connected by an edge, denoted (u, v) or equivalently (v, u), are called *adjacent* or *neighboring*. The *open neighborhood* of a vertex $v \in V$ is defined as $N(v) = \{u \in V : (v, u) \in E\}$, while the *closed neighborhood* is defined as $N[v] = \{v\} \cup N(v)$. For a subset of vertices $S \subseteq V(G)$, we expand the above definitions to $N(S) = \bigcup_{v \in S} N(v) \setminus S$ and $N[S] = N(S) \cup S$.

The degree of a vertex $v \in V$ is defined as $d(v) = |N(v)|$. The minimum, resp. maximum, degree of G is denoted by $\delta = \min_{v \in V} d(v)$, resp. $\Delta = \max_{v \in V} d(v)$.

Let us now consider the neighborhood of edges in terms of vertices. Given an edge $e = (v, u) \in E$, let $I(e) = \{v, u\}$ stand for the set containing its two incident vertices. We define the *neighborhood of an edge e* as $N[e] = \bigcup_{v \in I(e)} N[v]$. For a set of edges $E' \subseteq E$, we define $V(E') = \{v \in V \mid \exists e \in E' \text{ such that } v \in I(e)\}$. Then, we define the *edge-set neighborhood* as $N[E'] = N[V(E')]$. Here, we focus on a closed neighborhood definition, since it captures the number of vertices incident or adjacent to a set of edges in the standard edge-vertex domination paradigm (Definition 8 in [18]; originally introduced in [22]). That is, we say that a set of edges E' dominates $N[E']$.

Let us now proceed to formally define the problems studied in this paper.

Definition 1 (BUDGETED CONNECTED DOMINATING SET). *Given a graph $G = (V, E)$ and an integer k, select a subset $S \subseteq V$, where $|S| \le k$, such that the subgraph induced by S is connected and $|N[S]|$ is maximized.*

Definition 2 (BUDGETED EDGE-VERTEX DOMINATION). *Given a graph $G = (V, E)$ and an integer k, select a subset $E' \subset E$, where $|E'| \le k$, such that $|N[E']|$ is maximized.*

Definition 3 (PARTIAL EDGE-VERTEX DOMINATION). *Given a graph $G = (V, E)$ and an integer n', select a subset $E' \subseteq E$ of minimum size such that it holds $|N[E']| \ge n'$.*

3   Budgeted Connected Dominating Set

In this section, we consider the Budgeted Connected Dominating Set (BCDS) problem given in Definition 1. We initially present a summary of key aspects of the state of the art algorithm [12], which achieves a $(1-1/e)/13$ approximation factor. We then show how the analysis can be improved to achieve a $(1-1/e)/12$ guarantee via an alternative tree decomposition scheme; see Theorem 2. Then, we generalize the analysis of the greedy procedure in order to modify a call within the state of the art algorithm. This modification allows us to increase the approximation factor even further to $(1 - e^{-7/8})/11$; see Corollary 1. On the other hand, we conclude this section with a $(1 - 1/e + \epsilon)$, for any $\epsilon > 0$, inapproximability result; see Theorem 5.

3.1   Previous Approach

Khuller et al., see Algorithm 2 (Algorithm 5.1 in [12]), design the first constant factor approximation algorithm for BCDS with an approximation guarantee of $(1 - 1/e)/13$. Their approach comprises three method calls: (i) a call to an algorithm returning a greedy dominating set D and its corresponding profit function p; see Algorithm 1 (GDS), (ii) a call to a 2-approximation algorithm, which follows from [6,9], for the *Quota Steiner Tree* (QST) problem defined below, and (iii) a call to a dynamic programming scheme $Best_k(\cdot)$ to determine the maximum-profit subtree of size at most k within a bigger-size tree.

Algorithm 1: Greedy Dominating Set (GDS) [12]

    Input : A graph $G = (V(G), E(G))$
    Output: A dominating set $D \subseteq V(G)$ and a profit function
              $p : V(G) \to \mathbb{N} \cup \{0\}$
1 $D \leftarrow \emptyset$
2 $U \leftarrow V(G)$
3 **foreach** $v \in V(G)$ **do**
4 | $p(v) \leftarrow 0$
5 **end**
6 **while** $U \neq \emptyset$ **do**
7 | $w \leftarrow \arg\max_{v \in V(G) \setminus D} |N_U(v)|$ /* $N_U(v) = N[v] \cap U$ */
8 | $p(w) \leftarrow |N_U(w)|$
9 | $U \leftarrow U \setminus N_U(w)$
10 | $D \leftarrow D \cup \{w\}$
11 **end**
12 **return** (D, p)

Definition 4 (QUOTA STEINER TREE). *Given a graph G, a vertex profit function $p : V(G) \to \mathbb{N} \cup \{0\}$, an edge cost function $c : E(G) \to \mathbb{N} \cup \{0\}$ and a quota $q \in \mathbb{N}$, find a subtree T that minimizes $\sum_{e \in E(T)} c(e)$ subject to the condition $\sum_{v \in V(T)} p(v) \geq q$.*

Algorithm 2: Greedy Profit Labeling Algorithm for BCDS [12]

    Input   : A graph $G = (V(G), E(G))$ and $k \in \mathbb{N}$
    Output: A tree $\tilde{T}$ on at most k vertices
1 $(D, p) \leftarrow GDS(G)$
2 $T \leftarrow QST(G, (1 - 1/e)\text{OPT}, p)$
3 $\tilde{T} \leftarrow Best_k(T, p)$
4 **return** $\tilde{T}$

Theorem 1 (Follows from results in [6,9]**).** *There is a 2-approximation algorithm for QUOTA STEINER TREE.*

In their analysis, Khuller et al. [12] demonstrate that there exists a set $D' \subseteq D$ of size k which dominates at least $(1 - 1/e)\text{OPT}$ vertices, where OPT is the optimal number of dominated vertices achieved with a budget of k. Furthermore, D' can be connected by adding at most another $2k$ Steiner vertices, so giving a total of $3k$ vertices. Then, it suffices to call the 2-approximation algorithm for QST, see line 2 in Algorithm 2, with profit function p (returned by algorithm GDS at line 1), all edge costs equal to 1 and quota equal to $(1 - 1/e)\text{OPT}$. The value OPT can be guessed via a binary search between k and n. Overall, the returned tree has size at most $6k$ vertices and dominates at least $(1 - 1/e)\text{OPT}$ vertices: a $(6, 1 - 1/e)$ bicriteria approximation is attained (Lemma 5.2 [12]).

As a final step ($Best_k(\cdot)$ at line 3), a dynamic programming approach is used to identify the best-profit subtree with at most k vertices, such that the budget requirement is satisfied; see paragraph 5.2.2 in [12] for the relevant recurrences. To obtain a true approximation guarantee for the final solution, the following tree decomposition lemma is used recursively to prove that, for a sufficiently large value of k, a tree of size $6k$ can be decomposed into 13 trees; each of size at most k (Lemma 5.4 [12]).

Lemma 1 (Folklore). *Given any tree on n vertices, we can decompose it into two trees (by replicating a single vertex) such that the smaller tree has at most $\lceil \frac{n}{2} \rceil$ vertices and the larger tree has at most $\lceil \frac{2n}{3} \rceil$ vertices.*

3.2 Improvement to Previous Approach: Eligible Trees

An improvement to the analysis in [12] can be achieved by utilising a more refined tree decomposition (than the recursive application of Lemma 1) to provide the approximation guarantee at the final step. To do so, we consider a tree decomposition scheme based on the notion of *eligible trees* as introduced in [3].

Definition 5 ([3]). *Given a directed tree $T = (V_T, E_T)$, an eligible subtree T' is a subtree of T rooted at some vertex $i \in V_T$ such that the forest obtained by deleting the edges with both endpoints in T', and then all the remaining vertices of degree 0, consists of a single tree.*

Assuming T' is an eligible subtree not identical to T, after deleting all edges with both endpoints in T', the only vertex of T' with degree strictly greater than 0 is the root vertex of T'. That is, like in Lemma 1, a single vertex is replicated when removing T' from T; see Fig. 1. The following lemma suggests that, for any tree, there exists an eligible subtree within some specific size range.

Lemma 2 (Lemma 5 [3]). *For each directed tree $T = (V_T, E_T)$, and for each $p \in [1, |V_T|] \cap \mathbb{N}$, there exists an eligible subtree T' of T such that $p/2 \le |V_{T'}| \le p$.*

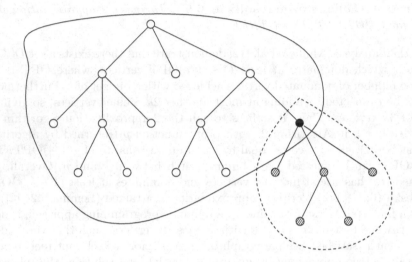

Fig. 1. An example eligible subtree of size 6 (enclosed within the dashed shape). After removing its edges and then all remaining vertices of degree 0 (vertices with lines), a single tree remains (enclosed within the solid shape). A single vertex is replicated in both trees, the black vertex.

We can now proceed to employ the above lemma iteratively toward a decomposition scheme for the tree of size at most $6k$ returned by the Quota Steiner Tree call in Algorithm 2.

Lemma 3. *Let k be an integer. Given any tree T on ak vertices, where $a \in \mathbb{N}$ is a constant, and $k \ge 4a - 2$, we can decompose it into $2a$ subtrees each on at most k vertices.*

Proof. To make T directed, we orient its edges away from some arbitrary vertex picked as the root. Now, we iteratively apply Lemma 2 with $p = k$, until we are left with a tree on at most k vertices.

First, let us show that after i iterations, the remaining tree has at most $ak - i \cdot (k/2 - 1)$ vertices. At the first iteration, there exists an eligible subtree T_1' such that $k/2 \le |V_{T_1'}| \le k$. After removing it from $T_1 := T$ we are left with T_2 of size $|V_{T_1}| - (|V_{T_1'}| - 1)$, since the root of T_1' remains in T_1. Hence, $|V_{T_1}| \le$

$ak - (k/2 - 1)$, since $k/2 \le |V_{T_1'}|$. Assume that after i iterations of the above procedure, it holds for the remaining tree T_{i+1} that $k < |V_{T_{i+1}}| \le ak - i \cdot (k/2-1)$. We inductively apply Lemma 2 with $p = k$ and get an eligible subtree T_{i+1}'. Removing T_{i+1}' from T_{i+1}, we get T_{i+2}, where $|V_{T_{i+2}}| = |V_{T_{i+1}}| - (|V_{T_{i+1}'}| - 1) \le ak - i \cdot (k/2 - 1) - (k/2 - 1) = ak - (i + 1) \cdot (k/2 - 1)$.

We proved that, after i removals of eligible subtrees from the original tree, for the remaining tree T_{i+1} it holds $|V_{T_{i+1}}| \le ak - i \cdot (k/2 - 1)$. For $i = 2a - 1$, we get $|V_{T_{2a}}| \le ak - (2a - 1) \cdot (k/2 - 1) = ak - ak + 2a + k/2 - 1 = k/2 + 2a - 1$, which is at most k for a sufficiently large value of k, i.e., $k \ge 4a - 2$. Overall, the original tree T_1 has been decomposed into $2a$ trees: $T_1', T_2', \ldots, T_{2a-1}'$ and T_{2a}, each of which has at most k vertices. □

Theorem 2. *Algorithm 2 is a $(1 - 1/e)/12$ approximation for BCDS.*

3.3 An Improved Modified Algorithm

In the following proof, we generalize the analysis given in Lemma 5.1 [12] regarding the existence of a greedily selected set (of at most k vertices) with a good intersection to the (neighborhood of the) optimal solution. Below, let D and p refer to the dominating set and profit function returned by GDS (line 1 in Algorithm 2).

Lemma 4. *There exists a set $D' \subseteq D$, $|D'| \le \lceil ck \rceil$, for some constant $0 < c \le 1$, such that $p(D') \ge (1 - e^{-c})$OPT. Furthermore, D' can be connected using at most another $k + \lceil ck \rceil$ Steiner vertices.*

Proof. We define the layers L_1, L_2, L_3 as follows. L_1 contains the (at most k) vertices of an optimal BCDS solution. Let $L_2 = N(L_1)$, meaning that the optimal number of dominated vertices is OPT $= |L_1 \cup L_2|$. Also, let $L_3 = N(L_2) \setminus L_1$ and $R = V \setminus (L_1 \cup L_2 \cup L_3)$, where R denotes the remaining vertices, i.e., those outside the three layers L_1, L_2, L_3. Let us now consider the intersection of these layers with the greedy dominating set D returned by GDS (Algorithm 1). Let $L_i' = D \cap L_i$ for $i = 1, 2, 3$ and $D' = \{v_1, v_2, \ldots, v_\lambda\}$ denote the first $\lambda = \lceil ck \rceil$ vertices from $L_1' \cup L_2' \cup L_3'$ in the order selected by the greedy algorithm. In order to bound the total profit in D', we define $g_i = \sum_{\mu=1}^{i} p(v_\mu)$ as the profit we gain from the first i vertices of D'. For the initial value, let $g_0 = 0$.

Proposition 1 (Claim 1 [12]). *For $i = 0, 1, \ldots, k - 1$, it holds $g_{i+1} - g_i \ge \frac{1}{k}(\text{OPT} - g_i)$.*

By solving the recurrence in Claim 1, we get $g_i \ge (1 - (1 - \frac{1}{k})^i)$OPT. Then, for D', we get $\sum_{v \in D'} p(v) = g_{\lceil ck \rceil} \ge \left(1 - (1 - \frac{1}{k})^{\lceil ck \rceil}\right)$OPT $\ge \left(1 - (1 - \frac{1}{k})^{ck}\right)$OPT $\ge \left(1 - \left((1 - \frac{1}{k})^{k}\right)^{c}\right)$OPT $\ge (1 - e^{-c})$OPT. Moreover, let us show that an extra $k + \lceil ck \rceil$ vertices are enough to ensure that D' is connected. We select a subset $D'' \subseteq L_2$ of size at most $|L_3 \cap D'| \le \lceil ck \rceil$ to dominate all vertices of $D' \cap L_3$. Then, we ensure that all vertices are connected by simply adding all the k vertices of

L_1. Thus, $\hat{D} = D' \cup D'' \cup L_1$ induces a connected subgraph that contains at most $k + 2\lceil ck \rceil$ vertices.   □

We can now make use of this generalized analysis and suggest a modified algorithm, parameterized by the parameter c, where the Quota Steiner Tree routine is called with a quota of $(1 - e^{-c})\text{OPT}$; see Algorithm 3 below.

Algorithm 3: Modified Greedy Profit Labeling Algorithm for BCDS(c)

    Input : A graph $G = (V(G), E(G))$, $k \in \mathbb{N}$
    Output: A tree $\tilde{T}$ on at most k vertices
1 $(D, p) \leftarrow GDS(G)$
2 $T \leftarrow QST(G, (1 - e^{-c})\text{OPT}, p)$
3 $\tilde{T} \leftarrow Best_k(T, p)$
4 **return** $\tilde{T}$

Theorem 3. *For some constant $0 < c \le 1$, there is a $(1 - e^{-c})/(\lceil 8c \rceil + 4)$ approximation for BCDS.*

Proof. By Lemma 4 and Theorem 1, it follows that Algorithm 3 (line 2) returns a tree of size at most $2k + 4\lceil ck \rceil \le 2k + 4(ck + 1) = (4c + 2)k + 4$ with profit at least $(1 - e^{-c})\text{OPT}$. For a final solution, it suffices to return a subtree of T, namely T', of size at most k which dominates the maximum number of vertices (call $Best_k(\cdot)$ in line 3 of Algorithm 3). This can be done in polynomial time via dynamic programming: see section 5.2.2 in [12].

To prove a lower bound on the number of vertices T' dominates, we decompose T into a set of subtrees via iteratively removing an eligible tree from T. To do so, we apply Lemma 2 with $p = k$. Like in the proof of Lemma 3, we can prove by induction that after i such removals of eligible subtrees of size at most k, the remaining tree has at most $|T| - i \cdot (k/2 - 1)$ vertices. For $i = \lceil 8c + 3 \rceil$, the remaining tree's size is upper bounded by $(4c + 2)k + 4 - \lceil 8c + 3 \rceil \cdot (k/2 - 1) \le (4c + 2)k + 4 - (8c + 3) \cdot (k/2 - 1) = k/2 + 8c + 7$, which is at most k for a sufficiently large choice of k, i.e., $k \ge 16c + 14$. Therefore, we can decompose T into $\lceil 8c + 3 \rceil + 1 = \lceil 8c \rceil + 4$ subtrees of size at most k, say $T_1, T_2, \ldots, T_{\lceil 8c \rceil + 4}$. Then, from pigeonhole principle and our decomposition, it follows $p(T') \ge \frac{1}{\lceil 8c \rceil + 4} \sum_{i=1}^{\lceil 8c \rceil + 4} p(T_i) \ge \frac{1}{\lceil 8c \rceil + 4} p(T) \ge \frac{1}{\lceil 8c \rceil + 4}(1 - e^{-c})\text{OPT}$.   □

For $c = 1$, Theorem 3 matches the approximation ratio already given in Theorem 2. Since the above ratio is a function of the parameter c, we numerically compute its maximum value to $1/11(1 - e^{-7/8})$ attained for $c = 7/8$.

Corollary 1. *There is a $1/11(1 - e^{-7/8})$-approximation for BCDS.*

3.4 Inapproximability

In this Subsection, we demonstrate a first inapproximability result for BCDS by identifying a reduction from the well known *Maximum Coverage* problem.

Definition 6 (MAX-k-COVER). *Given a positive integer k and a collection of sets $S = \{S_1, S_2, \ldots, S_m\}$, find a set $S' \subseteq S$, where $|S'| \leq k$, which maximizes the number of covered elements $|\bigcup_{S_i \in S'} S_i|$.*

Theorem 4 ([5,11]). *For any $\epsilon > 0$, there is no polynomial time approximation algorithm for MAX-k-COVER within a ratio of $(1 - 1/e + \epsilon)$ unless $P = NP$.*

Let us now demonstrate a *gap-preserving reduction* (Definition 10.2 [1]) which transforms an instance of MAX-k-COVER, namely MC(S, k), where $S = \{S_1, S_2, \ldots, S_m\}$ to an instance of BCDS, namely BCDS(G, k), where $G = (V, E)$. For an example illustration, see Fig. 2. For each set $S_i \in S$, we include a vertex s_i in V. Let the union of elements in the set system $\bigcup_{S_i \in S} S_i$ be represented as $\{x_1, x_2, \ldots, x_n\}$. For each element x_j, we include q vertices in V, namely $x_{j,1}, x_{j,2}, \ldots, x_{j,q}$, where q is a polynomial in m ($q \geq m^2$ suffices). Overall, $|V| = m + qn$. In the edge set E, we include edges (s_i, s_j), for each $i, j = 1, 2, \ldots, m$, $i \neq j$, and $(s_i, x_{j,z})$, for each i, j such that $x_j \in S_i$ and for each $z = 1, 2, \ldots, q$. Notice the size is polynomial in the input of MC(S, k), since we get $|E| \leq \binom{m}{2} + mqn$. In Lemma 5, let MC$(S, k)$, respectively BCDS$(G, k)$, also refer to the optimal solution for the corresponding MAX-k-COVER, resp. BCDS, instance.

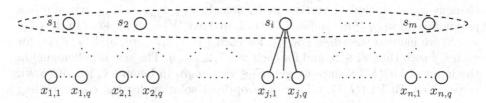

Fig. 2. The graph G constructed for the gap-preserving reduction employed in Lemma 5. Vertices s_i within the dashed ellipse form a clique. Vertex s_i is connected to vertices $x_{j,1}, x_{j,2}, \ldots, x_{j,q}$ in G if $S_i \ni x_j$ in MC(S, k).

Lemma 5. *There is a gap-preserving reduction from MAX-k-COVER to BCDS so that,*

(i) if MC$(S, k) \geq \lambda$, *then* BCDS$(G, k) \geq \Lambda$, *where* $\Lambda := m + q\lambda$, *and*
(ii) if MC$(S, k) < (1 - \frac{1}{e} + \epsilon) \cdot \lambda$, *then* BCDS$(G, k) < (1 - \frac{1}{e} + \frac{m}{e(m+q\lambda)} + \epsilon \cdot \frac{q\lambda}{m+q\lambda}) \cdot \Lambda$.

Theorem 5. *For any $\epsilon > 0$, there is no polynomial time approximation algorithm for BCDS within a ratio of $(1 - 1/e + \epsilon)$ unless $P = NP$.*

4   Edge-Vertex Domination

We now turn our attention to edge-vertex domination problems, where the goal is to identify a set of edges which dominate vertices of the graph. We consider both budgeted and partial cover cases.

4.1   Budgeted Edge-Vertex Domination

Let us consider the general case of BEVD (Definition 2), where the selected subset of edges does not need to be connected. We identify a strong connection to the classical MAX-k-COVER problem; see Definition 6 and Theorems 4, 6. On the positive side, in Theorem 7, we prove a $(1 - 1/e)$-approximation by reducing BEVD to an instance of MAX-k-COVER. On the negative side, we demonstrate a gap-preserving reduction from MAX-k-COVER to BEVD and therefore conclude that the above approximation is the best possible (Theorem 8).

Theorem 6 (Proposition 5.1 [5]). *There exists a $(1 - 1/e)$-approximation algorithm in polynomial time for MAX-k-COVER.*

Theorem 7. *There exists a $(1 - 1/e)$-approximation algorithm for BEVD.*

We now proceed and demonstrate a *gap-preserving reduction* (Definition 10.2 [1]) which transforms an instance of MAX-k-COVER, namely MC(S, k), where $S = \{S_1, S_2, \ldots, S_m\}$ to an instance of BEVD, namely BEVD(G, k), where $G = (V, E)$. For an illustration, see Fig. 3. The vertex set V contains a "root" vertex v_0. For each set $S_i \in S$, we include a vertex s_i in V. Let the union of elements in the set system $\bigcup_{S_i \in S} S_i$ be represented as $\{x_1, x_2, \ldots, x_n\}$. For each element x_j, we include q vertices in V, namely $x_{j,1}, x_{j,2}, \ldots, x_{j,q}$, where q is a polynomial in m ($q \geq m^2$ suffices) Overall, we have $|V| = m + 1 + qn$. In the edge set E, we include the edges (v_0, s_i), for each $i = 1, 2, \ldots, m$, and $(s_i, x_{j,z})$, for each i, j such that $x_j \in S_i$ and for each $z = 1, 2, \ldots, q$. The size is polynomial in the input of MC(S, k), since we get $|E| \leq m + mqn$. In Lemma 6, let MC(S, k), respectively BEVD(G, k), refer to the optimal solution for the corresponding max cover, resp. BEVD, instance.

Lemma 6. *There is a gap-preserving reduction from MAX-k-COVER to BEVD so that,*

(i) *if* MC$(S, k) \geq \lambda$, *then* BEVD$(G, k) \geq \Lambda$, *where* $\Lambda := m + 1 + q\lambda$, *and*
(ii) *if* MC$(S, k) < (1 - \frac{1}{e} + \epsilon) \cdot \lambda$, *then* BEVD$(G, k) < (1 - \frac{1}{e} + \frac{m+1}{e(m+1+q\lambda)} + \epsilon \frac{q\lambda}{m+1+q\lambda}) \cdot \Lambda$.

Theorem 8. *For any $\epsilon > 0$, there is no polynomial time approximation algorithm for BEVD within a ratio of $(1 - 1/e + \epsilon)$ unless $P = NP$.*

As a side note, consider the case where the selected edge set is required to be connected. That is, let BEVD$_C$ refer to the budgeted edge-vertex *connected* domination problem. Below, we prove that this problem is equivalent to the budgeted connected dominating set (BCDS) problem researched in Sect. 3.

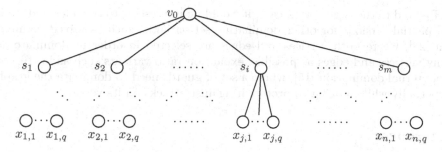

Fig. 3. Graph G constructed for the gap-preserving reduction employed in Lemma 6. Vertex s_i is connected to vertices $x_{j,1}, x_{j,2}, \ldots, x_{j,q}$ in G if $S_i \ni x_j$ in MC(S,k).

Proposition 2. *For any $G = (V, E)$ where $|V| \geq 2$, and integer $k \geq 2$, a feasible solution S to BCDS(G, k) can be transformed to a solution S_E to BEVD$_C(G, k-1)$, where $N[S] = N[S_E]$, and vice versa.*

4.2 Partial Edge-Vertex Domination

Herein, we prove an $O(\log n)$-approximation for Partial Edge-Vertex Domination (PEVD); refer to Definition 3. Given a graph $G = (V, E)$ and an integer n', we need to select a subset $E' \subseteq E$ of minimum size such that it holds $|N[E']| \geq n'$. To approximate the problem, we identify a reduction to *Partial Cover* (PC).

Definition 7 (PARTIAL COVER). *Given a universe (set) of elements $X = \{x_1, x_2, ..., x_n\}$, a collection of subsets of X, $S = \{S_1, S_2, ..., S_m\}$, and a real $0 < p \leq 1$, find a minimum-size sub-collection of S, say S', that covers at least a p-part of X, i.e., $|\bigcup_{S_i \in S'} S_i| \geq pn$.*

Theorem 9 (Theorems 3, 4 in [24]). *PARTIAL COVER is approximable within a factor $\min\{H(\lceil pn \rceil), H(D)\}$, where H is the Harmonic number $H(x) = \sum_{i=1}^{x} 1/x$ and D is the maximum size of a set in S.*

Theorem 10. *There exists a $\min\{H(n'), H(2\Delta)\}$-approximation for PEVD.*

5   Conclusion

We propose a new technique to obtain tree decompositions, and a generalized analysis, thus improving the approximation guarantee from $(1 - e^{-1})/13$ to $(1 - e^{-7/8})/11$ for BCDS. Furthermore, we prove a $(1 - 1/e + \epsilon)$ upper bound. Also, we introduce BEVD and PEVD, and provide (tight) approximation bounds.

Regarding future work on BCDS, the goal is to design an algorithm with an improved guarantee. Moreover, it would be interesting to capture the difficulty of the problem with a stronger inapproximability result. We believe that a tight bound lies somewhere between our currently established state of the art.

Related to the edge-vertex case, it would be interesting to consider budgeted and partial versions for other dominating set variants, such as mixed domination [28], where both vertices and edges are selected in order to dominate as many vertices and edges as possible, expansion ratio variants such as in [16], or even eternal domination [15], where a set of guards need to dominate the graph perpetually while moving to protect it against attacks on its vertices.

References

1. Arora, S., Lund, C.: Hardness of approximations. In: Approximation Algorithms for NP-Hard Problems, chap. 10, pp. 399–446. PWS Publishing Co., Boston (1997)
2. Boutrig, R., Chellali, M., Haynes, T.W., Hedetniemi, S.T.: Vertex-edge domination in graphs. Aequat. Math. **90**, 355–366 (2016). https://doi.org/10.1007/s00010-015-0354-2
3. Bermond, J.-C., et al.: Bin packing with colocations. In: Jansen, K., Mastrolilli, M. (eds.) WAOA 2016. LNCS, vol. 10138, pp. 40–51. Springer, Cham (2017). https://doi.org/10.1007/978-3-319-51741-4_4
4. Du, D.Z., Wan, P.J.: Connected Dominating Set: Theory and Applications. Springer Optimization and its Applications. Springer, New York (2013). https://doi.org/10.1007/978-1-4614-5242-3
5. Feige, U.: A threshold of $\ln n$ for approximating set cover. J. ACM **45**(4), 634–652 (1998)
6. Garg, N.: Saving an epsilon: a 2-approximation for the k-MST problem in graphs. In: Proceedings of the Thirty-Seventh Annual ACM Symposium on Theory of Computing (STOC), pp. 396–402 (2005)
7. Guha, S., Khuller, S.: Approximation algorithms for connected dominating sets. Algorithmica **20**(4), 374–387 (1998)
8. Horton, J.D., Kilakos, K.: Minimum edge dominating sets. SIAM J. Discrete Math. **6**(3), 375–387 (1993)
9. Johnson, D.S., Minkoff, M., Phillips, S.: The prize collecting Steiner tree problem: theory and practice. In: Proceedings of the Eleventh Annual ACM-SIAM Symposium on Discrete algorithms (SODA), pp. 760–769 (2000)
10. Khoa, N.M., Tung, D.D.: Locating fault on transmission line with static var compensator based on phasor measurement unit. Energies **11**, 2380 (2018)
11. Khuller, S., Moss, A., Naor, J.S.: The budgeted maximum coverage problem. Inf. Process. Lett. **70**(1), 39–45 (1999)
12. Khuller, S., Purohit, M., Sarpatwar, K.K.: Analyzing the optimal neighborhood: algorithms for budgeted and partial connected dominating set problems. In: Proceedings of the Twenty-Fifth Annual ACM-SIAM Symposium on Discrete Algorithms (SODA), pp. 1702–1713 (2014)
13. Khuller, S., Purohit, M., Sarpatwar, K.K.: Analyzing the optimal neighborhood: algorithms for budgeted and partial connected dominating set problems. SIAM J. Discrete Math. **34**(1), 251–270 (2020)
14. Kumar, R., Kaur, J.: Efficient beacon placement for network tomography. In: Proceedings of the 4th ACM SIGCOMM Conference on Internet Measurement, pp. 181–186 (2004)
15. Lamprou, I., Martin, R., Schewe, S.: Eternally dominating large grids. Theoret. Comput. Sci. **794**, 27–46 (2018)

16. Lamprou, I., Martin, R., Schewe, S., Sigalas, I., Zissimopoulos, V.: Maximum rooted connected expansion. In: 43rd International Symposium on Mathematical Foundations of Computer Science (MFCS), LIPIcs, vol. 117, pp. 25:1–25:14 (2018)
17. Lan, J.K., Chang, G.J.: On the mixed domination problem in graphs. Theoret. Comput. Sci. **476**, 84–93 (2013)
18. Lewis, J.: Vertex-edge and edge-vertex parameters in graphs. All Dissertations, 103 (2007)
19. Liu, Y., Liang, W.: Approximate coverage in wireless sensor networks. In: The IEEE Conference on Local Computer Networks 30th Anniversary (LCN 2005), pp. 68–75 (2005)
20. Liu, X., Wang, W., Kim, D., Yang, Z., Tokuta, A., Jiang, Y.: The first constant factor approximation for minimum partial connected dominating set problem in growth-bounded graphs. Wirel. Netw. **22**(2), 553–562 (2015). https://doi.org/10.1007/s11276-015-0981-5
21. Miyano, E., Ono, H.: Maximum domination problem. In: Proceedings of the Seventeenth Computing: the Australasian Theory Symposium (CATS), vol. 119, pp. 55–62 (2011)
22. Peters, K.W.: Theoretical and algorithmic results on domination and connectivity. Ph.D. thesis, Clemson University, Clemson (1986)
23. Sampathkumar, E., Kamath, S.S.: Mixed domination in graphs. Sankhya Indian J. Stat. **54**, 399–402 (1992)
24. Slavík, P.: Improved performance of the greedy algorithm for partial cover. Inf. Process. Lett. **64**, 251–254 (1997)
25. Venkatakrishnan, Y.B., Krishnakumari, B.: An improved upper bound of edge-vertex domination number of a tree. Inf. Process. Lett. **134**, 14–17 (2018)
26. Wang, B.: Coverage problems in sensor networks. ACM Comput. Surv. **43**(4), 32 (2011)
27. Yannakakis, M., Gavril, F.: Edge dominating sets in graphs. SIAM J. Appl. Math. **38**(3), 364–372 (1980)
28. Zhao, Y., Kang, L., Sohn, M.Y.: The algorithmic complexity of mixed domination in graphs. Theoret. Comput. Sci. **412**, 2387–2392 (2011)

Algorithms for Constructing Anonymizing Arrays

Erin Lanus[1](✉) and Charles J. Colbourn[2]

[1] Virginia Tech, Arlington, VA 22203, USA
lanus@vt.edu
[2] Arizona State University, Tempe, AZ 85281, USA
charles.colbourn@asu.edu

Abstract. Attribute-based methods are inherently identity-less as authorization decisions are made in terms of attributes possessed by the subject rather than identity. However, anonymity against the system is not guaranteed when attribute distribution allows for the composition of a policy that few subjects can satisfy. An anonymizing array ensures that any assignment of values to t attributes that appears in the array appears at least r times. When an anonymizing array is used for subjects registered to a system and policies contain conjunctions of at most t attributes, the system cannot identify the subject using the policy to to gain authorization with greater than $\frac{1}{r}$ probability. Anonymizing arrays are similar to covering arrays with higher coverage and constraints, but have an additional desired property, homogeneity, due to their application domain. In this paper, we develop constructions for anonymizing arrays and propose a post-optimization mechanism to reduce homogeneity.

Keywords: Combinatorial array · Construction algorithms · Anonymous authorization · Attribute-based methods

1 Introduction

In attribute-based systems used for access control, such as Attribute-Based Access Control and Ciphertext-Policy Attribute-Based Encryption (CP-ABE), decisions are made on the basis of attributes, or characteristics of a subject expressed as name-value pairs [1,5]. A feature of these systems is that they can achieve anonymous access control, granting access to authorized subjects and denying access to unauthorized subjects without knowledge of the subject's identity. This is not a guarantee that the identity cannot be deduced. CP-ABE encrypts a ciphertext in a policy, and decryption is performed by a private

Research of EL was supported by a National Physical Science Consortium Fellowship. Research of CJC was supported in part by the National Science Foundation under Grant No. 1421058 and Grant No. 1813729.

L. Gąsieniec et al. (Eds.): IWOCA 2020, LNCS 12126, pp. 382–394, 2020.
https://doi.org/10.1007/978-3-030-48966-3_29

key containing attributes satisfying the policy. CP-ABE is proposed to mediate authenticated key exchange with an anonymous mode [9]. Suppose a service broadcasts a session key encrypted by a policy. A subject whose private key contains attributes satisfying the policy decrypts the message, obtains the key, and begins communicating with the service via the session key. The service knows that the subject communicating with it is authorized based on possession of the required attributes to obtain the key. The authors claim that the service cannot uniquely identify the subject. All subjects must register with the service to receive a private key, and thus the service knows all attributes of the subjects in the system. If a policy can be composed so that only one subject's attributes satisfy the policy and this policy is used to encrypt the session key, the service knows the identity of the subject using this session key. "Anonymous ABE" uses hidden credentials which can be used to retrieve a session key anonymously, but the receiver anonymity is based on "plausible deniability" due to the fact that anyone can request the message, not just the intended recipient [6]. Plausible deniability fails if the message is decrypted to gain a session key to obtain authorization, proving that a subject with the correct credentials decrypted the message.

The contribution of this work is to achieve a guaranteed degree of anonymity by requiring that certain properties of attribute distribution hold given a maximum credential size. Policies can be considered disjunctions of conjunctions of attribute values with the most restrictive policy being a single conjunction of many attribute values. Let t be the largest number of attributes in a single conjunction. An *anonymizing array* ensures that any assignment of values to t attributes that appears in the array appears at least r times [7]. When an anonymizing array is used for subjects registered to a system and policies contain conjunctions of at most t attributes, the system cannot identify the subject using the policy for authorization with greater than $\frac{1}{r}$ probability.

An *access profile* is an assignment of values to attributes. When attributes are assigned to access profiles for the purpose of anonymous authorization, as in key distribution, rather than existing as real-world attributes of subjects, an anonymizing array is built from scratch. When the set of subject attributes registered to a system is fixed, an anonymizing array determines the largest conjunction that can be used while achieving the anonymity guarantee r or, equivalently, the guarantee achievable for the largest conjunction. When subject attributes are immutable but the set of access profiles can be appended, anonymizing arrays provide a mechanism to provide higher anonymity guarantees. Constructions for anonymizing arrays must account for constraints on attributes and, due to the security application, must not rely randomness in order to achieve the guarantee with a high probability. The rest of the paper is organized as follows. Definitions and relationship to covering arrays are in Sect. 2. Construction algorithms are in Sect. 3. Results are in Sect. 4, and conclusions are in Sect. 5.

2   Anonymizing Arrays

2.1   Definitions

Consider an array with N rows and k columns and each column i for $1 \leq i \leq k$ has entries from a set of v_i symbols. The rows of the array are access profiles, columns are attributes, and symbols in a column are the values for the attribute. To express the parameters of the array, write $\mathsf{AA}(N; r, t, k, (v_1, \ldots, v_k))$ or use exponential notation v_i^j when j columns share the same number of symbols v_i. Write $\mathsf{AA}(N; r, t, k, v)$ when the number is the same for all columns. Such an array is (r, t)-*anonymous* if, when choosing an $N \times t$ subarray, $1 \leq t \leq k$, each row that appears is repeated at least r times. A *credential* is a tuple of attribute-value pairs presented for an authorization decision. The *maximum credential size* is t and r is the *anonymity guarantee*. Given an $N \times k$ array $\mathbf{A}$, an $N' \times k$ array $\mathbf{A}'$ is (r, t)-*anonymizing* with respect to $\mathbf{A}$ if $\mathbf{A} \subseteq \mathbf{A}'$ and $\mathbf{A}'$ is (r, t)-anonymous. Interesting cases require $r > 1$ and $v_i > 1$ for all v_i. An access profile may represent a subject or it may be *padding*, a row added to reach an anonymity guarantee. Access profiles need not be unique. The rows or columns of the array can be shuffled to obtain an equivalent array on the same parameters.

Hard constraints are credentials that cannot appear, while *soft constraints* are credentials that need not appear, but are not illegal. That is, hard constraints and non-appearing soft constraints must appear 0 times, while soft constraints that appear and all unconstrained credentials must appear at least r times. Hard constraints may give rise to *implicit hard constraints* that cause there to be no feasible solution. Constraints must be considered when appending padding rows to an anonymizing array to reach the anonymity guarantee r.

Anonymizing arrays containing groups of highly similar access profiles may lead to affiliating subjects with and tracking subjects by their groups; [7] develops the following metrics to detect this similarity. *Local homogeneity* describes how often an access profile appears in small groups of similar access profiles, and *global homogeneity* is the average local homogeneity. The *neighborhood* of a credential is the set of access profiles possessing the credential. The *closeness* of a pair of access profiles is a sum of their *weight* over all credentials, and the weight of a pair of access profiles on a credential is inversely proportionate to the size of the neighborhood of the credential if the access profiles are in the neighborhood. Let $\mathcal{U}$ be a set of N access profiles and let $\mathcal{C}$ be the set of credentials. Define the neighborhood of a credential $c \in \mathcal{C}$ as $\rho(c) = \{u_i : u_i \text{ possesses } c, u_i \in \mathcal{U}\}$ and

$$weight(u_i, u_j, c) = \begin{cases} \frac{1}{|\rho(c)|} & \Longleftrightarrow \{u_i, u_j\} \subseteq \rho(c) \\ 0 & \text{otherwise} \end{cases}$$

$$closeness(u_i, u_j) = \sum_{c \in \mathcal{C}} weight(u_i, u_j, c)$$

$$neighbors(u_i) = \{\bigcup_{c \in \mathcal{C}} \rho(c) : u_i \in \rho(c)\}$$

$$homogeneity(u_i) = \frac{1}{|neighbors(u_i)|} \sum_{u_j \in \mathcal{U}, u_j \neq u_i} closeness(u_i, u_j)$$

2.2 Relationship to Covering Arrays

A *covering array* denoted $CA(N; t, k, v)$ is an $N \times k$ array on v symbols such that in every $N \times t$ subarray each of the v^t combinations of symbols, called *interactions*, appears as a row. When different columns can have different numbers of symbols, it is a *mixed-level covering array* $MCA(N; t, k, (v_1, \ldots, v_k))$. In rare cases when higher coverage is needed, interactions may be required to appear at least $\lambda > 1$ times. When not specified, $\lambda = 1$ is implied. Anonymizing arrays are similar to covering arrays with constraints and higher coverage. The primary difference due to application is in the desired homogeneity property, but also in how constraints are treated. For covering arrays, the norm is to define the interactions that must not appear (hard constraints), then to define the interactions that might appear (soft constraints, possibly further divided into "don't care" and "avoid"), and then to derive the interactions that must appear [3]. For anonymizing arrays, the access profiles provided define the unconstrained credentials. The system specification defines the hard constraints, and the soft constraints are defined to be the remaining credentials that are in neither set. Given an anonymizing array without a defined set of constraints, it may be impossible to distinguish the soft and hard constraints from the set of non-appearing credentials. The same difficulty arises distinguishing the soft constraints from the unconstrained credentials. Care must be taken when converting between covering arrays and anonymizing arrays that constraints are categorized correctly. Many construction algorithms exist for building covering arrays, though few explicitly include constraint handling or higher coverage requirements. The following non-exhaustive list of relationships elucidate how to use covering array constructions to build anonymizing arrays.

Any $MCA_\lambda(t, k, (v_1, \ldots, v_k))$ with hard constraint set $\mathcal{H}$ is also an $AA(\lambda, t, k, (v_1, \ldots, v_k))$ with hard constraint set $\mathcal{H}$ and all other credentials appearing. Every t-way interaction that appears in the covering array λ times is a credential that appears λ times in the corresponding anonymizing array. The interactions in $\mathcal{H}$ never appear in the covering array so they never appear in the anonymizing array. In the context of covering arrays, higher λ does not force a "don't care" or an "avoid" interaction to appear λ times if it appears once. Then soft constraints must not be present in a covering array used as an anonymizing array. There must also exist a mapping of soft constraints in the anonymizing array onto either hard constraints in the covering array if they do not appear or onto unconstrained interactions if they do.

If an $MCA(t, k, (v_1, \ldots, v_k))$ with hard constraint set $\mathcal{H}$ and soft constraint set $\mathcal{S}$ exists, then an $AA(r, t, k, (v_1, \ldots, v_k))$ with $\mathcal{H}$ and $\mathcal{S}$ exists. Copy the covering array vertically r times. No interaction of $\mathcal{H}$ appears in the covering array, so none of these credentials appear in the anonymizing array. Any interaction of $\mathcal{S}$ that appears in the covering array at least once appears in the anonymizing array at least r times, and the rest never appear. Unconstrained credentials appear at least once in the covering array and at least r times in the anonymizing array.

An $\mathsf{MCA}(t, k, (v_1, \ldots, v_k)), v = \min_{i=1}^{k}(v_i)$ with no constrained interactions is an $\mathsf{AA}(v, t - 1, k, (v_1, \ldots, v_k))$ with no constrained credentials. In the mixed-level covering array without constraints, a $(t - 1)$-way interaction appears at least v_i times, once with each of the v_i symbols in the t-th column of the t-way interaction including those $t - 1$ columns. Every $(t - 1)$-way interaction appears at least v times, v the minimum v_i. This is an anonymizing array for $r = v$.

If there exists a covering array $\mathsf{CA}(t, k, v)$ with a set of hard constraints $\{(c_1, \sigma_1), \ldots, (c_{t-1}, \sigma_{t-1}), (c_x, \sigma_y)\}$ for each column symbol pair (c_x, σ_y) with column $c_x \in \mathcal{K} \setminus \{c_1, \ldots, c_{t-1}\}$ and $\sigma_y \in \Sigma_x$, the symbol set of c_x, then there is an anonymizing array $\mathsf{AA}(v, t - 1, k, v)$ with $\{(c_1, \sigma_1), \ldots, (c_{t-1}, \sigma_{t-1})\}$ as a hard constraint. To guarantee that the constrained credential with $t - 1$ attributes never appears in the anonymizing array, it must be the case that no t-way interactions of which it is a subset appeared in the covering array. The coverage for all unconstrained credentials has already been shown. To extend this to soft constraints, there must be a mapping of soft constraints in the anonymizing array to either unconstrained interactions or hard constraints in the covering array.

Given an array $\mathbf{A}$ that is (r, t)-anonymous and not $(r + 1, t)$-anonymous, for every $t \leq t' \leq k$ for which $\mathbf{A}$ is (r', t')-anonymous, it must be the case that $r' \leq r$. Pick the credential c that appears the fewest number of times in $\mathbf{A}$ and let r be the number of times c appears. $\mathbf{A}$ is (r, t)-anonymous by definition and is not $(r + 1, t)$-anonymous. Choose any credential c' that contains c. The rows in which c' appears must be a subset of the rows in which c appeared. Then for $t' \geq t$, if $\mathbf{A}$ is (r', t')-anonymous, then $r' \leq r$. Similarly, an array that is (r, t)-anonymous is (r, t')-anonymous for $t' < t$. Any credential, c, of size t appears in at least r rows. Any t'-subset of c appears in at least these rows.

3   Construction Algorithms

3.1   Moser-Tardos-Style Column Resampling Algorithm

Algorithm 1 is a Moser-Tardos-style column resampling algorithm (MTCR) [8]. A *bad event* is either a violation of a hard constraint or lack of necessary coverage on unconstrained or soft constraints. A candidate is checked systematically until either no bad events are found or an iteration limit is reached. If any bad event is found, all involved columns are resampled. If $\mathcal{T}$ is the set of $\binom{k}{t}$ t-subsets of columns and there are $\sum_{T \in \mathcal{T}} \prod_{i \in T} v_i$ possible credentials, the position of T in colexicographic ordering of the sets is the *rank*. Estimating the number of rows is not obvious, so rows are added until coverage is met or an iteration limit is reached. When provided a set of rows, MTCR adds padding to meet the guarantee and forbids resampling of initial rows. When building from scratch, the candidate starts with no rows or an initial number of randomly populated rows is computed as r times the maximum number of non-constrained credentials of any rank. Adding rows too often may produce more rows than needed, while the iteration limit may be reached when adding conservatively. Too few rows

---

Algorithm 1: Moser-Tardos-style Column Resampling (MTCR)

---

input : $\mathbf{A}, r, t, k, (v_1, ..., v_k)$, and a set of constraints
output: $\mathbf{A}$ or $\emptyset$
begin
    while *iterations* < *limit* **do**
        foreach *rank while no bad event* **do**
            Check all credentials in rank
            if *coverage bad event* **then**
                Increment number of resamplings
                if *resamplings* > *rank* ∗ *threshold* **then**
                    Add a row to $\mathbf{A}$ and reset resamplings
        if *no bad event* **then**
            return A
        else
            Resample all columns of rank in $\mathbf{A}$
    return $\emptyset$

---

can contribute to lack of r coverage, but not to presence of a constraint, as more rows increase the likelihood of a constraint appearing. The candidate is checked by a fixed ordering, so it is expected, though not guaranteed, that fewer bad events exist in a candidate when checking a higher rank. The number of resamplings due to a lack of coverage bad event since adding the last row is used to estimate progress. To add rows readily when bad events occur early, the number of resamplings to add a row is proportional to the amount remaining to check.

3.2 Conditional Expectation Heuristic Search Algorithm

Algorithm 2, Conditional Expectation Heuristic Search (CEHS), is a greedy, one-row-at-a-time algorithm that combines ideas from conditional expectation with a heuristic to avoid constraints [2, 4]. Call a credential not-yet-r-covered if it is unconstrained appearing fewer than r times or a soft constraint appearing between 0 and r times. The expectation for a row is the number of not-yet-r-covered credentials that are covered if symbols are assigned to columns randomly. Given a row with $i-1$ columns fixed to symbols and the rest free, choose a column i randomly and consider the v_i symbols to place in column i. For the symbols of that column, there is a choice of symbol that does not reduce the expectation for the row. Let $\mathcal{I}_i$ be the set of $\binom{k-1}{t-1}$ sets of t columns involving i, and $\mathcal{C}_T$ the set of possible credentials for a t-set of columns, T. Suppose column i is fixed to symbol σ. If $P(c)$ is the probability of credential c appearing and $\Lambda(c)$ is related to the coverage status of c,

$$\Lambda(c) = \begin{cases} 1 \text{ if } c \text{ covered fewer than } r \text{ times,} \\ 0 \text{ if } c \text{ covered at least } r \text{ times or } c \text{ is a soft constraint,} \end{cases}$$

define

$$value(i, \sigma) = \sum_{T \in \mathcal{T}_i} \sum_{c \in \mathcal{C}_T} \Lambda(c) P(c).$$

The expected number of not-yet-r-covered credentials newly covered by placing σ in i is $value(i, \sigma)$. The best symbol is one that maximizes $value(i, \sigma)$ without violating a hard constraint. Ties can be broken randomly.

The heuristic lies in redefining Λ. Prioritizing credentials that have been covered fewer times over those that have been covered more may be more useful than the all-or-nothing approach that works well when $\lambda = r = 1$. To avoid fixing the last symbol σ in column i of a credential that violates a hard constraint when other not-yet-r-covered credentials require σ in i, define $\Lambda(c) = -\binom{k-1}{t-1}$ for this case. There are $\binom{k-1}{t-1} - 1$ other t sets involving column i. At most, a t-set contributes 1 to $value(i, \sigma)$, so the most positive value a symbol receives from the other credentials is $\binom{k-1}{t-1} - 1$. A lookahead attempts to drive the search away from fixing symbols leading to one or more eventual hard constraints without preventing covering unconstrained credentials. The lowest benefit of placing symbol σ in column j occurs when there is one credential to be covered one remaining time with the highest number of symbols, $v = \max_{i=1}^{k}(v_i)$. The probability of being placed is lowest when all other columns in the t-set are still free, assuming j is fixed to σ. Then $P(c) = \frac{1}{v^{t-1}}$ and $\Lambda(c) = \frac{1}{r}$, so the benefit is $\frac{1}{rv^{t-1}}$. The highest cost occurs when the other t-sets involving j have $\binom{k-1}{t-1} - 1$ potential hard constraints and one free column. For each t-set, let w be the number of symbols for the free column. There are w credentials with symbols matching the $t - 1$ fixed columns, and each is chosen with probability $P(c) = \frac{1}{w}$. Each t-set contributes at most $w\frac{1}{w}\Lambda$, so the total cost is $(\binom{k-1}{t-1} - 1)\Lambda$. The value of Λ must ensure that $|(\binom{k-1}{t-1} - 1)\Lambda| < \frac{1}{rv^{t-1}}$. Set $\Lambda = \frac{-1}{(\binom{k-1}{t-1}-1)ry^t}$, $y = \max_{i=1}^{k}(v_i)$. When $y \geq v$,

$$\left| \left(\binom{k-1}{t-1} - 1 \right) \frac{-1}{(\binom{k-1}{t-1}-1)ry^t} \right| = \left| \frac{-1}{ry^t} \right| < \left| \frac{1}{rv^{t-1}} \right|.$$

The full definition is then

$$\Lambda(c) = \begin{cases} \frac{r - \text{times } c \text{ covered}}{r} & \text{if } c \text{ is unconstrained or an appearing soft constraint,} \\ 0 \text{ if } c \text{ is a non-appearing soft constraint,} \\ \frac{-1}{(\binom{k-1}{t-1}-1)ry^t} : y = \max_{i=1}^{k} v_i, c \text{ a hard constraint with } \geq 1 \text{ free column,} \\ -\binom{k-1}{t-1}, c \text{ a hard constraint with 0 free columns.} \end{cases}$$

As with MTCR, a feasibility check should be conducted beforehand or an iteration limit used, as some scenarios can still result in infinite looping. CEHS lacks complete lookahead, so a series of local decisions based on the ordering of columns in an execution can lead to the placement of some hard constraint even if an anonymizing array exists. In this case, CEHS aborts and can be run again.

Algorithm 2: Conditional Expectation Heuristic Search (CEHS)

---

input : $r, t, k, (v_1, ..., v_k)$, and a set of constraints
output: **A** or $\emptyset$
begin

    Create an empty array, **A**, and set count of all credentials = 0

    while *some not-yet-r-covered credential remains* **do**

        Add a row to **A** with all columns free

        while *some column is free* **do**

            Randomly select a column i

            for *each symbol $\sigma \in [v_i]$* **do**

                Compute $value(i, \sigma) = \sum_{T \in \mathcal{T}_i} \sum_{c \in \mathcal{C}_T} \Lambda(c) P(c)$

                $P(c) = \dfrac{\text{ways to cover } c}{\text{ways to fix free columns of T}}$

                $\Lambda(c) =$

$$
\begin{cases}
\dfrac{r - \text{count of } c}{r}, & c \text{ unconstrained or appearing soft constraint,} \\
0, & c \text{ non-appearing soft constraint,} \\
\dfrac{-1}{(\binom{k-1}{t-1} - 1) r y^t} : y = \max_{i=1}^{k} v_i, & c \text{ hard constraint and } \geq 1 \text{ free column,} \\
-\binom{k-1}{t-1}, & c \text{ hard constraint with 0 free columns.}
\end{cases}
$$

            Place symbol σ in column i that maximizes $value(i, \sigma)$

        for *each of the credentials appearing in the row* **do**

            Update the count of the credential

            if *a hard constraint appears* **then**

                Return $\emptyset$

    Return **A**

3.3 Homogeneity Post-Optimization

We develop a post-optimization strategy in Algorithm 3 to reduce the homogeneity of an array by *crossover*, or swapping credentials between two access profiles. A first idea is to distance similar access profiles by identifying a high homogeneity access profile, u, and the access profile v with the largest $closeness(u, v)$. Then if credential c has the largest $weight(u, v, c)$, we might swap the symbols of u and access profile w in the columns of credential c for which $weight(u, w, c)$ is smallest. Computationally, this approach requires storage of the weight array whereas closeness can be computed as sums without the intermediary weights. Additionally, the view at the granularity level of weight does not inform how close u and w are on other credentials. They may be identical in all columns except some of c, and so crossover simply swaps u and w but the overall homogeneity of the array has not changed. Instead, select u and w such that $homogeneity(u)$ is highest and $closeness(u, w)$ is lowest. The key is to "decouple" u from u's group and create a link between u's group and w, an access profile outside the group, doing the same with w and w's group by swapping some credentials of u and w.

The weights give information about shared credentials so we could choose to swap any credentials c where $weight(u, w, c) = 0$. However, too many swaps results in swapping the entire row, and as u and w are chosen to have the smallest $closeness$ score, they may have no credentials in common. Swapping a single credential changes up to $\binom{k}{t} - \binom{k-t}{t}$ other credentials, so how to make the best

---

Algorithm 3: Homogeneity Post-optimization (HP)

---

input : $\mathbf{A}, r, t, k, (v_1, ..., v_k)$, and a set of constraints
output: $\mathbf{A}$
begin

    while *generations remain* **do**
        $mostFit = \mathbf{A}$
        Compute $homogeneity(i)$ for all rows in $\mathbf{A}$
        $u = \max_i^N (homogeneity(i))$
        for *each child in the generation* **do**
            Create a copy of $\mathbf{A}$ as *child*
            Mutate based on implementation choices
            Compute S, a set of s rows with smallest $closeness(u, w), w \in S$
            for *each block of attributes based on implementation* **do**
                Randomly select w with $\frac{1}{s}$ probability
                Swap w and u's attributes in the block
            if *child is (r, t)-anonymous with lowest global homogeneity* **then**
                Set $mostFit = child$
        Set $\mathbf{A} = mostFit$
    Return $\mathbf{A}$

---

decision without considering all possibilities is unclear. A middle path between random row resampling and computationally intensive search is to generate a set of child arrays by conducting crossover to probabilistically swap blocks of attributes between u and the S access profiles with the lowest closeness scores with u. The child with the lowest global homogeneity without violating hard constraints and meeting the anonymity guarantee becomes the parent of the next generation. As mentioned, swapping one credential changes up to $\binom{k}{t} - \binom{k-t}{t}$ other credentials in the same access profile. An affected credential that appears few times may fall below r coverage in all of the children allowed in a generation. In this case, the parent is retained and random resampling by *mutation* is conducted to allow additional appearances of the credential that is eliminated by resampling to occur elsewhere in the array to regain (r, t)-anonymity. It is not obvious how to set the mutation rate or how many and which rows to mutate. Additional tunable parameters include the set size of access profiles with which to swap, the blocksize of attributes to swap, the probability of swapping, and the number and size of generations. Stopping conditions include a generation limit, number of generations without reduced homogeneity, or meeting the expected global homogeneity.

4   Results

Comparison of MTCR and CEHS. In tests to construct $\mathsf{AA}(r, t, 10, (5^1 4^2 3^3 2^4))$, MTCR produces arrays with the same number of rows as CEHS when $t = 1$ without constraints if restricted to use the same number of rows produced by CEHS. When allowed to add additional rows, it typically adds more than needed.

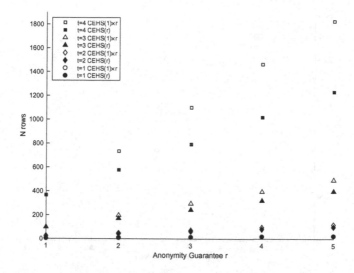

Fig. 1. CEHS versus CAcopy to build AAs with constraints

When $t = 2$ and MTCR is allowed 10^6 iterations, in general it requires more rows than CEHS to find a solution. For $t = 2$ with a hard constraint, MTCR requires about twice as many rows. For two hard constraints, MTCR does not complete in 10^6 iterations for any fixed number of rows or allowed unlimited rows. For two soft constraints, MTCR performs in fewer iterations and rows than for one hard constraint. These results suggest that randomized constructions perform poorly in the presence of hard constraints.

Comparison to Replicated Mixed-Level Covering Arrays with Constraints. A "from scratch" construction is used when attributes are assigned arbitrarily to subjects, as in key distribution. We compare the performance of CEHS against a covering array copy construction (CAcopy). CEHS is executed for $1 \leq r \leq 5$ for each $1 \leq t \leq 4$ with and 0, 6, 4, and 3 hard constraints for the values of t, respectively, to construct an $AA(r, t, 10, (5^1 4^2 3^3 2^4))$. The number of rows for this construction are plotted in Fig. 1 with closed markers and labels indicating t and "CEHS(r)." To obtain an arbitrary covering array with the same constraints, CEHS is used to construct an AA with $r = 1$. Next, AAs are made for $2 \leq r \leq 5$ by stacking r copies of each covering array. The number of rows for this construction is plotted in Fig. 1 with open markers and labeled by t and "CEHS(1) $\times r$." When $t = 1$, the number of rows needed is always r times the maximum number of levels, and both constructions produce the same number of rows. For $t > 1$, the redundancy of CAcopy clearly produces more rows than CEHS. A challenge in comparing these constructions by homogeneity is that additional rows increase the likelihood that access profiles have larger credential neighborhoods. In general, an anonymizing array with more rows is less

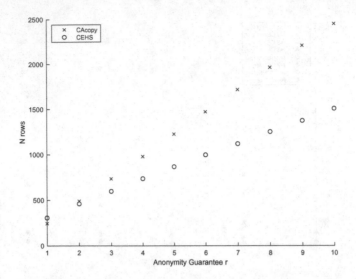

Fig. 2. CEHS versus CAcopy to build unconstrained AAs

homogeneous than one with fewer. When $t = 1$, the anonymizing arrays produced by both methods have the same number of rows for all values of r and so provide a good opportunity for comparison. In tests, the anonymizing arrays created by CEHS always have lower homogeneity scores than the copy constructed arrays. To attempt an ad-hoc comparison of the constructions in the absence of a standardized homogeneity metric that adequately compares arrays with differing numbers of rows, five rows are randomly selected from an $(r, 2)$-anonymizing array constructed by CEHS and appended to an $AA(43; 2, 2, 10, (5^1 4^2 3^3 2^4))$. The rows are not constructed randomly to ensure that no hard constraints are introduced. This method is not without bias due to the pool of rows from which they are selected and is not intended for practical use. The resulting array has lower global homogeneity than the $AA(48; 2, 2, 10, (5^1 4^2 3^3 2^4))$ created by CAcopy.

Comparison to Replicated Covering Arrays without Constraints. We construct a set of arrays, $AA(245r; r, 3, 10, 5)$ by making $1 \leq r \leq 10$ vertical copies of a $CA(245; 3, 10, 5)$ made by a conditional expectation algorithm shown to construct covering arrays with few rows efficiently [4]. As indicated in Fig. 2, when $r = 1$, the covering array has 62 fewer rows, but the CEHS algorithm produces anonymizing arrays with fewer rows for $r \geq 2$. Now, consider a row ρ in the covering array. After r copies, ρ appears (at least) r times, and this forms a cluster of rows sharing the same credentials and therefore neighborhoods. Instead, for each copy $i > 1$ and for each column j in the copy, choose a random permutation over the levels of a column, $p_{c_{i,j}} : v \mapsto v$. Each permuted copy is still a covering array, so the composed array is (r, t)-anonymous (CAperm). In this array, k independent permutations are applied to the columns of the ρth row in a copy, so the likelihood that this row closely matches ρ is reduced. In

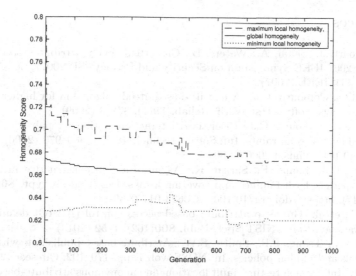

Fig. 3. HP on $AA(62; 3, 2, 19, (5^1 4^2 3^3 2^4))$ with 20 children and 1000 generations

all tests, the permuted arrays have lower average and maximum homogeneity scores than by CAcopy, and in all but one data point, they have lower minimum homogeneity scores. To compare homogeneity of CAperm to CEHS, randomly generated rows are appended to the CEHS array to equalize the number of rows. For $2 \leq r \leq 10$, CEHS produces lower minimum, global, and maximum homogeneity scores than CAperm. The one exception is that CAperm produced lower maximum homogeneity for $r = 10$. This suggests that CEHS typically produces arrays with fewer rows and lower homogeneity than by copying covering arrays, even when utilizing permutations.

Evaluation of Homogeneity Post-optimization (HP). HP contains a number of tunable parameters, and details for the implementation tested here are in [7]. An example of the reduction of global homogeneity on an $AA(62; 3, 2, 10, (5^1 4^2 3^3 2^2))$ generated by CEHS with six hard constraints is in Fig. 3.

5   Conclusion

Although anonymizing arrays differ from covering arrays in essential ways, constructive algorithms for covering arrays underlie useful algorithms for constructing anonymizing arrays. Indeed, this connection leads to copy constructions to produce arrays "from scratch" as well as two methods to add rows to a partial array (CEHS and MTCR). CEHS outperforms both MTCR and the copy constructions, both in terms of the number of rows generated and the homogeneity. Nevertheless, none of the construction methods examined ensures low homogeneity. To address this, we propose a "post-optimization" method (called HP) to reduce homogeneity, and provide preliminary evidence that HP is a reasonable first approach.

References

1. Bethencourt, J., Sahai, A., Waters, B.: Ciphertext-policy attribute-based encryption. In: 2007 IEEE Symposium on Security and Privacy (SP 2007), Los Alamitos, pp. 321–334. IEEE (2007)
2. Bryce, R.C., Colbourn, C.J.: A density-based greedy algorithm for higher strength covering arrays. Softw. Test. Verif. Reliab. **19**(1), 37–53 (2009)
3. Bryce, R.C., Colbourn, C.J.: Prioritized interaction testing for pair-wise coverage with seeding and constraints. Inf. Softw. Technol. **48**(10), 960–970 (2006). https://doi.org/10.1016/j.infsof.2006.03.004
4. Colbourn, C.J., Lanus, E., Sarkar, K.: Asymptotic and constructive methods for covering perfect hash families and covering arrays. Des. Codes Crypt. **86**(4), 907–937 (2017). https://doi.org/10.1007/s10623-017-0369-x
5. Hu, V.C., et al.: Guide to attribute based access control (ABAC) definition and considerations (draft). NIST Spec. Publ. **800**(162), 1–52 (2013)
6. Kapadia, A., Tsang, P.P., Smith, S.W.: Attribute-based publishing with hidden credentials and hidden policies. In: NDSS, vol. 7, pp. 179–192. Citeseer (2007)
7. Lanus, E.: Interaction testing, fault location, and anonymous attribute-based authorization. Ph.D. thesis, Arizona State University (2019)
8. Moser, R.A., Tardos, G.: A constructive proof of the general Lovász local lemma. J. ACM **57**(2), 11:1–11:15 (2010). https://doi.org/10.1145/1667053.1667060
9. Portnoi, M., Shen, C.C.: Location-enhanced authenticated key exchange. In: 2016 International Conference on Computing, Networking and Communications (ICNC), pp. 1–5. IEEE (2016)

Parameterized Algorithms for Partial Vertex Covers in Bipartite Graphs

Vahan Mkrtchyan[1], Garik Petrosyan[2], K. Subramani[3(✉)],
and Piotr Wojciechowski[3]

[1] School of Advanced Studies, Gran Sasso Science Institute, L'Aquila, Italy
vahan.mkrtchyan@gssi.it
[2] Department of Informatics and Applied Mathematics, Yerevan State University,
Yerevan, Armenia
garik.petrosyan.1@gmail.com
[3] West Virginia University, Morgantown, WV, USA
ksmani@csee.wvu.edu, pwojciec@mix.wvu.edu

Abstract. In this paper, we discuss parameterized algorithms for variants of the partial vertex cover problem. Recall that in the classical vertex cover problem (VC), we are given a graph $G = (V, E)$ and a number K and asked if we can cover all of the edges in E, using at most K vertices from V. In the partial vertex cover problem (PVC), in addition to the parameter K, we are given a second parameter K' and the question is whether we can cover at least K' of the edges in E using at most K vertices from V. The weighted generalizations of the VC and PVC problems are called the weighted vertex cover (WVC) and the partial weighted vertex cover problem (WPVC) respectively. In the WPVC problem, we are given two parameters R and L, associated respectively with the vertex set V and edge set E of the graph **G**. Additionally, we are given non-negative integral weight functions for the vertices and the edges. The goal then is to cover edges of total weight at least L, using vertices of total weight at most R. (In the WVC problem, the goal is to cover all the edges with vertices whose total weight is at most R). This paper studies several variants of the PVC problem and establishes new results from the perspective of fixed-parameter tractability and **W[1]-hardness**. We also introduce a new problem called the partial vertex cover with matching constraint and show that it is fixed-parameter tractable for a certain class of graphs.

1 Introduction

In this paper, we study several variants of the vertex cover problem (VC), from the perspectives of parameterized algorithm design and complexity. In particular,

The work of the first author has been partially supported by the Italian MIUR PRIN 2017 Project ALGADIMAR "Algorithms, Games, and Digital Markets." The research of the third author has been supported in part by the Air-Force Office of Scientific Research through Grant FA9550-19-1-017 and in part by the Air-Force Research Laboratory, Rome through Contract FA8750-17-S-7007.

© Springer Nature Switzerland AG 2020
L. Gąsieniec et al. (Eds.): IWOCA 2020, LNCS 12126, pp. 395–408, 2020.
https://doi.org/10.1007/978-3-030-48966-3_30

we consider the partial vertex cover problem (PVC), wherein the goal is to cover a certain threshold of edges (as opposed to all of the edges) using the fewest number of vertices. We also look into weighted variants of this problem. Our primary focus is on bipartite graphs. In this case, the problem is called the partial vertex cover problem on bipartite graphs (PVCB). The PVCB problem is an important problem with a number of applications in computer security [9] and risk assessment [6]. In the weighted partial vertex cover problem (WPVC), we are given a graph $G = (V, E)$, cost function $c : V \to \mathbb{N}$, profit function $p : E \to \mathbb{N}$, and positive integers R and L. The goal is to check whether there is a subset $V' \subseteq V$ of cost at most R, such that the total profit of edges covered by V' is at least L. In this paper we study the fixed-parameter tractability of WPVC in bipartite graphs (WPVCB). By extending the methods of Amini et al. [1], we show that WPVCB is Fixed-Parameter Tractable (FPT) with respect to R if $c \equiv 1$. On the negative side, it is **W[1]-hard** for arbitrary c, even when $p \equiv 1$. In particular, WPVCB is **W[1]-hard** parameterized by R. We complement this negative result by proving that for bounded-degree graphs WPVC is FPT with respect to R. Additionally, we show that WPVC is FPT with respect to L. Finally, we discuss a variant of PVCB in which the edges covered are constrained to include a large matching and derive a parameterized algorithm for this version of the problem.

The rest of this paper is organized as follows: The problems studied in this paper are formally described in Sect. 2. In Sect. 3, we discuss the motivation for our work and mention related approaches in the literature. Our main results are described in Sect. 4. A variant of the partial vertex cover problem with applications to computational social choice is detailed in Sect. 5. In Sect. 6, we examine the parameterized complexity of the WPVCB problem when there is a separate budget for each partition of the vertices. We conclude in Sect. 7, by summarizing our results and outlining avenues for future research.

2   Statement of Problems

We focus on finite, undirected graphs that have no loops or parallel edges. As usual, the degree of a vertex is the number of edges of the graph that are incident with it. The maximum degree of the graph G is just the maximum of all degrees of vertices of G. A graph $G = (V, E)$ is bipartite, if its vertex set can be partitioned into two sets V_1 and V_2, such that each edge of G connects a vertex from V_1 to one from V_2. Given a graph $G = (V, E)$, and a set $S \subset V$ of vertices, an edge $(i, j) \in E$ is *covered* by S if $i \in S$ or $j \in S$. Let $E(S)$ be the set of edges of G that are covered by S. The classical vertex cover problem (VC) is defined as finding the smallest set S of vertices of the input graph G, so that $E(S) = E$. The vertex cover problem is a well-known **NP-complete** problem [16]. In this paper, we study the following variants of VC:

1. The partial vertex cover problem (PVC) -

Definition 1. *Given an undirected graph $G = (V, E)$, vertex-cardinality parameter k_1, and edge-cardinality parameter k_2, is there a subset V' of V, such that $|V'| \leq k_1$ and V' covers at least k_2 edges?*

2. The weighted partial vertex cover problem (WPVC) -

Definition 2. *Given an undirected graph $G = (V, E)$, weight-functions $c : V \to N$ and $p : E \to N$, vertex-weight parameter R, and edge-weight parameter L, is there a subset S of V, such that $\sum_{v \in S} c(v) \leq R$ and $\sum_{e \in E(S)} p(e) \geq L$?*

3. The partial vertex cover problem on bipartite graphs (PVCB) - This is the restriction of the partial vertex cover problem (PVC) to bipartite graphs.
4. The weighted partial vertex cover problem on bipartite graphs (WPVCB) - This is the restriction of the weighted partial vertex cover problem (WPVC) to bipartite graphs.
5. The 2-budget partial vertex cover problem on bipartite graphs (2-PVCB) -

Definition 3. *Given an undirected bipartite graph $G = (V, U, E)$, vertex-cardinality parameters R_V and R_U, and an edge-cardinality parameter L, is there a subset S_V of V and a subset S_U of U, such that $|S_V| \leq R_V$, $|S_U| \leq R_U$, and the number of edges covered by $S_V \cup S_U$ is at least L?*

6. The partial vertex cover problem with matching constraint (PVCBM) - This is a variant of the PVCB problem, in which we are given a third parameter k_3 and the goal is to find a vertex subset of size at most k_1, covering at least k_2 edges, such that the edges covered include a matching of size at least k_3.

The principal contributions of this paper are as follows:

1. Fixed-parameter tractability of the WPVCB problem with respect to R when vertex weights are identically 1.
2. **W[1]-hardness** of the WPVCB problem with respect to R.
3. Fixed-parameter tractability of the weighted partial vertex cover problem in bounded degree graphs (not necessarily bipartite) with respect to R.
4. **W[1]-hardness** of the 2-PVCB problem with respect to R_U and R_V.
5. Fixed-parameter tractability of the WPVC problem with respect to L.
6. A parameterized algorithm for the matching variant of the PVCB problem with respect to k_1.

3   Motivation and Related Work

When the weight functions c and p are identically one (i.e. $c \equiv p \equiv 1$), we get the well-known partial vertex cover problem (PVC). PVC represents a natural theoretical generalization of VC and is motivated by practical applications. Flow-based risk-assessment models in computational systems, for example, can be viewed as instances of PVC [6]. In particular PVC has applications to computer security even when restricted to bipartite graphs [9].

VC is polynomial-time solvable in bipartite graphs. However, the computational complexity of PVC in bipartite graphs has remained open until it was recently shown to be **NP-hard** [3,9,10,14].

VC has also been extensively studied from the perspective of approximation algorithms. Many 2-approximation algorithms for VC are known [31]. There is an approximation algorithm for the VC problem which has an approximation factor of $(2 - \theta(\frac{1}{\sqrt{\log n}}))$ [15]. This is the best known algorithm. The VC problem is also known to be **APX-complete** [27]. Moreover, it cannot be approximated within a factor of 1.3606 unless **P = NP** [12]. This lower bound was recently improved to $(\sqrt{2} - \epsilon)$ for any $\epsilon > 0$ in [18]. If the unique games conjecture is true, then VC cannot be approximated within any constant factor smaller than 2 [17]. In [28], a $(\frac{4}{3} + \epsilon)$-approximation algorithm is designed for WPVC for each $\epsilon > 0$ when the input graph is bipartite.

All hardness results for the VC problem directly apply to the PVC problem because the PVC problem is an extension of the VC problem. Since the 1990's the PVC problem and the partial-cover variants of similar graph problems have been extensively studied [7,8,20,21,24,29]. In particular, there is an $O(n \cdot \log n + m)$-time 2-approximation algorithm for PVC based on the primal-dual method [21], as well as a combinatorial 2-approximation algorithm [22]. Both of these algorithms are for a more general soft-capacitated version of PVC. There are several older 2-approximations resulting from different approaches [5,8,13,19]. Let us also note that the WPVC problem for trees is studied in [23], which provides an FPTAS for the problem. Additionally, the paper provides a polynomial time algorithm in the case when the vertices are unweighted.

Another problem with a close relationship to WPVC is the budgeted maximum coverage problem (BMC). In this problem one tries to find a min-cost subset of vertices, such that the profit of covered edges is maximized. It can be shown that both problems are equivalent from the perspective of exact solvability. The BMC problem for sets (not necessarily graphs) admits a $(1 - \frac{1}{e})$-approximation algorithm [25]. However, special cases that beat this bound are rare. The pipage rounding technique gives a $\frac{3}{4}$-approximation algorithm for the BMC problem on graphs [2]. This is improved to $\frac{4}{5}$ for bipartite graphs [4]. Finally, in [9,10], an $\frac{8}{9}$-approximation algorithm for the problem is presented when the input graph is bipartite and the vertices are unweighted. The result is based on the linear-programming formulation of the problem, and the constant $\frac{8}{9}$ matches the integrality gap of the linear program used in the formulation. More recently, Vangelis Paschos has described a polynomial time approximation scheme for the edge-weighted maximum coverage problem on bipartite graphs [30].

In this paper, we address these problems from the perspective of fixed-parameter tractability. Recall that a combinatorial problem Π is said to be fixed-parameter tractable with respect to a parameter k, if there is an algorithm for solving Π, whose running time is bounded by $f(k) \cdot size^{O(1)}$. Here f is a computable function of k, and $size$ is the length of the input. From the perspective of FPT, PVC is in some sense more difficult than VC. For instance, PVC is **W[1]-complete** with respect to R, the number of vertices in the cover [11], while VC is FPT [11,26].

In [1], the decision version of WPVCB is considered. The authors show that this problem is FPT with respect to the vertex budget R, when the vertices and edges of the bipartite graph are unweighted [1]. In this paper, by extending the result of Amini et al. [1], we show that the decision version of WPVCB is FPT with respect to R, if the vertices have cost one, while the edges remain arbitrarily weighted. On the negative side, the problem is **W[1]-hard** for arbitrary vertex weights, even when edges have profit one. We complement this negative result by proving that for bounded-degree graphs WPVC is FPT with respect to R. The same result holds for WPVC with respect to L. We finish the paper by obtaining an FPT result for an extension of PVCB. Terms and concepts that we do not define can be found in [11].

4   Main Results

In this section, we present our results. Our goal is to investigate the fixed-parameter tractability of the WPVCB problem.

When c and p are identically one (i.e. $c \equiv p \equiv 1$), we get the PVCB problem. When c is identically one, we get EPVCB. Finally, when p is identically one, we get the VPVCB problem. We will also use the same scheme of notations when the input graph need not be bipartite. In [1], PVCB is considered and it is shown that the problem is FPT with respect to R. We strengthen this result.

Theorem 1. *EPVCB is FPT with respect to R.*

Proof. Roughly speaking, we obtain the result with the approach of [1] by considering the weighted degree instead of the usual degree. Below we present the technical details.

Assume that we have an instance I of EPVCB. For a vertex v of B, let $\delta(v)$ be the set of edges of B incident with v. Define the set S of vertices of B as follows:

$$S = \left\{ v \in V(B) : p(\delta(v)) \geq \frac{L}{R} \right\}.$$

Case 1: $|S| \geq 2 \cdot R$. Consider the subgraph H of B induced by S. Since B is bipartite, H is bipartite, too. Let (X, Y) be the bipartition of H, and assume that $|X| \geq |Y|$. Since $|X| + |Y| = |S| \geq 2 \cdot R$, we have $|X| \geq R$. Take any R vertices of X. Observe that X is an independent set in B, hence these R vertices will cover at least L edges. Thus, the total profit is at least L. This means that I is a yes-instance.

Case 2: $|S| < 2 \cdot R$. Observe that any feasible solution to I must intersect S. Hence, we do recursive guessing, that is, we try each vertex of S one by one as a possible vertex of the feasible solution.

In the Case 1, the algorithm will run in polynomial time, so the most expensive case is Case 2. Since the number of vertices in a feasible solution is at most R, we have that the depth of the recursion is at most R. Hence the total running time of our algorithm is $O((2 \cdot R)^R \cdot size^{O(1)})$. $\square$

Our next result shows that WPVCB and VPVCB are **W[1]-hard**. Our reduction is from the multi-colored clique problem [11]. It is formulated as follows:

Multi-colored Clique: Given a graph G, positive integer k, and a partition $(V_1, ..., V_k)$ of the vertices of G, the goal is to check whether G contains a k-clique Q, such that Q contains exactly one vertex from each V_j for $j = 1, ..., k$.

Multi-colored clique is a well-studied problem which is known to be **W[1]-hard** with respect to the number of partitions k. Observe that an edge e connecting two vertices from V_i $1 \le i \le k$ does not lie in a feasible clique. Thus, without loss of generality, we can assume that for $i = 1, ..., k$, V_i is an independent set of vertices.

Theorem 2. *WPVCB is* **W[1]-hard** *parameterized by R.[1]*

Proof. We construct an FPT-reduction from MULTI-COLORED CLIQUE. Let $G = (V, E)$ be an instance of this problem with vertices partitioned as $V = V_1 \cup ... \cup V_k$. We create a bipartite graph $B = (U' \cup V' \cup Z, E')$ as follows. Let U' and V' be two copies of V, and let $V = V' = \{v_1, ..., v_n\}$, $U' = \{u_1, ..., u_n\}$, where for each $i \in [n]$, u_i is a copy of v_i. Here as usual $[n] = \{1, ..., n\}$. For a vertex $v \in V$ let $\chi(v)$ be its color, i.e., $\chi(v) = i$ if $v \in V_i$, and extend this to $U' \cup V'$ so that $\chi(u_i) = \chi(v_i) + k$ (V' inherits the original χ values on $v_1, ..., v_n$). For a vertex $x \in U' \cup V'$, let the cost of x be $c(x) = 2^{\chi(x)}$. Add an edge $u_i v_j$ to B if either $\chi(u_i) = \chi(v_j) + k$ and $i \ne j$, or $\chi(u_i) \ne \chi(v_j) + k$ and $v_i v_j \notin E(G)$. Give all these edges profit 1. Observe that a selection of one vertex from every color class of $U' \cup V'$ forms an independent set in B, if and only if it corresponds to two copies of a k-clique in G. Add two additional vertices z_1 and z_2, let $Z = \{z_1, z_2\}$ and give both the cost $2^{2 \cdot k + 1}$. Finally, join every vertex $x \in U'$ to z_2, every vertex $x \in V'$ to z_1 and give these edges a profit value so that the *total* profit of all edges incident with x equals $(2^{\chi(x)} \cdot (n + 1) + 5^{\chi(x)})$. (This is clearly possible, since the total profit of all previously created edges incident with x is bounded by n.)

Set the budgets of the instance as vertex budget $R = \sum_{i=1}^{2 \cdot k} 2^i = 2^{2 \cdot k + 1} - 2$ and profit threshold $L = \sum_{i=1}^{2 \cdot k} (2^i \cdot (n + 1) + 5^i) = (n + 1) \cdot R + (\frac{5}{4}) \cdot (5^{2 \cdot k} - 1)$. This finishes the instance description. It is clear that the construction can be performed in polynomial time, and the budget R is a function of k.

In Fig. 1, a graph G and the bipartite graph B obtained after the reduction are given. In this example, $k = 2$, $V_1 = \{v_1, v_3\}$, $V_2 = \{v_2, v_4\}$ and $\chi(v_1) = \chi(v_3) = 1$, $\chi(v_2) = \chi(v_4) = 2$, $\chi(u_j) = \chi(v_j) + 2$ for $j = 1, 2, 3, 4$. We have $R = 30$ and $L = 930$. The costs of vertices and profits of edges in B are defined as follows: $c(v_1) = c(v_3) = 2$, $c(v_2) = c(v_4) = 4$, $c(u_1) = c(u_3) = 8$ and $c(u_2) = c(u_4) = 16$, and the profits of edges not incident with z_1 or z_2 are 1, finally, the edges incident with z_1 or z_2 are chosen so that the total profit of edges incident with any $x \in V(B) \backslash \{z_1, z_2\}$ is $(5 \cdot 2^{\chi(x)} + 5^{\chi(x)})$. For example, consider the edge $z_1 v_1$. Recall that $\chi(v_1) = 1$, thus the total profit of all edges incident with v_1 should be $5 \cdot 2^1 + 5^1 = 15$. Since the profit of $v_1 u_3$ is 1, the profit of the edge $z_1 v_1$ is 14.

---

[1] We are grateful to Magnus Wahlström for providing us with a proof of this theorem.

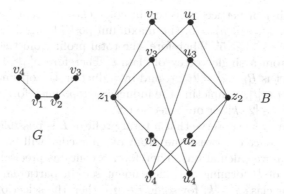

Fig. 1. A graph G and the bipartite graph B obtained from it in the reduction.

It remains to show that (B, R, L) is a positive instance of WPVCB if and only if G has a multi-colored clique.

From the multi-colored clique problem to the partial vertex cover problem. Let $X \subseteq V(G)$ be a multi-colored k-clique, and let $S = \{u_i \mid u_i \in X\} \cup \{v_i \mid u_i \in X\}$. Since S contains one vertex for every color class of B its total cost equals R, and since it induces an independent set in B the total profit of the edges covered equals L.

From the Partial Vertex Cover Problem to the Multi-colored Clique Problem. Now for the more challenging part of the argument. We need to argue that the costs and profits balance out so that the only way to select vertices to a total profit of L is to select one vertex from every color class of B. For this, first observe that for a vertex of class $i \in [2 \cdot k]$, the ratio of the total profit of its incident edges to its cost is

$$\frac{2^i \cdot (n+1) + 5^i}{2^i} = (n+1) + \left(\frac{5}{2}\right)^i. \tag{1}$$

Let S be a partial vertex cover of profit at least L and cost at most R, and for each $i \in [2 \cdot k]$ let n_i be the number of vertices of S with color class i. For each vertex, we can divide the contributions to the total profit into two parts. The first part corresponds to the $(n+1)$ term in (1), while the second corresponds to the $(\frac{5}{2})^i$ term. From the first part, each vertex v contributes a profit of at least $(n+1) \cdot c(v)$. It is clear that every selection of cost at most R contributes a profit of at least $(n+1) \cdot R$, regardless of the distribution n_i. Therefore we focus on the contribution of the second part of the formula, with target profit $L - (n+1) \cdot R$.

Considering this second part, define $R_t = \sum_{i=1}^{t} 2^i$ and $L_t = \sum_{i=1}^{t} 5^i$. We show by induction that for every $t \in [2 \cdot k]$, the largest possible contribution of a selection of cost at most R_t is L_t (which of course is achieved by making one selection per color class). For $t = 1$ this is trivial. Therefore, by induction, let $t > 1$ and assume that the claim holds for every value $t' < t$. Let n_i', $i \in [t]$, denote the number of vertices selected in color class i, for a selection of total cost at most R_t. Then if $n'_t = 0$, the maximum possible profit is $R_t \cdot (\frac{5}{2})^{t-1} < 2^{t+1} \cdot (\frac{5}{2})^{t-1} = 4 \cdot 5^{t-1} < 5^t$. Therefore, the total profit from the selection n'_i is less than that from a single vertex of class t. Therefore $n'_t \geq 1$. But then the remaining budget is $R_t - 2^t = R_{t-1}$, and by induction the optimal selection has $n'_i = 1$ for every $i \in [t]$, completing the induction step. Therefore we may assume $n_i = 1$ for every $i \in [2 \cdot k]$ for our selection S.

But then, finally, we observe that a total profit of L is possible only if S is an independent set, since otherwise the profit of some edge will have been double-counted in the above calculations. Therefore, S contains precisely one vertex of each color class of B forming an independent set. In particular, if $v_i \in S$ is a selection in color class $j \in [k]$ for some $i \in [n]$, then the selection in color class $j + k$ must be u_i. Also, for every pair of color classes $i, i' \in [k]$ the selections in classes i and $i' + k$ are independent in B and therefore the selections in classes i and i' are neighbors in G. Thus $X = S \cap V'$ is a multi-colored clique in G, as required. □

Corollary 1. *The problem is* **W[1]-hard** *also in the variant where all edge profits are 1, i.e., the VPVCB problem is* **W[1]-hard** *with respect to R.*

Proof. The only edges of weight more than 1 in the above reduction are the edges connecting to the special vertices z_i, and the largest edge weight used is bounded by a function $f(k) \cdot (n+1)$. Therefore, instead of using edge weights we can in FPT time simply create the corresponding number of pendant vertices for each vertex. These pendants can be given the same weight as the vertices z_i. □

A class of graphs is said to be bounded-degree, if there is a constant C, such that all graphs from the class have maximum degrees at most C. It turns out that when the input graphs have bounded-degree and need not be bipartite, we have the following result:

Theorem 3. *WPVC is fixed-parameter tractable with respect to R for bounded-degree graphs.*

Proof. Let I be an instance of the WPVC problem, where $G = (V, E)$ is a bounded-degree graph, $c : V \to \mathbb{N}$, $p : E \to \mathbb{N}$ are cost and profit functions, and L, R are constants. Assume that for every $v \in V$, we have $d(v) \leq d$. For $i = 1, 2, \ldots, R$, let $M_i(G) = \{v : c(v) = i\}$ (we disregard the vertices of cost greater than R). Choose a vertex $v_i \in M_i(G)$ which has the largest coverage, and let M be the set comprised of chosen vertices v_i. The set of vertices of G which have a neighbor from M, we denote by N. Consider the set $M \cup N$. It is obvious that $|M \cup N| \leq (d+1) \cdot R$, where d is the bound for the degree.

Let us show that if I is a yes-instance, then we can construct a feasible set, which intersects $M \cup N$. Indeed, let S be a feasible set, which does not have a vertex from $M \cup N$. Then we can replace any vertex $v \in S$ with $v_{c(v)}$ to get a new set S'. Since there is no vertex of S, which is a neighbor of $v_{c(v)}$, we have that the profit of S' does not decrease. Moreover, as $c(v) = c(v_{c(v)})$, S' is also feasible. Observe that S' intersects $M \cup N$.

Now we complete the proof by recursively guessing on $M \cup N$. Since the number of vertices in a feasible set is at most R, we have that the depth of the recursion is at most R, hence the total running time of the algorithm is $((d+1) \cdot R)^R \cdot size^{O(1)}$. $\qquad\square$

Exercise 5.11 from [11] implies that PVC is fixed-parameter tractable with respect to L. Below, we strengthen the statement of this exercise by showing that the WPVC problem can be parameterized with respect to L.

Theorem 4. *WPVC is fixed-parameter tractable with respect to L.*

Proof. Let I be an instance of the WPVC problem, where $G = (V, E)$ is a graph, $c : V \to \mathbb{N}$, $p : E \to \mathbb{N}$ are cost and profit functions, and L and R are constants. We can assume that no vertex of G is isolated. Moreover, without loss of generality, we can assume that for any vertex v, we have $c(v) \leq R$.

For every vertex v we denote $p(v) = p(\delta(v)) = \sum_{e:v \in e} p(e)$ the total profit of edges incident with v. We can assume that $p(v) \leq L - 1$ for all vertices, as otherwise we will have a feasible solution comprised of one vertex and, as a result, I is a yes-instance. This, in particular, means that $d(v) \leq L - 1$. For $i = 1, \ldots, L - 1$ let v_i be a vertex which has profit i and for any other vertex u, which has profit i, we have $c(v_i) \leq c(u)$. Let M be the set of those vertices v_i. The set of vertices of G which have a neighbor from M is denoted by N. It is obvious that $|M \cup N| \leq (L - 1) + (L - 1)^2 < L^2$.

Let us show that if I is a yes-instance and S is a feasible set in I, then we can construct a feasible set that intersects $M \cup N$. Indeed, assume that S does not contain any vertex from $M \cup N$. Then we can replace any vertex $v \in S$ by the vertex $v_{p(v)} \in M$. Since there is no vertex in S which is a neighbor of $v_{p(v)}$, it follows that the total vertex cost has not increased, and, as $p(v) = p(v_{p(v)})$, S' is also feasible. Now we complete the proof by recursively guessing on $M \cup N$. We remove any isolated vertex that may arise in each branch of recursion. Let us show that the depth of the recursion is less than $2 \cdot L$. For the sake of contradiction, assume that during the recursive guessing, the algorithm has considered the vertices $z_1, \ldots, z_{2 \cdot L}$. Let $Z = \{z_1, \ldots, z_{2 \cdot L}\}$. Since the algorithm has considered these vertices, we have that $c(Z) \leq R$. Then for the profit of edges covered by Z, we will have the following bound:

$$p(E(Z)) \geq |E(Z)| \geq \frac{d(z_1) + \ldots + d(z_{2 \cdot L})}{2} \geq \frac{2 \cdot L}{2} = L.$$

Thus, Z is a feasible set, hence I is a yes-instance. This means that there is no need to consider $2 \cdot L$ or more vertices during the recursive guessing. Hence the depth of the recursion is less than $2 \cdot L$. Since $|M \cup N| \leq L^2$, we have that the running time of our algorithm is bounded by $(L^2)^{2 \cdot L} \cdot size^{O(1)} = L^{4 \cdot L} \cdot size^{O(1)}$. $\qquad\square$

5   The Matching Problem

We now consider a variant of PVCB. In this variant, we are given a bipartite graph G and three integers k_1, k_2 and k_3. The goal is to check whether there is a subset of at most k_1 vertices, that covers at least k_2 edges, such that the covered edges contain a matching of size at least k_3. This variant is called PVCBM. Clearly, this problem is **NP-hard**, since when $k_3 = 0$ it results in PVCB. Since PVCB is FPT with respect to k_1, it would be interesting to parameterize this new version of the problem with respect to k_1. Observe that we can assume that $k_3 \leq k_1$ otherwise the problem is a trivial no-instance.

Theorem 5. *PVCBM is fixed-parameter tractable with respect to the parameter* k_1.

Proof. Let $PVCB(A, B)$ be an FPT algorithm for PVCB that checks whether there is a subset of A vertices that covers at least B edges of the input bipartite graph. Now, assume that the graph G and the parameters k_1, k_2 and k_3 are given in the matching problem. First, we run $PVCB(k_1, k_2)$. If there is no such subgraph, then the answer to the matching problem is also negative. So we can assume that $PVCB(k_1, k_2)$ returns such a subgraph. Next, by trying $R = 0, 1, ..., k_1$ we can find the smallest R for which $PVCB(R, k_2)$ is a yes-instance.

Let H be the edge-induced subgraph on these $\geq k_2$ edges. As usual, let $\nu(G)$ be the size of the largest matching in G, and let $\tau(G)$ be the size of the smallest vertex cover in G. By the classical König theorem we have $\nu(G) = \tau(G)$ for any bipartite graph G.

Observe that we can assume that $R < k_3 \leq k_1$. To see this, observe that R represents the number of vertices required to cover all the edges in H. In other words, it is a vertex cover of H. Thus, $R = \nu(H)$ and since H is a bipartite subgraph of G, R is also the size of a maximum matching in H. Thus, if $R \geq k_3$, then the edges in H, which number at least k_2 can be covered by $R \leq k_1$ vertices and a matching of size $R \geq k_3$ is contained in H. Also, observe that if $\tau(G) < k_3$, then we have trivial no-instance, as G contains no matching of size k_3. Thus, we can assume that $\nu(G) = \tau(G) \geq k_3$. Since $\tau(H) = R < k_3 \leq \tau(G)$, we have that $E(H) \neq E(G)$. Thus, there is as an edge e lying outside H. Add e to H. If $\tau(H)$ has increased by adding e, define $R := R + 1$, otherwise let R be the same. Repeat this process of adding edges outside H. Since $\tau(H) = R < k_3 \leq \tau(G)$, at some point we will arrive into H such that $R = \tau(H) = k_3 \leq k_1$. Observe that H can be covered with at most $R \leq k_1$ vertices, it has at least k_2 edges and it contains a matching of size k_3. Thus, the problem is a yes-instance.

Finally, let us observe that the running-time of this algorithm is FPT in k_1. We need at most k_1 calls of $PVCB(k_1, k_2)$. Since the latter is FPT with respect to k_1, we have the result.   □

6   The 2-PVCB Problem

In this section, we consider the problem of finding a partial vertex cover on a bipartite graph when we have a separate budget for each partition. In Theorem 6 we show that this problem is **W[1]-hard** with respect to both vertex budgets even when both the vertices and edges are unweighted.

Theorem 6. *2-PVCB is* **W[1]-hard** *with respect to R_V and R_U.*

Proof. We will show this by a reduction from clique on regular graphs. This problem is known to be **W[1]-complete** (Theorem 13.25 of [11]).

Let $G = (V, E)$ be an undirected graph where each vertex has degree r. From G we construct the bipartite graph $G' = (V', U', E')$ as follows:

1. For each vertex $v_i \in V$, add the vertex v_i' to V'.
2. For each edge $e_l = (v_i, v_j) \in E$ add the vertex u_l' to U'. Additionally add the edge (v', u_l') to E' for each $v' \in V' \setminus \{v_i', v_j'\}$. This connects the vertex u_l' to every vertex v_i' that does *not* correspond to an endpoint of e_l.

Figure 2 shows the original regular graph G and the corresponding bipartite graph G' constructed as in the reduction.

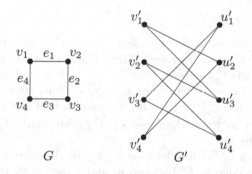

Fig. 2. A graph G and the bipartite graph G' obtained from it in the reduction.

Note that G has a clique of size $k = 2$ using vertices v_1 and v_2. This corresponds to the vertex cover using vertices v_1' and v_2' from V' and vertex u_1' from U' which covers $(4 - 2) \cdot \binom{2}{2} + (4 - 2) \cdot 2 = 6$ edges of G'.

We now show that G has a clique of size k if and only if we can cover at least $L = (n - k) \cdot \binom{k}{2} + k \cdot (m - r)$ edges of G' using at most $R_{V'} = k$ vertices from V' and at most $R_{U'} = \binom{k}{2}$ vertices from U'. Since checking if G has a clique of size n can be done in polynomial time, we assume without loss of generality that $k < n$.

First assume that G has a clique of size k. Let $S \subseteq V$ denote the vertices in the clique. We construct the desired partial vertex cover of G' as follows:

1. For each $v_i \in S$, add v_i' to $S_{V'}$.
2. For each edge $e_l = (v_i, v_j)$ such that $v_i, v_j \in S$, add the vertex u_l' to $S_{U'}$. Note that there are exactly $\binom{k}{2}$ such edges.

Observe the following:

1. Each vertex $v_i \in S$ is the endpoint of exactly r edges in E. Thus, there are $(m - r)$ edges in E which do not have v_i as an endpoint. This means that, by construction, $v_i' \in S_{V'}$ covers $(m - r)$ edges. Thus, the vertices in $S_{V'}$ cover a total of $k \cdot (m - r)$ edges.
2. Each vertex u_l' in $S_{U'}$ corresponds to an edge of G with both endpoints in S. Thus, there is an edge between u_l' and every vertex in $V' \setminus S_{V'}$. This means that u_l' covers an additional $(n - k)$ edges. Thus, the vertices in $S_{U'}$ cover a total of $(n - k) \cdot \binom{k}{2}$ additional edges.

Thus, the sets $S_{V'}$ and $S_{U'}$ cover a total of $L = (n - k) \cdot \binom{k}{2} + k \cdot (m - r)$ edges.

Now suppose that there exist sets $S_{V'} \subseteq V'$ and $S_{U'} \subseteq U'$ such that $|S_{V'}| \leq k$, $|S_{U'}| \leq \binom{k}{2}$, and the sets $S_{V'}$ and $S_{U'}$ cover at least $L = (n - k) \cdot \binom{k}{2} + k \cdot (m - r)$ edges of G'.

We can assume without loss of generality that $|S_{V'}| = k$ and $|S_{U'}| = \binom{k}{2}$ since adding additional vertices will not reduce the number of edges covered. Let $S = \{v_i : v_i' \in S_{V'}\}$.

Observe the following:

1. As before, each vertex in $S_{V'}$ has degree $(m - r)$. Thus, the vertices in $S_{V'}$ cover a total of $k \cdot (m - r)$ edges. Thus, the vertices in $S_{U'}$ must cover an additional $(n - k) \cdot \binom{k}{2}$ edges.
2. There are $(n - k) > 0$ vertices in $V' \setminus S_{V'}$. Thus, there must be an edge between each of the $\binom{k}{2}$ vertices in $S_{U'}$ and each of the $(n - k)$ vertices in $V' \setminus S_{V'}$. Otherwise, fewer than $(n - k) \cdot \binom{k}{2}$ additional edges will be covered by the vertices in $S_{U'}$.
3. Let u_l' be a vertex in $S_{U'}$ and let $e_l = (v_i, v_j)$ be the corresponding edge in G. Note there is no edge in E' between u_l' and v_i' nor is there an edge between u_l' and v_j'. Thus, both v_i' and v_j' are in $S_{V'}$. This means that every vertex $u_l' \in S_{U'}$ corresponds to an edge $e_l \in E$ with both endpoints in S.

Thus, G has $\binom{k}{2}$ edges with both endpoints in S. Since $|S| = k$, S must be a clique of size k.   □

7   Conclusion

In this paper, we studied the partial vertex cover problem from the perspective of parameterized tractability and **W[1]-hardness**. Although our primary focus was on bipartite graphs, we obtained new results for the general case as well. Our main contributions include showing that a restricted version of the WPVCB problem is fixed-parameter tractable and that this problem is **W[1]-hard**, with respect to the vertex-weight parameter. We also showed that the WPVC problem is fixed-parameter tractable in bounded degree graphs. Finally, we introduced a new variant of the partial vertex cover problem called PVCBM and showed that it is fixed-parameter tractable.

References

1. Amini, O., Fomin, F.V., Saurabh, S.: Implicit branching and parameterized partial cover problems. J. Comput. Syst. Sci. **77**, 1159–1171 (2011)
2. Ageev, A.A., Sviridenko, M.I.: Approximation algorithms for maximum coverage and max cut with given sizes of parts. In: Cornuéjols, G., Burkard, R.E., Woeginger, G.J. (eds.) IPCO 1999. LNCS, vol. 1610, pp. 17–30. Springer, Heidelberg (1999). https://doi.org/10.1007/3-540-48777-8_2
3. Apollonio, N., Simeone, B.: The maximum vertex coverage problem on bipartite graphs. Discrete Appl. Math. **165**, 37–48 (2014)
4. Apollonio, N., Simeone, B.: Improved approximation of maximum vertex coverage problem on bipartite graphs. SIAM J. Discrete Math. **28**(3), 1137–1151 (2014)
5. Bar-Yehuda, R.: Using homogeneous weights for approximating the partial cover problem. J. Algorithms **39**(2), 137–144 (2001)
6. Bilgin, C.C., Caskurlu, B., Gehani, A., Subramani, K.: Analytical models for risk-based intrusion response. Comput. Netw. (Special issue on Security/Identity Architecture) **57**(10), 2181–2192 (2013)
7. Bläser, M.: Computing small partial coverings. Inf. Process. Lett. **85**(6), 327–331 (2003)
8. Bshouty, N.H., Burroughs, L.: Massaging a linear programming solution to give a 2-approximation for a generalization of the vertex cover problem. In: Morvan, M., Meinel, C., Krob, D. (eds.) STACS 1998. LNCS, vol. 1373, pp. 298–308. Springer, Heidelberg (1998). https://doi.org/10.1007/BFb0028569
9. Caskurlu, B., Mkrtchyan, V., Parekh, O., Subramani, K.: Partial vertex cover and budgeted maximum coverage in bipartite graphs. SIAM J. Discrete Math. **31**(3), 2172–2184 (2017)
10. Caskurlu, B., Mkrtchyan, V., Parekh, O., Subramani, K.: On partial vertex cover and budgeted maximum coverage problems in bipartite graphs. In: Diaz, J., Lanese, I., Sangiorgi, D. (eds.) TCS 2014. LNCS, vol. 8705, pp. 13–26. Springer, Heidelberg (2014). https://doi.org/10.1007/978-3-662-44602-7_2
11. Cygan, M., et al.: Parameterized Algorithms, pp. 3–555. Springer, Cham (2015). https://doi.org/10.1007/978-3-319-21275-3. ISBN 978-3-319-21274-6
12. Dinur, I., Safra, S.: On the hardness of approximating minimum vertex cover. Ann. Math. **162**(1), 439–485 (2005)
13. Hochbaum, D.S.: The *t*-vertex cover problem: extending the half integrality framework with budget constraints. In: Jansen, K., Rolim, J. (eds.) APPROX 1998. LNCS, vol. 1444, pp. 111–122. Springer, Heidelberg (1998). https://doi.org/10.1007/BFb0053968
14. Joret, G., Vetta, A.: Reducing the rank of a matroid. Discrete Math. Theor. Comput. Sci. **17**(2), 143–156 (2015)
15. Karakostas, G.: A better approximation ratio for the vertex cover problem. ACM Trans. Algorithms **5**(4), 1–8 (2009)
16. Karp, R.: Reducibility among combinatorial problems. In: Miller, R., Thatcher, J. (eds.) Complexity of Computer Computations, pp. 85–103. Plenum Press, New York (1972)
17. Khot, S., Regev, O.: Vertex cover might be hard to approximate to within $2 - \epsilon$. J. Comput. Syst. Sci. **74**, 335–349 (2008)
18. Khot, S., Minzer, D., Safra, M.: Pseudorandom sets in Grassmann graph have near-perfect expansion. Electronic Colloquium on Computational Complexity, Report No. 6 (2018)

19. Khuller, S., Gandhi, R., Srinivasan, A.: Approximation algorithms for partial covering problems. J. Algorithms **53**(1), 55–84 (2004)
20. Kneis, J., Langer, A., Rossmanith, P.: Improved upper bounds for partial vertex cover. In: Broersma, H., Erlebach, T., Friedetzky, T., Paulusma, D. (eds.) WG 2008. LNCS, vol. 5344, pp. 240–251. Springer, Heidelberg (2008). https://doi.org/10.1007/978-3-540-92248-3_22
21. Mestre, J.: A primal-dual approximation algorithm for partial vertex cover: making educated guesses. Algorithmica **55**(1), 227–239 (2009)
22. Bar-Yehuda, R., Flysher, G., Mestre, J., Rawitz, D.: Approximation of partial capacitated vertex cover. In: Arge, L., Hoffmann, M., Welzl, E. (eds.) ESA 2007. LNCS, vol. 4698, pp. 335–346. Springer, Heidelberg (2007). https://doi.org/10.1007/978-3-540-75520-3_31
23. Mkrtchyan, V., Parekh, O., Segev, D., Subramani, K.: The approximability of partial vertex covers in trees. In: Steffen, B., Baier, C., van den Brand, M., Eder, J., Hinchey, M., Margaria, T. (eds.) SOFSEM 2017. LNCS, vol. 10139, pp. 350–360. Springer, Cham (2017). https://doi.org/10.1007/978-3-319-51963-0_27
24. Kneis, J., Mölle, D., Rossmanith, P.: Partial vs. complete domination: t-dominating set. In: van Leeuwen, J., Italiano, G.F., van der Hoek, W., Meinel, C., Sack, H., Plášil, F. (eds.) SOFSEM 2007. LNCS, vol. 4362, pp. 367–376. Springer, Heidelberg (2007). https://doi.org/10.1007/978-3-540-69507-3_31
25. Moss, A., Khuler, S., (Seffi) Naor, J.: The budgeted maximum coverage problem. Inf. Process. Lett. **70**(1), 39–45 (1999)
26. Guo, J., Niedermeier, R., Wernicke, S.: Parameterized complexity of generalized vertex cover problems. In: Dehne, F., López-Ortiz, A., Sack, J.-R. (eds.) WADS 2005. LNCS, vol. 3608, pp. 36–48. Springer, Heidelberg (2005). https://doi.org/10.1007/11534273_5
27. Papadimitriou, C.H., Yannakakis, M.: Optimization, approximation, and complexity classes. J. Comput. Syst. Sci. **43**(3), 425–440 (1991)
28. Parekh, O., Könemann, J., Segev, D.: A unified approach to approximating partial covering problems. Algorithmica **59**(4), 489–509 (2011)
29. Kneis, J., Mölle, D., Richter, S., Rossmanith, P.: Intuitive algorithms and t-vertex cover. In: Asano, T. (ed.) ISAAC 2006. LNCS, vol. 4288, pp. 598–607. Springer, Heidelberg (2006). https://doi.org/10.1007/11940128_60
30. Paschos, V.Th.: A polynomial time approximation schema for max k–vertex cover in bipartite graphs (2019). https://arxiv.org/abs/1909.08435v1
31. Vazirani, V.V.: Approximation Algorithms. Springer, New York (2001)

Acyclic Matching in Some Subclasses
of Graphs

B. S. Panda$^{(\boxtimes)}$ and Juhi Chaudhary

Computer Science and Application Group, Department of Mathematics,
Indian Institute of Technology Delhi, Hauz Khas 110016, New Delhi, India
bspanda@maths.iitd.ac.in, chaudhary.juhi5@gmail.com

Abstract. A subset $M \subseteq E$ of edges of a graph $G = (V, E)$ is called
a *matching* if no two edges of M share a common vertex. A matching
M in a graph G is called an *acyclic matching* if $G[V(M)]$, the subgraph
of G induced by the M-saturated vertices of G is acyclic. The ACYCLIC
MATCHING PROBLEM is the problem of finding an acyclic matching of
maximum size. The decision version of the ACYCLIC MATCHING PROB-
LEM is known to be NP-complete for general graphs as well as for bipartite graphs. In this paper, we strengthen this result by showing that
the decision version of the ACYCLIC MATCHING PROBLEM remains NP-
complete for comb-convex bipartite graphs and dually-chordal graphs.
On the positive side, we present linear time algorithms to compute an
acyclic matching of maximum size in split graphs and proper interval
graphs. Finally, we show that the ACYCLIC MATCHING PROBLEM is hard
to approximate within a factor of $n^{1-\epsilon}$ for any $\epsilon > 0$, unless $P = NP$ and
the ACYCLIC MATCHING PROBLEM is APX-complete for $2k + 1$-regular
graphs for $k \geq 3$, where k is a constant.

Keywords: Matching · Bipartite graphs · Chordal graphs · Graph
algorithm · NP-completeness · Approximation algorithm

1 Introduction

A subset $M \subseteq E$ of edges of a graph $G = (V, E)$ is called a *matching* if no two
edges of M share a common vertex. Vertices that are incident on the edges of
a matching M are called M-saturated vertices and are denoted by $V(M)$. In
this paper, we study an important variant of matching called *acyclic matching*
(see [3,5,9]). A matching M in G is called an *acyclic matching* if $G[V(M)]$,
the subgraph of G induced by the M-saturated vertices of G is acyclic. The
ACYCLIC MATCHING PROBLEM asks to find an acyclic matching of maximum

B. S. Panda—The author thanks the SERB, Department of Science and Technology
for their support vide Diary No. SERB/F/12949/2018-2019.
J. Chaudhary—The author has been supported by the Department of Science and
Technology through INSPIRE Fellowship for this research.

© Springer Nature Switzerland AG 2020
L. Gąsieniec et al. (Eds.): IWOCA 2020, LNCS 12126, pp. 409–421, 2020.
https://doi.org/10.1007/978-3-030-48966-3_31

size in a given graph G. The *acyclic matching number* of G, denoted by $\mu_{ac}(G)$ is the maximum size of an acyclic matching in G.

More formally, the ACYCLIC MATCHING PROBLEM and its decision version are defined as follows:

ACYCLIC MATCHING PROBLEM
Instance: A graph $G = (V, E)$.
Solution: An acyclic matching M in G.
Measure: Cardinality of the set M.

ACYCLIC MATCHING DECIDE PROBLEM
Instance: A graph $G = (V, E)$ and a positive integer k.
Question: Does there exist an acyclic matching M in G of size at least k?

Goddard et al. [5] introduced the concept of *acyclic matching* along with some other variants of the matching and proved that the ACYCLIC MATCHING DECIDE PROBLEM is NP-complete for general graphs. Later, Panda and Pradhan [9] strengthened this result by showing that the ACYCLIC MATCHING DECIDE PROBLEM remains NP-complete for bipartite graphs and even for perfect-elimination bipartite graphs, which is a subclass of bipartite graphs. They also gave a dynamic programming based algorithm to find an acyclic matching of maximum size in bipartite permutation graphs. Baste et al. [2] showed that finding a maximum size 1-degenerate matching in a graph G is equivalent to finding a maximum acyclic matching in G. They further proved that a maximum 1-degenerate matching could be found in polynomial time in chordal graphs, but the time complexity is very high. Recently, Fürst and Rautenbach showed that it is hard to decide whether a given bipartite graph of maximum degree at most four has a maximum matching that is acyclic [4]. They further characterized the graphs for which every maximum matching is acyclic and give linear time algorithms to compute a maximum acyclic matching in graph classes like P_4-free graphs and $2P_3$-free graphs [4]. There are no approximation results known for the ACYCLIC MATCHING PROBLEM till now.

In this paper, we study the complexity status of the ACYCLIC MATCHING PROBLEM and the ACYCLIC MATCHING DECIDE PROBLEM in some subclasses of graphs. The main contributions of this paper are summarized below.

1. We prove that the ACYCLIC MATCHING DECIDE PROBLEM is NP-complete for tree-convex bipartite graphs by showing that it is NP-complete for comb-convex bipartite graphs which is a subclass of tree-convex bipartite graphs.
2. We prove that the ACYCLIC MATCHING DECIDE PROBLEM is NP-complete for dually chordal graphs.
3. We prove that a maximum size acyclic matching can be computed in linear time in split graphs and proper interval graphs.
4. We prove that it is hard to approximate the ACYCLIC MATCHING PROBLEM within a factor of $n^{1-\epsilon}$ for any $\epsilon > 0$, unless $P = NP$.
5. We prove that the ACYCLIC MATCHING PROBLEM is APX-complete for $2k + 1$-regular graphs for $k \geq 3$, where k is a constant.

2   Preliminaries

We consider only simple and connected graphs. For a graph $G = (V, E)$, let n denotes the number of vertices and m denotes the number of edges in G. The open and closed neighborhood of a vertex $u \in V$ are denoted by $N(u)$ and $N[u]$ respectively, where $N(u) = \{w \mid wu \in E\}$ and $N[u] = N(u) \cup \{u\}$. The degree of a vertex u is $|N(u)|$ and is denoted by $d(u)$. For a graph $G = (V, E)$, the subgraph of G induced by $U \subseteq V$ is denoted by $G[U]$, where $G[U] = (U, E_U)$ and $E_U = \{xy \in E \mid x, y \in U\}$.

A graph $G = (V, E)$ is called a k-regular graph if $d(v) = k$ for every vertex v of G. A graph $G = (V, E)$ is called a *bipartite graph* if its vertex set V can be partitioned into two independent sets X and Y, such that every edge of G joins a vertex in X to a vertex in Y. A *comb* is a graph obtained by attaching a pendant vertex (tooth) to every vertex of a path (backbone). A bipartite graph $G = (X, Y, E)$ is said to be a *tree-convex bipartite* graph, if a tree $T = (X, E_X)$ can be defined on X such that for every vertex $y \in Y$, the vertices in $N_G(y)$ induces a subtree of T. It can be noted that tree-convex bipartite graphs are recognizable in linear time and the associated tree T can also be constructed in linear time [11]. If the tree T in a tree-convex bipartite graph is a comb, then G is called a *comb-convex bipartite graph*.

A graph $G = (V, E)$ is called a *chordal graph* if every cycle in G of length at least four has a *chord*, that is, an edge joining two non-consecutive vertices of the cycle. A graph $G = (V, E)$ is called a *split graph* if its vertex set V can be partitioned into two sets I and C such that I is an independent set and C is a clique. Let $\mathscr{F}$ be a family of sets. The *intersection graph* of $\mathscr{F}$ is obtained by taking each set in $\mathscr{F}$ as a vertex and joining two sets in $\mathscr{F}$ if and only if they have a nonempty intersection. A graph G is called a *proper interval graph* if it is the intersection graph of a family $\mathscr{F}$ of intervals on the real line such that no intervals in $\mathscr{F}$ contains another. A vertex $u \in N_G[v]$ in a graph G is called a maximum neighbor of v if for all $w \in N_G[v]$, $N_G[w] \subseteq N_G[u]$. An ordering $\alpha = (v_1, v_2, \ldots, v_n)$ of $V(G)$ is called a maximum neighborhood ordering, if v_i has a maximum neighbor in $G_i = G[\{v_i, \ldots, v_n\}]$ for all $i, 1 \leq i \leq n$. A graph G is called a *dually chordal graph* if it has a maximum neighborhood ordering.

3   NP-Completeness Results

3.1   Comb-Convex Bipartite Graphs

It has been shown in [9] that the ACYCLIC MATCHING DECIDE PROBLEM is NP-complete for bipartite graphs. In this subsection, we strengthen this result by showing that the ACYCLIC MATCHING DECIDE PROBLEM remains NP-complete for tree-convex bipartite graphs, which is a subclass of bipartite graphs by showing that it is NP-complete for comb-convex bipartite graphs.

Theorem 1. *The* ACYCLIC MATCHING DECIDE PROBLEM *is NP-complete for comb-convex bipartite graphs.*

Proof. Clearly, the ACYCLIC MATCHING DECIDE PROBLEM belongs to the class NP for comb-convex bipartite graphs. To show the NP-completeness, we give a polynomial reduction from the ACYCLIC MATCHING DECIDE PROBLEM for bipartite graphs, which is already known to be NP-complete [9].

Given a bipartite graph $G = (X, Y, E)$, we construct a comb-convex bipartite graph $H = (X_H, Y_H, E_H)$ as follows:

Let $X_H = X \cup X'$, where $X' = \{x_i' \mid x_i \in X\}$, $Y_H = Y$, and $E_H = E \cup E'$, where $E' = \{x_i'y \mid x_i' \in X' \text{ and } y \in Y\}$. The constructed graph H is a comb-convex bipartite graph if X' is taken as the backbone and X is taken as the teeth of a comb C. Further, note that given a bipartite graph G, the graph H can be constructed in polynomial time.

Now, the following claim is sufficient to complete the proof of the theorem.

Claim. G has an acyclic matching of size at least k if and only if H has an acyclic matching of size at least k.

Proof. Necessity: Let M be an acyclic matching in G of size at least k. Since G is a vertex induced subgraph of graph H, M is acyclic in H. Hence, H has an acyclic matching of size at least k.

Sufficiency: Let M' be an acyclic matching in H of size at least k. If M' does not have an edge from the edge set E' then M' is a required acyclic matching in G. Otherwise, note that M' can include at most one edge from the edge set E'. To the contrary, if $\{x_i'y_i, x_j'y_j\} \subseteq M'$ for some $x_i', x_j' \in X'$ then $G[\{x_i', y_i, x_j', y_j\}]$ forms a cycle, which is a contradiction. Thus, M' can include at most one edge from E'.

Next, let $x_i'y_i \in M'$ for some $x_i' \in X'$. Since G is connected, y_i will have a neighbor (say x_k) in X. Note that x_k must be unsaturated by M' because otherwise if $x_ky_k \in M'$ for some $y_k \in Y$, then $G[\{x_k, y_k, x_i', y_i\}]$ will form a cycle, which is a contradiction to the fact that M' is acyclic. Let $M = (M' \setminus \{x_i'y_i\}) \cup \{x_ky_i\}$. If $G[V(M)]$ is acyclic, then M is a required acyclic matching in G. Otherwise, let us assume that $G[V(M)]$ contains a cycle C''. If C' does not contain the vertex x_k, then C' is also a cycle in $G[V(M')]$. This contradicts the fact that M' is an acyclic matching. So, C' contains the vertex x_k. Let $x_ky_a, x_ky_b \in E(C')$. Since vertices of set X' are adjacent to every $y \in Y$, $x_i'y_a, x_i'y_b \in E_H$. Now, $C = (C' \setminus \{x_ky_a, x_ky_b\}) \cup \{x_i'y_a, x_i'y_b\}$ is also a cycle in $G[V(M')]$, which is a contradiction. Hence, M is acyclic and it is a required acyclic matching in G of size at least k.   ◇

Hence, the ACYCLIC MATCHING DECIDE PROBLEM is NP-complete for comb-convex bipartite graphs.   □

Corollary 1. *The* ACYCLIC MATCHING DECIDE PROBLEM *is NP-complete for tree-convex bipartite graphs.*

3.2   Dually Chordal Graphs

The ACYCLIC MATCHING PROBLEM is polynomial time solvable for chordal graphs [2] and hence for strongly chordal graphs. In this subsection, we show that the ACYCLIC MATCHING DECIDE PROBLEM is NP-complete for dually chordal graphs which is a superclass of strongly chordal graphs.

Theorem 2. *The* ACYCLIC MATCHING DECIDE PROBLEM *is NP-complete for dually chordal graphs.*

Proof. Clearly, the ACYCLIC MATCHING DECIDE PROBLEM belongs to the class NP for dually chordal graphs. To show the NP-completeness, we give a polynomial reduction from the ACYCLIC MATCHING DECIDE PROBLEM for general graphs, which is already known to be NP-complete [5].

Given a graph $G = (V, E)$, we construct a dually chordal graph $H = (V_H, E_H)$ as follows: $V_H = V \cup \{v_0\}$, $E_H = E \cup \{v_0 v \mid v \in V\}$.

Consider the ordering $\alpha = (v_1, v_2, \ldots, v_n, v_0)$. Since $N[v_j] \subseteq N[v_0]$ for all $v_j \in V$, v_0 is a maximum neighbor for all v_j in $G_j = G[\{v_j, \ldots, v_n\}]$. Therefore, it is easy to see that the constructed graph $H = (V_H, E_H)$ is a dually chordal graph. Also, note that given a graph G, the graph H can be constructed in polynomial time.

Now, the following claim is sufficient to complete the proof of the theorem.

Claim. G has an acyclic matching of size at least k if and only if H has an acyclic matching of size at least k, where $k > 1$.

Proof. Necessity: Let M be an acyclic matching in G of size at least k. Since G is a vertex induced subgraph of H, so M is an acyclic matching in graph H of size at least k.

Sufficiency: Let M' be an acyclic matching in graph H of size at least k, $k > 1$. Observe that if the vertex v_0 is saturated by M', that is $v_0 v_i \in M'$ for some $v_i \in V$, then $|M'| = 1$. To the contrary, if there exists another edge $v_j v_k \in M'$, then the graph $H[\{v_0, v_j, v_k\}]$ forms a cycle, which is a contradiction.

As $|M'| \geq k > 1$, vertex v_0 is not saturated by M', that is, M' does not have any edge of the form $v_0 v_i$ for any $v_i \in V$. Thus, M' is a required acyclic matching in graph G of size at least k.   ◇

Hence, the ACYCLIC MATCHING DECIDE PROBLEM is NP-complete for dually chordal graphs.   □

4   Polynomial Time Algorithms

4.1   Split Graphs

In this subsection, we show that an acyclic matching of maximum size can be computed in linear time for split graphs which is a subclass of chordal graphs, where the complexity of computing a maximum size acyclic matching is $O(n^7)$.

Let $G = (V, E)$ be a split graph. Throughout this section, $I \cup C$ represents a given partition of the vertex set V, where I is an independent set and C is a clique in G. Now, the following lemma shows that the cardinality of an acyclic matching in a split graph $G = (V, E)$ can be either 1 or 2 only.

Lemma 1. *Let $G = (V, E)$ be a split graph. If M is an acyclic matching in G, then $1 \leq |M| \leq 2$.*

Proof. Let M be an acyclic matching in G and let $|M| \geq 3$. Let $\{a_1 b_1, a_2 b_2, a_3 b_3\} \subseteq M$ for some $a_i, b_i \in V$, $1 \leq i \leq 3$. Since I is an independent set, we can assume without loss of generality that $b_1, b_2, b_3 \in C$. This leads to a contradiction as $G[\{b_1, b_2, b_3\}]$ forms a cycle. Thus, $|M| \leq 2$.   ◇

Next, we will characterize the split graphs depending on the size of an acyclic matching in G. For this purpose, let us recall the definition of *threshold graphs*, which is a proper subclass of split graphs.

A split graph $G = (V, E)$ is called a *threshold graph* if the vertices in I can be linearly ordered, say $(v_1, v_2, \ldots v_{|I|})$, such that $N(v_1) \subseteq N(v_2) \subseteq \ldots \subseteq N(v_{|I|})$. This linear ordering of a threshold graph can be computed in linear time [6].

Lemma 2. *Let $G = (V, E)$ be a split graph and let M be a maximum acyclic matching in G. Then, $|M| = 2$ if and only if there exist a pair of vertices $v_i, v_j \in I$ such that $N(v_i) \setminus N(v_j) \neq \emptyset$ and $N(v_j) \setminus N(v_i) \neq \emptyset$.*

Proof. Necessity: Let M be an acyclic matching in G and let $M = \{a_i b_i, a_j b_j\}$. Since C is a clique and I is an independent set, exactly two vertices from the set $\{a_i, a_j, b_i, b_j\}$ belong to C and the other two belongs to I. Without loss of generality, let us assume that $a_i, a_j \in C$ and $b_i, b_j \in I$. Since $G[\{a_i, a_j, b_i, b_j\}]$ is acyclic and $a_i b_i, a_j b_j, a_i a_j \in E$, so $a_i \notin N(b_j)$ and $a_j \notin N(b_i)$. Hence, $b_i, b_j \in I$ is the required pair of vertices.

Sufficiency: Let us assume that there exist two vertices $v_1, v_2 \in I$ such that $N(v_1) \setminus N(v_2) \neq \emptyset$ and $N(v_2) \setminus N(v_1) \neq \emptyset$. Let $c_1 \in N(v_1) \setminus N(v_2)$ and $c_2 \in N(v_2) \setminus N(v_1)$. Now, define a matching $M = \{v_1 c_1, v_2 c_2\}$. It is easy to see that $G[\{v_1, c_1, v_2, c_2\}]$ contains no cycle, and hence M is acyclic.   ◇

Theorem 3. *Let $G = (V, E)$ be a split graph and let M be a maximum acyclic matching in G. Then, $|M| = 1$ if and only if G is a threshold graph.*

Proof. Necessity: Let $G = (V, E)$ be a split graph and let M be a maximum acyclic matching in G such that $|M| = 1$. For the sake of contradiction, let us suppose that G is not a threshold graph. Then, there will exist a pair of vertices $v_i, v_j \in I$ such that $N(v_i) \nsubseteq N(v_j)$ and $N(v_j) \nsubseteq N(v_i)$. Define a matching $M' = \{c_i v_i, c_j v_j\}$, where $c_i \in N(v_i) \setminus N(v_j)$ and $c_j \in N(v_j) \setminus N(v_i)$. It is easy to see that M' is acyclic as $G[\{v_i, v_j, c_i, c_j\}]$ is a path graph. Since $|M'| = 2$, this leads to a contradiction to the fact that M is a maximum acyclic matching in G. Hence, G is a threshold graph.

Sufficiency: Let $G = (V, E)$ be a threshold graph and let $(v_1, v_2, \ldots v_{|I|})$ be an ordering of I, such that $N(v_1) \subseteq N(v_2) \subseteq \ldots \subseteq N(v_{|I|})$. Clearly, there does not exist any pair of vertices $v_i, v_j \in I$ such that $N(v_i) \setminus N(v_j) \neq \emptyset$ and $N(v_j) \setminus N(v_i) \neq \emptyset$. Hence, by Lemma 1 and Lemma 2, it is easy to see that $|M| = 1$.   □

Based on the above discussions we have the following theorem.

Theorem 4. *A maximum size acyclic matching in a split graph $G = (V, E)$ can be computed in $O(|V| + |E|)$ time.*

Proof. Due to space restriction, the proof has been deferred to the longer version of the paper.   □

4.2   Proper Interval Graphs

In this subsection, we show that an acyclic matching of maximum size can be computed in linear time for proper interval graphs which is a subclass of chordal graphs, where the complexity of computing a maximum size acyclic matching is $O(n^7)$.

Let $G = (V, E)$ be a given graph. A vertex $v \in V$ is called a simplicial vertex, if $N[v]$ induces a clique in G. An ordering $\alpha = (v_1, v_2, \ldots, v_n)$ of vertices is called a perfect elimination ordering (PEO) of G if v_i is a simplicial vertex in $G_i = G[\{v_i, v_{i+1}, \ldots, v_n\}]$ for all $1 \le i \le n$. A PEO $\alpha = (v_1, v_2, \ldots, v_n)$ of a graph G is called a bi-compatible elimination ordering (BCO) if $\alpha^{-1} = (v_n, v_{n-1}, \ldots, v_1)$ i.e., the reverse of α, is also a PEO of G. It has been characterized in [7] that a graph is proper interval if and only if it has a BCO.

Observation 5. *[8] Let $\sigma = (v_1, v_2, \ldots, v_n)$ be a BCO of a proper interval graph G. If $v_i v_j \in E$, then $v_k v_j \in E$ for all $k, i \le k \le j - 1$.*

Observation 6. *Let $\sigma = (v_1, v_2, \ldots, v_n)$ be a BCO of a proper interval graph G and let $L[v_i]$ denotes the last neighbor of a vertex v_i in σ. If $v_i < v_j$ in σ, then $L[v_i] \le L[v_j]$.*

Proof. Let us suppose that there exists v_i and v_j such that $v_i < v_j$ in σ and $L[v_i] > L[v_j]$. Then by Observation 5, $v_j L[v_i] \in E$ but since $L[v_j] < L[v_i]$, we reach at a contradiction.   ◇

Observation 7. *Let $\sigma = (v_1, v_2, \ldots, v_n)$ be a BCO of a proper interval graph G and let $L[v_i]$ denotes the last neighbor of a vertex v_i in σ. If M is an acyclic matching in G, then at most two vertices from the set $\{v_i, v_{i+1}, \ldots, L[v_i]\}$ can be saturated by M.*

Proof. The result easily follows from Observation 5.   ◇

Lemma 3. *Let G be a proper interval graph with a BCO $\sigma = (v_1, v_2, \ldots, v_n)$ and let M be an acyclic matching in G. If the edges $u_1 w_1, u_2 w_2 \in M$ such that $u_1 < w_1$ and $u_2 < w_2$ in σ, then either $w_1 < u_2$ or $w_2 < u_1$.*

Proof. Let us assume without loss of generality that there exist two edges $e_1 = u_1 w_1$ and $e_2 = u_2 w_2$ such that $u_1 < w_2 < w_1$ in σ. Now, the $G[\{u_1, w_2, w_1\}]$ forms a cycle, which is a contradiction. Thus, either $w_1 < u_2$ or $w_2 < u_1$.   ◇

Lemma 4. *Let G be a proper interval graph with a BCO $\sigma = (v_1, v_2, \ldots, v_n)$ and let M be a maximum acyclic matching in G. Then, there exists an acyclic matching M' in G such that $v_1 v_2 \in M'$ and $|M'| = |M|$.*

Proof. Let G be a proper interval graph with a BCO $\sigma = (v_1, v_2, \ldots, v_n)$ and let M be a maximum acyclic matching in G. Let $v_a v_b$ be the first edge with respect to σ that belongs to M. Let us assume without loss of generality that

$v_a < v_b$. It is easy to see that $v_a \leq L[v_2]$ in σ. To the contrary, if $v_a > L[v_2]$ in σ, then $G[\{v_1, v_2, v_a, v_b\}]$ is acyclic and hence the edge $v_1 v_2$ can be added to M. This leads to a contradiction to the fact that M is a maximum acyclic matching in G. Hence, $v_a \leq L[v_2]$ in σ. Now, if $v_a = v_2$, then replace $v_2 v_b$ by $v_2 v_1$ in M. If $v_a \neq v_2$, then replace $v_a v_b$ by $v_a v_2$ in M. If $v_a = v_1$, then we are done. Otherwise, again replace the edge $v_a v_2$ by $v_1 v_2$ in M.

By Observation 6 and Lemma 3, it is easy to see that we can replace the edge $v_a v_b$ with the desired edges in the cases mentioned above.   ◇

Let $\sigma = (v_1, v_2, \ldots, v_n)$ be a BCO of a proper interval graph G and let $\sigma' = (v_a, v_b, \ldots, v_k)$ be an ordering obtained from σ by removing some vertices from σ. Then, σ' is also a BCO of some proper interval graph G', where G' is a subgraph of G. Hence, we have the following corollary to Lemma 4.

Corollary 2. *If $\sigma' = (v_a, v_b, \ldots, v_k)$ is a BCO of a subgraph G' of a proper interval graph G, then the edge $v_a v_b$ is contained in some maximum acyclic matching of G'.*

Based on the above lemmas, we now present a linear time algorithm AM-PIG(G), which computes an acyclic matching of maximum size in a given proper interval graph G. The pseudocode of the algorithm is given below:

Algorithm 1. AM-PIG(G)

Input: A proper interval graph G_1 with BCO $\sigma(G_1) = (v_1, v_2, \ldots, v_n)$;
Output: A Maximum Acyclic Matching M;
$M = \emptyset$, $i = 1$, $v + 1 =$ vertex next to vertex v in $\sigma(G_i)$ for $i \geq 1$;
   $F[G_i] =$ first vertex in the BCO $\sigma(G_i)$ of graph G_i for $i \geq 1$;
   $L[v] =$ last neighbor of vertex v in $\sigma(G_i)$ for $i \geq 1$;
   while $(|G_i| \geq 2)$ **do**
   | $M = M \cup \{F[G_i](F[G_i] + 1)\}$;
   | **if** $(L[F[G_i]] = L[F[G_i] + 1])$ *or* $(L[F[G_i]] + 1 = L[F[G_i] + 1])$ **then**
   | | $i = i + 1$;
   | └ $G_i = G_{i-1} \setminus \{F[G_{i-1}], \ldots, L[F[G_{i-1}]]\}$;
   | **else if** $((L[F[G_i]] + 1 = v_k) < L[F[G_i] + 1] < L[v_k])$ **then**
   | | $i = i + 1$;
   | └ $G_i = G_{i-1} \setminus (\{F[G_{i-1}], \ldots, L[F[G_{i-1}]]\} \cup \{v_{k+1}, \ldots, L[F[G_{i-1}] + 1]\})$;
   | **else if** $(L[F[G_i]] + 1 = v_k) < L[F[G_i] + 1])$ *and* $L[v_k] = L[F[G_i] + 1]$ **then**
   | | $i = i + 1$;
   | | $G_i = G_{i-1} \setminus \{F[G_{i-1}], \ldots, L[F[G_{i-1}]]\}$;
   | | **while** $(L[v_k] = L[F[G_i] + 1])$ **do**
   | | | $temp = v_k$;
   | | | $v_k = v_k + 1$;
   | | └ $G_i = G_i \setminus temp$;
   | └ $G_i = G_i \setminus \{v_{k+1}, \ldots, L[F[G_i] + 1]\}$;
return M;

Theorem 8. *Given a proper interval graph G_1 with BCO $\sigma(G_1)$, AM-PIG(G_1) correctly computes a maximum size acyclic matching in G_1.*

Proof. Due to space restriction, the proof has been deferred to the longer version of the paper. $\qquad\square$

5   Inapproximation Results

Let $G = (V, E)$ be a graph with n vertices. It is easy to note that the maximum size of an acyclic matching in G can be at most $\frac{n}{2}$. So, the ACYCLIC MATCHING PROBLEM can be approximated within a factor of n in polynomial time. In this section, we show that for any $\epsilon > 0$, it is hard to approximate the ACYCLIC MATCHING PROBLEM within a factor of $n^{1-\epsilon}$, unless $P = NP$.

To prove the result, we will need the following theorem for the MAXIMUM INDEPENDENT SET PROBLEM.

Theorem 9. *[12] The Maximum Independent Set Problem for a graph G cannot be approximated within a factor of $n^{1-\epsilon}$ for any $\epsilon > 0$, unless $P = NP$.*

Now, consider the following construction:

Construction 1. *Let $G = (V, E)$, where $V = \{v_1, v_2, \ldots, v_n\}$, be an instance of the MAXIMUM INDEPENDENT SET PROBLEM. We construct a graph $H = (V_H, E_H)$, an instance of the ACYCLIC MATCHING PROBLEM, in the following way:*

- *$V_H = V \cup V'$, where $V' = \{v_i' \mid v_i \in V\}$.*
- *$E_H = E \cup \{v_i v_i' \mid 1 \le i \le n\} \cup \{v_i v_j' \mid v_i v_j \in E\} \cup \{v_i' v_j' \mid v_i v_j \in E\}$.*

Clearly, H can be constructed in polynomial time as $|V_H| = 2|V|$ and $|E_H| = 4|E| + |V|$.

Also, note that the edges in H can be one of the following four types:

1. *Type-I* $= \{v_i v_i' \mid v_i \in V \text{ and } v_i' \in V'\}$.
2. *Type-II* $= \{v_i v_j \mid v_i, v_j \in V\}$.
3. *Type-III* $= \{v_i' v_j \mid v_i' \in V' \text{ and } v_j \in V\}$.
4. *Type-IV* $= \{v_i' v_j' \mid v_i', v_j' \in V'\}$.

Now, we will discuss some lemmas that will be used in the proof of the main theorem of this section. Let us recall that $V_H(M)$ denotes the set of M-saturated vertices of graph H.

Lemma 5. *Let H be the graph obtained from a given graph G by Construction 1. If M is an acyclic matching in H, then there exists an acyclic matching M' in H such that $|M'| = |M|$ and M' contains edges of Type-I and Type-II only.*

Proof. First, let us suppose that M is an acyclic matching in H and let M contains an edge (say $v_i' v_j'$) of *Type-IV*. Since M is acyclic and $v_i' v_j' \in M$, both v_i and v_j are unsaturated by M. Let $M' = (M \setminus \{v_i' v_j'\}) \cup \{v_i' v_j\}$. If $G[V_H(M')]$ is acyclic, then we are done. So, assume that $G[V_H(M')]$ contains a cycle C. If C does not contain the vertex v_j, then C is also a cycle in $G[V_H(M)]$. This contradicts the fact that M is an acyclic matching. So, C contains the vertex v_j. Let $v_j u_a, v_j u_b \in E(C)$. Since $N[v_j] = N[v'_j]$, $v'_j u_a, v'_j u_b \in E_H$. Now, $C' = (C \setminus \{v_j u_a, v_j u_b\}) \cup \{v'_j u_a, v'_j u_b\}$ is a cycle in $G[V_H(M)]$, which is a contradiction. Hence, M' is acyclic. In this way, an acyclic matching of same size can be obtained by replacing an edge of $Type\text{-}IV$ with a corresponding $Type\text{-}III$ edge.

Using the similar arguments, we can show that an acyclic matching of same size can be obtained by replacing an edge of $Type\text{-}III$ with a corresponding $Type\text{-}II$ edge. $\Diamond$

Lemma 6. *Let H be the graph obtained from a given graph G by Construction 1. If M' is an acyclic matching in H containing edges of $Type\text{-}I$ and $Type\text{-}II$ only, then there exists an acyclic matching M'' in H such that $|M''| = |M'|$ and M'' contains edges of $Type\text{-}I$ only.*

Proof. Due to space restriction, the proof has been deferred to the longer version of the paper. $\Diamond$

The following lemma shows that the described reduction is exactly what we need.

Lemma 7. *Let H be the graph obtained from a given graph G by Construction 1. Then, G has an independent set of size at least k if and only if H has an acyclic matching of size at least k.*

Proof. Necessity: Let $I = \{v_1, v_2, \ldots, v_l\}$ be an independent set in G of size at least k. Define a matching $M = \{v_1 v'_1, v_2 v'_2, \ldots, v_l v'_l\}$ in H. It is easy to see that M is an acyclic matching as $G[V(M)]$ is a disjoint union of $K'_2 s$.

Sufficiency: Let $M = \{e_1, e_2, \ldots, e_l\}$ be an acyclic matching in H of size at least k. By Lemma 6, there exists an acyclic matching M' in H such that $|M'| = |M|$ and M' contains edges of $Type\text{-}I$ only. Define a set $I = \{v_i \mid v_i v'_i \in M'\}$. It is easy to see that I is an independent set of graph G. $\Diamond$

Corollary 3. *G has a maximum independent set of size k if and only if H has a maximum acyclic matching of size k.*

Theorem 10. *The Acyclic Matching Problem for a graph G cannot be approximated within a factor of $n^{1-\epsilon}$ for any $\epsilon > 0$, unless $P = NP$.*

Proof. Let $G = (V, E)$ be a graph with n vertices. Construct a graph $H = (V_H, E_H)$ with $|V_H| = \bar{n}$ from G using Construction 1. Let I^* denotes a maximum independent set in G and M^* denotes a maximum acyclic matching in H.

Now, let us suppose that the Acyclic Matching Problem can be approximated within a ratio $\alpha \geq 1$ by using an algorithm ALG, where $\alpha = \bar{n}^{1-\epsilon'}$ for some fixed $\epsilon' > 0$.

If $M_{ALG}(H)$ is an acyclic matching in H obtained by applying algorithm ALG, then $|M^*(H)| \leq \alpha |M_{ALG}(H)|$.

By Corollary 3, $|I^*(G)| = |M^*(H)|$. By Lemma 7, we can construct an independent set I_{ALG} of G corresponding to M_{ALG} of H such that $|M_{ALG}(H)| = |I_{ALG}(G)|$.

Hence, we obtain, $|I^*(G)| \leq \alpha|I_{ALG}(G)| = \overline{n}^{1-\epsilon'}|I_{ALG}(G)| = (2n)^{1-\epsilon'}|I_{ALG}(G)| = (2)^{1-\epsilon'}(n)^{1-\epsilon'}|I_{ALG}(G)|$.

If we choose ϵ, such that $2^{1-\epsilon'} < n^{\epsilon'-\epsilon}$, then $|I^*(G)| < (n)^{\epsilon'-\epsilon}(n)^{1-\epsilon'}|I_{ALG}(G)| = (n)^{1-\epsilon}|I_{ALG}(G)|$.

Hence, $|I^*(G)| < (n)^{1-\epsilon}|I_{ALG}(G)|$, which leads to a contradiction to Theorem 9. Therefore, the ACYCLIC MATCHING PROBLEM cannot be approximated within a factor of $n^{1-\epsilon}$ for any $\epsilon > 0$, unless $P = NP$. $\qquad\square$

6 APX-Completeness

In this section, we show that the ACYCLIC MATCHING PROBLEM is APX-complete for $2k + 1$-regular graphs for $k \geq 3$, where k is a constant.

To prove the result, we first show that the ACYCLIC MATCHING PROBLEM is approximable within a constant factor when restricted to k-regular graphs for $k \geq 3$, where k is a constant. For the purpose, consider the following algorithm:

Algorithm 2. APPROX-AM(G)

Input: A graph $G = (V, E)$;
Output: An acyclic matching M_{ac} in G;
$M_{ac} = \emptyset$;
  while ($E \neq \emptyset$) **do**
  | Choose an edge $e = uv$ from E;
  | $M_{ac} = M_{ac} \cup \{uv\}$;
  | $V = V \setminus (N_G(u) \cup N_G(v))$;
return M_{ac}.

Lemma 8. *The algorithm* APPROX-AM(G) *produces an acyclic matching of G in polynomial time.*

Proof. For any pair of edges in M_{ac}, say $e_i = a_ib_i$ and $e_j = a_jb_j$, $G[\{a_i, b_i, a_j, b_j\}]$ is a disjoint union of $K_2's$. $\qquad\diamond$

Lemma 9. *The* ACYCLIC MATCHING PROBLEM *for a k-regular graph G can be approximated with an approximation ratio of $\frac{2k(k-1)+1}{k}$, where k is a constant.*

Proof. Given a k-regular graph G, construct an acyclic matching M_{ac} of G by using algorithm APPROX-AM(G). In each step, after adding an edge in the matching M_{ac}, we are removing at most k^2 edges, hence $\frac{kn}{2[2k(k-1)+1]} \leq |M_{ac}|$. Moreover, it is easy to see that the size of any matching can be at most $\frac{n}{2}$.

Hence, the ACYCLIC MATCHING PROBLEM is approximable within a factor of $\frac{2k(k-1)+1}{k}$ in k-regular graphs, where k is a constant. $\qquad\diamond$

To prove the result, we will need the following theorem for the MAXIMUM INDEPENDENT SET PROBLEM.

Theorem 11. *[1, 10] The* MAXIMUM INDEPENDENT SET PROBLEM *is APX-complete for k-regular graphs for $k \geq 3$.*

Observation 12. *If G is a k-regular graph in Construction 1, then the constructed graph H is a $2k + 1$-regular graph for $k \geq 3$.*

Now, we are ready to prove the APX-completeness of the ACYCLIC MATCHING PROBLEM for $2k + 1$-regular graphs for $k \geq 3$, where k is a constant. For this purpose, we recall the concept of L-reduction. Given two NP optimization problems π_1 and π_2 and a polynomial time transformation f from instances of π_1 to instances of π_2, we say that f is an L-reduction if there are positive constants α and β such that for every instance x of π_1:

1. $opt_{\pi_2}(f(x)) \leq \alpha.opt_{\pi_1}(x)$;
2. for every feasible solution y of $f(x)$ with objective value $m_{\pi_2}(f(x), y) = c_2$, we can find a solution y' of x in polynomial time with $m_{\pi_1}(x, y') = c_1$ such that $|opt_{\pi_1}(x) - c_1| \leq \beta.|opt_{\pi_2}(f(x)) - c_2|$.

Theorem 13. *The* ACYCLIC MATCHING PROBLEM *is APX-complete for $2k+1$-regular graphs for $k \geq 3$, where k is a constant.*

Proof. By Lemma 9, it is clear that the ACYCLIC MATCHING PROBLEM for $2k + 1$-regular graphs for $k \geq 3$ belongs to the class APX. By Theorem 11, it is enough to construct an L-reduction from the instances of the MAXIMUM INDEPENDENT SET PROBLEM for k-regular graphs to the instances of the ACYCLIC MATCHING PROBLEM for $2k + 1$-regular graphs. Given a k-regular graph $G = (V, E)$, where $V = \{v_1, v_2, \ldots, v_n\}$. We construct a graph $H = (V_H, E_H)$, an instance of the ACYCLIC MATCHING PROBLEM by Construction 1. It is easy to see by Lemma 7 and Corollary 3 that the reduction described in Construction 1 is an L-reduction with $\alpha = 1$ and $\beta = 1$.

Therefore, the ACYCLIC MATCHING PROBLEM is APX-complete for $2k + 1$-regular graphs for $k \geq 3$, where k is a constant.   □

7   Conclusion

In this paper, we have shown that the ACYCLIC MATCHING DECIDE PROBLEM is NP-complete for comb-convex bipartite graphs and dually chordal graphs. On the positive side, we have shown that the ACYCLIC MATCHING PROBLEM can be solved in linear time in split graphs and proper interval graphs. Apart from these, we have shown that the ACYCLIC MATCHING PROBLEM cannot be approximated within a factor of $n^{1-\epsilon}$ for any $\epsilon > 0$, unless $P = NP$. We have also shown that the ACYCLIC MATCHING PROBLEM is APX-complete for $2k+1$-regular graphs for $k \geq 3$, where k is a constant. Further, it will be interesting to study better approximation algorithms for this problem for bipartite graphs and other important graph classes.

References

1. Alimonti, P., Kann, V.: Some APX-completeness results for cubic graphs. Theoret. Comput. Sci. **237**(1–2), 123–134 (2000)
2. Baste, J., Rautenbach, D.: Degenerate matchings and edge colorings. Discrete Appl. Math. **239**, 38–44 (2018)
3. Fürst, M., Rautenbach, D.: A lower bound on the acyclic matching number of subcubic graphs. Discrete Math. **341**(8), 2353–2358 (2018)
4. Fürst, M., Rautenbach, D.: On some hard and some tractable cases of the maximum acyclic matching problem. Ann. Oper. Res. **279**(1), 291–300 (2019). https://doi.org/10.1007/s10479-019-03311-1
5. Goddard, W., Hedetniemi, S.M., Hedetniemi, S.T., Laskar, R.: Generalized subgraph-restricted matchings in graphs. Discrete Math. **293**(1), 129–138 (2005)
6. Heggernes, P., Kratsch, D.: Linear-time certifying recognition algorithms and forbidden induced subgraphs. Nord. J. Comput. **14**(1–2), 87–108 (2007)
7. Jamison, R.E., Laskar, R.: Elimination orderings of chordal graphs. In: Combinatorics and Applications, pp. 192–200 (1982)
8. Panda, B.S., Das, S.K.: A linear time recognition algorithm for proper interval graphs. Inf. Process. Lett. **87**(3), 153–161 (2003)
9. Panda, B.S., Pradhan, D.: Acyclic matchings in subclasses of bipartite graphs. Discrete Math. Algorithms Appl. **4**(04), 1250050 (2012)
10. Papadimitriou, C.H., Yannakakis, M.: Optimization, approximation, and complexity classes. J. Comput. Syst. Sci. **43**(3), 425–440 (1991)
11. Bao, F.S., Zhang, Y.: A review of tree convex sets test. Comput. Intell. **28**(3), 358–372 (2012)
12. Zuckerman, D.: Linear degree extractors and the inapproximability of max clique and chromatic number. In: Proceedings of the Thirty-Eighth Annual ACM Symposium on Theory of Computing, pp. 681–690 (2006)

Author Index

Printed in the United States
By Bookmasters